Navigated Transcranial Magnetic Stimulation in Neurosurgery

Sandro M. Krieg • Thomas Picht
Editors

Navigated Transcranial Magnetic Stimulation in Neurosurgery

Second Edition

Springer

Editors
Sandro M. Krieg
Department of Neurosurgery
Heidelberg University Hospital
Heidelberg, Germany

Thomas Picht
Department of Neurosurgery
Charité – Universitätsmedizin Berlin
Berlin, Germany

ISBN 978-3-031-97154-9 ISBN 978-3-031-97155-6 (eBook)
https://doi.org/10.1007/978-3-031-97155-6

This Springer imprint is published by the registered company Springer Nature Switzerland AG
The registered company address is: Gewerbestrasse 11, 6330 Cham, Switzerland

Foreword

It is a great pleasure for me to write the foreword to the second edition of this book dedicated to navigated transcranial magnetic stimulation (nTMS) in neurosurgery. The first edition, led by Sandro Krieg in 2017, was very successful, establishing this text as the definitive reference on the topic. Due to a rapid development in this exciting field, a second edition was needed, as a continuation of the breadth of the knowledge essential to fully apprehend this extraordinarily complex subject.

With the goal to achieve optimal precision and safety in the operating theater, a neurosurgeon must investigate not only the structure and vasculature of the brain but also its neural functions. In this setting, in order to preserve the quality of life of patients who will undergo cerebral surgery, it is crucial to study the organization of neural circuits before to remove a part of the central nervous system (CNS) affected by a cerebral disease, e.g., epilepsy or tumor. Due to a major inter-individual anatomo-functional variability, especially in the event of brain lesions which can induce mechanisms of neuroplasticity, mapping techniques are very helpful to understand the distribution of cortical hubs and subcortical pathways underlying motor, language, cognitive, and emotional functions at the individual level. To this end, intraoperative direct electrical stimulation (DES) in awake patients is considered the gold standard to optimize the extent of resection while minimizing neurological morbidity. However, even though this method allows real-time anatomo-functional correlations throughout the surgical procedure, it is also important to benefit from complementary techniques that permit a non-invasive preoperative mapping. Functional neuroimaging has been extensively used in the past decades, but its main limitation is the inability to differentiate critical areas which should not be removed during surgery to avoid permanent deficit, versus regions involved in a neural network but which can be compensated—and thus which can be surgically resected with no neurological consequence.

In this state of mind, nTMS represents an original tool opening new avenues in the exploration of the CNS, both in healthy volunteers as well in brain-damaged patients. Indeed, as intrasurgical DES, nTMS offers the unique opportunity to create a transient virtual disruption of neural networks, with the aim to identify the cortical areas essential for brain functions. However, contrary to DES, nTMS is a non-invasive technique that can be used before surgery to map the eloquent cortex and to plan the resection accordingly. Led by the editors, Sandro Krieg and Thomas Picht, this collective body of work will serve as a comprehensive textbook for all physicians with an individualized approach to brain surgery. What makes this book so unusual is that it contains all the required information to use nTMS in a department of neurosurgery and outlines the pros and cons of other techniques. The approach the authors have taken in defining this rapidly evolving technology and its implications for neurosurgical management is quite unique and innovative, to say the least.

The second edition of the book is organized in a very logical and informative fashion, starting off with critically important chapters covering the basic principles of nTMS. The clinical aspects are further evoked in chapters on preoperative motor and language mapping. To this end, Drs. Krieg and Picht are masters at explaining and detailing how to use nTMS for surgical planning and how to combine this method with other techniques, especially with fiber tracking. I particularly like the way in which further brain functions can be mapped by nTMS, and in which this methodology may be used in children—knowing that it may be difficult to achieve awake surgery in pediatric population. New chapters about the contribution of nTMS in epilepsy and in cervical myelopathy have been added in this second edition, showing the increased role of this technique in the surgical armamentarium. Interestingly, the fact that nTMS is also able to modulate neural networks for neurosurgical applications is depicted in a series of detailed chapters on these subjects. For example, nTMS can be helpful to treat chronic pain, depression, and may have a value in spinal cord injury. In the future, this technique could also be considered to induce and canalize neuroplasticity, allowing an increase of the extent of resection or even an improvement of the neurological status—for example, by combining it with specific programs of rehabilitation in patients with neurological deficits. Finally, in the field of cognitive neurosciences, nTMS may represent a valuable tool to investigate CNS processing in humans. Indeed, thanks to recent advances in the new science of connectomics, which aims to comprehensively map neural connections at both structural and functional levels, coordination of cognitive and behavioral domains is now attributed to parallel and intersecting large-scale neural circuits that contain interconnected cortical and subcortical components. In this context, a technique based on the concept of transitory disruption of neural circuits will undoubtedly provide new insights into the organization of such a networking brain. Yet, it is worth noting that nTMS can achieve only a mapping of the cortex, but it is not able to map directly the white matter tracts, which are nonetheless constituting a crucial part of the connectome. From a clinical point of view, preservation of subcortical pathways is essential during brain surgery, because the white matter connectivity is a well-known limitation of neuroplasticity. This is the reason why, currently, nTMS should still be considered in combination with other mapping techniques, especially preoperative advanced tractography as well as intraoperative DES.

It is crucial for modern clinical neuroscience, and especially for neurosurgery, to incorporate advances in the topic of brain mapping in as timely a fashion as possible, so that patient care becomes guided by the latest increments of relevant technology and knowledge with regard to neural processing. I have no doubt that this comprehensive second edition will serve this purpose with considerable distinction. Once again, this text represents the definitive source on this complex field. I congratulate Dr. Krieg, Dr. Picht, and their colleagues for bringing so much critical information about nTMS, its current and future applications in neurosurgery, in the confines of this book. This is a must-read for everybody in this subject!

Department of Neurosurgery
Gui de Chauliac Hospital
Montpellier University Medical Center
Institute for Neurosciences of Montpellier
INSERM U1191
Montpellier, France

Hugues Duffau

Introduction

I am delighted and honored to comment on the potential role of nTMS as we move into the future. The co-authors of this textbook, Drs. Picht and Krieg, have done a magnificent job in this second edition dedicated to nTMS to define and emphasize the current indications for nTMS such as preoperative functional localization of cortical regions of interest and how these functional areas can be used to delineate more accurately subcortical white matter tracts subserving these cortical regions of interest with DTI. However, as nTMS becomes a mainstay in Neurosurgical practice in conjunction with intraoperative stimulation mapping, we are beginning to understand the immense potential of this technology, pioneered by Drs. Picht and Krieg to enhance our understanding of functional pathways in the brain, brainstem and spinal cord but to also highlight its untapped potential to modulate plasticity and influence complex disorders such as chronic pain, depression, and psychological disorders.

In San Francisco, we have begun to follow the lead of Drs. Picht and Krieg to utilize nTMS to modulate postoperative deficit recovery following removal of tumors with intraoperative mapping, and the early experience and impression is dramatic in terms of enhancing recovery following post-surgical motor deficits. We intend to further pursue this strategy to overcome postoperative language deficits to determine if nTMS is able to promote early reorganization of language circuits. If our results are anything like what the authors have demonstrated with their stellar studies using nTMS in stroke patients, we as a specialty are in for a great benefit. Neuromodulation is the new frontier in neurosurgery for such complex disorders as pain, movement disorders, and perhaps behavioral entities such as obesity, depression, and psychological illnesses. The virtual sky is the limit in terms of what nTMS can accomplish. We need to continue to explore the role of nTMS for neuromodulation and promoting reorganization of cortical function. We may be able to push our extent of resection even further out into territories that result in deficits that can be reduced or abolished with nTMS during physical rehabilitation. We are entering a new age in our field and we must have nTMS as a means to enhance safe maximal resection and expand avenues to promote neuromodulation of disorders we once thought were impossible to influence. The impossible is now possible with nTMS, and neurosurgeons in all disciplines in our field must embrace this essential technology.

UCSF Brain Tumor Center
San Francisco, CA, USA

Mitchel S. Berger

Words from the Editors

Headed by the neurosurgical departments in Heidelberg and Berlin, the second edition of n*TMS in Neurosurgery* was a joint effort of a fast-growing international community of brain stimulation enthusiasts and other neuroscientists. We are amazed and delighted at how the field has developed since the first edition, and we are pleased to be able to present the findings of many of the major players in the field in one comprehensive yet concise volume. This book shows how far we've come, with cutting-edge developments that have opened new frontiers in neurosurgical planning and treatment.

One of the most remarkable aspects of the nTMS journey has been the unparalleled spirit of collaboration that has driven this progress. From the first edition to this one, we have seen how working together across borders can produce ground-breaking research, and how our international partnerships have translated nTMS research into tangible and meaningful clinical applications. By bringing together minds from around the world, we've been able to refine and improve the way nTMS is used—far beyond pre-operative mapping, where it all started. Tractography and connectomics, as the natural counterparts of TMS, have enabled us to deepen our understanding of the brain's complex networks and provide powerful tools, for example, to guide TMS stimulation and therapeutic interventions with greater precision.

This second edition highlights the groundbreaking integration of these technologies, allowing us to probe brain function in ways we never thought possible. We've extended our reach from isolated brain regions to complex networks, understanding their relevance and vulnerability in the context of individual patient care. We are privileged to work in a field where scientists, engineers, physicists, clinicians, and many others work together in an environment of camaraderie and shared vision, advancing the field at an unprecedented rate.

We hope this issue will continue to serve as a testament to the evolution of the field of navigated TMS in neurosurgery, and we invite you to be a part of this growing community, contributing your insights and experiences as we all work towards the same goal.

Let's keep moving forward together.

Sandro (Krieg) and Thomas (Picht)

The Editors

Contents

Part VI Further Potential of nTMS

Part I

General Aspects of nTMS

Basic Principles of Navigated TMS

1

Henri Hannula and Risto J. Ilmoniemi

1.1 Introduction

Neurons in the human brain can be triggered to fire action potentials by transcranial magnetic stimulation (TMS) (Barker et al. 1985; Ilmoniemi et al. 1999): a strong, quickly changing magnetic field is generated by a coil outside the head to induce, according to Faraday's law, an electric field within the brain. The method is noninvasive: no contact to the patient is needed; no electric current is fed through the scalp or skull. The strong magnetic field itself has no direct biological consequences; the effect is entirely due to the electric field-driven electric current, which accumulates charge at conductivity boundaries such as cell membranes, depolarizing or hyperpolarizing them. Sufficient depolarization initiates synchronous action potentials that propagate in the axon just like naturally occurring ones; both anterograde and retrograde propagation is possible. The physiological consequences of TMS are mainly the result of action potentials triggered by sodium-channel opening, although, e.g., dendritic calcium channels might be activated by TMS as well.

At first, TMS was produced with nonfocal round coils, but soon the electric field-focusing figure-of-eight coil was introduced (Ueno et al. 1988). However, positioning of the coil was and still is often performed based merely on external landmarks

H. Hannula (✉)
Department of Neuroscience and Biomedical Engineering, Aalto University School of Science, Espoo, Finland

Nexstim Plc., Helsinki, Finland
e-mail: henri.j.hannula@aalto.fi

R. J. Ilmoniemi
Department of Neuroscience and Biomedical Engineering, Aalto University School of Science, Espoo, Finland

BioMag Laboratory, Helsinki University Hospital, Helsinki, Finland
e-mail: risto.ilmoniemi@aalto.fi

S. M. Krieg, T. Picht (eds.), *Navigated Transcranial Magnetic Stimulation in Neurosurgery*, https://doi.org/10.1007/978-3-031-97155-6_1

on the head or by trial and error, meaning that the anatomical target in the cortex remains inaccurately known. This limitation was eventually eliminated by the introduction of navigated TMS, the so-called nTMS (*see* Ruohonen and Karhu 2010), also called navigated brain stimulation (NBS; introduced by Nexstim Ltd. in 2003) or stereotactic magnetic stimulation (Krings et al. 2001). Navigated targeting based on magnetic resonance imaging (MRI) was originally proposed by Ilmoniemi and Grandori in the context of multi-coil stimulation (e.g., Ilmoniemi and Grandori 1993).

Initially, TMS navigation was performed simply by placing a figure-of-eight coil over the target area on the basis of the subject's individual MR image. The maximal activation was assumed to be located on the line that passes through the center of the coil perpendicularly to the surface of the coil bottom; this methodology is called "line navigation." However, it was realized (Fig. 1.1) and later confirmed that line navigation can result in inaccurate targeting if the coil is not perfectly tangential with respect to the skull (Sollmann et al. 2016).

Subsequently developed "electric field navigation," on the other hand, takes the geometry of the head into account, resulting in sufficient targeting accuracy for the most demanding clinical applications like presurgical mapping of eloquent cortex (Picht et al. 2011). In addition to the precise definition of the stimulator parameters such as coil shape, location, and orientation, individual head size and shape as well as the orientation of cortical folds have to be taken into account in order to know

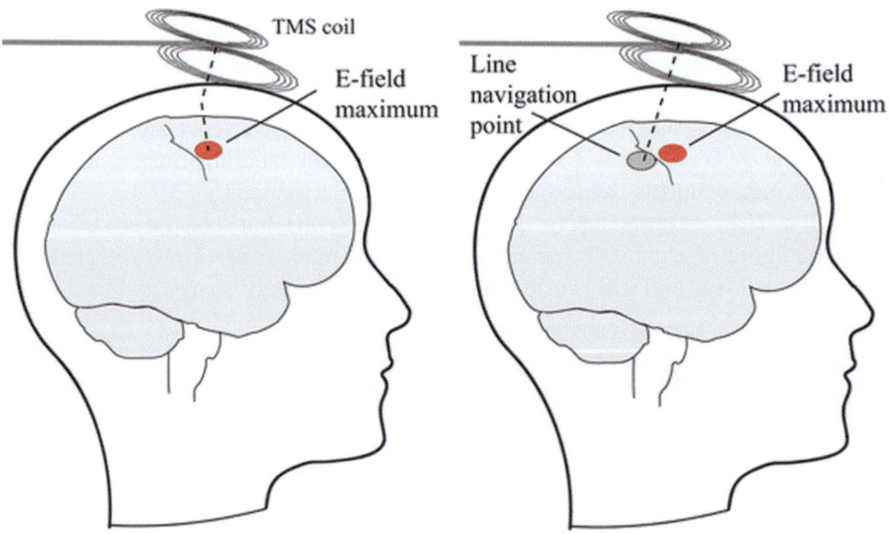

Fig. 1.1 The difference between line and electric field (E-field) navigation. Somewhat analogously to light bending (refraction) at water–air boundaries, the electromagnetic influence of nTMS depends on air–tissue, skull–intracranial and other conductivity boundaries. *Left*: Electric field navigation takes into account the conductivity boundaries and computes the electric field maximum where the cortex is best stimulated. *Right*: Line navigation does not visualize the spot of maximal stimulation if there is any coil tilt

which sulcus or gyrus is stimulated. This is particularly important in clinical applications, where the location and dose of stimulation should be accurately defined.

This chapter will provide an introduction to the basic physical, physiological, and technical principles of nTMS.

1.2 Basic Principles of TMS

The TMS pulse is generated by feeding a strong electric current to a coil placed over the stimulated cortical area. Figure 1.2 shows basic stimulator electronics: a capacitor, a charging circuit, and an electric switch that connects the capacitor to the coil. First, the switch is in the nonconducting state and the capacitor is charged to a voltage V_c of typically several kV. Then, the switch is closed, and the capacitively stored energy, $E_c = ½CV_c^2$ is transformed to the energy of the magnetic field B: $E_B = ½\,\mu_0^{-1} \int B^2 dV = ½LI^2$, where μ_0 is the permeability of free space, L is the inductance of the TMS coil, and I is the coil current. The capacitor C and the inductance L form a resonant circuit with time constant $\tau = \sqrt{LC}$, resulting in a sinusoidal current waveform with frequency $f = (2\pi\tau)^{-1}$.

In biphasic operation (Fig. 1.2, upper row), one oscillatory cycle is completed if the switch is opened at the end of the cycle.

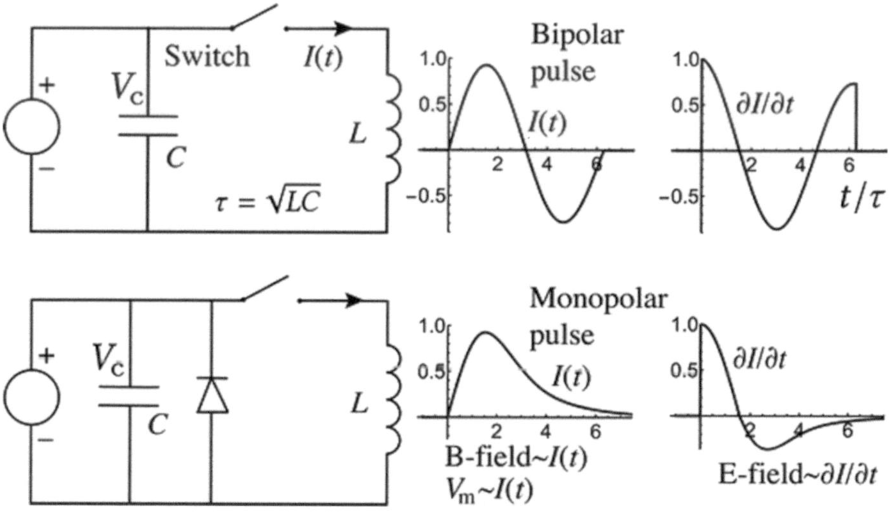

Fig. 1.2 Basic TMS circuits and corresponding waveforms. The electric field (E-field) in the brain is proportional to the rate of change of the magnetic field while the change in membrane potential V_m, being proportional to the time integral of the induced current, is proportional to the magnetic field (B-field). However, for slow pulses, leakage currents through the cellular membrane and current flow within the cell reduce the change in membrane voltage relative to faster pulses with the same peak magnetic field. *Upper row*: Bipolar circuit. *Lower row*: Monopolar circuit

The lower row of Fig. 1.2 shows a circuit very similar to that of the upper row, but a diode has been added. This essentially prevents oppositely directed current, producing monophasic stimulation.

The spatial pattern of the magnetic field is defined by the shape and size of the coil. Two coil types are common: round coil and, the so-called, figure-of-eight coil. When the coil is placed over the scalp, the electric field induced in the brain is a blurred mirror image of the coil current (Koponen et al. 2015), the electric field strength being proportional to the time derivative of the magnetic field (Fig. 1.2). The round coil produces an unfocused annulus of current while the figure-of-eight coil produces a focal electric field. In the latter case, the electric fields of the two wings are additive under the center of the coil, doubling the electric field there. As shown quantitatively by Deng et al. (2013), small coils produce more focal and more superficial electric fields in the brain than large similarly shaped ones.

The TMS-induced intracranial current causes transient membrane potential changes, which, if these changes are sufficiently depolarizing, trigger action potentials. The induced electric field of a figure-of-eight coil being focal, a small target patch of cortex (of the order of 0.2–2 cm^2) can be selectively activated. It should be noted, however, that the focality depends on the TMS intensity: the stronger the pulse, the larger the area of cortex that is influenced (Fig. 1.3).

The effect of TMS is manifested in multiple ways. (1) The direct effect is triggering action potentials in cortical excitatory and inhibitory neurons. (2) Pyramidal cells send their signals to connected brain areas and to the spinal cord; the spread of

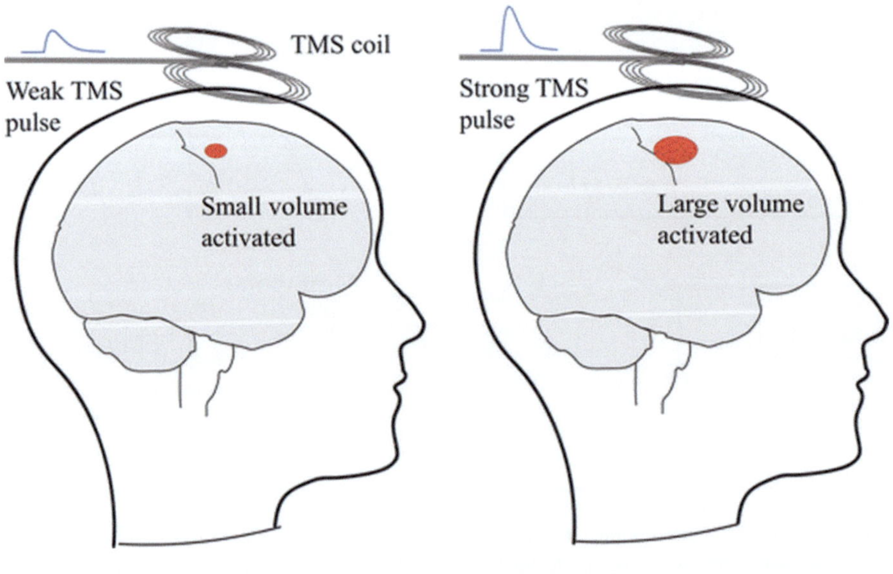

Fig. 1.3 Pulse intensity and stimulated volume. If a weak pulse is given, only a small volume near the peak electric field is activated. However, for strong pulses, the threshold of activation is exceeded in a large volume. Thus, in addition to a focal coil, maximal focality also requires that the stimulation intensity only slightly exceeds the activation threshold

the activity can be observed with electroencephalography (EEG) (Ilmoniemi et al. 1997; Massimini et al. 2005; Ilmoniemi and Kicic 2010), functional MRI (fMRI) (Mizutani-Tiebel et al. 2022), near infrared spectroscopy (Näsi et al. 2011), or, if the motor cortex is stimulated, with electromyography (EMG) in the form of MEPs from muscles (Rothwell et al. 1999). (3) The activated inhibitory cells quiet down the stimulated site, causing the so-called cortical silent period (CSP), which in the case of the motor cortex is visible as dampened EMG signal lasting about 100–150 ms after the MEP (Säisänen et al. 2008). (4) TMS may disturb the processing or transfer of information in the target area, producing a "transient lesion"; thus, stimulation of the somatosensory or language areas can degrade sensory perception (Hannula et al. 2005) or language functions (Picht et al. 2013), respectively. (5) High-frequency (HF, ~10 Hz or more) repetitive TMS (rTMS) can increase cortical excitability while low frequency (LF, ~1 Hz or less) rTMS can reduce excitability; such changes can outlast the stimulation period. Repeated sessions of rTMS can have therapeutic effects (Lefaucheur et al. 2014). (6) TMS can modulate ongoing oscillatory activity, which, in turn, may influence effective connectivity at least transiently.

Thus, the basic principle of TMS is simple: action potentials in the target area are triggered by electric field-driven currents that depolarize cell membranes. Based on precise knowledge of the shape of the coil and its location and orientation with respect to the head and its internal structure, the electric field pulse can be directed accurately to the target area. Such targeting, when an anatomical image of the brain is used as the map, is called electric field-navigated TMS (En-TMS).

1.3 Principles of Navigated TMS

1.3.1 General Considerations

In En-TMS, the induced intracranial electric field or the location of the maximum electric field is determined online and is usually displayed on anatomical brain images in order to allow targeting the pulses to desired cortical structures. The accurate determination of the electric field requires knowledge of (1) the precise geometry of coil windings, (2) coil current as a function of time, (3) the location and orientation of the coil with respect to the head, (4) the size and shape of the head and its compartments and, if detailed analysis of the electric field within the brain is investigated, the conductivity values and profiles of different head compartments such as cerebrospinal fluid, gray matter and white matter. As the magnetic field penetrates tissues without distortion and virtually no current is fed to the brain through the scalp or skull, the conductivity values of the scalp and the skull do not play a big role in determining the induced current flow in the brain. This is in stark contrast to transcranial electrical stimulation (TES) where these conductivity values as well as the details of skull thickness and shape play a crucial role.

Focal En-TMS is comparable to DES of the brain in that it produces a stimulating electric field that is confined to a small area of tissue. The benefit of En-TMS over DES is of course its noninvasiveness that allows one to perform presurgical mapping without risks or time pressure and therapeutic stimulation in populations where invasive procedures cannot be considered. Although En-TMS is administered transcranially, the lack of distortion of the magnetic field by intervening tissues has allowed it to be developed to a level of accuracy that is at par with DES in clinical applications (Takahashi and Vajkoczy 2013).

The three-dimensional (3D) structure of the head is often approximated with the so-called spherical model, in which the conductivity is assumed to be spherically symmetric, i.e., being only a function of the distance from the center of the sphere. In areas of the head where the local curvature is approximately spherical, this approximation has been shown to provide quite accurate results both in practice (Picht et al. 2011) and in simulations (Nummenmaa et al. 2013). However, the spherical model is accurate only if the sphere is fitted to the local curvature of the head, not the whole head (Fig. 1.4). This is precisely what is done by the En-TMS systems designed for the demanding clinical work such as presurgical localization of eloquent cortex.

It is particularly straightforward to calculate the TMS-induced electric fields inside a spherically symmetric conductor. Consider a triangular path within a (locally) spherically symmetric conductor as in Fig. 1.4, where one of the vertices is at the center of symmetry. At the short edge the triangle, the electric field component $\boldsymbol{E}_l$ in the short-edge direction, $\boldsymbol{e}_l = \boldsymbol{\Delta l}/\Delta l$, is

$$\boldsymbol{E}_l(t) = \boldsymbol{E}(t)\cdot\boldsymbol{e}_l = M\left(dI(t)/dt\right)/\Delta l,$$

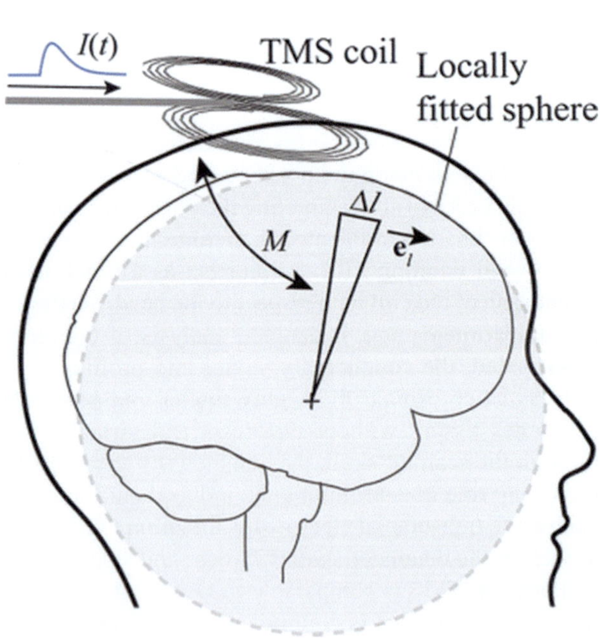

Fig. 1.4 Computing the TMS-Induced Electric Field. The triangle construction can be used to compute the TMS-induced electric field. The electric field component in the direction $\boldsymbol{e}_l$ at the short edge of the (imagined) triangle is $E_l = dI/dt\ M/\Delta l$, where M is the mutual inductance between the TMS coil and the triangle, $I(t)$ is the current fed into the TMS coil and Δl the length of the short edge of the triangle. The vertex opposite to the short edge is at the center of the sphere that is fitted to the local radius of head curvature near the location of the coil or the short edge

where M is the mutual inductance between the triangular path and the TMS coil; Δl is the length of the short edge of the triangle; $I(t)$ is the current fed into the coil. Easily programmable formulae for mutual inductances have been presented by Grover (1946).

It may be worth mentioning another relationship (Ilmoniemi et al. 1996) that can be used to compute electric field values, not only in the case of the spherical model, but in the general case. This relationship ties TMS to what is common knowledge in MEG. Namely, the electric field $\mathbf{E}(\mathbf{r}')$ at location $\mathbf{r}'$ induced by TMS coil current I is proportional to the "lead field" $\mathbf{L}$ of the coil

$$\mathbf{E}(\mathbf{r}') = dI / dt\, \mathbf{L}(\mathbf{r}').$$

In MEG, the lead field expresses the sensitivity of a pickup coil to primary or source currents $\mathbf{J}^{\mathrm{p}}$ in the brain so that the flux Φ threading the pickup coil is a weighted integral of the source currents:

$$\Phi = \int \mathbf{J}^{\mathrm{p}}(\mathbf{r}')\cdot\mathbf{L}(\mathbf{r}')\mathrm{d}\mathbf{r}'.$$

If a head model (spherical, realistically-shaped, or more detailed realistic model) and a solution to the forward problem are available, i.e., if one can calculate $\Phi(\mathbf{Q}, \mathbf{r}')$ threading the coil due to current dipole $\mathbf{Q}$ at location $\mathbf{r}'$, then $\mathbf{L}(\mathbf{r}')\cdot\mathbf{Q} = \Phi(\mathbf{Q}, \mathbf{r}')$. The Cartesian components of the lead field at $\mathbf{r}'$, $\mathrm{L}_x(\mathbf{r}')$, $\mathrm{L}_y(\mathbf{r}')$, and $\mathrm{L}_z(\mathbf{r}')$ can thus be calculated by computing $\Phi(\mathbf{e}_x, \mathbf{r}')$, $\Phi(\mathbf{y}_x, \mathbf{r}')$, and $\Phi(\mathbf{e}_z, \mathbf{r}')$, respectively.

As pointed out above, the local sphere model in En-TMS is sufficiently accurate for state-of-the-art presurgical localization of motor and language areas. However, the accuracy can be further increased if more realistic models of the head are available. The boundary-element method (BEM) (Nummenmaa et al. 2013) can be used if the head conductivity structure is approximated by compartments of uniform conductivity. Often, 3- or 4-compartment models are used in these calculations, but even a single-layer model (intracranial space with uniform conductivity) is quite accurate because current flow within the skull, which has poor electrical conductivity, has a very limited effect on the induced currents in the intracranial space. For the three-compartment model, head MRI is segmented to intracranial, scalp, and skull volumes. In the four-compartment model, the intracranial cavity is further divided into two parts: the cerebrospinal fluid and the brain. A limiting factor of BEM in practical real-time En-TMS has been the computational load, but this limitation is already subsiding with better algorithms and faster computers.

In principle, even more accurate modeling can be achieved with the finite-element method (FEM), in which the conductivity is described at the level of small voxels, each of which can be assigned an individual conductivity value; these values can even be anisotropic. Unfortunately, FEM suffers from several practical issues. First, the computation time is far too long for real-time navigation. Second, with typical voxel sizes of 3–5 mm, conductivity boundaries and conductivity gradients become very poorly represented and the results from FEM can be less accurate than those from BEM.

A practical, real-time solution to En-TMS was developed by Nexstim Ltd. (Helsinki, Finland). Keeping track of the stimulator parameters and the coil's location, orientation, and tilting, the system calculates the electric field using the local sphere model fitted to the individual head shape in a region below the coil. This local-sphere approach is also called multiple-sphere approach. Local-sphere modeling has turned out to be highly successful and is now approved for clinical use (U.S. Food and Drug Administration = FDA Predicate 510(k) number for the Nexstim NBS system K091457 for motor mapping and K112881 for the combination of the Nexstim NBS system used with NexSpeech® for language mapping).

1.3.2 How Does nTMS Work?

The technique of nTMS was developed to make targeting based on individual MRI (the navigation map) possible and visible. For the accurate navigation needed in diagnostics or in modulation of brain activity in targeted therapy, one has to eliminate uncertainty in the technical (nonphysiological) factors that affect the stimulation level at the neurons. For both, nTMS accuracy and visibility, the physical parameters of the coil (location, orientation, tilt, size, and shape of the copper windings) and the stimulator pulse waveform and amplitude need to be accurately determined and taken into account.

In addition to the technical TMS parameters, we need physical parameters of the individual head and brain for the calculation of the stimulating intracortical electric fields; structural information for this purpose can be obtained from MRI. The MRI scan needs to include the whole head so that the nose, ears, and head surface are all visible and thus accessible as landmarks. Typically, T1-weighted, 3D MRI scans with 1 × 1 × 1-mm voxels are used, as they are also required for intraoperative neuronavigation. With the MRI, individual brain morphology with patient-specific structural forms such as cortical convolutions becomes visible and thus applicable for nTMS targeting.

Navigation tools needed to locate the TMS coil and to align the MRI scan with the patient's head include the head and coil trackers and the digitizer pen; these are located in real time with a 3D position sensor (Figs. 1.5 and 1.6).

The head tracker, acting as a reference tool, is used to locate the head and to monitor its movements. The coil tracker can then indicate the coil's location and 3D orientation in relation to the head, i.e., in the head coordinate system. The digitizer pen is used to determine arbitrary location coordinates; e.g., to align the MRI scan and the head by pointing at landmarks that are visible both in the MRI and on the head. After the registration of the MRI and the individual head coordinate systems (aligning the MRI and the real head), the stimulating electric field in the brain can be calculated from the knowledge of coil and stimulation parameters and visualized in real time on the MRI (Ruohonen and Ilmoniemi 2005).

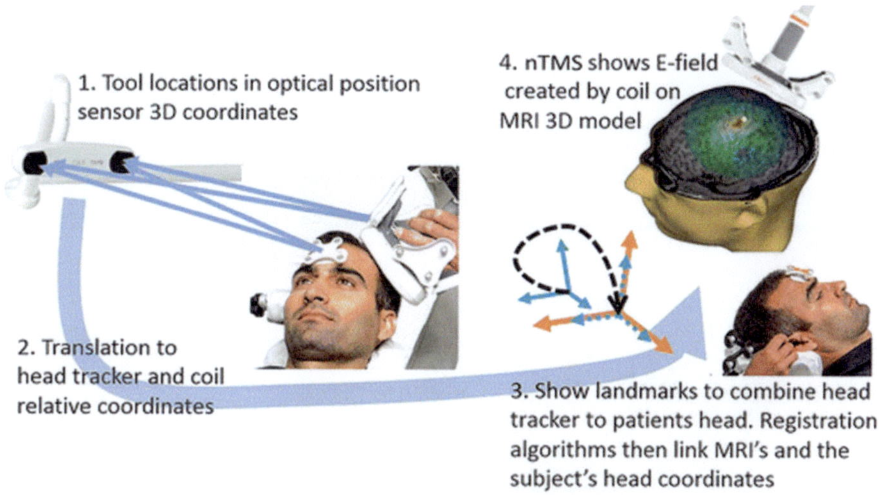

Fig. 1.5 3D navigation. The principle of En-TMS with online calculation and targeting based on electric fields (E-field). The stimulation session is started with registration to align the MRIs to the individual head. After successful registration, the stimulating electric field is displayed online over the 3D MRI brain topology to guide stimulation targeting when the coil is moved

1.3.3 Contributors to Accuracy

Being critical for many clinical applications, the accuracy of nTMS targeting has raised many questions and discussions as long as this technique has been available. Although it is crucial for operators to understand the determinants of spatial accuracy of nTMS, there is very limited literature or published data on this topic. Some of the publications report nTMS precision (variability or repeatability) instead of accuracy; this can lead to misinterpretations. It is therefore important to clarify the meaning of these terms. The U.S. FDA (2014) gives a clear definition of accuracy by referring to Institute of Electrical and Electronics Engineers (IEEE) standards: "The measure of an instrument's capability to approach a true or absolute value" and "A quantitative measure of the magnitude of error." Precision is a quite different measure: "The relative degree of repeatability, i.e., how closely the values within a series of replicate measurements agree."

Based on the FDA definition, nTMS accuracy is a measure of the typical distances from the visualized stimulation point to the real in vivo location of the maximal electric field. This targeting error is of course a random variable and usually unknown; a useful way to describe this error is its standard deviation. In a good En-TMS system, the standard deviation of the error may be 5–6 mm, comparable to that in determining eloquent cortical areas during surgery by direct electrical stimulation (DES), includes stimulation of both the cortex (DCS) and subcortical structures.

The navigation system displays the location of the electric field maximum on the individual MRI of the cortex. The accuracy of determining the visualized electric field depends on how accurately the multiple parameters and variables used in the calculations are known: intensity of stimulus, stimulator parameters, coil shape and

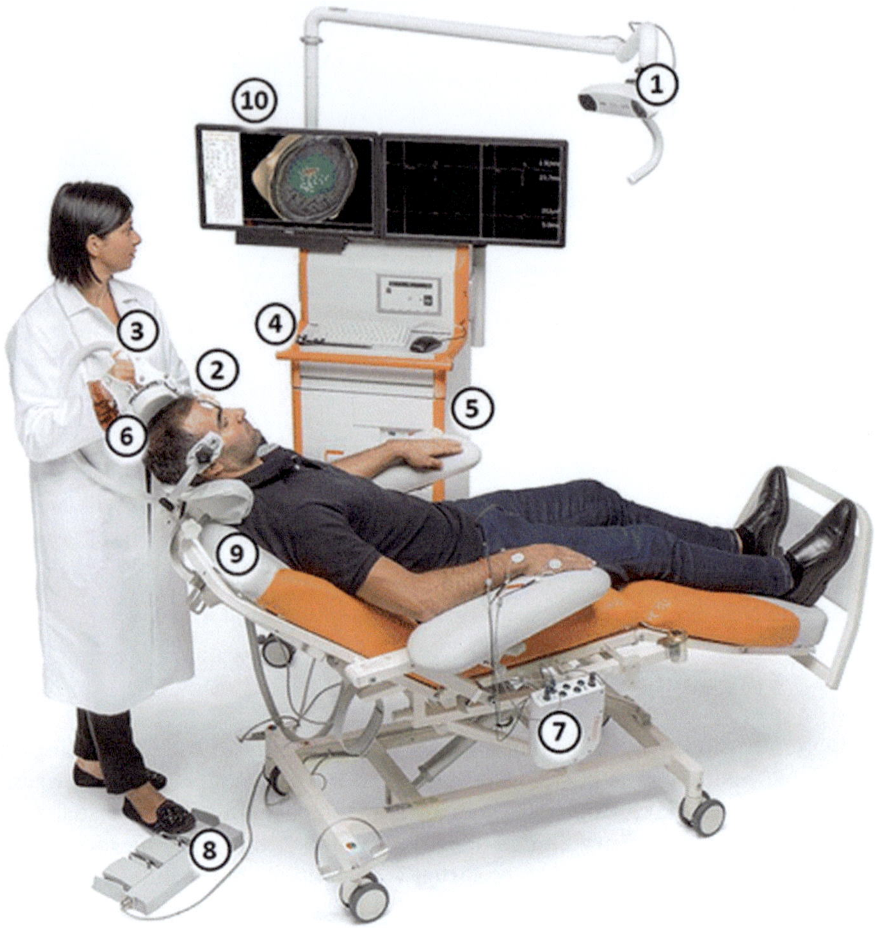

Fig. 1.6 Electric Field-Navigated TMS Hardware. 1. Positioning sensor (optical)—tracking the tools; 2. Head tracker—location and movement of the head; 3. Coil tracker—location, orientation, tilting of the coil; 4. Digitizer pen—enables recording of individual points, e.g., for MRI registration; 5. TMS stimulator—producing the controlled pulse adjusted by the navigation software; 6. Coil—producing the TMS pulse, output controlled by the navigation software; 7. EMG—response measurement for MEPs, synchronized to each stimulus and enabling the coloring of the stimulation spot based on the MEP amplitude; 8. Foot pedal—to apply stimulus when the electric field is targeted at the right spot; 9. Chair with head rest—good ergonomics for operator and subject. 10. nTMS system display for controls and targeting the stimulating electric field

location, orientation and tilting of the coil, head shape, size, and electrical conductivities. The position sensor used for tracking the navigation tools must have a high spatial accuracy. Indeed, a typical error for a modern optical position sensor is only 0.2 mm, but this is only a minimal part of the total error for an nTMS system. Unfortunately, in some of the published nTMS literature, the accuracy of optical position sensor is reported instead of the overall accuracy. Also sometimes only the precision with which a previous nTMS coil location can be determined is mentioned; precision is typically much better than accuracy, since many of the errors are constant across the repeated stimulations like inaccuracies in MRI, uncertainties in the conductivity model, or errors in tracker calibration. To properly characterize the En-TMS accuracy, it should be divided into different groups (here, A–D) that independently impact the reliability of the determination of the visualized stimulation spot location:

A. Determination of coil location and 3D orientation
   - Manufacturing tolerances for the coil and coil trackers, digitizer pen, and head tracker
   - 3D localization of the reference (head) tracker and the coil by optical (or electromagnetic) techniques
B. Computational model of the stimulating electric field
   - Fitting of the conductor model to head-shape information
   - Modeling the stimulation coil (shape, size, details of the copper windings)
   - Head conductor model (sphere, multiple-compartment model)
   - Computational method (spherical-model formulas, BEM, FEM)
C. Errors in alignment of MRI scan to individual head
   - Registration algorithm that aligns the individual MRI and the real head
   - Operator errors in alignment procedures, when pointing of head landmarks with the digitizer pen
   - MRI resolution and quality
   - MRI susceptibility distortions
D. Movement of the reference (head) tracker during examinations:
   - Head tracker movement (reference tool) during the stimulation session, affecting the calculated coil position and the electric field pattern

If the system is built from individually designed standalone components, including a TMS with a coil not initially designed to be navigated, navigation software, position sensor with trackers to be installed to the coil and an EMG device, all connected together as an nTMS system, the overall accuracy can be determined only by testing the particular combination. In addition to this, there exist nTMS systems that operate without electric field navigation. For the integrated En-TMS system that calculates the stimulating electric field, the accuracy can be derived from independent uncertainties by the square root of the sum of squared errors. Table 1.1 gives an example on how the accuracy of one En-TMS system (Nexstim Plc., previously Nexstim Ltd.) has been calculated; this system was clinically validated for the localization and assessment of the motor cortex and motor tract integrity for preoperative planning. The specifications of other manufacturers' equivalent accuracy specifications are not published or accessible yet.

Table 1.1 Error sources

Error source	Mean error in electric field stimulation spot (mm)
Coil localization: Manufacturing tolerances for the coil and coil trackers 3D localization with optical position sensor	1.6
Head tracker movement (reference tool) during the stimulation session	3.1
Computation of the stimulating electric field: The output and characteristics of stimulation coil Model of the intracranial electric field Fitting of the model to the individual head	3.8
Registration to anatomical MRIs: MR image imperfections Registration algorithms	2.5
System accuracy with root square sum of all error sources: $\sqrt{(1.6)^2+(3.1)^2+(3.8)^2+(2.5)^2}$	5.7

Error sources contributing to the accuracy of an En-TMS system (in this example: Nexstim eXimia NBS)

To integrate En-TMS into clinical workflow, dedicated En-TMS protocols are developed to ensure consistent targeting accuracy. The multiple factors affecting the results include the definition of individualized stimulation intensity, electric field direction, and response detection. The nTMS Workshop Group (2016) finalized a summary of their protocol meeting, describing the motor mapping protocol that has been demonstrated to work in presurgical planning. Section 1.5 of this chapter includes a summary of the protocol's key factors; it also lists an nTMS system's general software and hardware features that are essential for the clinical use of En-TMS.

1.3.4 Electric Field Navigation Versus Line Navigation

Currently, there are two basic nTMS principles. En-TMS is a technique where the stimulating electric field is calculated and navigated at the cortex, while in the so-called "line navigation" the stimulation spot is assumed to be on the line that passes through the geometric center of the coil and is perpendicular to the coil surface. Line navigated TMS (Ln-TMS) is prone to increased errors when the coil is not held continuously tangentially against the head. In practice, there is a continuous need to adjust coil tilting when moving the coil along the head surface. The effect of the head in EN-TMS versus Ln-TMS can be seen in primary motor cortex (M1) hand area stimulation of a healthy subject (Figs. 1.7 and 1.8).

As highlighted in Figs. 1.7 and 1.8, electric field navigation and line navigation may indicate the stimulation spot in different gyri depending on local head curvature and coil tilting. There is very limited literature available where these two

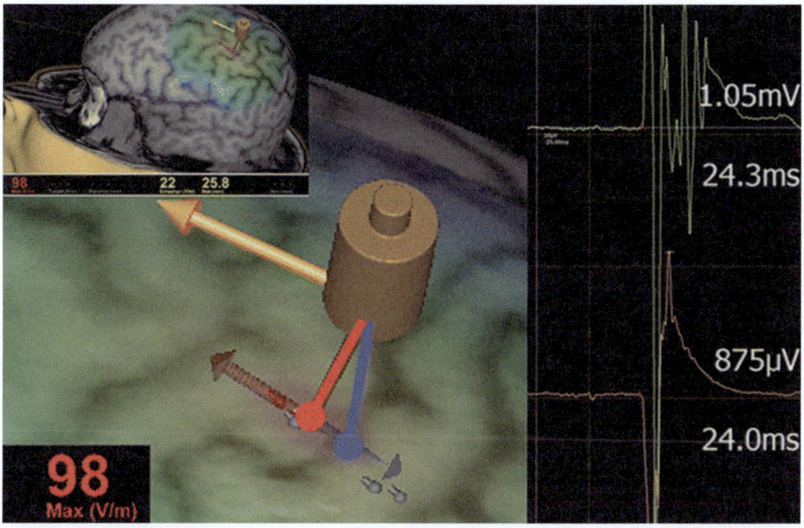

Fig. 1.7 M1 stimulation with electric field navigation. *Right*: EMG view from the nTMS software: the top green line indicates a MEP from the abductor pollicis brevis (APB) with 24.3 ms latency and 1.05 mV amplitude; the bottom purple line shows a MEP from abductor digiti minimi (ADM) with 24.0 ms latency and 875 µV amplitude. *Left*: 3D stimulation targeting view from the nTMS: view from electric field-targeted stimulation of M1. The red line and dot indicate the maximum of the electric field in M1. The electric field in this hotspot, 98 V/m, caused the 1.05 mV MEP; the electric field orientation was perpendicular to the sulcus at the APB representation area. The blue line and dot visualize line navigation, which points misleadingly to the primary sensory cortex (S1), which has no motor function to produce the MEPs from APB or ADM. This was demonstrated with the electric field targeted S1 stimulation visible in Fig. 1.8

methods have been compared. A published study by Sollmann et al. (2016) was designed to compare stimulus targeting in tumor patients with an En-TMS system and an Ln-TMS system. In this clinical study, only the data generated by the clinically validated FDA-approved En-TMS system could be used for tumor-patient diagnostics (Picht et al. 2011; Picht 2014; Forster et al. 2011; Krieg et al. 2012, 2013; Tarapore et al. 2012). However, the clinical team compared motor mapping with a broadly distributed Ln-TMS system also marketed for preoperative motor mapping in 12 patients suffering from brain tumors in order to examine potential differences in clinical applicability, workflow, and mapping results between the two navigation techniques.

The nTMS-naive operator who had no previous nTMS experience was trained for both devices by the manufacturers. Motor mappings were performed with both nTMS systems on each patient in an alternating pattern starting with either En-TMS or Ln-TMS. Study outcomes comparing En-TMS and Ln-TMS were as follows:

1. The upper and the lower extremity motor maps generated by En-TMS and Ln-TMS were different, resulting in only partial overlap of the results.

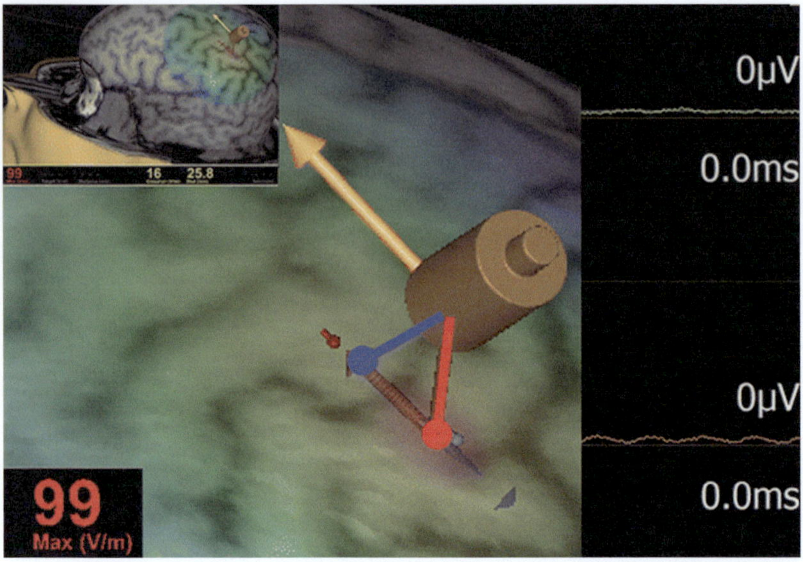

Fig. 1.8 S1 stimulation with electric field navigation. *Right*: EMG view from the nTMS software: the top green line indicates the absence of MEP from APB and the bottom purple line the absence of MEP from ADM. *Left*: 3D stimulation targeting view: The red line and dot indicate the maximum of the electric field in S1 (postcentral gyrus). Here, the electric field of 99 V/m causes no MEP. The blue line and dot point to assumed stimulation location in M1 according to line navigation. Line navigation is misleading in this case due to the fact that no MEP (amplitude 0 μV) is coming from the APB or ADM muscles

2. In two patients, Ln-TMS and En-TMS identified the motor hotspot in different gyri. It is important to note that this is initial data for the comparison due to the fact that only with 6 out of 12 patients the ipsilateral mappings and 8 out of 12 contralateral mappings could be completed due to technical hardware problems in the Ln-TMS device (these problems were unrelated to the stimulation targeting method).
3. Distances between En-TMS and Ln-TMS motor hotspots were 8.3 ± 4.4 mm on the ipsilesional and 8.6 ± 4.5 mm on the contralesional hemisphere.
4. The number of those (motor-positive) spots where nTMS succeeded to elicit MEPs was significantly higher for En-TMS (En-TMS vs. Ln-TMS: 128.3 ± 35.0 vs. 41.3 ± 26.8, $p < 0.0001$).
5. The ratio of motor-positive spots and the number of stimulations was higher for En-TMS: En-TMS vs. Ln-TMS: 38.0 ± 9.2%vs. 20.0 ± 14.4%, $p = 0.0031$.
6. Both technical approaches are quick to learn and fast in operation; it allows mapping of a large area of the cortex in a given time. The learning process for En-TMS and Ln-TMS was studied by comparing overall mapping durations and speed for the first and the last mapping conducted: In En-TMS, the time per stimulation was reduced by 75.9% while the number of stimulations per mapping was increased by 66.5%. In Ln-TMS, a reduction of time per stimulation

was only 29.0% and the number of stimulations per mapping was decreased by 9.1% meaning that the necessity of manual placing the coil on the head tangentially makes the mapping itself more time consuming despite learning effects.

According to the authors, the above-mentioned significantly higher ratio of positive responses could be seen favorable toward En-TMS regarding more elaborate capabilities to optimize coil location, tilting, and orientation during the stimulation, as coil orientation has been reported to be one of the key factors for optimized stimulation (Ruohonen and Karhu 2010; Schmidt et al. 2015). En-TMS is able to calculate and visualize the electric field online with its orientation and dose, allowing the continuous optimization of coil positioning. The authors stated that Ln-TMS shows the location of the coil without providing information about coil angulation with respect to the subject's head. The way the Ln-TMS indicates the coil's orientation with the line projection may lead to nonoptimal coil orientation and tilting with decreased electric field at the cortex; this could explain the lower rate of overall stimulations that led to a positive response.

Conclusions from this first En-TMS and Ln-TMS motor mapping comparison in neurosurgical context: *Although both nTMS systems tested in the present study are explicitly designed for application during motor mapping in patients with brain lesions, there are differences in applicability, workflow, and results between En-TMS and Ln-TMS, which should be distinctly considered during clinical use of the technique. However, to draw final conclusions about accuracy, confirmation of motor-positive Ln-TMS spots by intraoperative stimulation is crucial within the scope of upcoming investigations. As already done for En-TMS, future studies using intraoperative DES have to examine whether Ln-TMS is also able to detect valid and sufficiently accurate motor maps in the neurosurgical context.*

1.3.5 Clinical Validation of Accuracy

The accuracy in clinical practice of En-TMS systems has been evaluated and clinically validated in brain tumor patients only with the Nexstim system so far. Corresponding clinically validated data are not currently available for Ln-TMS.

A meta-analysis of Takahashi and Vajkoczy (2013) reviewed articles where the En-TMS system was used to map patients with rolandic brain tumors. They reported: "The mean distance between motor cortex identified on nTMS and DES by using the mean distance in 81 patients described in 6 quantitatively evaluated studies was 6.18 mm" (Takahashi and Vajkoczy 2013). The review article also concludes that nTMS "is the only modality that is analogous to DES in that it allows for electrical stimulation of the brain and observation of the induced effect" due to the fact that the nTMS "technique spatially correlates well with the gold standard of DES." Chapter 7 outlines the details of the different studies analyzed in this meta-analysis.

Since the FDA motor mapping approval of En-TMS in December 2009, 132 original neurosurgical articles for preprocedural planning, with 5645 patients have been published. These include 13 presurgical motor mapping articles, with 1231 patients reporting the impact of nTMS motor maps on better clinical outcomes. Similar clinically validated data basis are not currently available for Ln-TMS.

1.4 Practical Issues and Examples of the Use of nTMS

1.4.1 General Consideration

TMS is based on the fact that a rapidly changing magnetic field induces an electric field to cause neuronal activation or transient alterations of neuronal excitability in order to diagnose or modify brain functions. Electric field-guided navigation was developed to rule out uncertainties and to improve repeatability. Diagnostics of the motor cortex can be done by single-pulse TMS to probe corticospinal excitability (CSE). However, CSE measurements have variability due to physical and physiological factors. An integrated En-TMS system controls, guides, and records physical parameters of TMS like coil tilting, orientation, location, and stimulator output and estimate the stimulating electric field at the individual brain anatomy. A system setup with an integrated EMG has enabled investigations in which these parameters are partitioned with stepwise regression to find out the variance of physical parameters in CSE measurement (Schmidt et al. 2015). This study also indicated the validity of calculated and visualized electric fields with the model used by an En-TMS system. Results of the study highlight the following:

- CSE variability is reduced when the physical parameters can be controlled with the help of navigation.
- The calculated and visualized electric field of the En-TMS system is valid.
- The spatial accuracy of En-TMS is better than 5 mm.
- Small fluctuations of physical parameters can influence statistical comparisons of CSE measurements.

1.4.2 Factor—Electric Field Location

Schmidt et al. (2015) conclude that in their En-TMS study, the spatial definition of En-TMS is possibly as small as 2 mm. Location changes larger than 2 mm resulted in significantly increased CSE variability, meaning that CSE measurements are much more susceptible to small changes in physical parameters than might be expected from previous studies.

1.4.3 Factor—Orientation of the Stimulating Electric Field

The orientation of the electric field has a clear impact on TMS responses (Schmidt et al. 2009; Kallioniemi et al. 2015). The importance of the optimal orientation perpendicular to the gyrus is reported in En-TMS and DES comparison studies (Picht et al. 2011; Krieg et al. 2012; Tarapore et al. 2012), suggesting that suprathreshold nTMS mapping may lead to combined activation of transsynaptic pathways and direct stimulation of axonal pathways deeper in the gray matter and in the bends of white matter axons (Amassian et al. 1992). Action potentials are created when large enough membrane depolarization is caused by the TMS-induced electric field. TMS is most effective when the stimulating electric field is oriented longitudinally and orthodromically with respect to the largest population of neurons at the stimulation target, which in most stimulation spots means the direction perpendicular to the gyrus. This is why the orientation of the coil has to be confirmed before individual resting motor threshold (rMT) definition. The rMT is required to find the optimal mapping intensity by repeated stimulation of the optimal hot spot location while turning the coil in 20° steps in both directions as long as responses are seen. When doing motor mapping close to a lesion or when obtaining motor responses far away from the assumed primary motor cortex, it is important to repeat the stimulations with electric field orientation changed by +45° and −45° to test the population of the neurons at the stimulation target and to confirm consistency of the motor responses.

1.4.4 Factor—rMT-Guided Stimulation Intensity

To create a focal stimulation for high-resolution mappings, the stimulation intensity must be defined individually by determining the rMT and by adjusting the stimulation intensity to 110% rMT to get responses every time when a motor-relevant area is targeted. It is important to realize that stimulator output in percent of the maximum output from the capacitor does not tell much about the stimulation intensity at the cortex due to the impact of coil distance from the cortex, coil tilting, coil location, and coil orientation. While nTMS motor mapping with 110% rMT is regarded standard, some teams find the mapping result even more focal when using 105% rMT as stimulation intensity. If weaker pulses are given, e.g., 105% rMT, a smaller volume of neurons near the peak electric field is activated. However, the use of too low intensity harbors a severe risk to miss some functional areas during the mapping. Moreover, all validation studies were performed using a stimulation intensity of 110% rMT. In contrast, the case of strong pulses, e.g., 130% rMT, the threshold of activation is exceeded in a large volume, thus leading to more unspecific maps. For accurate rMT definition, the hot spot and optimal orientation must be determined before running the rMT algorithm. Suboptimal definition of the hotspot and electric field orientation may lead to higher mapping intensities due to an erroneously high rMT value, compromising the resolution of mapping due to activation of larger neuronal volumes. Thus, in addition to a focal coil, a pulse only slightly above activation threshold can be used for good focality.

1.4.5 Factor—Electric Field Strength and Location

Schmidt et al. (2015) stimulated the primary the motor cortex of healthy subjects, providing evidence for the importance and validity of En-TMS motor mapping. The key factor here is that the navigation system knows and controls the electronics and hardware for the stimulation delivery and response measurement. Schmidt et al. (2015) demonstrated that changes in electric field strength and location result in associated changes in MEP amplitudes, showing MEP susceptibility to small changes of electric field values calculated and visualized at the brain. Coil tilting and distance from the individual anatomy among the other parameters mentioned before affect the electric field. When moving the coil during the mapping session, it is important to follow the electric field values, location, and orientation.

1.4.6 Factor—Preactivation

Continuous triggered EMG monitoring of target muscles during the motor mapping is essential. Preinnervation in the target muscle is known to strongly modify the MEP amplitudes; in the Schmidt et al. (2015) study, more than tenfold effects were observed. To keep the mapping session stable, the preinnervation needs to be minimized by keeping the patient comfortable and thus the muscles relaxed. Temporary tension of the target muscle between the relaxations during the mapping helps to minimize preinnervation during stimulation.

1.4.7 Factor—Anatomy

To map the individual brain, the morphology with its unique shapes, structures, and cortical convolutions of each person needs to be taken into account. The orientation of the electric field must be defined and tested individually according to anatomical structures. A standardized coil orientation with respect to external landmarks of the skull is not sufficient for preoperative mapping of eloquent cortical areas or for delivery of therapeutic stimulation. The mapping also needs to exceed the area of interest, especially when individual patterns of sulci and gyri cannot be determined by any morphometric landmarks. In presurgical planning, the anatomical structures are often smeared by tumor growth, edema, bleeding, and/or vascular alterations. The orientation of the electric field needs to be changed again by +45° and −45° in these locations in order to check for the consistency of the motor responses.

1.5 Current Motor Mapping Protocol

The above-mentioned factors of accuracy are crucial for precise use of nTMS; especially En-TMS. In order to put these abstract technical data into a clinical context, this paragraph explains the different steps of nTMS motor mapping by using the protocol of the nTMS Workshop Group (2016) and by referring to the factors outlined in paragraph 4.

1.5.1 Preparation of the Session

1. Upload the patient's MRI to the system; based on the individual MRI, the software generates a 3D head model.
2. Prepare the patient (interview, checking for contraindications (Chap. 6), explaining the procedure, attaching EMG leads).
3. Align the patient's head with the MRI-based 3D head: (a) attach the head tracker to the patient's forehead as reference tool for tracking by the infrared camera; the head tracker will allow free head movement by the patient; (b) point with the digitizer pen at the three points predefined as MRI landmarks (left ear crus of helix, nasion, right ear crus of helix); (c) touch with the pen nine points around the head, guided by the software. Typical tool-tracking accuracy is better than 1.5 mm (mean); the head as a whole can be aligned with the MRI at 2 mm accuracy when the software-guided registration process with 3 + 9 points is followed.

1.5.2 Hot Spot Identification

1. Identify the hand knob at 20–25-mm depth from the scalp and adjust the stimulator output to 80–100 V/m.
2. Adjust the stimulation intensity to be suitable for mapping (muscle at rest, MEP latencies 15–25 ms).
   (a) If MEP amplitudes are >500 μV, lower the intensity 1–2% of the stimulator output.
   (b) If MEP amplitudes are <100 μV, increase the intensity until abductor pollicis brevis (APB) MEP amplitude is 100–500 μV.
3. Rough mapping (Fig. 1.9): Stimulate along the central sulcus toward the midline and toward the Sylvian fissure, keeping the electric field orientation perpendicular to the central sulcus, until no more MEPs are evoked. Select the largest APB response and set the corresponding location as the stimulation target.
4. Test the coil orientation: Check the EMG amplitude in this 0° orientation. Repeat stimulations on the marker with the aiming tool (Fig. 1.10, lower right corner panel), while turning the coil 20° in both directions after each stimulus. Check the amplitude (μV) after each stimulus. Select the coil orientation evoking the largest MEP for defining the rMT.

The En-TMS software provides an aiming tool that allows the operator to stimulate only if the location is within 2 mm and coil orientation and tilting within 2° of those of the target position to ensure that rMT determination is performed with the coil at the predetermined location and orientation (Fig. 1.10). The stimulation of the

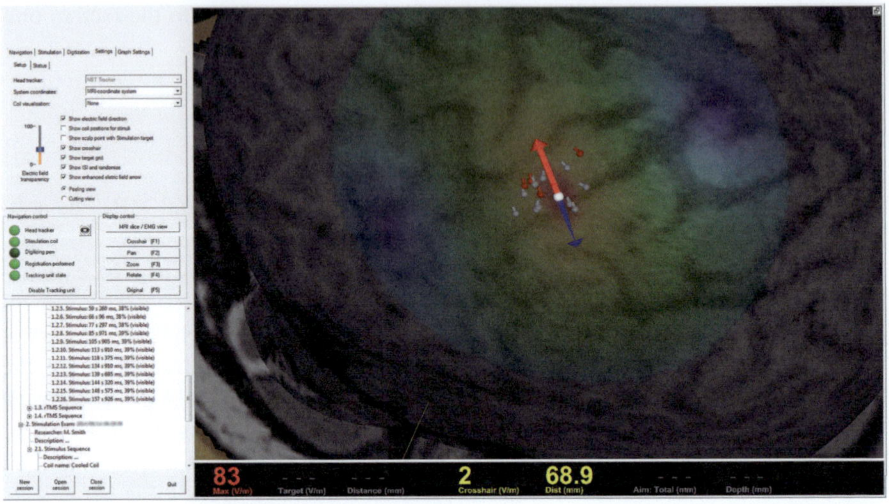

Fig. 1.9 Hotspot identification. Colors correspond to motor response in each location: white: MEP ≥1 mV peak-to-peak amplitude; red: MEP = 50–500 µV; gray: MEP ≤50 µV (no response). The direction of the stimulating electric field is highlighted with the red and blue arrows. The stronger stimulation direction is displayed as a red arrow and the weaker stimulation direction as a blue arrow. The arrows start at the stimulation hotspot. The En-TMS software calculates the electric field at the chosen target location based on a validated head model, which utilizes a predefined set of over 40,000 locally fitted spheres; the coil design and manufacturing specifications enable focal and precise delivery of TMS pulses that generate electric field patterns that corresponds to the one calculated by the software; the recording and display of the TMS-evoked EMG responses is synchronized with stimulus delivery, providing instant feedback; the software guides the user to place the nTMS coil in the right place and orientation in relation to brain anatomy to focus the electric field at the target and to enable electric field to be perpendicular to the stimulated sulcus

same anatomical location with the coil in different orientations is allowed by the system when the location control is not activated to ensure that rMT determination is performed with the coil orientation producing the largest responses.

1.5.3 rMT Determination

1. Select the optimal stimulation from Sect. 1.5.2, step 4 as reference stimulation to be repeated with exactly the same coil location, tilting and orientation with the help of the targeting tool (Fig. 1.11 lower right corner window).
2. Start the rMT determination software by selecting the target muscle EMG channel and setting the ending parameters: Integrated online EMG monitoring and response control for each stimulus enables computer-aided rMT determination for automated control of mapping intensity (*see* Fig. 1.11 for details).
3. Verify the rMT by repeating the stimulation with the suggested value 10 times. If fewer than 5 responses out of 10 stimuli elicit a proper response, increase the intensity by 1% and start over again. If you get 10 responses out of 10 stimuli, decrease the stimulation intensity by 1% and start over again.

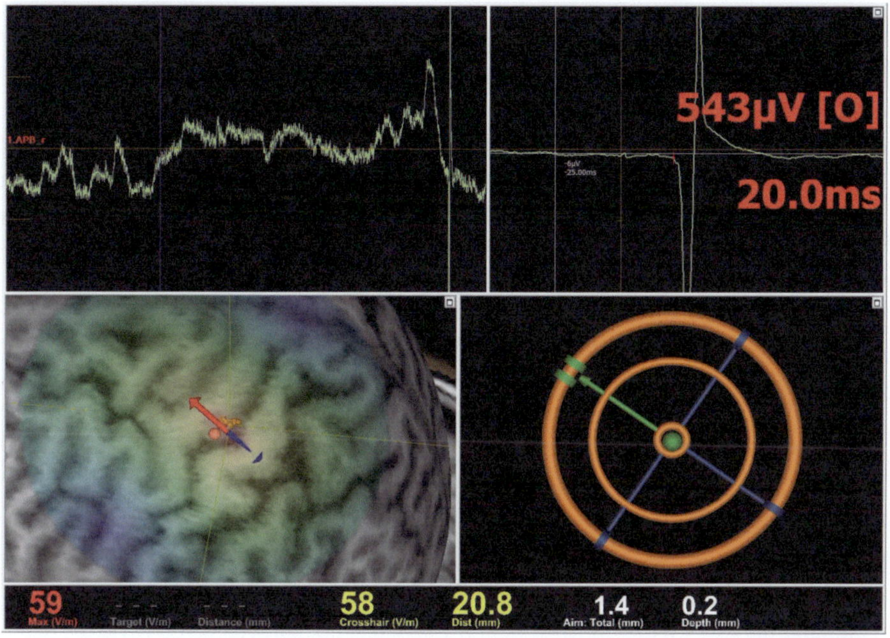

Fig. 1.10 Visualizing electric field orientation with En-TMS. *Left lower corner*: Location of the electric field maximum and electric field orientation at the maximum. *Right lower corner*: The aiming tool that helps to reproduce stimulations with identical coil location and 3D orientation

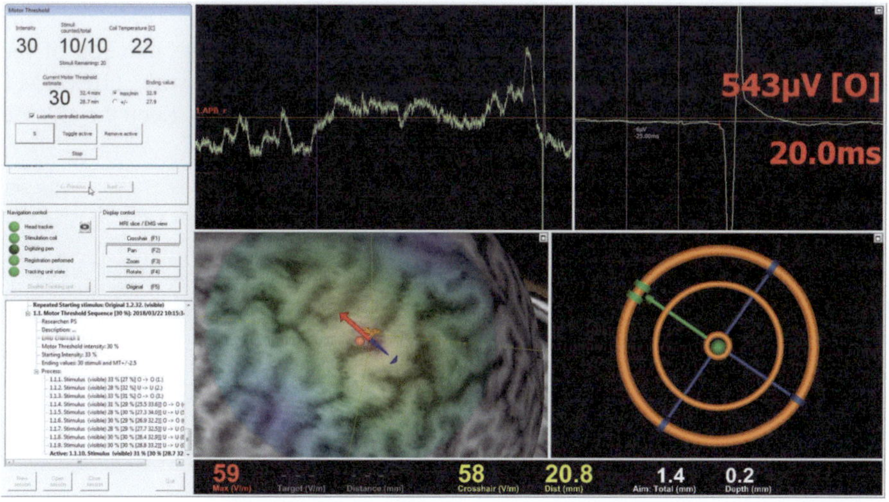

Fig. 1.11 EMG response. Integrated online EMG monitoring and response control for each stimulus enables computer-aided rMT determination. After each stimulus, the peak-to-peak amplitude of the muscle response is checked and the software classifies the response as "over" or "under" the amplitude threshold (character "O" or "U"). Responses with nonphysiological latency or amplitude as well as those with preceding muscle tension is rejected. After each stimulus, the algorithm shows and adjusts the stimulator output for the next stimulus. The rMT determination is finished when the error bounds of the rMT estimate fall below the set value or when the number of stimuli reaches a prespecified maximum

1.5.4 Mapping of Hand Motor Area with 110% rMT

1. Set the stimulator output to 110% of the rMT defined previously (Optional: Refined mapping at 105% rMT).

   Stimulate along the central sulcus, precentral gyrus, postcentral sulcus, and precentral sulcus with 2–3 mm spacing while keeping the electric field orientation perpendicular to the local sulcus (Fig. 1.12). Expand the mapping area as long as motor responses are detected. There should be a rim of 1 or 2 rows surrounding the MEP-positive area, in which nTMS was not able to elicit MEPs. When obtaining motor responses far away from the assumed primary motor cortex or close to the lesion (e.g., tumor), repeat stimulation with +45° and −45° in these locations to test the consistency of the motor responses.
2. Verify registration integrity after mapping. The nTMS system enables 3D visualization of the digitizer pen with the 3D MRI after the 12-point MRI-to-head registration. The operator can take at any time of the session the digitizer pen and move it over the scalp to check from the nTMS display that the tip of the pen moves correspondingly over the scalp in the 3D head view. A separate digitization examination enables digitization of extra scalp landmarks that can be seen in the 3D head view and checked at any time on the system 3D view by pointing at landmarks with the digitizer pen. If the pen is not shown correctly on the head, the operator must redo the 12-point MRI-to-head registration.

1.5.5 Mapping of Leg Motor Area (TA, Soleus)

1. The tibialis anterior muscle (TA) representation area near the junction of the central sulcus and longitudinal fissure is stimulated first with 110% rMT +20 V/m intensity. Leg muscles should be kept initially at rest. Keep the electric

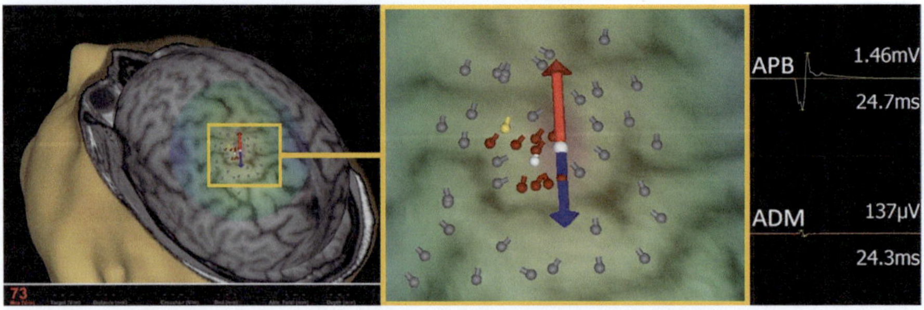

Fig. 1.12 Electric field orientation for hand area mapping. This screenshot shows a hand area mapping with 110% rMT, with APB and ADM as target muscles. The stimulating electric field orientation is to be maintained perpendicular to the central sulcus. The direction of the electric field is indicated in each stimulus location with a small sphere and a cylinder. The cylinder direction is the same as the red arrow during targeting (middle view). Colors correspond to motor response strength as in Fig. 1.9

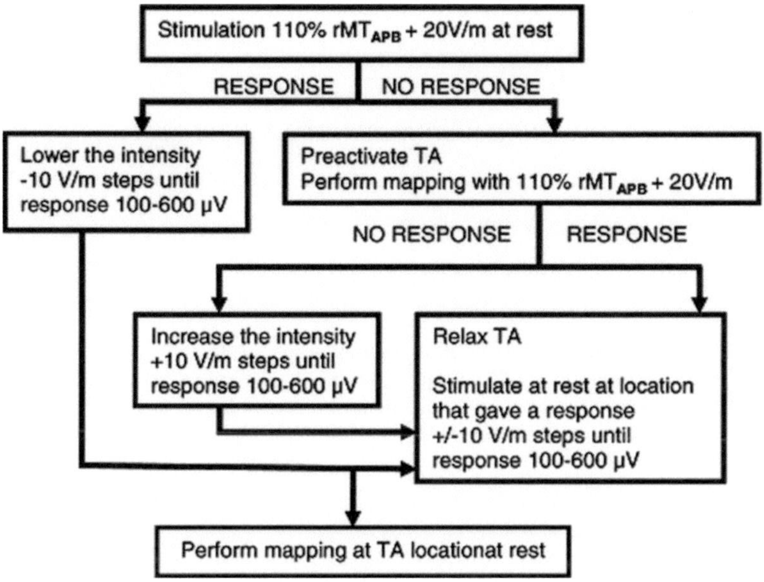

Fig. 1.13 Algorithm for lower extremity mapping. The chart describes the different steps of lower extremity mapping

field orientation perpendicular to the longitudinal fissure, the electric field pointing laterally.

2. If motor responses are obtained, continue stimulation by following the longitudinal fissure 2 cm anteriorly and 2 cm posteriorly. Stimulate also perpendicular to the medial parts of central sulcus, up to 3 cm from the longitudinal fissure as far as responses emerge.
3. If no responses are obtained, use the following procedure as outlined in Figs. 1.13 and 1.14.
4. Verify the registration after mapping. Move the digitizer pen over the scalp to check from the nTMS display that the tip of the pen moves correspondingly over the scalp and touches the scalp landmarks in the 3D head view.

1.5.6 Mapping of Facial Muscles (Mentalis, Orbicularis Oris)

1. Facial representation area in the lateral parts of the central sulcus in the precentral gyrus is stimulated first with 110% rMT +20 V/m stimulation intensity (or perform an rMT determination for the face area separately). Facial muscles should be kept initially at rest. Keep the coil orientation first perpendicular to central sulcus.

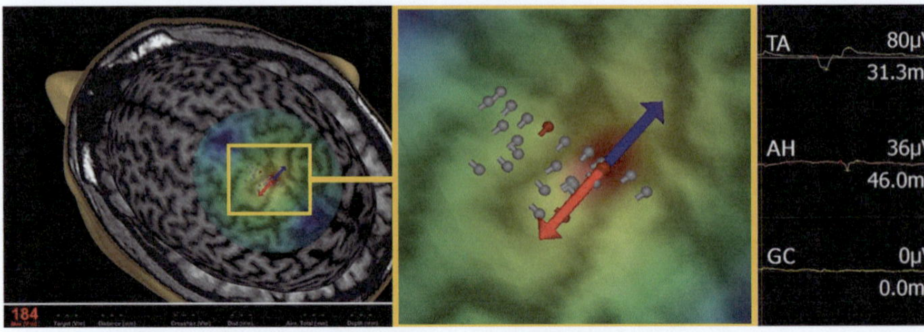

Fig. 1.14 Electric field orientation for lower extremity mapping. This screenshot shows a leg area mapping of three target muscles (TA), abductor hallucis (AH), and gastrocnemius muscle (GC). The stimulation is done with an electric field orientation perpendicular to the longitudinal fissure with the electric field oriented laterally at the junction of the central sulcus and the longitudinal fissure. Colors as described in Fig. 1.9

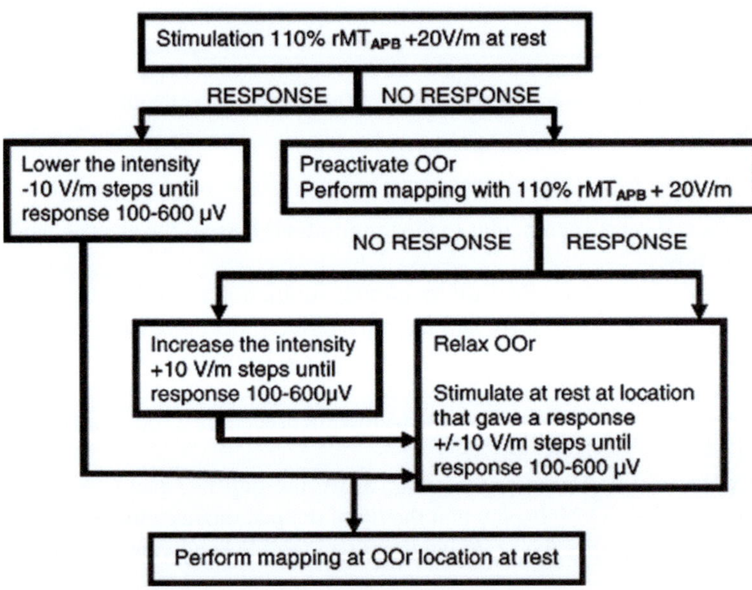

Fig. 1.15 Algorithm for face mapping. The chart describes the different steps of face mapping

2. If motor responses are obtained, continue stimulation by following the longitudinal fissure 2 cm anteriorly and 2 cm posteriorly as far as responses emerge.
3. If no responses are obtained, use the following procedure as outlined in Figs. 1.15 and 1.16.
4. Verify the registration after mapping.

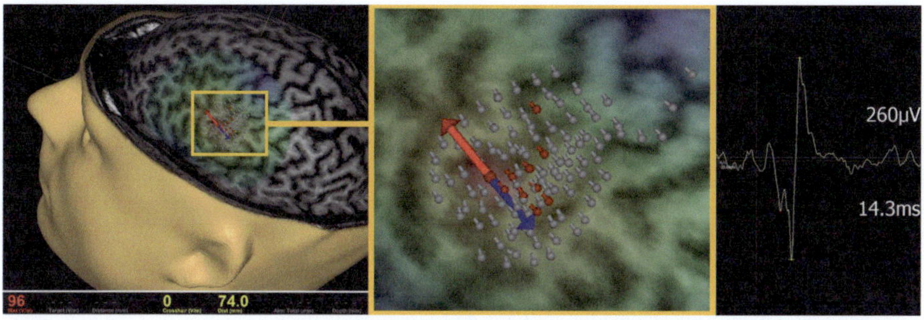

Fig. 1.16 Electric field orientation for face area mapping. This screenshot shows a face area mapping of the orbicularis oris muscle (OrO). The stimulation should be done with electric field orientation perpendicular to the central sulcus. Colors as in Fig. 1.9

1.6 Conclusion

A strong, quickly changing magnetic field generated by a coil outside the head induces, according to Faraday's law, a stimulating electric field within the brain. The navigation of the stimulating electric field was developed to solve many of the issues associated with nTMS accuracy, reliability, reproducibility, and usability. In the clinically validated En-TMS system, the accurate targeting of the stimulating electric field is enabled by the focal figure-of-eight coil design, reproducible stimulator pulse delivery, and individual MRIs with accurate head alignment and integrated controlled response measurements via triggered EMG. Such system design minimizes the influence of nonphysiological factors on the accuracy of nTMS so that clinical application for preoperative motor and language mapping became feasible.

By facing all the factors impairing accuracy as outlined in this chapter, the En-TMS technology with individual electric field modeling can now provide the level of accuracy, reliability, and reproducibility sufficient for neurosurgical applications.

Acknowledgments This work was supported by the Academy of Finland. We thank Dr. Jaakko Nieminen for helpful comments on the manuscript.

References

Amassian VE, Eberle L, Maccabee PJ, Cracco RQ. Modelling magnetic coil excitation of human cerebral cortex with a peripheral nerve submersed in a brain shaped volume conductor: the significance of fiber-bending excitation. Electroencephalogr Clin Neurophysiol. 1992;77:390–401.

Barker AT, Jalinous R, Freeston IL. Non-invasive magnetic stimulation of human motor cortex. Lancet. 1985;1(8437):1106–7. https://doi.org/10.1016/s0140-6736(85)92413-4. PMID: 2860322.

Deng Z-D, Lisanby SH, Peterchev AV. Electric field depth–focality tradeoff in transcranial magnetic stimulation: Simulation comparison of 50 coil designs. Brain Stimul. 2013;6:1–13.
FDA. Glossary of computer system software development terminology (8/95). 2014. www.fda.gov/iceci/inspections/inspectionguides/ucm074875.htm. Accessed 22 Feb 2024.
Forster MT, Hattingen E, Senft C, Gasser T, Seifert V, Szelenyi A. Navigated transcranial magnetic stimulation and functional magnetic resonance imaging: advanced adjuncts in preoperative planning for central region tumors. Neurosurgery. 2011;68:1317–24, discussion 1324–1315.
Grover FW. Inductance calculations. New York: Van Norstrand Company inc.; 1946
Hannula H, Ylioja S, Pertovaara A, Korvenoja A, Ruohonen J, Ilmoniemi RJ, Carlson S. Somatotopic blocking of sensation with navigated transcranial magnetic stimulation of the primary somatosensory cortex. Hum Brain Mapp. 2005;26:100–9.
Ilmoniemi RJ, Grandori F. Device for applying a programmable excitation electric field to a target. European Patent Application 94203134.5; EP0709115 A1; 1993.
Ilmoniemi RJ, Kicic D. Methodology for combined TMS and EEG. Brain Topogr. 2010;22:233–48.
Ilmoniemi RJ, Ruohonen J, Virtanen J. Relationships between magnetic stimulation and MEG/EEG. In: Nilsson J, Panizza M, Grandori F, editors. Advances in occupational medicine and rehabilitation, vol 2, No 2. Pavia: Fondazione Salvatore Maugeri Edizioni; 1996. p. 65–72.
Ilmoniemi RJ, Virtanen J, Ruohonen J, Karhu J, Aronen HJ, Näätänen R, Katila T. Neuronal responses to magnetic stimulation reveal cortical reactivity and connectivity. Neuroreport. 1997;8(16):3537–40. https://doi.org/10.1097/00001756-199711100-00024. PMID: 9427322.
Ilmoniemi RJ, Ruohonen J, Karhu J. Transcranial magnetic stimulation—a new tool for functional imaging of the brain. Crit Rev Biomed Eng. 1999;27(3–5):241–84. PMID: 10864281.
Kallioniemi E, Könönen M, Julkunen P. Repeatability of functional anisotropy in navigated transcranial magnetic stimulation—coil-orientation versus response. Neuroreport. 2015;26(9):515–21.
Koponen LM, Nieminen JO, Ilmoniemi RJ. Minimum-energy coils for transcranial magnetic stimulation: application to focal stimulation. Brain Stimul. 2015;8:124–34.
Krieg SM, Shiban E, Buchmann N, Gempt J, Foerschler A, Meyer B, Ringel F. Utility of presurgical navigated transcranial magnetic brain stimulation for the resection of tumors in eloquent motor areas. J Neurosurg. 2012;116:994–1001.
Krieg SM, Shiban E, Buchmann N, Meyer B, Ringel F. Presurgical navigated transcranial magnetic brain stimulation for recurrent gliomas in motor eloquent areas. Clin Neurophysiol. 2013;124:522–7.
Krings T, Chiappa KH, Foltys H, Reinges MHT, Cosgrove GR, Thron A. Introducing navigated transcranial magnetic stimulation as a refined brain mapping methodology. Neurosurg Rev. 2001;24:171–9.
Lefaucheur JP, André-Obadia N, Antal A, Ayache SS, Baeken C, Benninger DH, Cantello RM, Cincotta M, de Carvalho M, De Ridder D, Devanne H, Di Lazzaro V, Filipovic SR, Hummel FC, Jääskeläinen SK, Kimiskidis VK, Koch G, Langguth B, Nyffeler T, Oliviero A, Padberg F, Poulet E, Rossi S, Rossini PM, Rothwell JC, Schönfeldt-Lecuona C, Siebner HR, Slotema CW, Stagg CJ, Valls-Sole J, Ziemann U, Paulus W, Garcia-Larrea L. Evidence-based guidelines on the therapeutic use of repetitive transcranial magnetic stimulation (rTMS). Clin Neurophysiol. 2014;125:2150–206.
Massimini M, Ferrarelli F, Huber R, Esser SK, Singh H, Tononi G. Breakdown of cortical effective connectivity during sleep. Science. 2005;309(5744):2228–32. https://doi.org/10.1126/science.1117256. PMID: 16195466.
Mizutani-Tiebel Y, Tik M, Chang KY, Padberg F, Soldini A, Wilkinson Z, Voon CC, Bulubas L, Windischberger C, Keeser D. Concurrent TMS-fMRI: technical challenges, developments, and overview of previous studies. Front Psychiatry. 2022;13:825205. https://doi.org/10.3389/fpsyt.2022.825205. PMID: 35530029; PMCID: PMC9069063.
Näsi T, Mäki H, Kotilahti K, Nissilä I, Haapalahti P, Ilmoniemi RJ. Magnetic-stimulation-related physiological artifacts in hemodynamic near-infrared spectroscopy signals. PLoS One. 2011;6(8):e24002.
nTMS Workshop Group. Protocol hand-out, 8th international nTMS symposium, Berlin; 2016.

Nummenmaa A, Stenroos M, Ilmoniemi RJ, Okada YC, Hämäläinen MS, Raij T. Comparison of spherical and realistically shaped boundary element head models for transcranial magnetic stimulation navigation. Clin Neurophysiol. 2013;124:1995–2007.

Picht T. Current and potential utility of transcranial magnetic stimulation in the diagnostics before brain tumor surgery. CNS Oncol. 2014;3:299–310.

Picht T, Schmidt S, Brandt S, Frey D, Hannula H, Neuvonen T, Karhu J, Vajkoczy P, Suess O. Preoperative functional mapping for rolandic brain tumor surgery: comparison of navigated transcranial magnetic stimulation to direct cortical stimulation. Neurosurgery. 2011;69(3):581–8, discussion 588.

Picht T, Krieg SM, Sollmann N, Rösler J, Niraula B, Neuvonen T, Savolainen P, Lioumis P, Mäkelä JP, Deletis V, Meyer B, Vajkoczy P, Ringel F. A comparison of language mapping by preoperative navigated transcranial magnetic stimulation and direct cortical stimulation during awake surgery. Neurosurgery. 2013;72:808–19.

Rothwell JC, Hallett M, Berardelli A, Eisen A, Rossini P, Paulus W. Magnetic stimulation: motor evoked potentials. Electroencephalogr Clin Neurophysiol Suppl. 1999;52:97–103.

Ruohonen J, Ilmoniemi RJ. Basic physics and design of TMS devices and coils. In: Hallett M, Chokroverty S, editors. Magnetic stimulation in clinical neurophysiology. Butterworth: Philadelphia; 2005. p. 17–30.

Ruohonen J, Karhu J. Navigated transcranial magnetic stimulation. Neurophysiol Clin. 2010;40:7–17.

Säisänen L, Pirinen E, Teitti S, Könönen M, Julkunen P, Määttä S, Karhu J. Factors influencing cortical silent period: optimized stimulus location, intensity and muscle contraction. J Neurosci Methods. 2008;169:231–8.

Schmidt S, Cichy RM, Kraft A, Brocke J, Irlbacher K, Brandt SA. An initial transient-state and reliable measures of corticospinal excitability in TMS studies. Clin Neurophysiol. 2009;120:987–93.

Schmidt S, Bathe-Peters R, Fleischmann R, Ronnefarth M, Scholz M, Brandt SA. Nonphysiological factors in navigated TMS studies; confounding covariates and valid intracortical estimates. Hum Brain Mapp. 2015;36:40–9.

Sollmann N, Goblirsch-Kolb MF, Ille S, Butenschoen VM, Boeckh-Behrens T, Meyer B, Ringel F, Krieg SM. Comparison between electric-field-navigated and line-navigated TMS for cortical motor mapping in patients with brain tumors. Acta Neurochir. 2016;158:2277–89.

Takahashi S, Vajkoczy P, Picht T. Navigated transcranial magnetic stimulation for mapping the motor cortex in patients with rolandic brain tumors. Neurosurg Focus. 2013;34(4):E3.

Tarapore PE, Tate MC, Findlay AM, Honma SM, Mizuiri D, Berger MS, Nagarajan SS. Preoperative multimodal motor mapping: a comparison of magnetoencephalography imaging, navigated transcranial magnetic stimulation, and direct cortical stimulation. J Neurosurg. 2012;117:354–62.

Ueno S, Tashiro T, Harada K. Localized stimulation of neural tissues in the brain by means of a paired configuration of time-varying magnetic fields. J Appl Phys. 1988;64(10):5862–4.

TMS, MEG, and fMRI: Comparing and Contrasting Three Functional Mapping Techniques

2

Marianne Kanaris, Srikantan S. Nagarajan, and Phiroz E. Tarapore

2.1 Introduction

Neuroimaging is a cornerstone of modern practice in the fields of neurology, neuro-oncology, and neurosurgery. Techniques such as computerized tomography (CT) and magnetic resonance imaging (MRI) have revolutionized the clinician's ability to diagnose, treat, and follow patients with neoplastic processes of the central nervous system (CNS). The last two decades have seen the introduction of additional techniques that go beyond simple anatomic and structural description, offering insights into the function of CNS structures themselves. These functional neuroimaging techniques enable clinicians to characterize a neoplastic lesion not just in an anatomical context, but also in the context of the overall neurological function of the patient, thus reducing treatment-related morbidity. Should a lesion require surgical management, intraoperative imaging techniques have enhanced the neurosurgeon's ability to maximize extent of resection without compromising neurological function. Finally, in following patients long term, serial functional imaging plays an integral role in defining the evolution of the functional relationship between an identified neoplastic lesion and its surrounding structures. In short, the significant

M. Kanaris · S. S. Nagarajan
Department of Radiology and Biomedical Imaging, University of California at San Francisco, San Francisco, CA, USA
e-mail: sri@ucsf.edu

P. E. Tarapore (✉)
Department of Neurological Surgery, University of California at San Francisco, San Francisco, CA, USA

Veterans Affairs Medical Center, Department of Veterans Affairs, San Francisco, CA, USA
e-mail: Phiroz.Tarapore@ucsf.edu

S. M. Krieg, T. Picht (eds.), *Navigated Transcranial Magnetic Stimulation in Neurosurgery*, https://doi.org/10.1007/978-3-031-97155-6_2

improvements in morbidity and mortality associated with neoplastic lesions of the CNS are due, in large part, to the advancement of neuroimaging techniques.

This chapter briefly summarizes the available functional imaging techniques that are in common clinical use. For most of these modalities, a detailed discussion of underlying physics is beyond the scope of the text; however, details will be provided if they are directly relevant to understanding the clinical application. The techniques of navigated transcranial magnetic stimulation (nTMS), magnetoencephalography (MEG), and functional magnetic resonance imaging (fMRI) will be covered, as well as the strengths and weaknesses of each. Finally, the reasons for selecting each study in clinical practice will be discussed.

2.2 Functional Imaging Techniques

Functional imaging is being utilized increasingly in the characterization of intracranial lesions, particularly in the preoperative setting to evaluate eloquent brain region in patients with brain tumors. It is most useful in the management of lesions within or near eloquent cortex such as motor or visual cortex because, by defining the function of perilesional parenchyma, it allows for optimization of surgical strategy by, for example, identifying the safest route of access for a deep-seated lesion. It also allows clinicians to discuss the morbidity associated with a given lesion in specific and accurate terms, thereby giving patients clearer understanding of the ramifications of various management strategies (e.g., observation vs. subtotal resection vs. aggressive resection). Functional imaging is therefore an invaluable modality for identifying the clinical sequelae associated with a given lesion.

2.2.1 Functional MRI

Functional magnetic resonance imaging (fMRI) uses MRI to measure changes in the blood-oxygen-level dependence (BOLD) signal, which is a reflection of the changing ratios of oxyhemoglobin and deoxyhemoglobin in functionally active brain regions. Because functionally active brain has a higher energy requirement than resting brain, the metabolic activity within functional brain is increased. This increased metabolic demand results in greater oxygen consumption, thereby altering the ratio of oxyhemoglobin to deoxyhemoglobin within the tissue.

Since its advent in 1991, fMRI has become the dominant modality for functional brain imaging in both the clinical and the research community for a number of reasons. It is safe, involves no ionizing radiation, and completely painless for the subject. It allows for whole-brain coverage, including the ability to examine activity in deep brain structures. Importantly, the widespread availability of commercial and open-source tools for analysis of fMRI data has enabled many researchers to embrace this technology easily. Finally, the acquisition of MRI data in standard formats allows for seamless integration of fMRI datasets with existing picture archiving and communication system (PACS) workstations and intraoperative frameless stereotactic navigational systems.

2.2.1.1 Clinical Application

In the clinical setting, particularly in patients with brain tumors, fMRI has been used to identify sensorimotor cortex (Fig. 2.1), as well as language and visual cortices (DeYoe and Raut 2014; Hirsch et al. 2000; Kapsalakis et al. 2012; Gupta 2014). Currently, the use of fMRI to delineate regions associated with specific language tasks (i.e., word repetition, word reading, and object naming) is in the experimental stage. Similarly, language lateralization with fMRI is a subject of significant continuing research, and is rapidly reaching equivalent sensitivity and specificity to Wada testing, but has not achieved a sensitivity or specificity sufficient to become the gold standard for language lateralization (Abou-Khalil 2007).

2.2.1.2 Limitations

fMRI is limited by the fact that the BOLD signal is an indirect measure of neural activity. In effect, the BOLD signal is three steps removed from the process it is meant to quantify: oxyhemoglobin/deoxyhemoglobin ratio is a proxy for oxygen demand, which is a proxy for metabolic activity, neuronal activity, and function. Moreover, because it is limited by the rate of oxygen consumption and subsequent blood flow mechanism, fMRI lacks the temporal resolution required to image the dynamic and oscillatory spatiotemporal patterns that are associated with cognitive processes. Furthermore, it might not accurately reflect true neuronal processes, especially in regions of altered vasculature, as frequently occurs in patients with neurosurgical pathology that are undergoing fMRI for the purposes of mapping eloquent cortices. In fact, the exact frequency band of neuronal processes that corresponds to the BOLD signal is still being actively debated (Logothetis et al. 2002;

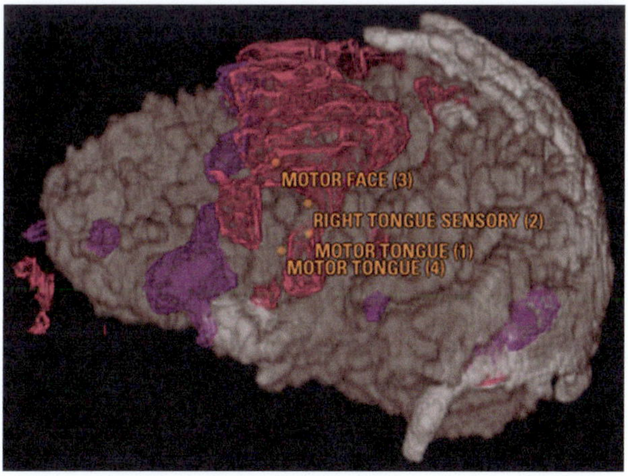

Fig. 2.1 Functional MRI of motor (pink) and pre-motor areas (purple). Intraoperative direct cortical mapping results (orange dots) are overlaid on the three-dimensional volume. The motor region (pink) defined by fMRI is more extensive than that defined by direct cortical mapping; this lack of specificity makes fMRI suboptimal as a stand-alone technique for the identification of eloquent cortical regions

Niessing et al. 2005). Finally, in the context of speech and language studies, because fMRI measurements involve loud scans caused by fast forces on MR gradient coils, the scans themselves invoke auditory responses that have to be deconvolved from the signals in order to examine other stimulus related activity. Hence, to image brain activity non-invasively on a neurophysiologically relevant timescale and to observe neurophysiological processes more directly, silent imaging techniques that have high temporal and spatial resolution are needed.

2.2.2 Magnetoencephalography

Magnetoencephalography (MEG) measures tiny magnetic fields outside of the head that are generated by neural activity. Because it measures these fields directly, in contrast to fMRI, MEG offers excellent temporal resolution (<1 ms). Furthermore, magnetic fields are unimpeded by biological tissue, so MEG recordings offer an undistorted signature of underlying neural activity.

Biomagnetic fields detected by MEG are extremely small, in the tens-to-hundreds of femto-Tesla (fT) range—seven orders of magnitude smaller than the earth's magnetic field. As a result, appropriate data collection necessitates a magnetically shielded room and highly sensitive detectors called superconducting quantum interference devices (SQUIDs) (Vrba and Robinson 2002). The fortuitous anatomical arrangement of cortical pyramidal cells allows the noninvasive detection of their activity by MEG. The long apical dendrites of these cells are arranged parallel to each other and often perpendicular to the cortical surface, and their electromagnetic (EM) fields sum up to magnitudes large enough to be detected at the scalp. Synchronously fluctuating dendritic currents result in electric and magnetic dipoles that produce these electromagnetic fields (Okada et al. 1987, 1999). These dendritic currents from the brain are typically sensed using detection coils called flux transformers or magnetometers, which are positioned closely to the scalp and connected to SQUIDs. SQUIDs act as magnetic-field-to-voltage converters, and their typically non-linear response is linearized by flux-locked loop electronic circuits. SQUIDs have a sensitivity of approximately 10 femto-Tesla per square root of Hz which is adequate for detection of the brain's magnetic fields (Vrba and Robinson 2002; Vrba and Robinson 2001).

Modern MEG systems often consist of simultaneous recordings from many differential sensors that cover the whole head, with the total number of SQUIDs varying from 100 to 300. Typical MEG systems have sensors that are spaced approximately 2.2–3.6 cm apart. Although the maximum sampling rate is approximately 12 kHz, most MEG data is usually recorded at about 1 kHz, thereby maintaining excellent temporal resolution for measuring the dynamics of cortical neuronal activity at the millisecond level.

MEG scanners, however, are expensive and relatively rare, so MEG is less widespread than MR-based techniques. MEG studies are useful for localization of sensory, motor (Nagarajan et al. 2008) and language regions (Fig. 2.2) (Edwards et al. 2010). They are also used to localize seizure foci, which is often helpful in the

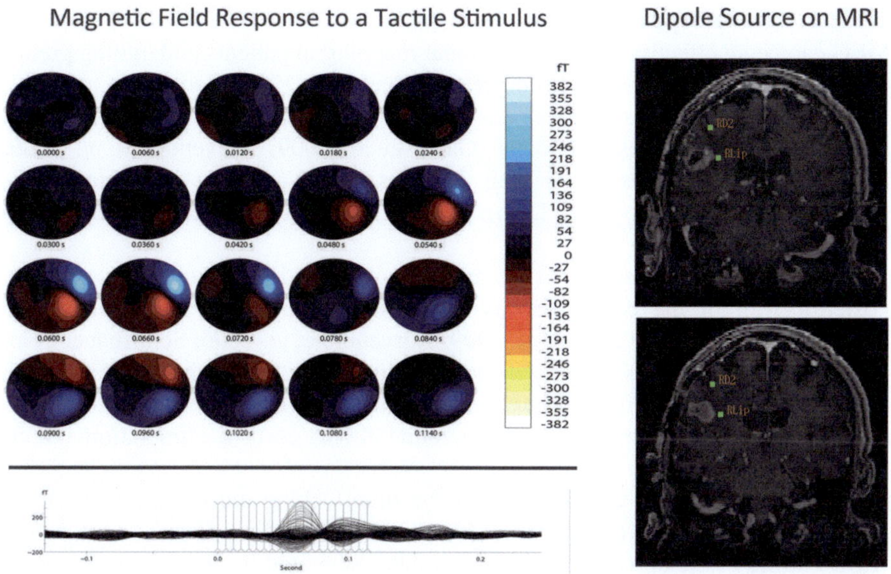

Fig. 2.2 Example MEG of somatosensory stimuli to the right lip (RLip) and right index finger (RD2). Multiple stimulus trials are performed for each site and cortical magnetic fields are recorded. The trials are averaged and a single dipole is reconstructed for each site using the least square fit method. On the left are depicted the results in 0.006 s intervals, with the magnetic dipoles represented as red-orange (negative femto-Tesla) to blue-white (positive femto-Tesla). On the right, the resulting dipoles are then displayed on a coregistered, T1-weighted post-gadolinium coronal MR slice. On the bottom of the figure is the overall timecourse of magnetic field with all trials superimposed

management of epileptogenic tumors such as oligodendrogliomas. Finally, a recently described MEG-based algorithm for language lateralization successfully identify the language-dominant hemisphere with accuracy approaching that of Wada testing, the current gold standard (Findlay et al. 2012).

MEG signals must be reconstructed using complex algorithms to quantify the underlying brain activity from observed sensor data. This reconstruction of brain activity from MEG data typically involves two major components—a forward model and an inverse model. The details of these algorithms are complex and beyond the scope of this chapter. Depending on the type of source reconstruction chosen, the quality of the signal data, and the particular characteristics of the machine itself, the spatial resolution of MEG can vary. As a rule of thumb, for typical datasets, newer beamforming methods can reconstruct tens-to-hundreds of sources at about 5 mm distances (assuming time-frequency separation and detectability). This estimate can be considered an approximate spatial resolution for MEG, keeping in mind that under certain circumstances the spatial resolution can be even greater.

2.2.2.1 Clinical Application

MEG can be an effective tool for functional mapping in patients undergoing preoperative workup for brain tumor surgery (Gallen et al. 1995; Kamada et al. 1993; Makela et al. 2001). This modality has been used for localization of the sensorimotor cortex along the central sulcus (Gaetz et al. 2009; Korvenoja et al. 2006; Taniguchi et al. 2004; Ossenblok et al. 2003) as well as mapping the primary auditory (Rowley and Roberts 1995; Lutkenhoner et al. 2003) and visual cortices (Plomp et al. n.d.).

The primary motor cortex and the sensory cortex are located on the anterior and posterior wall of the central sulcus, respectively. Identifying the hand region (Nagarajan et al. 2008; Gaetz et al. 2009; Korvenoja et al. 2006; Schiffbauer et al. 2003; Ishibashi et al. 2001) and the mouth region (Schiffbauer et al. 2003; Kirsch et al. 2007) of the primary sensorimotor cortex has been useful for presurgical evaluation, and also confirmed with intracranial direct cortical stimulation (DCS) mapping.

Motor evoked fields can be recorded by time-locking the MEG signal corresponding to movement (Rezai et al. 1996), and single equivalent current dipole (ECD) fitting of the corresponding evoked field generated from the average sensor data (Korvenoja et al. 2006; Schiffbauer et al. 2003; Ishibashi et al. 2001; Kirsch et al. 2007; Kober et al. 2001). Using this approach, Schiffbauer et al. compared MEG to intraoperative mapping in tumor patients receiving painless tactile somatosensory stimulation to the lip, hand, and foot and found that both approaches had a favorable degree of quantitative correlation (Schiffbauer et al. 2003). Similarly, a favorable degree of quantitative correlation was also seen from utilizing dipole fitting with MEG versus fMRI (Kober et al. 2001). Confirmed with ECOG, dipole fitting of evoked magnetic fields to median nerve stimulation proved to be superior to fMRI for 15 patients in identifying the sensorimotor cortex (Korvenoja et al. 2006). Following dipole fitting of the mouth motor cortex, electrocortical stimulation (ECS) sites were usually anterior and lateral to MEG localization of the lip somatosensory cortex (Kirsch et al. 2007).

The use of MEG spatial filtering holds promise for a more robust method for mapping the motor cortex in presurgical patients (Nagarajan et al. 2008; Gaetz et al. 2009; Cheyne et al. 2006). The use of a spatial filter beamformer while subjects performed a self-paced index finger movement can generate high-resolution imaging of the spatiotemporal patterns of premotor and motor cortex activity (Fig. 2.3) (Cheyne et al. 2006; Tarapore et al. 2012a). Peaks of the tomographic distribution of beta-band event-related desynchronization sources reliably localized the hand motor cortex in a group of 66 patients, which was confirmed with DCS (Nagarajan et al. 2008).

Location of the language cortex (i.e., Broca's and Wernicke's areas) also holds clinical value as mass lesions can distort the anatomy and also because of interindividual anatomic variation among patients. Recently, Hirata et al. used synthetic aperture magnetometry (SAM, an improvement over MEG dipole methods) to prospectively determine language lateralization, and found high concordance with Wada testing and intraoperative cortical stimulation results (Hirata et al. n.d.).

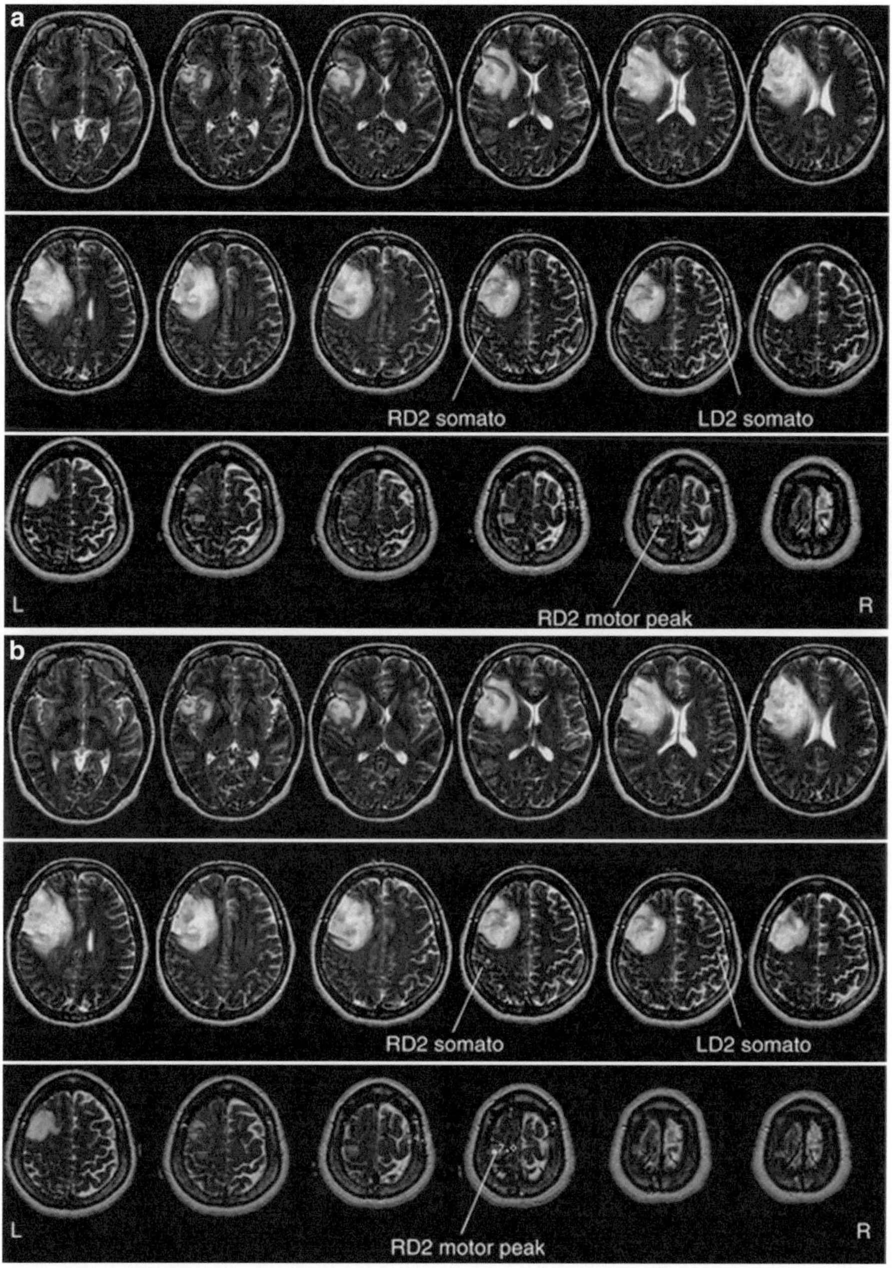

Fig. 2.3 (**a**) Localization of β-band desynchronization preceding right index finger flexion for a subject with a frontal tumor. The location of hand motor cortex relative to a single dipole localization of hand somatosensory cortex is also shown. (**b**) Localization of β-band desynchronization due to left index finger flexion in the same subject, showing contralateral hand motor cortical activation in the right hemisphere

Extending this approach, SAM has been used to accurately characterize dynamics of language dominance using MEG (Guggisberg et al. 2008).

MEG has also been used to quantify functional connectivity in peri-tumoral regions, a technique that may have implications regarding the risk of postoperative complications. The term functional connectivity essentially defines the complex functional interaction between local and more remote brain areas. When compared with healthy controls, all patients with brain tumors had diffuse brain areas with decreased alpha coherence, as well as a decrease in high-frequency bands for long distance connections and an increase in slower frequency bands for more local connections (Bartolomei et al. 2006). Two follow-up studies have demonstrated that by quantifying the functional connectivity in the peri-tumoral region, connectivity maps help to predict regions of positive intraoperative stimulation (Martino et al. 2011). Furthermore, these connectivity maps demonstrate that patients with high peri-tumoral connectivity are at greater risk of postoperative deficit than are patients with low peri-tumoral connectivity (Tarapore et al. 2012b).

2.2.2.2 Limitations

In addition to being relatively rare, MEG can be limited by its relative preference for specifically oriented field sources: it senses primarily the tangential currents in the brain closer to the surface (Hamalainen 1991). In addition, the post-processing requirements for MEG are substantial, both from the standpoint of technical knowledge of complex signal processing algorithms and from the standpoint of raw computational power. MEG datasets are large, and processing them requires substantial resources, often distributed over a computer cluster involving several servers. Because of its high sensitivity, MEG can also be impeded by magnetic noise: occasionally, a few dental fillings will render a patient's scan unusable. As data acquisition and signal processing algorithms improve, MEG will contribute increasingly valuable functional imaging data in the clinical setting.

2.2.3 Transcranial Magnetic Stimulation

Transcranial magnetic stimulation (TMS) is a technique that has demonstrated great promise in functional mapping of the human cortex. Based on Faraday's principle of electromagnetic induction, TMS applies a brief pulse of high-strength magnetic field over the scalp which passes through the skull and induces an electrical current in the underlying brain region (Wagner et al. 2007). These pulses of current, if applied appropriately, are sufficient to depolarize a population of neurons, inducing an action potential (Mills et al. 1987). Single TMS pulses, when delivered over the cortex, will thus briefly stimulate the underlying cortical region. Repetitive trains of these pulses, so-called repetitive TMS (rTMS), can have either an inhibitory or a stimulatory effect on cortical excitability, depending on the frequency of the rTMS trains (Kobayashi and Pascual-Leone 2003). Therefore, by altering the protocol of stimulation, TMS can cause either a temporary excitation or lesion effect in the cortex. The technique has been used for decades, although its application in neurosurgical patients has been more recent. Nevertheless, it has an excellent safety profile (Tarapore et al. 2016a, b).

Furthermore, the development of *navigated* TMS (nTMS) has allowed for this technology to arrive rapidly at the forefront of noninvasive mapping modalities (Fig. 2.4). What sets apart the nTMS system from a non-navigated TMS system is its demonstration, in real-time, of the precise location and strength of the magnetic pulse. By integrating a frameless stereotactic navigational system (such as those used commonly utilized in neurosurgical and other procedures) with a TMS coil, one can co-register a structural MRI or computed tomographic (CT) brain scan to a subject's anatomy using fiducial markers or anatomical landmarks. This advancement allows the investigator to deliver TMS pulses with unprecedented precision under image guidance (Julkunen et al. 2009; Krings et al. 2001a, b; Picht et al. 2009). Furthermore, in some nTMS systems, the strength and directionality of each TMS pulse is calculated on-the-fly according to a dynamic spherical model which takes into account the preset parameters of stimulation as well as the subject's scalp/skull thickness (Sarvas 1987; Tarkiainen et al. 2003). As a result, when the stimulation coil is positioned over the subject's scalp, the investigator can visualize the targeted cortical region, the strength and orientation of the magnetic dipole, and the cone of activation generated by the magnetic pulse.

nTMS therefore allows the investigator, for the first time, to pinpoint precisely the cortical region that is being targeted. This capacity for accurate offers the

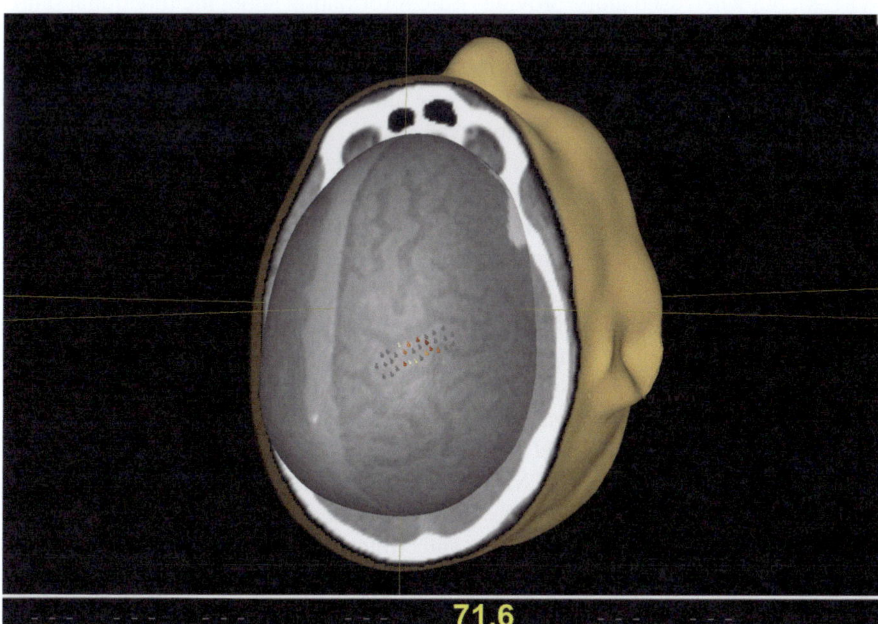

Fig. 2.4 A motor map of the precentral gyrus generated with TMS. Variably colored pins indicate the amplitude of motor evoked potentials in abductor pollicis brevis during the mapping procedure

possibility of mapping essential cortical regions associated with motor and language function. Prior studies (Pascual-Leone et al. 1991; Michelucci et al. 1994; Jennum et al. 1994; Wassermann et al. 1999; Epstein et al. 1996, 1999, 2000) have examined the use of repetitive TMS to cause speech arrest and lateralize language (see below for details). However, these efforts were not stereotactically guided, and showed repetitive TMS to be an unreliable technique for determining language laterality, largely because of a high false-positive rate for speech arrest sites on the supposedly non-dominant hemisphere.

2.2.3.1 TMS Combined with Fiber Tractography

Preoperative functional imaging modalities like those described aim to identify eloquent areas of the brain that control critical functions such as language or motor. The combination of different imaging modalities, including structural imaging, is now being used to yield further insights into the visualization of functional brain areas. One growing area of interest is the use of diffusion tensor imaging (DTI) in conjunction with functional imaging modalities such as TMS and fMRI.

DTI is a form of diffusion-weighted MRI that allows for tractography of the white matter of the brain. fMRI and TMS are predominantly used to measure activity on the surface of cerebral cortex, and so the addition of white matter tractography through modalities like DTI may strengthen the understanding of functional connectivity, and thus pose benefits for surgical planning. DTI is used to detect the magnitude and direction of water diffusion in brain tissue along white matter tracts (WMTs), but when used alone provides little information about the functional status of the white matter tract (Kekhia et al. 2011). Combining DTI with functional imaging, by using fMRI and TMS as seeding regions for DTI tractography, may make it possible to spatially detect eloquent cortical and subcortical structures involved in language or motor function (Sollmann et al. 2018).

As postoperative deficits may originate from either cortical damage or injury to subcortical structures, a more complete preoperative functional planning should include a reconstruction of subcortical white matter tracts (Raffa et al. 2017). This can provide insight into which areas of the brain should be preserved during surgery via more robust preoperative imaging and thereby inform surgical planning to reduce postoperative deficits. Raffa et al. demonstrated how the use of preoperative nTMS-based DTI tractography in a small cohort of low-grade glioma patients undergoing surgery increased surgeon's awareness of functional anatomy and surgical risks in 75% of cases, and was associated with a decrease in postoperative motor and/or language deficits (Raffa et al. 2017).

Early studies into fMRI and TMS-seeded DTI tractography have compared the accuracy of the two modalities in predicting permanent neurological deficit. Muir et al., in 2022, investigated the accuracy and clinical relevance of TMS tractography (Muir et al. 2022a). They overlaid postoperative MRI onto preoperative imaging to visualize TMS points in the context of the surgical cavity of resected gliomas. Meaningful subsets of TMS points were identified through generated DTI tractography at different fractional anisotropic (FA) thresholds of 25%, 50%, and 75%.

Muir et al. discovered that the resection of TMS-positive points within the glioma was significantly predictive of permanent neurological deficits. Moreover, resections of TMS-seeded white matter tracts (WMTs) identified at 50% and 75% FA thresholds were significantly predictive of permanent deficits, with tractography generated at 75% FA thresholds providing the highest positive predictive value (PPV) and negative predictive value (NPV). However, previous studies on TMS tractography have demonstrated different conclusions regarding the optimal FA threshold, particularly in reference to motor-associated versus language-associated WMTs. Of note, Sollman et al. found that all relevant language-associated fibers were not identified at an FA threshold of 75%, thus necessitating DTI fiber tracking at 25% and 50% FA thresholds (Sollmann et al. 2019). Future research on optimal FA thresholds may elucidate different thresholds for specific tracts, but overall the combination of TMS with DTI tractography shows great promise for prediction of postoperative neurological deficits.

In a more recent study in 2023, Muir et al. directly compared the predictive value of TMS tractography and fMRI tractography in a cohort of glioma patients across a variety of FA thresholds (Muir et al. 2022b). They found that TMS tractography provided a significantly greater predictive value of permanent postoperative deficits across all metrics of sensitivity, specificity, PPV, and NPV when compared to fMRI tractography, regardless of the FA threshold. One possible mechanistic explanation is that TMS may better capture clinically relevant neuroplasticity, as was demonstrated in 35% of patients, compared to only 2.5% of patients who had a BOLD signal displaced from precentral gyrus during fMRI. Both modalities may be part of the toolkit for a comprehensive presurgical functional imaging workup for patients with glioma, but TMS-based DTI tractography is emerging as a powerful tool to identify eloquent cortical and subcortical structures, and may better inform presurgical planning to reduce permanent neurological deficits in patients.

2.2.3.2 Clinical Application

The clinical application of nTMS is discussed in detail throughout this text, and will not be summarized in detail in this chapter. Briefly, nTMS is commonly used in the management of peri-Rolandic tumors, where the pyramidal tract is at highest risk of disruption from surgical resection, and has shown to correlate well with intraoperative direct cortical stimulation (Tarapore et al. 2012a; Picht et al. 2011; Krieg et al. 2012). Additionally, nTMS is used increasingly for preoperative language mapping in patients with neoplastic lesions in peri-eloquent language regions (Krieg et al. 2013; Tarapore et al. 2013).

The functional maps generated with nTMS can be used for more than just localizing function. They also offer insights into the relative vulnerability of surround cortices, and can be used for risk stratification (Lefaucheur and Picht 2016; Picht et al. 2012). Clinical studies are increasingly demonstrating the value of adding nTMS to the surgical management of brain tumors, with improvements seen in functional outcomes and extent of resection (Picht et al. 2016; Krieg et al. 2014).

2.2.3.3 Limitations

While navigated TMS mapping is a promising new modality, there are several limitations that bear mentioning. The first of these relate to the precision of the navigation itself. The tolerance of registration on most modern nTMS systems is estimated at 2–3 mm; given that the guidance system is not frame-based it is possible that the actual error is higher. Care must therefore be taken to ensure that preoperative maps match up with anatomical landmarks in the operating room: gyri and sulci, for example, and blood vessels can be used to ensure an accurate co-registration between the preoperative and the intraoperative findings.

Single pulse TMS for motor mapping, as described earlier, has been shown to be highly accurate. Limitations with this modality are similar to that of intraoperative cortical stimulation: namely that cortical and subcortical lesions can interfere with the cortico-spinal tract, making MEPs difficult or impossible to obtain. In the majority of these cases, patients will demonstrate signs of clinical weakness as well. This limitation can in fact be of great use; if preoperative mapping yields a difficult or unobtainable map, intraoperative cortical stimulation will likely encounter the same difficulty. Additionally, if motor maps of inferior motor cortex are required, patients may complain of discomfort from associated temporalis contractions. In general, this discomfort is less than with rTMS because of the differences in protocol.

An inherent limitation in the specificity of nTMS for speech mapping is transsynaptic excitation of downstream (and, possibly, upstream) neuronal units. Neurophysiological and neuroimaging studies have shown that repetitive TMS of a given brain region induces distributed activation of neural circuitry via transsynaptic spread, which follows established functional networks (Paus et al. 1997; Valero-Cabre and Pascual-Leone 2005; Bestmann 2008). Thus, behavioral effects may be a result of activation not in the target region but in a distant, functionally connected region. The "over-calling" seen in many nTMS validation studies is likely a result of this limitation. Nevertheless, it should be pointed out that the overall nTMS language map does reflect the distribution of DCS language sites in the three published series to date (Tarapore et al. 2013; Sanai et al. 2008; Picht et al. 2013). Additionally, it seems to correlate with prior studies of un-navigated rTMS mapping of language sites.

Another inherent limitation to both motor and speech maps is the spread of the magnetic field itself. The figure-of-eight coil used in most modern systems generates a conical magnetic field. The field is therefore roughly circular at the cortical surface with a diameter of about 2 cm (and greatest intensity at the center with a sharp fall-off at the edges) and tapers toward its apex which occurs approximately 4.5 cm from the coil surface. As a result, the magnetic pulses might disrupt subcortical white matter tracts, while the overlying cortex is inappropriately identified as a site of motor function or language disruption.

Furthermore, a given neuron's orientation, volume, axonal and dendritic organization, and innate threshold affect the likelihood of a magnetic pulse generating an action potential (Pashut et al. 2011). Thus, nTMS-positive sites are not "points"—that is a misnomer; it is more accurate to say that they are regions, and to be aware that closely approximated regions of positivity on the map may all be associated with a single eloquent cortical site.

Finally, the basic parameters of nTMS stimulation, particularly with regard to speech mapping, should be examined more systematically and thoroughly. It is possible that relatively small adjustments in the frequency or number of pulses could improve results. Similarly, the timing of the onset of the pulse train is potentially important. In one published series, the pulse train was initiated just before the presentation of the stimulus in an object naming task, largely because that is the protocol used for DCS (Tarapore et al. 2013). It is possible, however, that initiating the pulse train with, or just after the stimulus might improve the specificity of nTMS mapping. These variations in task and parameter must be methodically explored in future studies of nTMS protocols.

2.3 Comparing and Contrasting Functional Imaging Techniques

Functional MRI, magnetoencephalography, and navigated TMS each have particular strengths which make them clinically useful in the correct situation. Similarly, each of these modalities has limitations that the clinician must keep in mind when ordering and interpreting results.

2.3.1 Temporal Considerations

fMRI is, by far, the most widely used of these three functional imaging modalities. To many clinicians, the term "functional imaging" is synonymous with that of fMRI. It is in use worldwide, largely because of its accessibility—anyone with a reasonably modern MRI scanner can perform an fMRI study. Ironically, fMRI is one of the less well-suited functional studies for the purposes of pre-surgical mapping. The BOLD signal is 3 degrees removed from neuronal activity and, as a result, the fMRI suffers from poor temporal resolution. While this limitation may not be of particular significance if the objective is simple localization of the primary motor cortex, it becomes much more important in localization of language cortex, where multiple cortical regions are involved and the interplay between those regions defines their functionality. MEG, on the other hand, is particularly well suited for defining cortical regions associated with language function. Because of its high temporal resolution, it can identify the temporal relationship between various cortical sites as well as the spatial relationship.

2.3.2 Spatial Considerations

One advantage to fMRI is its relative insensitivity to the location or orientation of neuronal sources. fMRI can resolve deep brain sources, such as those in the basal ganglia, as well as it can superficial sources. MEG, on the other hand, has difficulty with deep sources because of the low signal to noise ratio inherent to MEG recording. Similarly, fMRI has no preference for the orientation of neuronal groups, while MEG picks up radially oriented sources than tangentially oriented sources. Of critical importance, also, is the fact that fMRI only detects cortical involvement in the functional brain. White matter connections, which are equally important, are not detected with fMRI.

Similar to MEG, nTMS is most effective at identifying relatively superficial cortical sites. The magnetic pulse of the nTMS coil is cone shaped and tends to penetrate to a depth of 2–3 cm with most standard machines. Beyond this depth, the magnetic field of an nTMS coil is usually too weak to generate an inductive current in the target neurons.

2.3.3 Vascular and Electromagnetic Artifacts

fMRI data can be affected by abnormal/asymmetric vasculature. In the neurosurgical patient, abnormal vasculature is commonly encountered, especially in higher grade brain lesions. Therefore, in those patients for whom accurate functional mapping is most vital, fMRI can be subject to artifact. This artifact can significantly reduce the reliability of the fMRI results and, on occasion, make them useless. Neither MEG nor nTMS is subject to these vascular artifacts.

MEG, on the other hand, is particularly sensitive to electromagnetic artifact. Dental fillings, implanted hardware in the head/neck, pacemakers, deep brain stimulators, and aneurysms clips can all generate enough electromagnetic interference that the MEG recording is not useful for clinical purposes. It should be noted that some of these devices also preclude patients from obtaining an MRI; in contrast to an MRI, however, there is no danger to the patient if an implanted device is not compatible with MEG.

nTMS, in contradistinction to fMRI and MEG, is largely resistant to artifact. Because it involves direct stimulation of the underlying brain, it is not subject to inaccuracies from vascular abnormalities or from implanted devices. It should be noted, however, that certain implanted devices such as cochlear implants and programmable shunt valves are a relative contraindication for TMS.

2.3.4 Side Effects and Patient Participation

fMRI and MEG have similar side-effect profiles, in that they are completely non-invasive and do not involve any patient contact. Patients experience no additional sensation as a direct result of these studies. Patients can experience claustrophobia in an fMRI and, to a lesser extent, in an MEG, but these effects can usually be ameliorated with small doses of anxiolytic medication. If motor and/or language mapping is being performed, patient participation is required.

nTMS, although technically non-invasive, does require patient contact during the course of the study, and the patient experiences the sensation of stimulation throughout. Depending on the intensity, frequency, and location of stimulation, the experience can range from benign to quite painful. In particular, repetitive nTMS at high frequency and intensity over the temporalis muscle can generate significant patient discomfort.

Because motor mapping with nTMS is accomplished with integrated EMG, no patient participation is required for this study. It can be performed in sleeping or anesthetized patients, as well as in patients with decreased cognitive status secondary to intracranial pathology. It is also useful in small children. Language mapping with nTMS, similar to MEG and fMRI, requires patient participation in the form of picture naming or other language task.

2.3.5 Accuracy

In quantifying the accuracy of fMRI, MEG, and nTMS, one must first define the gold standard by which these techniques are judged. In the case of neurosurgical patients, this standard is direct cortical stimulation (DCS) in which an electrified electrode is placed directly on the cortical surface during surgery. In all other patients for whom DCS maps are not available, the techniques can only be judged against one another, and their relative accuracy is more subjective.

Several studies have compared the accuracy of fMRI, MEG, and nTMS with DCS, although, to our knowledge, no study has compared all three modalities with DCS. fMRI-based maps of motor function have been extensively validated against DCS, with high concordance rates (Tomczak et al. 2000; Lehericy et al. 2000; Jack Jr. et al. 1994). In maps of language function, on the other hand, fMRI has correlated poorly with DCS-based cortical speech arrest sites, but it has demonstrated excellent correlation with the Wada test for language lateralization (Binder et al. 1996; Deblaere et al. 2004).

Similar to fMRI, MEG-based maps of motor and sensory function demonstrate relatively good concordance with intraoperative DCS data. On average, the error distance between MEG-based and DCS-based motor sites is approximately 8 mm, which is well within the tolerances of coregistration and spatial resolution of most MEG sensor arrays (Tarapore et al. 2012a). MEG also correlates poorly with DCS-based cortical speech arrest sites, but can differentiate language laterality with accuracy comparable to the Wada test (Findlay et al. 2012; Tarapore et al. 2013).

Of these three functional techniques, nTMS-based motor maps have the highest concordance rates with intraoperative DCS motor maps, with error distances from 3 to 6 mm (Picht 2014). Not only is the colocation extremely reliable, but the sensitivity and specificity of nTMS-based motor maps approach 100%. Therefore, if a site appears positive on nTMS, it will be positive on DCS, and if a site is negative on nTMS, it will be negative on DCS. This is of particular value when performing limited craniotomies that minimize cortical exposure, as it allows the surgeon to maintain confidence in a negative intraoperative map and obviates the need for exposing a positive DCS control site. In comparing nTMS-based language maps with those of DCS, the results are less concordant. nTMS-based maps have high sensitivity (i.e., few false negative sites) but have poor specificity (many false positive sites). Although this profile is better than the opposite, it demonstrates the continued need for refinement of nTMS-based language mapping protocols.

2.3.6 Integration of Noninvasive Imaging and Intraoperative Neurophysiology

The combination of structural and functional imaging modalities can serve to spatially enclose eloquent cortical and subcortical brain structures. This not only can provide a preoperative reconstruction of motor and language pathways, but also can be integrated with intraoperative neurophysiology as a guide for resection.

Currently, intraoperative direct electrical stimulation is the gold standard technique for functional brain mapping during awake surgery (Aaronson et al. 2021). However, DES can be supplemented by preoperative modalities, of which fMRI and MEG have previously been applied in clinical routine, to support resection planning and intraoperative guidance (Sollmann et al. 2018). Negative nTMS language mapping has been shown to correlate well with negative DES, detecting language-related cortical sites that fMRI or MEG did not detect (Tarapore et al. 2013; Sollmann et al. 2013). In a study by Sollmann et al., nTMS-based DTI FT was routinely applied as an adjunct to awake surgery to assess clinical treatment and outcomes of highly language-eloquent gliomas. In addition to using preoperative nTMS-based DTI FT functional data for surgical planning, they also used it for intraoperative application, directing DES through nTMS and nTMS-based DTI FT data made available on navigational screens during tumor removal (Sollmann et al. 2018). In the 2017 study by Raffa et al., intraoperative use of the nTMS-based functional mapping helped guide direct cortical stimulation (DCS) and direct subcortical stimulation (DSS) during awake surgery (Raffa et al. 2017).

By providing further methods to distinguish between true-eloquent and non-eloquent areas, particularly in subcortical white matter tracts, nTMS-based DTI FT can extend surgical indications and lead to increased surgical resection. In particular, Picht et al. demonstrated the high negative predictive value of nTMS language mapping, which can be valuable when managing surgical strategy for patients with peri-eloquent lesions (Picht et al. 2013). The intraoperative use of DTI-nTMS can allow surgeons to orient to intraoperative neurophysiological mapping in order to more confidently and quickly identify functional structures (Raffa et al. 2017). This may potentially allow for more radical resections as surgeons can better identify the individual functional anatomy of a patient and adjust surgical strategy using a comprehensive reconstruction of both cortical and subcortical structures.

The results of the 2017 Raffa et al. study demonstrate that intraoperative use of nTMS-based DTI FT allowed for real-time verification of the CST location through neuronavigation, increasing the confidence of surgeons in identifying anatomo-functional landmarks and also significantly reducing the DSS intensity at which resection was terminated as compared to the control group (Raffa et al. 2017). In addition, TMS-based DTI tractography can reduce the incidence of postoperative functional deficits. The study further shows the use of nTMS DTI FT led to an increase in the surgeon's awareness of functional anatomy and surgical risks in 75% of cases, and was associated with greater stability of motor and language outcomes at discharge as well as a decrease of postoperative deficits (Raffa et al. 2017). Preoperative planning and intraoperative guidance from nTMS-based DTI FT also resulted in good clinical outcome in the Sollmann study, including parameters such as extent of resection and permanent deficit (Sollmann et al. 2018).

nTMS DTI tractography provides practical information for preoperative functional reconstruction of motor and language pathways as well as the ability to supplement neurophysiology-based image-guided surgery through increased surgeon's confidence in detecting functional structures. The potential for more radical surgical resections and the demonstrated reduction of postoperative deficits suggests great utility for the application of nTMS DTI tractography in clinical neurosurgical procedures.

2.4 Conclusion

Functional neuroimaging is an integral component in the diagnosis and management of CNS neoplasms. In addition to the standard anatomical techniques, modern functional imaging provides insight into the functional status of the peritumoral environment. These techniques allow the treating clinicians to predict the neurological sequelae associated with the observation or treatment of a specific lesion, and to optimize treatment strategies to minimize neurological morbidity. In comparing the techniques of fMRI, MEG, and nTMS, care should be taken in study selection, as each of these studies has unique strengths and weaknesses. Thorough knowledge of the limitations inherent to these techniques is crucial to accurate interpretation of the results. Each of these modalities, applied singly or in combination, will help to improve clinical outcomes in patients with brain tumor.

References

Aaronson DM, Martinez Del Campo E, Boerger TF, Conway B, Cornell S, Tate M, Mueller WM, Chang EF, Krucoff MO. Understanding variable motor responses to direct electrical stimulation of the human motor cortex during brain surgery. Front Surg. 2021;8:730367. https://doi.org/10.3389/fsurg.2021.730367.

Abou-Khalil B. An update on determination of language dominance in screening for epilepsy surgery: the Wada test and newer noninvasive alternatives. Epilepsia. 2007;48(3):442–55.

Bartolomei F, Bosma I, Klein M, Baayen JC, Reijneveld JC, Postma TJ, et al. Disturbed functional connectivity in brain tumour patients: evaluation by graph analysis of synchronization matrices. Clin Neurophysiol. 2006;117(9):2039–49.

Bestmann S. The physiological basis of transcranial magnetic stimulation. Trends Cogn Sci. 2008;12(3):81–3.

Binder JR, Swanson SJ, Hammeke TA, Morris GL, Mueller WM, Fischer M, et al. Determination of language dominance using functional MRI: a comparison with the Wada test. Neurology. 1996;46(4):978–84.

Cheyne D, Bakhtazad L, Gaetz W. Spatiotemporal mapping of cortical activity accompanying voluntary movements using an event-related beamforming approach. Hum Brain Mapp. 2006;27(3):213–29.

Deblaere K, Boon PA, Vandemaele P, Tieleman A, Vonck K, Vingerhoets G, et al. MRI language dominance assessment in epilepsy patients at 1.0 T: region of interest analysis and comparison with intracarotid amytal testing. Neuroradiology. 2004;46(6):413.

DeYoe EA, Raut RV. Visual mapping using blood oxygen level dependent functional magnetic resonance imaging. Neuroimaging Clin N Am. 2014;24(4):573–84.

Edwards E, Nagarajan SS, Dalal SS, Canolty RT, Kirsch HE, Barbaro NM, et al. Spatiotemporal imaging of cortical activation during verb generation and picture naming. NeuroImage. 2010;50(1):291–301.

Epstein CM, Lah JJ, Meador K, Weissman JD, Gaitan LE, Dihenia B. Optimum stimulus parameters for lateralized suppression of speech with magnetic brain stimulation. Neurology. 1996;47(6):1590–3.

Epstein CM, Meador KJ, Loring DW, Wright RJ, Weissman JD, Sheppard S, et al. Localization and characterization of speech arrest during transcranial magnetic stimulation. Clin Neurophysiol. 1999;110(6):1073–9.

Epstein CM, Woodard JL, Stringer AY, Bakay RA, Henry TR, Pennell PB, et al. Repetitive transcranial magnetic stimulation does not replicate the Wada test. Neurology. 2000;55(7):1025–7.

Findlay AM, Ambrose JB, Cahn-Weiner DA, Houde JF, Honma S, Hinkley LB, et al. Dynamics of hemispheric dominance for language assessed by magnetoencephalographic imaging. Ann Neurol. 2012;71(5):668–86.

Gaetz W, Cheyne D, Rutka JT, Drake J, Benifla M, Strantzas S, et al. Presurgical localization of primary motor cortex in pediatric patients with brain lesions by the use of spatially filtered magnetoencephalography. Neurosurgery. 2009;64(3 Suppl):177–85. discussion 86.

Gallen CC, Schwartz BJ, Bucholz RD, Malik G, Barkley GL, Smith J, et al. Presurgical localization of functional cortex using magnetic source imaging. J Neurosurg. 1995;82(6):988–94.

Guggisberg AG, Honma SM, Findlay AM, Dalal SS, Kirsch HE, Berger MS, et al. Mapping functional connectivity in patients with brain lesions. Ann Neurol. 2008;63(2):193–203.

Gupta SS. fMRI for mapping language networks in neurosurgical cases. Indian J Radiol Imaging. 2014;24(1):37–43.

Hamalainen MS. Basic principles of magnetoencephalography. Acta Radiol Suppl. 1991;377:58–62.

Hirata M, Goto T, Barnes G, Umekawa Y, Yanagisawa T, Kato A, et al. Language dominance and mapping based on neuromagnetic oscillatory changes: comparison with invasive procedures. J Neurosurg. n.d.;112(3):528–38.

Hirsch J, Ruge MI, Kim KH, Correa DD, Victor JD, Relkin NR, et al. An integrated functional magnetic resonance imaging procedure for preoperative mapping of cortical areas associated with tactile, motor, language, and visual functions. Neurosurgery. 2000;47(3):711–21. discussion 21–2

Ishibashi H, Morioka T, Nishio S, Shigeto H, Yamamoto T, Fukui M. Magnetoencephalographic investigation of somatosensory homunculus in patients with peri-Rolandic tumors. Neurol Res. 2001;23(1):29–38.

Jack CR Jr, Thompson RM, Butts RK, Sharbrough FW, Kelly PJ, Hanson DP, et al. Sensory motor cortex: correlation of presurgical mapping with functional MR imaging and invasive cortical mapping. Radiology. 1994;190(1):85–92.

Jennum P, Friberg L, Fuglsang-Frederiksen A, Dam M. Speech localization using repetitive transcranial magnetic stimulation. Neurology. 1994;44(2):269–73.

Julkunen P, Saisanen L, Danner N, Niskanen E, Hukkanen T, Mervaala E, et al. Comparison of navigated and non-navigated transcranial magnetic stimulation for motor cortex mapping, motor threshold and motor evoked potentials. NeuroImage. 2009;44(3):790–5.

Kamada K, Takeuchi F, Kuriki S, Oshiro O, Houkin K, Abe H. Functional neurosurgical simulation with brain surface magnetic resonance images and magnetoencephalography. Neurosurgery. 1993;33(2):269–72. discussion 72–3

Kapsalakis IZ, Kapsalaki EZ, Gotsis ED, Verganelakis D, Toulas P, Hadjigeorgiou G, et al. Preoperative evaluation with FMRI of patients with intracranial gliomas. Radiol Res Pract. 2012;2012:727810.

Kekhia H, Rigolo L, Norton I, Golby AJ. Special surgical considerations for functional brain mapping. Neurosurg Clin N Am. 2011;22(2):111–32. vii

Kirsch HE, Zhu Z, Honma S, Findlay A, Berger MS, Nagarajan SS. Predicting the location of mouth motor cortex in patients with brain tumors by using somatosensory evoked field measurements. J Neurosurg. 2007;107(3):481–7.

Kobayashi M, Pascual-Leone A. Transcranial magnetic stimulation in neurology. Lancet Neurol. 2003;2(3):145–56.

Kober H, Nimsky C, Moller M, Hastreiter P, Fahlbusch R, Ganslandt O. Correlation of sensorimotor activation with functional magnetic resonance imaging and magnetoencephalography in presurgical functional imaging: a spatial analysis. NeuroImage. 2001;14(5):1214–28.

Korvenoja A, Kirveskari E, Aronen HJ, Avikainen S, Brander A, Huttunen J, et al. Sensorimotor cortex localization: comparison of magnetoencephalography, functional MR imaging, and intraoperative cortical mapping. Radiology. 2006;241(1):213–22.

Krieg SM, Shiban E, Buchmann N, Gempt J, Foerschler A, Meyer B, et al. Utility of presurgical navigated transcranial magnetic brain stimulation for the resection of tumors in eloquent motor areas. J Neurosurg. 2012;116(5):994–1001.

Krieg SM, Schaffner M, Shiban E, Droese D, Obermuller T, Gempt J, et al. Reliability of intraoperative neurophysiological monitoring using motor evoked potentials during resection of metastases in motor-eloquent brain regions. J Neurosurg. 2013;118:1269.

Krieg SM, Sabih J, Bulubasova L, Obermueller T, Negwer C, Janssen I, et al. Preoperative motor mapping by navigated transcranial magnetic brain stimulation improves outcome for motor eloquent lesions. Neuro-Oncology. 2014;16(9):1274–82.

Krings T, Chiappa KH, Foltys H, Reinges MH, Cosgrove GR, Thron A. Introducing navigated transcranial magnetic stimulation as a refined brain mapping methodology. Neurosurg Rev. 2001a;24(4):171–9.

Krings T, Foltys H, Reinges MH, Kemeny S, Rohde V, Spetzger U, et al. Navigated transcranial magnetic stimulation for presurgical planning – correlation with functional MRI. Minim Invasive Neurosurg. 2001b;44(4):234–9.

Lefaucheur JP, Picht T. The value of preoperative functional cortical mapping using navigated TMS. Neurophysiologie clinique = Clin Neurophysiol. 2016;46(2):125–33.

Lehericy S, Duffau H, Cornu P, Capelle L, Pidoux B, Carpentier A, et al. Correspondence between functional magnetic resonance imaging somatotopy and individual brain anatomy of the central region: comparison with intraoperative stimulation in patients with brain tumors. J Neurosurg. 2000;92(4):589–98.

Logothetis N, Merkle H, Augath M, Trinath T, Ugurbil K. Ultra high-resolution fMRI in monkeys with implanted RF coils. Neuron. 2002;35(2):227–42.

Lutkenhoner B, Krumbholz K, Lammertmann C, Seither-Preisler A, Steinstrater O, Patterson RD. Localization of primary auditory cortex in humans by magnetoencephalography. NeuroImage. 2003;18(1):58–66.

Makela JP, Kirveskari E, Seppa M, Hamalainen M, Forss N, Avikainen S, et al. Three-dimensional integration of brain anatomy and function to facilitate intraoperative navigation around the sensorimotor strip. Hum Brain Mapp. 2001;12(3):180–92.

Martino J, Honma SM, Findlay AM, Guggisberg AG, Owen JP, Kirsch HE, et al. Resting functional connectivity in patients with brain tumors in eloquent areas. Ann Neurol. 2011;69(3):521–32.

Michelucci R, Valzania F, Passarelli D, Santangelo M, Rizzi R, Buzzi AM, et al. Rapid-rate transcranial magnetic stimulation and hemispheric language dominance: usefulness and safety in epilepsy. Neurology. 1994;44(9):1697–700.

Mills KR, Murray NM, Hess CW. Magnetic and electrical transcranial brain stimulation: physiological mechanisms and clinical applications. Neurosurgery. 1987;20(1):164–8.

Muir M, Prinsloo S, Michener H, Traylor JI, Patel R, Gadot R, de Almeida Bastos DC, Kumar VA, Ferguson S, Prabhu SS. TMS seeded diffusion tensor imaging tractography predicts permanent neurological deficits. Cancers (Basel). 2022a;14(2):340. https://doi.org/10.3390/cancers14020340.

Muir M, Gadot R, Prinsloo S, Michener H, Traylor J, Athukuri P, Tummala S, Kumar VA, Prabhu SS. Comparative study of preoperative functional imaging combined with tractography for prediction of iatrogenic motor deficits. J Neurosurg. 2022b;139(1):65–72. https://doi.org/10.3171/2022.10.JNS221684.

Nagarajan S, Kirsch H, Lin P, Findlay A, Honma S, Berger MS. Preoperative localization of hand motor cortex by adaptive spatial filtering of magnetoencephalography data. J Neurosurg. 2008;109(2):228–37.

Niessing J, Ebisch B, Schmidt KE, Niessing M, Singer W, Galuske RA. Hemodynamic signals correlate tightly with synchronized gamma oscillations. Science. 2005;309(5736):948–51.

Okada Y, Lauritzen M, Nicholson C. MEG source models and physiology. Phys Med Biol. 1987;32(1):43–51.

Okada Y, Lahteenmaki A, Xu C. Comparison of MEG and EEG on the basis of somatic evoked responses elicited by stimulation of the snout in the juvenile swine. Clin Neurophysiol. 1999;110(2):214–29.

Ossenblok P, Leijten FS, de Munck JC, Huiskamp GJ, Barkhof F, Boon P. Magnetic source imaging contributes to the presurgical identification of sensorimotor cortex in patients with frontal lobe epilepsy. Clin Neurophysiol. 2003;114(2):221–32.

Pascual-Leone A, Gates JR, Dhuna A. Induction of speech arrest and counting errors with rapid-rate transcranial magnetic stimulation. Neurology. 1991;41(5):697–702.

Pashut T, Wolfus S, Friedman A, Lavidor M, Bar-Gad I, Yeshurun Y, et al. Mechanisms of magnetic stimulation of central nervous system neurons. PLoS Comput Biol. 2011;7(3):e1002022.

Paus T, Jech R, Thompson CJ, Comeau R, Peters T, Evans AC. Transcranial magnetic stimulation during positron emission tomography: a new method for studying connectivity of the human cerebral cortex. J Neurosci. 1997;17(9):3178–84.

Picht T. Current and potential utility of transcranial magnetic stimulation in the diagnostics before brain tumor surgery. CNS Oncol. 2014;3(4):299–310.

Picht T, Mularski S, Kuehn B, Vajkoczy P, Kombos T, Suess O. Navigated transcranial magnetic stimulation for preoperative functional diagnostics in brain tumor surgery. Neurosurgery. 2009;65(6 Suppl):93–8. discussion 8–9

Picht T, Schmidt S, Brandt S, Frey D, Hannula H, Neuvonen T, et al. Preoperative functional mapping for rolandic brain tumor surgery: comparison of navigated transcranial magnetic stimulation to direct cortical stimulation. Neurosurgery. 2011;69(3):581–8. discussion 8.

Picht T, Strack V, Schulz J, Zdunczyk A, Frey D, Schmidt S, et al. Assessing the functional status of the motor system in brain tumor patients using transcranial magnetic stimulation. Acta Neurochir. 2012;154(11):2075–81.

Picht T, Krieg SM, Sollmann N, Rosler J, Niraula B, Neuvonen T, et al. A comparison of language mapping by preoperative navigated transcranial magnetic stimulation and direct cortical stimulation during awake surgery. Neurosurgery. 2013;72:808.

Picht T, Frey D, Thieme S, Kliesch S, Vajkoczy P. Presurgical navigated TMS motor cortex mapping improves outcome in glioblastoma surgery: a controlled observational study. J Neuro-Oncol. 2016;126(3):535–43.

Plomp G, Leeuwen C, Ioannides AA. Functional specialization and dynamic resource allocation in visual cortex. Hum Brain Mapp. n.d.;31(1):1–13.

Raffa G, Conti A, Scibilia A, Sindorio C, Quattropani MC, Visocchi M, Germanò A, Tomasello F. Functional reconstruction of motor and language pathways based on navigated transcranial magnetic stimulation and DTI fiber tracking for the preoperative planning of low grade glioma surgery: a new tool for preservation and restoration of eloquent networks. Acta Neurochir Suppl. 2017;124:251–61.

Rezai AR, Hund M, Kronberg E, Zonenshayn M, Cappell J, Ribary U, et al. The interactive use of magnetoencephalography in stereotactic image-guided neurosurgery. Neurosurgery. 1996;39(1):92–102.
Rowley HA, Roberts TP. Functional localization by magnetoencephalography. Neuroimaging Clin N Am. 1995;5(4):695–710.
Sanai N, Mirzadeh Z, Berger MS. Functional outcome after language mapping for glioma resection. N Engl J Med. 2008;358(1):18–27.
Sarvas J. Basic mathematical and electromagnetic concepts of the biomagnetic inverse problem. Phys Med Biol. 1987;32(1):11–22.
Schiffbauer H, Berger MS, Ferrari P, Freudenstein D, Rowley HA, Roberts TP. Preoperative magnetic source imaging for brain tumor surgery: a quantitative comparison with intraoperative sensory and motor mapping. Neurosurg Focus. 2003;15(1):E7.
Sollmann N, Picht T, Mäkelä JP, Meyer B, Ringel F, Krieg SM. Navigated transcranial magnetic stimulation for preoperative language mapping in a patient with a left frontoopercular glioblastoma. J Neurosurg. 2013;118(1):175–9. https://doi.org/10.3171/2012.9.JNS121053.
Sollmann N, Kelm A, Ille S, Schröder A, Zimmer C, Ringel F, Meyer B, Krieg SM. Setup presentation and clinical outcome analysis of treating highly language-eloquent gliomas via preoperative navigated transcranial magnetic stimulation and tractography. Neurosurg Focus. 2018;44(6):E2.
Sollmann N, Fratini A, Zhang H, Zimmer C, Meyer B, Krieg SM. Associations between clinical outcome and tractography based on navigated transcranial magnetic stimulation in patients with language-eloquent brain lesions. J Neurosurg. 2019;132:1033–42. https://doi.org/10.3171/2018.12.JNS182988.
Taniguchi M, Kato A, Ninomiya H, Hirata M, Cheyne D, Robinson SE, et al. Cerebral motor control in patients with gliomas around the central sulcus studied with spatially filtered magnetoencephalography. J Neurol Neurosurg Psychiatry. 2004;75(3):466–71.
Tarapore PE, Tate MC, Findlay AM, Honma SM, Mizuiri D, Berger MS, et al. Preoperative multimodal motor mapping: a comparison of magnetoencephalography imaging, navigated transcranial magnetic stimulation, and direct cortical stimulation. J Neurosurg. 2012a;117:354.
Tarapore PE, Martino J, Guggisberg AG, Owen J, Honma SM, Findlay A, et al. Magnetoencephalographic imaging of resting-state functional connectivity predicts postsurgical neurological outcome in brain gliomas. Neurosurgery. 2012b;71(5):1012–22.
Tarapore PE, Findlay AM, Honma S, Mizuiri D, Houde JF, Berger MS, et al. Language mapping with navigated repetitive TMS: proof of technique and validation. Neuroimage. 2013;82:260.
Tarapore PE, Picht T, Bulubas L, Shin Y, Kulchytska N, Meyer B, et al. Safety and tolerability of navigated TMS for preoperative mapping in neurosurgical patients. Clin Neurophysiol. 2016a;127(3):1895–900.
Tarapore PE, Picht T, Bulubas L, Shin Y, Kulchytska N, Meyer B, et al. Safety and tolerability of navigated TMS in healthy volunteers. Clin Neurophysiol. 2016b;127(3):1916–8.
Tarkiainen A, Liljestrom M, Seppa M, Salmelin R. The 3D topography of MEG source localization accuracy: effects of conductor model and noise. Clin Neurophysiol. 2003;114(10):1977–92.
Tomczak RJ, Wunderlich AP, Wang Y, Braun V, Antoniadis G, Gorich J, et al. fMRI for preoperative neurosurgical mapping of motor cortex and language in a clinical setting. J Comput Assist Tomogr. 2000;24(6):927–34.
Valero-Cabre A, Pascual-Leone A. Impact of TMS on the primary motor cortex and associated spinal systems. IEEE engineering in medicine and biology magazine: The quarterly magazine of the Engineering in Medicine & Biology Society. 2005;24(1):29–35.
Vrba J, Robinson SE. Signal processing in magnetoencephalography. Methods. 2001;25(2):249–71.
Vrba J, Robinson SE. SQUID sensor array configurations for magnetoencephalography applications. Supercond Sci Technol. 2002;15:51–89.
Wagner T, Valero-Cabre A, Pascual-Leone A. Noninvasive human brain stimulation. Annu Rev Biomed Eng. 2007;9:527–65.
Wassermann EM, Blaxton TA, Hoffman EA, Berry CD, Oletsky H, Pascual-Leone A, et al. Repetitive transcranial magnetic stimulation of the dominant hemisphere can disrupt visual naming in temporal lobe epilepsy patients. Neuropsychologia. 1999;37(5):537–44.

3 Basic and Advanced Intraoperative Neurophysiological Monitoring and Mapping in Supratentorial Brain Tumor Surgery

Kathleen Seidel, Pablo Alvarez Abut, Jonathan Wermelinger, and Andreas Raabe

3.1 Introduction

3.1.1 The Surgical Context

Brain tumor surgery has two main goals: maximizing the extent of tumor resection to improve survival, and avoiding postoperative deficits to maintain quality of life (Chang et al. 2011a; De Witt Hamer et al. 2012). This is why many surgical teams also guide resection based on function of the nervous system rather than anatomical landmarks only (Chang et al. 2011a; De Witt Hamer et al. 2012; Raabe et al. 2014; Bello et al. 2014; Sala and Lanteri 2003; Seidel et al. 2013; Duffau et al. 2003a). It has been shown that complete removal or gross total resection leads to prolonging of progression-free and overall survival in both lower grade gliomas and glioblastomas (Smith et al. 2008; Stummer et al. 2008; Senft et al. 2011; Sanai et al. 2011a; McGirt et al. 2008, 2009; Lacroix et al. 2001; Kamp et al. 2012). However, more than 50% of these tumors are classified as being located in "presumably eloquent" brain regions based on preoperative imaging (Chang et al. 2011a; De Witt Hamer et al. 2012; Stummer et al. 2006). Preoperative and especially intraoperative mapping techniques increase the success rate of radical tumor resection by establishing whether a region presumed to be eloquent on preoperative magnetic resonance imaging (MRI) is actually eloquent or not (Chang et al. 2011a). Where this is not the case, resection can proceed. However, removing the final and deepest part of a tumor near eloquent tracts puts neurologic function at risk. The oncologic advantage of a more radical resection (McGirt et al. 2009; Lacroix et al. 2001; Stummer

K. Seidel (✉) · P. A. Abut · J. Wermelinger · A. Raabe
Department of Neurosurgery, Inselspital, Bern University Hospital, and University of Bern, Bern, Switzerland
e-mail: Kathleen.Seidel@insel.ch; pabloabel.alvarezabut@insel.ch; jonathan.wermelinger@insel.ch; andreas.raabe@insel.ch

S. M. Krieg, T. Picht (eds.), *Navigated Transcranial Magnetic Stimulation in Neurosurgery*, https://doi.org/10.1007/978-3-031-97155-6_3

et al. 2006; Jakola et al. 2012; Sanai et al. 2011b; Schucht et al. 2014a; Salvati et al. 2012) may be offset by an increased risk of a neurologic deficit (Schucht et al. 2014b). A "*maximal safe resection*" can be achieved with the guidance of intraoperative neurophysiologic mapping methods.

Intraoperative neurophysiological monitoring methods of evoked potentials allow for real-time assessment of functional integrity of the pathway being monitored by the specific evoked potential technique (Deletis et al. 2020). It has been shown that evoked potentials can predict postoperative long-term outcome of the patient in certain situations (Asimakidou et al. 2021). Ischemia due to injury of perforating vessels and vasospasm can cause harm to the patient, especially in tumors around the insula. Monitoring of evoked potentials can help preventing irreversible injury by providing a warning, which is supposed to lead to an intervention ideally resulting in a restored evoked potential by the end of the surgery (Asimakidou et al. 2021; Neuloh et al. 2007a).

In the following, we explain key aspects of *intraoperative neurophysiological monitoring and mapping* (IONM) methods, and discuss how to apply them in the surgical context.

3.1.2 Principles of Recording and Stimulation in IONM

3.1.2.1 Origin of Neurophysiological Signals

The neuron constitutes not only the structural but also the functional unit of the nervous tissue. The cell membrane has selective permeability to the passage of sodium, potassium, and chloride ions. At rest, positive charges predominate on the outside and negative charges on the inside, with a membrane potential of approximately 70 mV. In this case, the cell is said to be polarized. When a neuron is stimulated, there is a change in the permeability of its membrane resulting in a flow of current from outside to inside the cell, with a change in its potential during which it becomes positive inside compared to the outside. This is known as depolarization of the cell membrane. If depolarization is significant enough, an electric pulse called an action potential is generated, which self-propagates to the axonic terminal. All electrophysiological signals are due to these movements of ions across the cell membrane (Møller. 2006; Kandel et al. 2000). Neurophysiological signals can be divided into two main categories: *spontaneous activity* in absence of external stimuli and *evoked responses* elicited via external stimuli along a sensory or motor pathway.

3.1.2.2 Basic Principles of Stimulation for Evoked Responses

Electric stimulation requires two electrodes, one *positive* (*anode*) and one *negative* (*cathode*). Stimulation can be monopolar or bipolar. Using a *monopolar* IONM stimulation probe, the stimulating electrode is placed on the structure of interest to stimulate, whereas the other electrode is placed at an indifferent location ("far" away from the area of interest). *Bipolar* IONM stimulation probes, on the other hand, contain both electrodes. Stimulation can be *cathodic* (cathode closer to the structure to stimulate), *anodic* (anode closer to the structure to stimulate), or *biphasic*, in which the polarity of the electric current alternates, with the current of the first phase flowing in one direction and the current of the second phase flowing in the opposite direction. During the application of the stimulus, there is a depolarization of the neuronal membrane such that an action potential is generated, which propagates along the nerve and, after passing through the corresponding synapses in the involved nerve pathways, generates a new action potential that is obtained via the recording electrodes (Papanicolaou. 2000).

3.1.2.3 Basic Principles of Recording of Evoked Responses

Recording neurophysiological signals requires two electrodes, one called the active electrode and the other called the reference electrode. The *active electrode* is placed close to the structure to be evaluated, whereas the *reference electrode* is placed at an indifferent location ("far" away from the active recoding site). In case of muscle motor evoked potentials (MEP) recordings in the context of IONM, usually both electrodes are placed in the belly of the muscle. Active and reference electrodes are connected to the input of an amplifier. Together, the amplifier and the two electrodes constitute a *recording channel*, whose resulting recording is displayed on a screen to be visualized and analyzed by the neurophysiologist.

An important principle in IONM is to be "*specific*" (selective) at the stimulation and/or recording site (Deletis et al. 2020). This is to make sure that the structure of interest is being monitored. An advanced example is the recording of responses from the facial (orbicularis oris muscles) compared to responses from the trigeminal muscles (masseter muscle) of the face. If the surgical exposure allows for it, the cranial nerve 5 (innervating masseter) or 7 (innervating orbicularis oris) can be stimulated separately in the resection cavity. If not, and transcranial electrical stimulation (TES) is performed, isolated needles or hook wire electrodes would need to be placed in both muscles to make sure not to record a combined response. Such a mixed response of both muscles makes the interpretation of an alarm difficult, since it would be unclear whether a signal drop was the result of injury to the cranial nerve 5 or 7.

3.1.3 Monitoring Versus Mapping Techniques

Neurophysiological monitoring techniques can be divided into two basic, complementary, and non-exclusive groups. *Mapping techniques* are used for identification of structures at risk and can guide the surgical tumor resection to indicate if a certain structure is eloquent. *Monitoring techniques* provide continuous feedback on the functional integrity of nervous structures throughout the surgical procedure and can facilitate immediate detection of potential surgical damage (Deletis et al. 2020).

Mapping techniques allow to identify eloquent tissue before resection alters the structure of interest. However, it is important to understand that mapping only provides information at, and distal to, the stimulation site. If feedback about the whole integrity of the motor or sensory system in an anesthetized patient is needed,

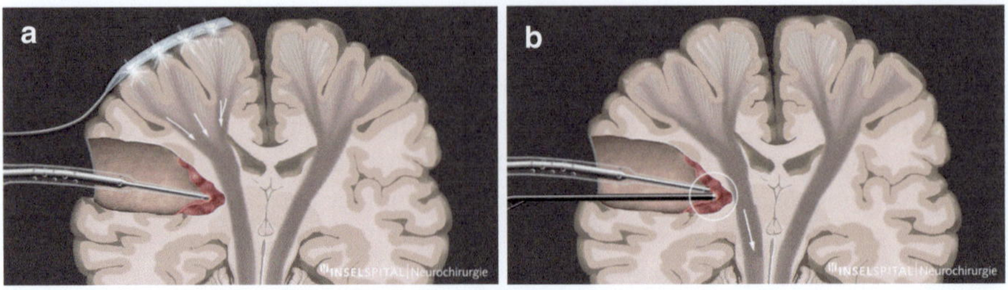

Fig. 3.1 Principles of monitoring and mapping techniques. (**a**) Illustration of *monitoring* of motor evoked potentials (MEPs) via stimulation through a strip electrode placed on the motor cortex. This allows assessment of the functional integrity of the corticospinal tract (CST) including detection of vascular injury or direct resection into the CST. (**b**) Subcortical *mapping* allows to estimate the distance to the CST before direct surgical injury to the tract occurs. However, only the CST distal to the stimulation site can be assessed. (© Inselspital, Bern University Hospital, Dept. of Neurosurgery)

monitoring methods are recommended for a continuous evaluation (Sala and Lanteri 2003; Seidel et al. 2013; Landazuri and Eccher 2013). For example, remote vascular injury due to coagulation of vessel perforators can only be detected by monitoring methods (Neuloh et al. 2004, 2007a; Szelenyi et al. 2010a). Therefore, mapping should always be combined with monitoring techniques (Fig. 3.1).

3.1.4 Basics of Anesthesia

Anesthetic agents affect evoked responses. The effects of intravenous and inhalational agents depend on the mechanism of action and the dose. The biggest impact results at the level of the synaptic transmission. Inhalation agents have a greater effect on evoked responses and are more likely to suppress signal recordings during monitoring (Sloan and Heyer 2002). Common inhalation agents include desflurane, isoflurane, sevoflurane, halothane, and nitrous oxide, whereas common intravenous anesthetics include propofol, barbiturates, dexmedetomidine, etomidate, ketamine, and benzodiazepines.

Among the evoked responses used in supratentorial procedures, visual evoked potentials (VEP) are most influenced by anesthetics, followed by MEP, which in turn are more affected than somato-sensory evoked potentials (SSEP). MEPs recorded from the muscles are susceptible to anesthetic agents at three sites: the motor cortex, the anterior horn cell, and the neuromuscular junction (Pechstein et al. 1998). Inhalational anesthetic agents have a dose-dependent effect on SSEPs. They lead to an increase in signal latency and decrease in amplitude of cortical responses (Sloan and Heyer 2002).

The use of sedative intravenous agents and analgesics without inhalation agents (namely, total intravenous anesthesia) has proved to be the most effective anesthetic protocol to obtain evoked responses in the context of IONM (Malcharek et al. 2015). At low doses, ketamine and etomidate increase the cortical amplitude of SSEP and MEP. Hence, they can be used as alternatives in case of low amplitude recordings.

Muscle relaxants work through their action on the neuromuscular junction and consequently impair the obtainment of MEPs. Therefore, short-acting muscle relaxants should only be used during intubation if monitoring of MEPs is planned. To test whether there is still residual neuromuscular blockade, one can carry out the *train of four* stimulation on a peripheral nerve. This method consists of four successive stimuli delivered at a constant intensity and a specified frequency, typically 2 Hz. The pulses are applied to a peripheral nerve, giving four responses recorded at a muscle innervated by the selected nerve (e.g., stimulating the median nerve and recording from the thenar muscle). If the amplitude of the final response matches that of the initial response, it indicates that the patient is not under muscle relaxation. Conversely, if there is a decrease in the amplitude of the final response compared to the initial response, the greater the decrement, the deeper the ongoing muscle relaxation.

3.2 The Integration of IONM and TMS Techniques

Advances in imaging techniques, such as functional MRI and fiber tracking algorithms, as well as preoperative *navigated transcranial magnetic stimulation (nTMS)* examinations substantially help to plan the surgical approach and to anticipate intraoperative challenges (Krieg et al. 2017). nTMS is especially valuable for identifying motor (and potential speech) eloquent regions in cases where the brain tumor has distorted the cortical anatomy (Krieg et al. 2017). Using positive nTMS points as seeds for fiber tracking allows to isolate and visualize the most important fibers (e.g., of the corticospinal tract, CST) which should be preserved. Nowadays, all of this information can be visualized in the operating room through *neuronavigation*, the computer-assisted transfer of three-dimensional image data of a patient to the current or future surgical field before or during an operation. However, after opening the dura and especially after corticotomy and tumor debulking, the resulting brain shift leads to a loss of accuracy in neuronavigation (Stieglitz et al. 2013). The use of intraoperative imaging to update navigation is currently still limited by restricted availability and time requirements (in case of intraoperative MRI) or accuracy and challenges of image data fusion (in case of navigated ultrasound). In order to resect the last remaining deep tumor parts, it is essential to identify functional tissue in real time, especially when approaching the tumor boundaries or even aiming for supra-marginal resection (Duffau 2023; Lutz et al. 2023). Over the last decades, intraoperative neurophysiological mapping has been established as the main tool to guide resection during these critical steps (Seidel et al. 2022).

In addition, nTMS after surgery can help predict recovery of the patients in case of intraoperative events and loss of MEPs (Seidel et al. 2019) or other evoked potentials. It may even guide the treatment strategy of the patients after surgery regarding whether to favor early rehabilitation or early adjuvant treatment (Seidel et al. 2020a). A promising avenue for future developments is (n)TMS-based rehabilitation (Ille et al. 2021; Vitulli et al. 2023).

To date, little research has been done combining and comparing IONM and nTMS data. This would, however, be desirable, as it might reinforce both methods and lead to an overall better planning and safer monitoring in the operating room. One such study has been published by Rosenstock et al., where they investigate how a previously introduced preoperative nTMS motor risk stratification might improve the interpretation of IONM phenomena (Rosenstock et al. 2017, 2021, 2022).

3.3 IONM in Surgeries Under Anesthesia

3.3.1 Evoked Potential Monitoring

3.3.1.1 Somato-Sensory Evoked Potentials (SSEPs)

SSEPs assess the entire dorsal somatosensory system from the limbs to the cerebral cortex, producing quantifiable, reproducible responses with defined amplitude, latency, and morphology. Multiple recording sites can be used to evaluate different segments of the neuroaxis (Cruccu et al. 2008). To obtain *SSEPs*, stimulation is applied to a peripheral nerve with pairs of electrodes typically placed distally on the wrists and ankles. Commonly used nerves include the median, ulnar, and posterior tibial nerves. SSEPs can then be recorded at peripheral nerve, spinal, subcortical, and cortical level. Each recording site helps assessing the integrity of the sensory pathway at a specific level (Cruccu et al. 2008). Loss of response indicates injury at, or distally from, that level. However, these signals are susceptible to anesthetic and hemodynamic factors. Further, due to a low signal-to-noise ratio, they require signal averaging (MacDonald et al. 2019), which can sometimes make it difficult to give a real-time feedback.

To identify neurological lesions through SSEPs, it is crucial to obtain a baseline measurement of responses before starting the surgical manipulation that might cause neurological injury. Warning criteria take into account signal parameters relative to baseline values (MacDonald et al. 2019). During supratentorial surgery, the most widely used criterion is a decrease of more than 50% in amplitude. However, an adaptive criterion has been recommended, corresponding to a visually obvious amplitude reduction from recent pre-change values and clearly exceeding variability, particularly when abrupt and focal (MacDonald et al. 2019).

When a significant change in a potential occurs, it is important to identify whether it was caused by technical failure, anesthetic or hemodynamic factors, or an actual neurological injury. For this purpose, it is essential to understand the surgical steps, to maintain constant communication with the surgical team, and to coordinate with the anesthesia team. Surgical feedback to the IONM team at critical steps is equally important to better interpret signal changes in the surgical context.

3.3.1.2 Motor Evoked Potentials (MEPs)

MEPs recorded from the muscles (so-called muscle MEPs) require trains of stimuli to overcome anesthetic inhibition over the anterior spinal horn. Signal averaging is not necessary. Standard stimulation parameters are a train of 5 stimuli, interstimulus interval of 4 ms, pulse duration of 500 μs and a 0.5–1.0 Hz repetition rate (Taniguchi et al. 1993). MEPs can be elicited via *transcranial electric stimulation* (TES) through corkscrew-electrodes from various montages (transcranially elicited MEP or *TES MEP*) (Szelenyi et al. 2007). For a more focal stimulation, a strip electrode can be placed on the motor cortex (if exposed) or slid under the dura toward the motor cortex (Fig. 3.1 Panel A). This method is called *direct cortical stimulation* (DCS) and the resulting MEPs are called *DCS MEPs* (Seidel et al. 2013; Szelenyi et al. 2005).

Muscle MEPs provide specific information about the motor pathway and muscle groups of interest. However, they are influenced by muscle relaxation, as well as anesthetic and hemodynamic factors. Stimulation, especially at high intensity, can induce patient movement. In case of patient movement, the surgeon should be informed and asked to pause resection before stimulation.

It is crucial to obtain baseline MEPs before the surgical maneuvers. During surgery, significant MEP deterioration compared to an earlier baseline should lead to a warning (Asimakidou et al. 2021). Several muscle MEP warning criteria have been proposed, including the following.

- *Threshold criterion:* an increase in the stimulus threshold to elicit a response (variable according to type of surgery and different authors).
- *Amplitude criterion:* a decrease in the amplitude of responses (variable, generally ranging from 50% to 80%, according to different authors).
- *All-or-nothing criterion:* the presence or absence of responses, which should not be used in supratentorial surgery.
- *Morphological criterion:* a change in morphology with a reduction in the number of phases of the evoked response (less established).

While monitoring MEPs, inadvertent bolus injections of anesthetics or muscle relaxants should be avoided, since they significantly affect amplitudes. So does a drop in blood pressure and body temperature. In particular, one should beware a slow, gradual decrease of blood pressure and body temperature. The MEP parameters are not linearly related to cerebral perfusion but can change abruptly and in a stepwise manner. However, as a rule of thumb, a mean arterial pressure below 70 mmHg can become critical (Neuloh et al. 2007b; Neuloh and Seidel 2023).

After checking for nonsurgical causes, a warning must be issued to the surgeon. Typically, resection or dissection is halted at this point. The surgeon should inspect potential technical causes for MEP changes. These include displacement of stimulation electrodes, poor contact of the electrode resulting in high impedance (in which case subdural irrigation and wet cottonoids on top of the electrode can be helpful), subdural air collection, or a shift of the motor cortex away from the stimulation electrodes after removal of a mass lesion (Neuloh et al. 2007b; Neuloh and Seidel 2023).

If MEP changes were potentially caused by surgical action, the surgeon's attention should be directed at the specific maneuver at the origin of the monitoring alarm event. Some straightforward causes include resection and electrocoagulation in close vicinity to the CST as revealed by subcortical mapping. A typical cause of (seemingly) inexplicable MEP deterioration is extensive manipulation of remote blood vessels supplying the motor tract which happened at some previous step of dissection. A temporary halt of dissection and readjustment of the brain retraction is often sufficient to enable MEP recovery and the continuation of safe resection

(Neuloh et al. 2007a). It may be useful to place pieces of gelfoam or cottonoids soaked with papaverine or nimodipine at sites of (preceding) vascular manipulation.

In some cases, there can be a false positive alarm. Stopping the resection prematurely might indirectly harm the patient. However, if signal changes remain inexplicable during tumor resection, one should always keep in mind the possibility of doing a second surgery on a following day. Thus, all MEP alarms should be taken seriously. Indeed, if irreversible MEP alterations occur, a postoperative motor deficit is expected in most cases (Seidel et al. 2013; Asimakidou et al. 2021).

Irreversible MEP alterations are associated with a higher number of transient deficits compared to reversible MEP changes, and a higher likelihood that these motor deficits persist. In most studies, MEPs show a high specificity and negative predictive value (Asimakidou et al. 2021). Thus, in the absence of an irreversible alteration, the surgeon may be reassured that the patient will not be suffering from a motor deficit in the short-term and long-term follow-up. In contrast, reports of sensitivity estimates and the positive predictive value are less consistent. This might be due to the low prevalence of reported alarm events in most series. MEPs seem to perform well as surrogate markers, since successful intervention followed by a reversal of MEP deterioration indicates postoperative motor function preservation (Asimakidou et al. 2021).

However, to date only a limited number of warning criteria are well established and have been implemented in the clinical routine. The traditional MEP parameters can vary within and across patients. This can affect MEP monitoring of individual patients during surgery as well as limit generalizability for different patients and monitoring procedures (Macdonald 2006). The use of machine learning methods as a way to cope with these difficulties is an active field of recent research (Park and Kim 2022) and might lead to a better understanding of the important properties of MEPs and other evoked potentials (Parduzi et al. 2025; Wermelinger et al. 2023; Jamaludin et al. 2022; Mirallave-Pescador et al. 2023).

3.3.1.3 Visual Evoked Potentials (VEPs)

When light hits the eye, the activity generated in the retina is transmitted through the optic nerve to the brain. Information from both optic nerves converges at the optic chiasm, and subsequently continues through the optic tracts (via the thalamus, optic radiation, and other subcortical structures) to the visual cortex. After stimulation of the visual pathway, the resulting signal at the visual cortex is called a *visual evoked potential (VEP)*. They primarily allow to assess central vision (Papanicolaou. 2000).

VEPs are reported to be unreliable due to their high variability and instability. This is partly attributed to the visual pathway's greater sensitivity to contrast changes (induced by checkboard stimulus) compared to luminous stimulation (performed with flash, the only feasible option in the operating room). Another potential reason for low VEP reliability is inadequate flash stimulation due to stimulator displacement during surgical maneuvers (especially rotation of the skin flap during the surgical approach). This is addressed by combining the cortical recording of VEPs with electroretinography (ERG). In case of VEP alteration, the ERG response can indicate whether the stimulus adequately reaches the retina to activate the visual pathway (Kurozumi et al. 2017). This has increased the reliability of VEP monitoring (Gutzwiller et al. 2018).

VEPs are very sensitive to anesthetic agents, especially inhalational agents. Propofol, fentanyl, and remifentanil can be used if VEP monitoring is planned, as they have less effect on these signals (Uribe et al. 2017). The following factors can also affect VEPs: preoperative visual deficit, low body temperature, hypoxia, or hypotension. If all those factors are excluded, there is an increased likelihood that the visual pathway is damaged. The most commonly used alarm criterion for VEPs is a decrease in response amplitude by more than 50%, although the loss of waveform morphology and an increase in latency by more than 2 standard deviations have also been described (Kamio et al. 2014; Luo et al. 2015).

Intraoperative VEPs are used to assess the functional integrity of the visual pathway in surgeries involving tumors affecting the optic nerves, chiasm, and potentially also optic tracts, or occipital visual cortex (Kamio et al. 2014; Kodama et al. 2010). To date, VEPs have mainly been reported to be useful in cases involving retroorbital or parasellar regions (Kurozumi et al. 2017). The use of VEP for monitoring the post-chiasmatic pathway still needs validation in larger series regarding postoperative patient outcome.

3.3.1.4 Electroencephalography (EEG)

Electroencephalography (EEG) is the real-time measurement of cortical activity. It can be recorded on the scalp or with intracranial electrodes, which is then called *electrocorticography* (ECoG). In IONM, it is useful for monitoring the anesthetic depth, detecting ischemic changes of the cerebral cortex, or visualizing epileptiform activity. The activity represented in EEG stems from the postsynaptic potentials of pyramidal cells. Brain wave frequencies are commonly divided into four categories: delta (≤4 Hz), theta (4–7 Hz), alpha (8–12 Hz), and beta (13–30 Hz). The anesthetic depth influences frequency and amplitude of brain activity captured by EEG through modifications in the synapses. During anesthetic induction, beta activity is the first to appear, followed by a successive lowering in frequency. If the anesthetic level continues to deepen, it eventually generates a *burst-suppression pattern* (periods of electrical activity interspersed by flattening of the EEG) and later electroencephalographic silence. In addition to direct visual observation of the EEG, algorithms such as the bispectral index or spectral frequency analysis can provide simplified continuous assessment of the anesthetic depth. However, these methods have not replaced the real-time visual analysis of “raw” EEG (Jameson et al. 2007).

3.3.2 Cortical Mapping

When monitoring TES MEPs during surgery close to the primary motor cortex, a critical concern is the site of MEP activation, particularly when high stimulation intensities are used. With high stimulation intensities, MEPs may be elicited more caudally in the CST (see also previous paragraphs). That is, the current might “jump” to levels below the resection cavity, resulting in a missed cortical or superficial subcortical injury and a false-negative result (Rothwell et al. 1994; Neuloh and Schramm 2009; Boex et al. 2016). Surgery at the level of the precentral gyrus therefore requires lower stimulation intensities and a more focal stimulation method.

However, placing a strip electrode for DCS MEP monitoring is sometimes impossible because it can interfere with the surgical approach, especially at the cortical incision site. Therefore, for tumors involving the precentral gyrus, frequent or continuous motor mapping with step-by-step tumor removal can be the most reliable and practical technique to timely identify motor eloquent cortex (Seidel et al. 2012). Cortical mapping is an efficient method to identify the primary motor area (M1) in the precentral gyrus. Subcortical motor mapping is used to provide information about the spatial relation of the CST and U-shaped fibers emerging from M1 (Rossi et al. 2019).

Recently, new IONM approaches have been introduced to probe other motor areas (in addition to M1 and the CST) under general anesthesia. By adapting the number of pulses within the train stimulation, Rossi et al. suggested that responses elicited from the fast and direct projecting fibers of M1 can be differentiated from other motor areas of the precentral gyrus (Rossi et al. 2019). To define their tumor resection limits within the precentral gyrus, they measured latency and amplitude of their MEPs (Rossi et al. 2019). To assess projections from the parietal cortex to M1, Cattaneo et al. introduced a novel approach to evaluate brain connectivity (Cattaneo et al. 2020). They applied a conditioning stimulus over the parietal cortex at defined intervals before the monitoring stimulus over the motor cortex. They found inhibition and a reduction in MEP amplitude in dorsal projections, while observing facilitation and an increased MEP amplitude in ventral projections (Cattaneo et al. 2020). Although these techniques are promising, they need to be clinically validated on larger cohorts.

3.3.3 Subcortical and Dynamic Mapping

Intraoperative subcortical mapping can guide the surgery toward the goal of maximizing tumor resection and minimizing neurologic deficits (Chang et al. 2011a; De Witt Hamer et al. 2012). One important concern during tumor removal is estimating how distant a certain point in the resection cavity is from the CST. Thus, elaborating warning signs for impending mechanical damage to the CST remains a challenge for the surgeon and the neurophysiologist. Eliciting an MEP depends on the charge applied to the tissue (Szelenyi et al. 2011). Charge is the integral of current or voltage (up to a factor) over time, and for rectangular pulses, it is equal to current times pulse duration. Moreover, the charge density decreases with distance. The higher the charge, the larger the area MEPs will be generated from, and vice versa (Seidel et al. 2013). This also implies that with higher stimulus intensity or charge, MEPs will be elicited through electrodes located at a greater distance from the CST. Many groups have investigated this "stimulation-strength-to-CST-distance" relationship by correlating the stimulation intensity in mA needed to elicit MEPs with distance in mm to the CST using different pulse durations (Kamada et al. 2009; Maesawa et al. 2010; Nossek et al. 2011; Prabhu et al. 2011; Ohue et al. 2012; Shiban et al. 2015a).

One of the first studies of this kind was published by Kamada et al. (2009). They found a convergent nonlinear correlation of distance and threshold current, and determined 1.8 mA as the threshold of direct CST contact during monopolar cathodal subcortical stimulation using trains of five 0.2 ms pulses. Nossek et al. (2011) compared the thresholds of subcortical monopolar mapping using trains of 5–7 with 0.5 ms pulse duration at 300 Hz to the CST displayed by intraoperative navigation, which was corrected for brain shift by intraoperative ultrasonography. They showed an approximately linear correlation between MEP thresholds and distance to the CST with a relationship of 0.97 mA for every 1 mm of brain tissue, which has subsequently been established as the "rule-of-thumb" of 1 mA = 1 mm (Seidel et al. 2012). A linear correlation between stimulus intensity and distance was also demonstrated in a study by Ohue et al. (2012). They used trains of five 0.2 ms pulses at 500 Hz and analyzed the distance of the postoperative resection cavity to the reconstructed CST in early postoperative diffusion tensor imaging MRI. Maesawa et al. (2010) and Prabhu et al. (2011) looked at subcortical stimulation and intraoperative MRI tractography. Maesawa et al. (bipolar probe, train of 5 stimuli, 0.4 ms pulse duration, 500 Hz) also found an approximately linear correlation. Further, Prabhu et al. (monopolar probe, train of 5 stimuli, 0.3–0.5 ms pulse duration, 500 Hz) described a trend toward worsening neurologic deficits if the distance from the probe to the CST was <5 mm. On the other hand, Shiban et al. (2015a) found a nonlinear correlation between stimulation intensity and the distance to the CST. They marked the intraoperative stimulation site at the end of tumor resection with a titanium clip and executed different stimulation paradigms with cathodal or anodal stimulation using pulse durations of 0.3, 0.5, and 0.7 ms. After the surgery, the distance between the CST as outlined by postoperative diffusion tensor imaging fiber tracking and the location of the titanium clip was measured. In their study, cathodal 0.5 ms pulse duration stimulation appeared best suited for subcortical stimulation.

When comparing these different studies, it is important to consider the applied number of stimuli, the pulse duration, the polarity, and especially the charge. For subcortical motor mapping, it may be advisable to keep the number of stimuli and pulse duration constant, and to apply constant-current cathodal stimulation (Raabe et al. 2014; Seidel et al. 2012, 2013; Plans et al. 2017). It is important to note that, to date, no definitive consensus exists on the relationship between stimulation current and distance to the CST. Yet, the approximate rule of thumb "1 mA correlates to 1 mm" is being used more and more when performing subcortical short train monopolar stimulation with five 0.5 ms cathodal constant-current pulses.

The remaining question now is: what is a reasonable lowest mapping threshold (in mA) to stop tumor resection? This question is crucial since the main goal is to achieve as much tumor removal as is safely possible (Chang et al. 2011a; De Witt Hamer et al. 2012; Raabe et al. 2014; Seidel et al. 2013). Several studies have shown that decreasing MEP thresholds correlates with an increasing risk of mechanical injury to the CST (Sala and Lanteri 2003; Seidel et al. 2013; Prabhu et al. 2011; Plans et al. 2017; Sanmillan et al. 2017). Applying monopolar subcortical cathodal short train stimulation at low stimulation intensities, Seidel et al. (2013) showed a correlation between motor thresholds and the risk of CST injury. However, it was demonstrated that mapping thresholds ≤3 mA can be safe if DCS MEP monitoring remains stable at the same time and mapping is repeated frequently (Seidel et al. 2013; Landazuri and Eccher 2013). However, the stimulation intensity limit during subcortical short train mapping at which tumor removal is supposed to stop depends on various factors like tumor histopathology, planned goal of tumor resection, infiltration of other eloquent tracts, or method of hemostasis. Consequently, different surgical centers have different safety limits (Raabe et al. 2014; Bello et al. 2014; Sala and Lanteri 2003; Seidel et al. 2013; Nossek et al. 2011; Prabhu et al. 2011; Ohue et al. 2012; Shiban et al. 2015a; Plans et al. 2017; Sanmillan et al. 2017; Mikuni et al. 2007; Moiyadi et al. 2018).

Moreover, using the sporadic conventional mapping with insufficient spatial and temporal coverage of the surgical site might not effectively mitigate motor deficits even when resection is halted at seemingly secure mapping thresholds, such as 3–6 mA using monopolar short train stimulation (Seidel et al. 2013). A significant progress would be achieved by mapping continuously with the highest possible temporal coverage, and directly at the site of tumor removal. This would be especially important at very low mapping thresholds of ≤3 mA in close proximity to the CST. In keeping with this concept, Raabe et al. (2014) described a new mapping protocol and developed a device allowing continuous subcortical stimulation without interrupting the surgical workflow of tumor removal. This was achieved by integrating cathodal short-train stimulation (five 0.5 ms pulses, 250 Hz) into the tip of the surgical suction device at a repetition rate of 0.4–2 Hz. Positive MEP responses were coupled with an alarm sound to facilitate real-time feedback to the surgeon. In a proof-of-concept analysis of this *continuous "dynamic" mapping* method, its safety was confirmed with a 3% rate of permanent motor deficits (due to remote vascular injury of the CST). In this series, no patient had a permanent motor deficit caused by a mechanical injury of the CST. The intraoperative rate of seizures was 4%, all detected by simultaneous electrocorticography (ECoG) recordings and easily terminated without alteration of the surgical workflow. Applying this continuous dynamic mapping technique in surgeries of glioblastoma, Schucht et al. (2014b) demonstrated the feasibility of complete resection of enhancing tumor in 69% of patients despite high proximity to the CST and a permanent motor deficit rate of 4% (caused by vascular injury). Continuous "dynamic" mapping will not completely avoid so-called "mechanical" injury of the CST. However, enabling mapping during every surgical step at every surgical site increases the mapping coverage and provides valuable real-time feedback (Seidel et al. 2020b).

Additionally, continuous subcortical stimulation through the cavitron ultrasonic surgical aspirator (CUSA) was shown to be safe (Shiban et al. 2015b; Roth et al. 2017). However, CUSA activity might interfere with mapping results on rare occasions (Carrabba et al. 2008). In our opinion, integration of mapping into a standard suction device is more ergonomic because it allows for continuous mapping during all steps of tumor removal, including subpial dissection and hemostasis which are performed with a variety of instruments (Raabe et al. 2014; Schucht et al. 2014b, 2015; Moiyadi et al. 2018; Seidel et al. 2020b). The integration of the stimulation probe into any surgical instrument (CUSA or suction probe) might improve the reliability, surgical acceptance, and clinical handling of subcortical mapping.

A patient case which illustrates the neurophysiological strategies of preoperative nTMS, intraoperative cortical mapping, DCS MEP monitoring, subcortical mapping and postoperative nTMS described above is presented in Figs. 3.2 and 3.3.

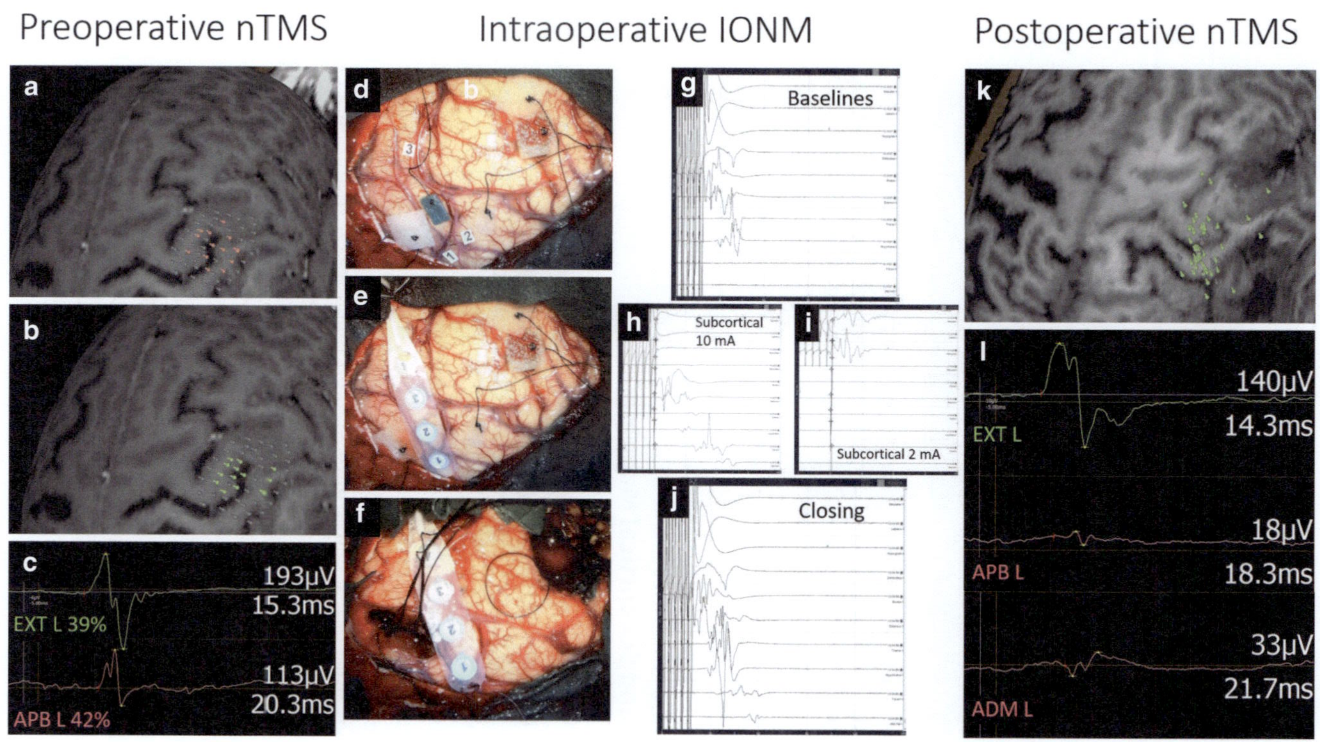

Fig. 3.2 Illustrative patient case with a suspected high-grade glioma in the peri-central region on the right hemisphere (see Fig. 3.3 for the MRI). (**a–c**) Preoperative navigated transcranial magnetic stimulation (nTMS) mapping data of extensor digitorum communis (EXT) and abductor pollicis brevis (APB). (**d**) Microscope image with paper marks 1 and 2 for hand and 3 for face muscle MEP responses after cortical monopolar short train mapping. The white cottonoids indicate the planed cortical entry sites. (**e**, **f**) Strip electrode position on the precentral gyrus for continuous MEP surveillance before and during tumor resection. (**g**) Baseline MEP recordings of different muscles. (**h**, **i**) Subcortical short train monopolar mapping with 10 mA and later during the resection with 2 mA. (**k**) Preserved MEP responses after tumor resection. (**l**, **m**) Postoperative nTMS mapping results 7 days after resection for arm and hand muscles. Directly after surgery the patient presented with a motor weakness of the arm and hand muscles, which completely recovered to the baseline performance within 2 weeks. (© Inselspital, Bern University Hospital, Dept. of Neurosurgery)

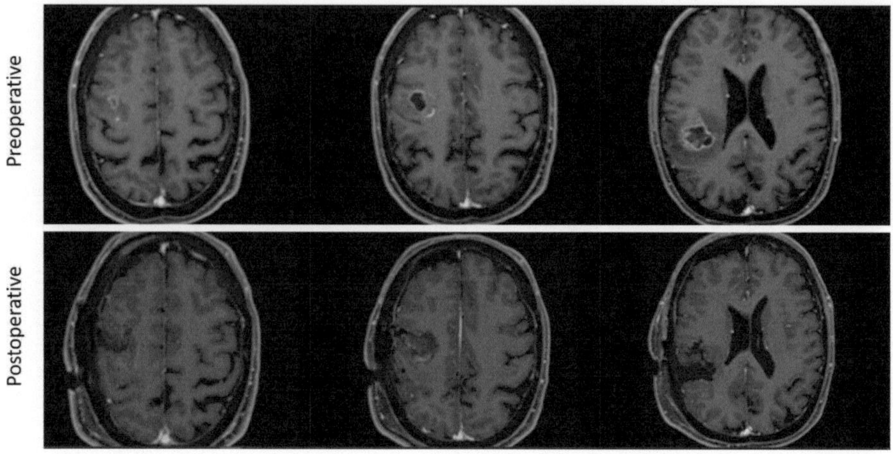

Fig. 3.3 Preoperative (upper row) and postoperative (lower row) axial T1 MR images with contrast. The postoperative MR images show a complete resection of contrast enhancing tumor (CRET) of a glioblastoma IDH wild type. (© Inselspital, Bern University Hospital, Dept. of Neurosurgery)

3.4 IONM in Awake Craniotomies

3.4.1 Cortical and Subcortical Mapping

In 1937, Wilder Penfield pioneered a paradigm which is commonly called the 50–60 Hz or low frequency stimulation, but it is also referred to as the *Penfield method* (Seidel et al. 2022; Szelenyi et al. 2010b; Penfield 1937). This method is commonly employed in awake craniotomies. A bipolar probe is used with a continuous train of consecutive stimuli at an interstimulus interval of 16.7–20 ms (50–60 Hz pulse frequency) and stimulation is applied for 1–4 seconds (Szelenyi et al. 2010b). Due to the stimulation duration of at least 1 second, the stimulation paradigm does not elicit a single time-locked response. It results in a continuum of positive or negative responses. If the primary motor cortex is stimulated, it can result in tonic-clonic movement of the patient. If the premotor area is activated, it may lead to a dystonia-like movement. If the supplementary motor area is stimulated, negative motor phenomena such as interruption of voluntary movement have been observed (Seidel et al. 2022). Simultaneous recording of muscle EMG may help analyzing these different motor phenomena (Seidel et al. 2022). Stimulation of the primary somatosensory cortex elicits the subjective feeling of dysesthesia. Stimulation of cortical and subcortical structures within the language and cognitive network can result in the partial or complete inhibition of the performed task.

3.4.2 ECOG and Language/Cognitive Testing

3.4.2.1 Electrocorticography (ECoG)

When EEG is recorded using subdural electrodes placed directly on the brain's surface, it is referred to as *electrocorticography (ECoG)* (Jameson et al. 2007). In the context of language and cognitive mapping, many centers employ ECoG alongside with mapping of the eloquent cortex (Spena et al. 2017). ECoG is used to record epileptiform discharges potentially occurring during cortical Penfield stimulation, called *after-discharges* (Berger and Ojemann 1992). These after-discharges are rhythmic, transient epileptiform activities. Mapping cortical function in the presence of after-discharges may lead to an inaccurate assessment of the stimulated cortex, as the clinical manifestation from the stimulation might have resulted from propagated cortical activation. Furthermore, the identification of after-discharges enables the rapid detection of subclinical epileptic seizures, allowing for their cessation before propagation leads to a clinical epileptic seizure. This might be achieved by stopping the stimulation, irrigating with cold physiological solution, or eventually administering antiepileptic medication (Szelenyi et al. 2010b; Spena et al. 2017).

An advanced ECoG technique is measuring high gamma activity when the patient is performing a certain task. It has been shown that activation of cortical regions is associated with high gamma activity. This activation is triggered during cognitive tasks, such as language tasks, providing another mapping tool without the need for electrical stimulation (Vakani and Nair 2019). However, further studies are needed before this technique can be integrated into the clinical routine.

3.4.2.2 Language and Cognitive Mapping Tasks

Language and cognitive mapping is recommended in patients with intrinsic tumors involving language or cognitive pathways. In current practice, the evaluation of language and other cognitive functions requires the patient to be awake and cooperative during the surgery. In addition, during the preoperative phase, the degree of functional reorganization achieved by the patient's brain should be evaluated. This preoperative assessment is meant to estimate the patients' ability to endure the awake phase of surgery and correctly select the intraoperative tests (Mandonnet et al. 2017; Rofes et al. 2017). Neuropsychologists and/or speech therapists are mandatory for such complex evaluations consisting of neuropsychologic tests and language examination. Various neurologic functions can be explored with neuropsychologic evaluations, like memory, language, praxis, executive functions, and fluid intelligence (Mandonnet et al. 2017; Rofes et al. 2017). In particular, language examinations are used to establish spoken language proficiency, language dominance, language production, and comprehension according to the educational level of the patient (Seidel et al. 2022).

"Naming" is the most commonly used intraoperative language test. Many centers combine conventional naming with semantic association tasks and other tests. Indeed, this is important for cortical and especially subcortical mapping since more than one tract contributes to language. It has been suggested that the arcuate fascicle is associated with phonological encoding (Duffau et al. 2003b; Giampiccolo and

Duffau 2022; Usui et al. 2003), the inferior fronto-occipital fascicle with semantic control (Duffau et al. 2005), and the superior longitudinal fascicle III with anarthria (Fridriksson et al. 2013; Gajardo-Vidal et al. 2021). During subcortical tumor resection, it is recommended to alternate between tests and spontaneous speech (Chang et al. 2011a, b; Duffau and Taillandier 2015).

Furthermore, Penfield stimulation might be applied to the posterior temporal lobe to identify the inferior longitudinal fasciculus, to the parietal lobe to identify the latter and visual tracts, or to the lateral frontal lobe to identify peri-insular connectivity involved in higher cognitive functions (Duffau 2023).

3.4.3 Cortico-Cortical Evoked Potentials (CCEPs)

In the neurosurgical context, Matsumoto et al. were the first to demonstrate the feasibility of recording evoked potentials from temporal language sites when stimulating frontal language hubs and vice versa (Matsumoto et al. 2004). They called these signals *cortico-cortical evoked potentials (CCEPs)* (Matsumoto et al. 2004; Vincent et al. 2017). While initially only performed in awake epilepsy patients with implanted subdural grids (Matsumoto et al. 2017), some groups have been studying the applicability of this technique in glioma surgery in awake (Yamao et al. 2014, 2017, 2021; Saito et al. 2014, 2022; Tamura et al. 2016; Suzuki et al. 2019; Vega-Zelaya et al. 2023) and even asleep patients (Yamao et al. 2014, 2017, 2021; Saito et al. 2014; Suzuki et al. 2019; Vega-Zelaya et al. 2023; Giampiccolo et al. 2021; Kanno et al. 2018; Nakae et al. 2020; Kang et al. 2023). However, intraoperative real-time recordings of CCEPs are still technically challenging, and often require careful signal processing to be differentiated from artifacts. Their reliability and predictive power have to be tested and established before CCEPs can be introduced into clinical routine (Seidel et al. 2024).

References

Asimakidou E, Abut PA, Raabe A, Seidel K. Motor evoked potential warning criteria in supratentorial surgery: a scoping review. Cancer. 2021;13(11):2803.

Bello L, Riva M, Fava E, Ferpozzi V, Castellano A, Raneri F, et al. Tailoring neurophysiological strategies with clinical context enhances resection and safety and expands indications in gliomas involving motor pathways. Neuro Oncol. 2014;16(8):1110–28.

Berger MS, Ojemann GA. Intraoperative brain mapping techniques in neuro-oncology. Stereotact Funct Neurosurg. 1992;58(1–4):153–61.

Boex C, Haemmerli J, Momjian S, Schaller K. Prognostic values of motor evoked potentials in insular, precental, or postcentral resections. J Clin Neurophysiol. 2016;33(1):51–9.

Carrabba G, Mandonnet E, Fava E, Capelle L, Gaini SM, Duffau H, et al. Transient inhibition of motor function induced by the Cavitron ultrasonic surgical aspirator during brain mapping. Neurosurgery. 2008;63(1):E178–9; discussion E9.

Cattaneo L, Giampiccolo D, Meneghelli P, Tramontano V, Sala F. Cortico-cortical connectivity between the superior and inferior parietal lobules and the motor cortex assessed by intraoperative dual cortical stimulation. Brain Stimul. 2020;13(3):819–31.

Chang EF, Clark A, Smith JS, Polley MY, Chang SM, Barbaro NM, et al. Functional mapping-guided resection of low-grade gliomas in eloquent areas of the brain: improvement of long-term survival. Clinical article. J Neurosurg. 2011a;114(3):566–73.
Chang EF, Wang DD, Perry DW, Barbaro NM, Berger MS. Homotopic organization of essential language sites in right and bilateral cerebral hemispheric dominance. J Neurosurg. 2011b;114(4):893–902.
Cruccu G, Aminoff MJ, Curio G, Guerit JM, Kakigi R, Mauguiere F, et al. Recommendations for the clinical use of somatosensory-evoked potentials. Clin Neurophysiol. 2008;119(8):1705–19.
De Witt Hamer PC, Robles SG, Zwinderman AH, Duffau H, Berger MS. Impact of intraoperative stimulation brain mapping on glioma surgery outcome: a meta-analysis. J Clin Oncol. 2012;30(20):2559–65.
Deletis V, Shils J, Sala F, Seidel K. Neurophysiology in neurosurgery- a modern approach. 2nd ed. Boston: Academic Press, Elsevier; 2020.
Duffau H. Oncological and functional neurosurgery: perspectives for the decade regarding diffuse gliomas. Rev Neurol. 2023;179(5):437–48.
Duffau H, Taillandier L. New concepts in the management of diffuse low-grade glioma: proposal of a multistage and individualized therapeutic approach. Neuro Oncol. 2015;17(3):332–42.
Duffau H, Capelle L, Denvil D, Sichez N, Gatignol P, Taillandier L, et al. Usefulness of intraoperative electrical subcortical mapping during surgery for low-grade gliomas located within eloquent brain regions: functional results in a consecutive series of 103 patients. J Neurosurg. 2003a;98(4):764–78.
Duffau H, Gatignol P, Denvil D, Lopes M, Capelle L. The articulatory loop: study of the subcortical connectivity by electrostimulation. Neuroreport. 2003b;14(15):2005–8.
Duffau H, Gatignol P, Mandonnet E, Peruzzi P, Tzourio-Mazoyer N, Capelle L. New insights into the anatomo-functional connectivity of the semantic system: a study using cortico-subcortical electrostimulations. Brain. 2005;128(4):797–810.
Fridriksson J, Guo D, Fillmore P, Holland A, Rorden C. Damage to the anterior arcuate fasciculus predicts non-fluent speech production in aphasia. Brain. 2013;136(11):3451–60.
Gajardo-Vidal A, Lorca-Puls DL, Team P, Warner H, Pshdary B, Crinion JT, et al. Damage to Broca's area does not contribute to long-term speech production outcome after stroke. Brain. 2021;144(3):817–32.
Giampiccolo D, Duffau H. Controversy over the temporal cortical terminations of the left arcuate fasciculus: a reappraisal. Brain. 2022;145(4):1242–56.
Giampiccolo D, Parmigiani S, Basaldella F, Russo S, Pigorini A, Rosanova M, et al. Recording cortico-cortical evoked potentials of the human arcuate fasciculus under general anaesthesia. Clin Neurophysiol. 2021;132(8):1966–73.
Gutzwiller EM, Cabrilo I, Radovanovic I, Schaller K, Boex C. Intraoperative monitoring with visual evoked potentials for brain surgeries. J Neurosurg. 2018;130(2):654–60.
Ille S, Kelm A, Schroeder A, Albers LE, Negwer C, Butenschoen VM, et al. Navigated repetitive transcranial magnetic stimulation improves the outcome of postsurgical paresis in glioma patients – a randomized, double-blinded trial. Brain Stimul. 2021;14(4):780–7.
Jakola AS, Myrmel KS, Kloster R, Torp SH, Lindal S, Unsgard G, et al. Comparison of a strategy favoring early surgical resection vs a strategy favoring watchful waiting in low-grade gliomas. JAMA: J Am Med Assoc. 2012;308(18):1881–8.
Jamaludin MR, Lai KW, Chuah JH, Zaki MA, Hasikin K, Abd Razak NA, et al. Machine learning application of transcranial motor-evoked potential to predict positive functional outcomes of patients. Comput Intell Neurosci. 2022;2022:2801663.
Jameson LC, Janik DJ, Sloan TB. Electrophysiologic monitoring in neurosurgery. Anesthesiol Clin. 2007;25(3):605–30, x.
Kamada K, Todo T, Ota T, Ino K, Masutani Y, Aoki S, et al. The motor-evoked potential threshold evaluated by tractography and electrical stimulation. J Neurosurg. 2009;111:785.
Kamio Y, Sakai N, Sameshima T, Takahashi G, Koizumi S, Sugiyama K, et al. Usefulness of intraoperative monitoring of visual evoked potentials in transsphenoidal surgery. Neurol Med Chir. 2014;54(8):606–11.

Kamp MA, Dibue M, Niemann L, Reichelt DC, Felsberg J, Steiger HJ, et al. Proof of principle: supramarginal resection of cerebral metastases in eloquent brain areas. Acta Neurochir. 2012;154(11):1981–6.

Kandel ER, Schwartz JH, Jessel TM. Principles of neural science. 4th ed. New York: McGraw-Hill; 2000.

Kang KM, Kim KM, Kim IS, Kim JH, Kang H, Ji SY, et al. Functional magnetic resonance imaging and diffusion tensor imaging for language mapping in brain tumor surgery: validation with direct cortical stimulation and cortico–cortical evoked potential. Korean J Radiol. 2023;24(6):553.

Kanno A, Enatsu R, Ookawa S, Noshiro S, Ohtaki S, Suzuki K, et al. Interhemispheric asymmetry of network connecting between frontal and temporoparietal cortices: a corticocortical-evoked potential study. World Neurosurg. 2018;120:e628–e36.

Kodama K, Goto T, Sato A, Sakai K, Tanaka Y, Hongo K. Standard and limitation of intraoperative monitoring of the visual evoked potential. Acta Neurochir. 2010;152(4):643–8.

Krieg SM, Lioumis P, Makela JP, Wilenius J, Karhu J, Hannula H, et al. Protocol for motor and language mapping by navigated TMS in patients and healthy volunteers; workshop report. Acta Neurochir. 2017;159(7):1187–95.

Kurozumi K, Kameda M, Ishida J, Date I. Simultaneous combination of electromagnetic navigation with visual evoked potential in endoscopic transsphenoidal surgery: clinical experience and technical considerations. Acta Neurochir. 2017;159(6):1043–8.

Lacroix M, Abi-Said D, Fourney DR, Gokaslan ZL, Shi W, DeMonte F, et al. A multivariate analysis of 416 patients with glioblastoma multiforme: prognosis, extent of resection, and survival. J Neurosurg. 2001;95(2):190–8.

Landazuri P, Eccher M. Simultaneous direct cortical motor evoked potential monitoring and subcortical mapping for motor pathway preservation during brain tumor surgery: is it useful? J Clin Neurophysiol. 2013;30(6):623–5.

Luo Y, Regli L, Bozinov O, Sarnthein J. Clinical utility and limitations of intraoperative monitoring of visual evoked potentials. PLoS One. 2015;10(3):e0120525.

Lutz K, Hani L, Kissling C, Raabe A, Schucht P, Seidel K. Resection of low-grade gliomas in the face area of the primary motor cortex and neurological outcome. Cancer. 2023;15(3):781.

Macdonald DB. Intraoperative motor evoked potential monitoring: overview and update. J Clin Monit Comput. 2006;20:347–77.

MacDonald DB, Dong C, Quatrale R, Sala F, Skinner S, Soto F, et al. Recommendations of the International Society of Intraoperative Neurophysiology for intraoperative somatosensory evoked potentials. Clin Neurophysiol. 2019;130(1):161–79.

Maesawa S, Fujii M, Nakahara N, Watanabe T, Wakabayashi T, Yoshida J. Intraoperative tractography and motor evoked potential (MEP) monitoring in surgery for gliomas around the corticospinal tract. World Neurosurg. 2010;74(1):153–61.

Malcharek MJ, Loeffler S, Schiefer D, Manceur MA, Sablotzki A, Gille J, et al. Transcranial motor evoked potentials during anesthesia with desflurane versus propofol--a prospective randomized trial. Clin Neurophysiol. 2015;126(9):1825–32.

Mandonnet E, Sarubbo S, Duffau H. Proposal of an optimized strategy for intraoperative testing of speech and language during awake mapping. Neurosurg Rev. 2017;40(1):29–35.

Matsumoto R, Nair DR, LaPresto E, Najm I, Bingaman W, Shibasaki H, et al. Functional connectivity in the human language system: a cortico-cortical evoked potential study. Brain. 2004;127(10):2316–30.

Matsumoto R, Kunieda T, Nair D. Single pulse electrical stimulation to probe functional and pathological connectivity in epilepsy. Seizure. 2017;44:27–36.

McGirt MJ, Chaichana KL, Attenello FJ, Weingart JD, Than K, Burger PC, et al. Extent of surgical resection is independently associated with survival in patients with hemispheric infiltrating low-grade gliomas. Neurosurgery. 2008;63(4):700–7; author reply 7–8.

McGirt MJ, Chaichana KL, Gathinji M, Attenello FJ, Than K, Olivi A, et al. Independent association of extent of resection with survival in patients with malignant brain astrocytoma. J Neurosurg. 2009;110(1):156–62.

Mikuni N, Okada T, Nishida N, Taki J, Enatsu R, Ikeda A, et al. Comparison between motor evoked potential recording and fiber tracking for estimating pyramidal tracts near brain tumors. J Neurosurg. 2007;106(1):128–33.

Mirallave-Pescador A, Lavrador JP, Lejarde A. Bayesian networks for risk assessment and postoperative deficit prediction in intraoperative neurophysiology for brain surgery. Research Square [Internet]; 2023.

Moiyadi A, Velayutham P, Shetty P, Seidel K, Janu A, Madhugiri V, et al. Combined motor evoked potential monitoring and subcortical dynamic mapping in motor eloquent tumors allows safer and extended resections. World Neurosurg. 2018;120:e259.

Møller. Intraoperative neurophysiological monitoring. Humana Press; 2006.

Nakae T, Matsumoto R, Kunieda T, Arakawa Y, Kobayashi K, Shimotake A, et al. Connectivity gradient in the human left inferior frontal gyrus: intraoperative cortico-cortical evoked potential study. Cereb Cortex. 2020;30(8):4633–50.

Neuloh G, Schramm J. Are there false-negative results of motor evoked potential monitoring in brain surgery? Cen Eur Neurosurg. 2009;70(4):171–5.

Neuloh G, Seidel K. Intraoperative monitoring and mapping of the motor system during surgery for supratentorial mass lesions under general anesthesia. In: Seubert C, Balzer JR, editors. Monitroing of the nervous system for anesthesiologists and other health care professionals. 3rd ed. Springer; 2023.

Neuloh G, Pechstein U, Cedzich C, Schramm J. Motor evoked potential monitoring with supratentorial surgery. Neurosurgery. 2004;54(5):1061–70; discussion 70–2.

Neuloh G, Pechstein U, Schramm J. Motor tract monitoring during insular glioma surgery. J Neurosurg. 2007a;106(4):582–92.

Neuloh G, Pechstein U, Cedzich C, Schramm J. Motor evoked potential monitoring with supratentorial surgery. Neurosurgery. 2007b;61(1 Suppl):337–46; discussion 46–8.

Nossek E, Korn A, Shahar T, Kanner AA, Yaffe H, Marcovici D, et al. Intraoperative mapping and monitoring of the corticospinal tracts with neurophysiological assessment and 3-dimensional ultrasonography-based navigation. Clinical article. J Neurosurg. 2011;114(3):738–46.

Ohue S, Kohno S, Inoue A, Yamashita D, Harada H, Kumon Y, et al. Accuracy of diffusion tensor magnetic resonance imaging-based tractography for surgery of gliomas near the pyramidal tract: a significant correlation between subcortical electrical stimulation and postoperative tractography. Neurosurgery. 2012;70(2):283–93; discussion 94.

Papanicolaou. Guide to intraoperative monitoring. Boca Raton: Taylor & Francis Group; 2000.

Parduzi Q, Wermelinger J, Koller SD, Sariyar M, Schneider U, Raabe A, Seidel K. Explainable AI for intraoperative motor-evoked potential muscle classification in neurosurgery: bicentric retrospective study. J Med Internet Res. 2025;27:e63937. https://doi.org/10.2196/63937. PMID: 40127441.

Park D, Kim I. Application of machine learning in the field of intraoperative neurophysiological monitoring: a narrative review. Appl Sci. 2022;12(15):7943.

Pechstein U, Nadstawek J, Zentner J, Schramm J. Isoflurane plus nitrous oxide versus propofol for recording of motor evoked potentials after high frequency repetitive electrical stimulation. Electroencephalogr Clin Neurophysiol. 1998;108(2):175–81.

Penfield W. Somatic motor and sensory representation in the cerebral cortex of man as studied by electrical stimulation. Brain J Neurol. 1937;60:389–443.

Plans G, Fernandez-Conejero I, Rifa-Ros X, Fernandez-Coello A, Rossello A, Gabarros A. Evaluation of the high-frequency monopolar stimulation technique for mapping and monitoring the corticospinal tract in patients with supratentorial gliomas. A proposal for intraoperative management based on neurophysiological data analysis in a series of 92 patients. Neurosurgery. 2017;81(4):585–94.

Prabhu SS, Gasco J, Tummala S, Weinberg JS, Rao G. Intraoperative magnetic resonance imaging-guided tractography with integrated monopolar subcortical functional mapping for resection of brain tumors. Clinical article. J Neurosurg. 2011;114(3):719–26.

Raabe A, Beck J, Schucht P, Seidel K. Continuous dynamic mapping of the corticospinal tract during surgery of motor eloquent brain tumors: evaluation of a new method. J Neurosurg. 2014;120(5):1015–24.
Rofes A, Mandonnet E, Godden J, Baron MH, Colle H, Darlix A, et al. Survey on current cognitive practices within the European Low-Grade Glioma Network: towards a European assessment protocol. Acta Neurochir. 2017;159(7):1167–78.
Rosenstock T, Grittner U, Acker G, Schwarzer V, Kulchytska N, Vajkoczy P, et al. Risk stratification in motor area-related glioma surgery based on navigated transcranial magnetic stimulation data. J Neurosurg. 2017;126(4):1227–37.
Rosenstock T, Tuncer MS, Munch MR, Vajkoczy P, Picht T, Faust K. Preoperative nTMS and intraoperative neurophysiology – a comparative analysis in patients with motor-eloquent glioma. Front Oncol. 2021;11:676626.
Rosenstock T, Hani L, Grittner U, Schlinkmann N, Ivren M, Schneider H, et al. Bicentric validation of the navigated transcranial magnetic stimulation motor risk stratification model. J Neurosurg. 2022;136(4):1194–206.
Rossi M, Conti Nibali M, Vigano L, Puglisi G, Howells H, Gay L, et al. Resection of tumors within the primary motor cortex using high-frequency stimulation: oncological and functional efficiency of this versatile approach based on clinical conditions. J Neurosurg. 2019:1–13.
Roth J, Korn A, Bitan-Talmor Y, Kaufman R, Ekstein M, Constantini S. Subcortical mapping using an electrified Cavitron UltraSonic Aspirator in pediatric supratentorial surgery. World Neurosurg. 2017;101:357–64.
Rothwell J, Burke D, Hicks R, Stephen J, Woodforth I, Crawford M. Transcranial electrical stimulation of the motor cortex in man: further evidence for the site of activation. J Physiol. 1994;481(Pt 1):243–50.
Saito T, Tamura M, Muragaki Y, Maruyama T, Kubota Y, Fukuchi S, et al. Intraoperative cortico-cortical evoked potentials for the evaluation of language function during brain tumor resection: initial experience with 13 cases. J Neurosurg. 2014;121(4):827–38.
Saito T, Muragaki Y, Tamura M, Maruyama T, Nitta M, Tsuzuki S, et al. Monitoring cortico-cortical evoked potentials using only two 6-strand strip electrodes for gliomas extending to the dominant side of frontal operculum during one-step tumor removal surgery. World Neurosurg. 2022;165:e732–e42.
Sala F, Lanteri P. Brain surgery in motor areas: the invaluable assistance of intraoperative neurophysiological monitoring. J Neurosurg Sci. 2003;47(2):79–88.
Salvati M, Pichierri A, Piccirilli M, Floriana Brunetto GM, D'Elia A, Artizzu S, et al. Extent of tumor removal and molecular markers in cerebral glioblastoma: a combined prognostic factors study in a surgical series of 105 patients. J Neurosurg. 2012;117(2):204–11.
Sanai N, Snyder LA, Honea NJ, Coons SW, Eschbacher JM, Smith KA, et al. Intraoperative confocal microscopy in the visualization of 5-aminolevulinic acid fluorescence in low-grade gliomas. J Neurosurg. 2011a;115(4):740–8.
Sanai N, Polley MY, McDermott MW, Parsa AT, Berger MS. An extent of resection threshold for newly diagnosed glioblastomas. J Neurosurg. 2011b;115(1):3–8.
Sanmillan JL, Fernandez-Coello A, Fernandez-Conejero I, Plans G, Gabarros A. Functional approach using intraoperative brain mapping and neurophysiological monitoring for the surgical treatment of brain metastases in the central region. J Neurosurg. 2017;126(3):698–707.
Schucht P, Knittel S, Slotboom J, Seidel K, Murek M, Jilch A, et al. 5-ALA complete resections go beyond MR contrast enhancement: shift corrected volumetric analysis of the extent of resection in surgery for glioblastoma. Acta Neurochir. 2014a;156(2):305–12; discussion 12.
Schucht P, Seidel K, Beck J, Murek M, Jilch A, Wiest R, et al. Intraoperative monopolar mapping during 5-ALA-guided resections of glioblastomas adjacent to motor eloquent areas: evaluation of resection rates and neurological outcome. Neurosurg Focus. 2014b;37(6):E16.
Schucht P, Beck J, Seidel K, Raabe A. Extending resection and preserving function: modern concepts of glioma surgery. Swiss Med Wkly. 2015;145:w14082.
Seidel K, Beck J, Stieglitz L, Schucht P, Raabe A. Low-threshold monopolar motor mapping for resection of primary motor cortex tumors. Neurosurgery. 2012;71(1 Suppl Operative):104–14; discussion 14–5.

Seidel K, Beck J, Stieglitz L, Schucht P, Raabe A. The warning-sign hierarchy between quantitative subcortical motor mapping and continuous motor evoked potential monitoring during resection of supratentorial brain tumors. J Neurosurg. 2013;118(2):287–96.

Seidel K, Hani L, Lutz K, Zbinden C, Redmann A, Consuegra A, et al. Postoperative navigated transcranial magnetic stimulation to predict motor recovery after surgery of tumors in motor eloquent areas. Clin Neurophysiol. 2019;130(6):952–9.

Seidel K, Raabe A, Schucht P. Reply to "Technology in context: a holistic care approach". Clin Neurophysiol. 2020a;131(2):579.

Seidel K, Schucht P, Beck J, Raabe A. Continuous dynamic mapping to identify the corticospinal tract in motor eloquent brain tumors: an update. J Neurol Surg A Cent Eur Neurosurg. 2020b;81(2):105–10.

Seidel K, Szelenyi A, Bello L. Intraoperative mapping and monitoring during brain tumor surgeries. Handb Clin Neurol. 2022;186:133–49.

Seidel K, Wermelinger J, Alvarez-Abut P, Deletis V, Raabe A, Zhang D, Schucht P. Cortico-cortical evoked potentials of language tracts in minimally invasive glioma surgery guided by Penfield stimulation. Clin Neurophysiol. 2024;161:256–67. https://doi.org/10.1016/j.clinph.2023.12.136. Epub 2024 Mar 5. PMID: 38521679.

Senft C, Bink A, Franz K, Vatter H, Gasser T, Seifert V. Intraoperative MRI guidance and extent of resection in glioma surgery: a randomised, controlled trial. Lancet Oncol. 2011;12(11):997–1003.

Shiban E, Krieg SM, Haller B, Buchmann N, Obermueller T, Boeckh-Behrens T, et al. Intraoperative subcortical motor evoked potential stimulation: how close is the corticospinal tract? J Neurosurg. 2015a;123(3):711–20.

Shiban E, Krieg SM, Obermueller T, Wostrack M, Meyer B, Ringel F. Continuous subcortical motor evoked potential stimulation using the tip of an ultrasonic aspirator for the resection of motor eloquent lesions. J Neurosurg. 2015b;123(2):301–6.

Sloan TB, Heyer EJ. Anesthesia for intraoperative neurophysiologic monitoring of the spinal cord. J Clin Neurophysiol. 2002;19(5):430–43.

Smith JS, Chang EF, Lamborn KR, Chang SM, Prados MD, Cha S, et al. Role of extent of resection in the long-term outcome of low-grade hemispheric gliomas. J Clin Oncol. 2008;26(8):1338–45.

Spena G, Schucht P, Seidel K, Rutten GJ, Freyschlag CF, D'Agata F, et al. Brain tumors in eloquent areas: a European multicenter survey of intraoperative mapping techniques, intraoperative seizures occurrence, and antiepileptic drug prophylaxis. Neurosurg Rev. 2017;40(2):287–98.

Stieglitz LH, Fichtner J, Andres R, Schucht P, Krahenbuhl AK, Raabe A, et al. The silent loss of neuronavigation accuracy: a systematic retrospective analysis of factors influencing the mismatch of frameless stereotactic systems in cranial neurosurgery. Neurosurgery. 2013;72(5):796–807.

Stummer W, Pichlmeier U, Meinel T, Wiestler OD, Zanella F, Reulen HJ. Fluorescence-guided surgery with 5-aminolevulinic acid for resection of malignant glioma: a randomised controlled multicentre phase III trial. Lancet Oncol. 2006;7(5):392–401.

Stummer W, Reulen HJ, Meinel T, Pichlmeier U, Schumacher W, Tonn JC, et al. Extent of resection and survival in glioblastoma multiforme: identification of and adjustment for bias. Neurosurgery. 2008;62(3):564–76; discussion -76.

Suzuki Y, Enatsu R, Kanno A, Yokoyama R, Suzuki H, Tachibana S, et al. The influence of anesthesia on corticocortical evoked potential monitoring network between frontal and temporoparietal cortices. World Neurosurg. 2019;123:e685–e92.

Szelenyi A, Kothbauer K, de Camargo AB, Langer D, Flamm ES, Deletis V. Motor evoked potential monitoring during cerebral aneurysm surgery: technical aspects and comparison of transcranial and direct cortical stimulation. Neurosurgery. 2005;57(4 Suppl):331–8; discussion -8.

Szelenyi A, Kothbauer KF, Deletis V. Transcranial electric stimulation for intraoperative motor evoked potential monitoring: stimulation parameters and electrode montages. Clin Neurophysiol. 2007;118(7):1586–95.

Szelenyi A, Hattingen E, Weidauer S, Seifert V, Ziemann U. Intraoperative motor evoked potential alteration in intracranial tumor surgery and its relation to signal alteration in postoperative magnetic resonance imaging. Neurosurgery. 2010a;67(2):302–13.

Szelenyi A, Bello L, Duffau H, Fava E, Feigl GC, Galanda M, et al. Intraoperative electrical stimulation in awake craniotomy: methodological aspects of current practice. Neurosurg Focus. 2010b;28(2):E7.

Szelenyi A, Senft C, Jardan M, Forster MT, Franz K, Seifert V, et al. Intra-operative subcortical electrical stimulation: a comparison of two methods. Clin Neurophysiol. 2011;122(7):1470–5.

Tamura Y, Ogawa H, Kapeller C, Prueckl R, Takeuchi F, Anei R, et al. Passive language mapping combining real-time oscillation analysis with cortico-cortical evoked potentials for awake craniotomy. J Neurosurg. 2016;125(6):1580–8.

Taniguchi M, Cedzich C, Schramm J. Modification of cortical stimulation for motor evoked potentials under general anesthesia: technical description. Neurosurgery. 1993;32(2):219–26.

Uribe AA, Mendel E, Peters ZA, Shneker BF, Abdel-Rasoul M, Bergese SD. Comparison of visual evoked potential monitoring during spine surgeries under total intravenous anesthesia versus balanced general anesthesia. Clin Neurophysiol. 2017;128(10):2006–13.

Usui K, Ikeda A, Takayama M, Matsuhashi M, Yamamoto JI, Satoh T, et al. Conversion of semantic information into phonological representation: a function in left posterior basal temporal area. Brain. 2003;126(3):632–41.

Vakani R, Nair D. Electrocorticography and functional mapping. In: Levin K, Chauvel P, editors. Handbook of clinical neurology. Elsevier; 2019.

Vega-Zelaya L, Pulido P, Sola RG, Pastor J. Intraoperative cortico-cortical evoked potentials for monitoring language function during brain tumor resection in anesthetized patients. J Integr Neurosci. 2023;22(1):17.

Vincent M, Guiraud D, Duffau H, Mandonnet E, Bonnetblanc F. Electrophysiological brain mapping: basics of recording evoked potentials induced by electrical stimulation and its physiological spreading in the human brain. Clin Neurophysiol. 2017;128(10):1886–90.

Vitulli F, Kalaitzoglou D, Soumpasis C, Diaz-Baamonde A, Mosquera JDS, Gullan R, et al. Cortical-subcortical functional preservation and rehabilitation in neuro-oncology: tractography-MIPS-IONM-TMS proof-of-concept study. J Pers Med. 2023;13(8):1278.

Wermelinger J, Parduzi Q, Sariyar M, Raabe A, Schneider UC, Seidel K. Opportunities and challenges of supervised machine learning for the classification of motor evoked potentials according to muscles. BMC Med Inform Decis Mak. 2023;23(1):198.

Yamao Y, Matsumoto R, Kunieda T, Arakawa Y, Kobayashi K, Usami K, et al. Intraoperative dorsal language network mapping by using single-pulse electrical stimulation. Hum Brain Mapp. 2014;35(9):4345–61.

Yamao Y, Suzuki K, Kunieda T, Matsumoto R, Arakawa Y, Nakae T, et al. Clinical impact of intraoperative CCEP monitoring in evaluating the dorsal language white matter pathway. Hum Brain Mapp. 2017;38(4):1977–91.

Yamao Y, Matsumoto R, Kunieda T, Nakae T, Nishida S, Inano R, et al. Effects of propofol on cortico-cortical evoked potentials in the dorsal language white matter pathway. Clin Neurophysiol. 2021;132(8):1919–26.

Integration of Functional Data in the Clinical Workflow

4

Haosu Zhang and Maximilian Schwendner

This chapter aims to present an application of functional data in clinical practice by integrating navigated transcranial magnetic stimulation (nTMS) and magnetic resonance imaging (MRI) data into the routine workflow. This chapter provides insights into the necessary requirements, suggested workflows, and the range of clinical applications. It provides an easy-to-implement reference for departments interested in implementing this workflow.

4.1 Introduction

4.1.1 Brief Overview of the Importance of Integrating Functional Data in Neurosurgery

In recent years, integrating functional data into the clinical workflow has emerged as a pivotal advancement in neurosurgical practice to fulfill the concept of individualized surgical treatment (Bagarinao et al. 2018; Subbiah 2023). Functional data mostly results from functional tests combined with structural localization, for example, eloquent cortical areas and neural tracts responsible for language and motor function (van de Ven et al. 2004; Smith et al. 2013). The functional data can be applied in the neurosurgical practice to achieve accurate and comprehensive preoperative planning, precise intraoperative navigation, and the preservation of critical eloquent brain regions (Pillai 2010).

H. Zhang · M. Schwendner (✉)
Department of Neurosurgery, Heidelberg University Hospital, Ruprecht-Karls-University Heidelberg, Heidelberg, Germany
e-mail: haosu.zhang@med.uni-heidelberg.de; maximilian.schwendner@med.uni-heidelberg.de

S. M. Krieg, T. Picht (eds.), *Navigated Transcranial Magnetic Stimulation in Neurosurgery*, https://doi.org/10.1007/978-3-031-97155-6_4

Navigated transcranial magnetic stimulation (nTMS) is an established means to facilitate the integration of functional data into neurosurgery (Frey et al. 2014; Krieg et al. 2014). By non-invasively stimulating specific cortical regions and assessing the corresponding functional performance, nTMS enables neurosurgeons to localize eloquent brain areas with an accuracy similar to intraoperative direct cortical stimulation (DCS) (Jeltema et al. 2021). This allows neurosurgeons to develop an optimal surgical strategy and to avoid damage to critical functional regions to minimize postoperative deficits (Krieg et al. 2014). The application of functional data in neurosurgical practice needs the teamwork of professionals from different disciplines, which meets the advocacy for the integration between medical and engineering disciplines to facilitate the development, translation, and clinical application of advanced healthcare technologies (Khoo 2012; Taşgetiren et al. 2015). The investigation of cross-discipline theories can also be strengthened through the integration process of functional data. This capitalizes on each field's unique strengths and expertise, creating a multi-disciplinary approach to address challenges in healthcare. Engineers bring innovative technologies, data analytics, and problem-solving methods, while medical professionals contribute their clinical insights, patient care experience, and in-depth understanding of diseases and treatments. So, building a team with different professional backgrounds is mandatory for integrating functional data in clinical practice (Lennox-Chhugani 2023). The combined effort from various professionals can reduce the time required to translate scientific and technological advancements into clinical practice and systematically improve the clinical process for a better prognosis.

4.1.2 Role of nTMS in Functional Data

As elucidated in previous chapters, nTMS employs magnetic fields to activate neurons in targeted brain regions or to induce temporary "virtual-lesions," to identify the regions related to alterations in functional performances (Silvanto and Muggleton 2008; Krieg 2017).

The primary advantage of nTMS is the noninvasive delineation of functional distributions for preoperative planning and navigation (Ottenhausen et al. 2015; Raffa et al. 2023). nTMS is applied to map cortical regions. Combining this data with diffusion tensor imaging (DTI) data in fiber tracking (FT), subcortical structures associated with certain cerebral functions can be identified (Krieg 2017; Krieg et al. 2012; Picht et al. 2014). The mapped cortical regions and the corresponding tracked subcortical fibers compose the structural networks, which allow neurosurgeons to have a comprehensive view of eloquent structures critical for functions such as speech, motor control, and sensory perception, enabling them to minimize functional damage while achieving the surgical objectives (Krieg 2017; Krieg et al. 2012; Picht et al. 2014).

The wide acceptance of nTMS in acquiring functional data is due to its unique approach to exploring functionality, which is similar to the gold standard of intraoperative DCS. Primarily, nTMS induces observable functional changes, providing intuitiveness that surpasses other techniques (Sun et al. 2023), an attribute crucial for planning neurosurgical interventions. This direct effect offers a more immediate and tangible understanding of neurophysiological processes (Rossi et al. 2009). Those understandings of cerebral functions are independent of the classical neuroanatomical theory, which fails to understand each individual's cerebral functions.

4.1.3 Purpose and Scope of the Chapter

This chapter illustrates the integration of functional data into clinical routine, including cortical nTMS mapping and subcortical analysis. Our workflow to apply functional data in neurosurgical practice results from a decade of experience and will be demonstrated as a reference. Given that each hospital's conditions vary and legal requirements differ from one country to another, the detailed implementation process of the workflow should be adapted accordingly.

4.2 Integrating Functional Data into a Clinical Data Management System

4.2.1 Basic Requirements

In our workflow, nTMS plays a central role in bridging functional performance and anatomical structures. This workflow involves multimodal data analysis and cooperation among disciplines. Even though many research centers have unique methods for data collection and analysis, the fundamental procedures are similar, including equipment and analytical techniques.

The integration of functional data involves two main aspects: (1) hardware, such as picture archiving and communication systems (PACS) with its servers, systems for intraoperative neuronavigation, nTMS devices, and MRI scanners. A well-functioning server system is recommended as one of the prerequisites for efficient operation throughout the entire workflow, and the initial investment will significantly pay off during subsequent clinical use; (2) software, including hospital information system (HIS), nTMS systems, and neuronavigation systems, such Brainlab iPlan Net® and Medtronic StealthStation®; (3) a team of professionals with backgrounds in the medical and technical background. The recommended team should include the following roles:

- nTMS operators and software experts: These scientists and technicians offer medical and technical assistance, manage the machine's operation and maintenance, and schedule examinations according to clinical needs.
- Analysts with medical background for image and functional data: These analysts can analyze medical images, extract critical information from images, and provide technical support for clinical and research purposes. They play a vital role in troubleshooting potential image issues.

- Neurosurgeons familiar with functional data: Surgeons are introduced into the functional mapping process and its application in neurosurgery to understand its indications, technical implications, benefits, and limitations in clinical work and research.

4.2.2 Integration in the Hospital Integration Systems

Hospital integration systems are mostly modularly designed and customizable to meet healthcare providers' specific needs. We created a specific interface for nTMS examinations in our HIS. It allows entering patient and tumor-specific information essential for the examination as well as the opportunity to schedule the examination (Krieg 2017; Krieg et al. 2017), such as

- Basic patient details, including name, age, gender, and handedness.
- Tumor location and preliminary diagnosis.
- Clinical data based on the type of the planned nTMS examination. For example, the clinical status of muscle strength should be recorded if motor mapping is required. If language testing is conducted, the patient's native language and spoken languages, as well as the type and level of aphasia must be documented.
- Information about epilepsy and current antiepileptic medications.
- Any contraindications for nTMS, such as pacemakers, cochlear implants, or deep brain stimulation electrodes.

All recorded information is stored for future reference (Fig. 4.1).

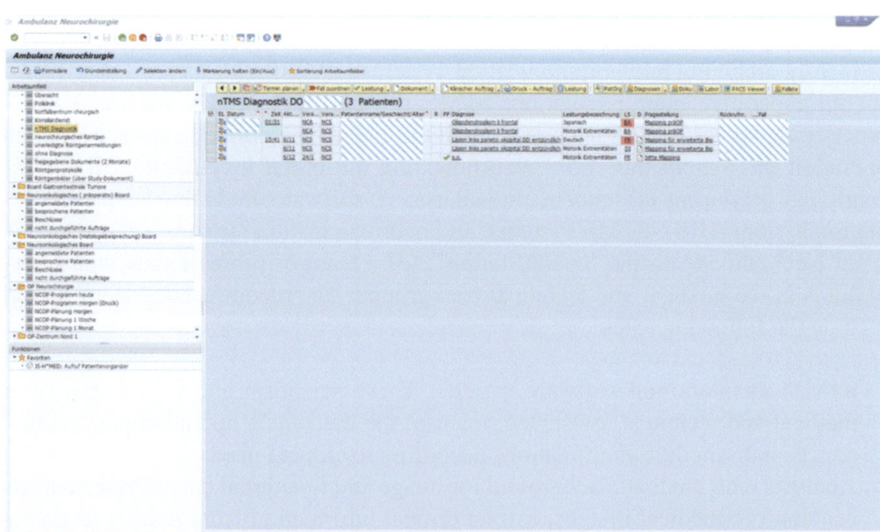

Fig. 4.1 Integration of the requests for nTMS examinations in the Hospital Information System

Performing nTMS examinations based on the latest MRI sequences is recommended to provide optimal results, as the brain can be potentially altered in a short time when suffering from pathological conditions. Therefore, the nTMS examination should be scheduled after the preoperative MRI scans to apply the latest MRI sequences for navigation. The HIS should open a portal to schedule nTMS mapping to avoid overlapping appointments for different examinations. In the HIS system, the documentation for the nTMS mapping and following fiber tracking should be standardized for future studies among subjects or centers since the multicenter research on nTMS mapping is increasingly emphasized.

- Mapping parameters: encompassing details of the nTMS examination, such as the mapping specification (motor, language, calculation), and stimulation parameters, including the resting motor thresholds (rMT), stimulation paradigms, baselines, and the duration of stimulation.
- Medication, especially antiepileptic drugs (Darmani et al. 2019).
- History of epilepsy (Walton et al. 2021).
- Adverse events: these include adverse events or side effects encountered during mapping (Abler et al. 2005) and the overall examination experience of patients, including rating of the discomfort on a numeric rating scale.
- Brief summary of the overall findings on cortical mapping and fiber tracking. This might include the cortical location of eloquent brain areas and lesion-to-tract distances. For mapping of the language function, an error rate (ER) can be applied to assess the frequency of stimulation-related language impairments (Picht et al. 2013). If navigated repetitive transcranial magnetic stimulation (nrTMS) language mapping is conducted on both hemispheres, a hemispheric dominance ratio (HDR) can be obtained by dividing the left-sided ER by the right-sided ER to assess the hemispheric language dominance (Krieg et al. 2014; Ille et al. 2016).

The documentation of the nTMS mapping is archived within the electronic patient file, and the nTMS-based tractography is saved together with the MRI data on the imaging server for later access, for example, during surgery but also radiotherapy treatment planning (Schwendner et al. 2018; Diehl et al. 2019).

4.2.3 Integrating Functional Data to Picture Archiving and Communication Systems

Several data storage solutions, such as Vendor Neutral Archives (VNAs) (Sirota-Cohen et al. 2019) and Radiology Information Systems (RIS) (Roberson and Shieh 1998), are commonly used in clinical settings. Our center uses the Picture Archiving and Communication System (PACS) (Roberson and Shieh 1998; Strickland 2000). This system opens access to data for imaging analysis software, such as Elements (Brainlab AG, Germany; Version 3.2) (Krieg et al. 2012; Caverzasi et al. 2016; Rosenstock et al. 2017; Zhang et al. 2022). It provides the source of MRI data for navigation during preoperative mapping. Subsequently, nTMS mapping data sets are uploaded to the PACS for long-term storage. Data is stored in the Digital Imaging and Communications in Medicine (DICOM) format.

However, due to safety regulations, only the radiology department can upload imaging data such as nTMS or data from external institutions, utilizing tailored software such as Uploader 1.0 within the PACS integrator software (Gemed, Ulm, Germany). Alternatively, nTMS functional images can be transmitted to the patient imaging dataset by opening upload permissions through a temporary order in the HIS system. Some institutions have established a direct connection between the nTMS devices and the PACS (Makela et al. 2015), facilitating a significantly smoother clinical workflow and seamless data connectivity. Nevertheless, due to legal variations in different regions and disparities in technical specifications of related equipment, achieving such data transmission might not be universally feasible. Therefore, implementing data transfer services through third-party software, such as Elements (Brainlab, Munich, Germany) to access and temporarily upload data, remains a practical solution. This software is compatible with the PACS system and can access data from servers for image reconstruction and analysis (Kaspar et al. 2020).

However, storing and retrieving data within the system poses significant challenges to the PACS server (Kalyanpur et al. 2010; Abbas and Singh 2019). Data stored for extended periods may encounter slow or even failed data retrieval, a common problem in big data management (Kalyanpur et al. 2010; Abbas and Singh 2019). Furthermore, an increasing volume of data over time may lead to a decline in the performance of the data storage system. This could result in longer data retrieval times, as the storage system needs to process vast amounts of information and locate specific data (Hindel 1990). Hence, guaranteeing data storage and retrieval efficiency is crucial when procuring and installing relevant hospital equipment and services.

4.2.4 Integrating Functional Data to Intraoperative Navigation Systems

Applying functional data in neurosurgery is helpful in surgical planning and intraoperative guidance, particularly concerning eloquent brain areas. Most navigation systems available on the market, such as Brainlab iPlan Net® and Medtronic StealthStation®, can visualize functional data, such as nTMS data (see Fig. 4.2).

Since all stimulation targets during mapping are constructed based on the coordinates of the patient's MR structural images, these images shall be well aligned to avoid co-registration inaccuracies to ensure that their exact anatomical locations can be identified (Krieg et al. 2014). Other functional imaging modalities, like functional MRI and positron emission tomography (PET) (Catana et al. 2012), can be co-registered to the same structural MR image, which enables multimodal analysis of the distribution of functional brain areas based on identical structural coordinates (Markello et al. 2022; Cheng et al. 2020). When registering imaging data sets, it is

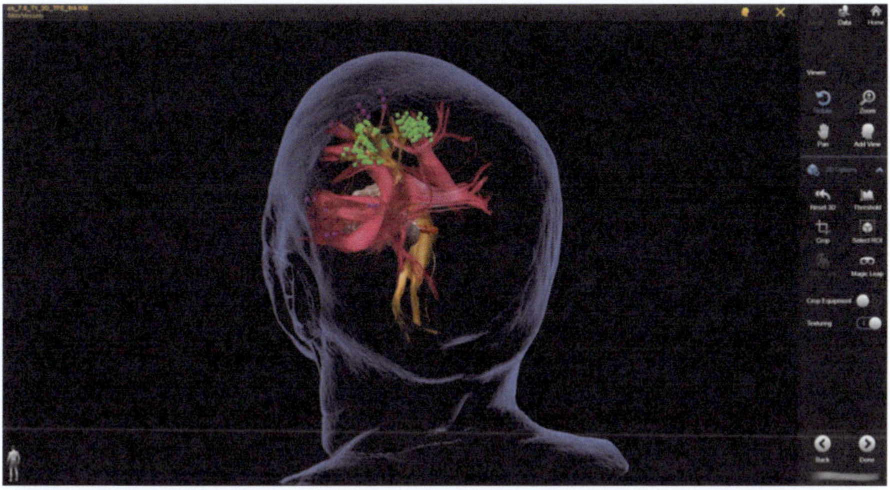

Fig. 4.2 Mapping results of language and motor mappings presented in different colors, representing different information

essential to note that all functional imaging data sets are directly registered to a high-resolution anatomical MRI scan because direct co-registration among functional images may lead to significant fusion inaccuracies due to their low resolution.

Automatic registration and fusion between structural and functional images is recommended, and most processes can be accurately fulfilled. Furthermore, most navigation systems allow manual image fusion if inaccuracies are identified after automatic registration and fusion. This process involves co-registering different functional image sequences onto the same structural image, typically utilizing enhanced T1-weighted images. In nTMS data, structural markers are exported alongside functional localization points, namely makers at the nasion and both crura of the helix from the anatomical image. Post-fusion validation is crucial; those anatomical landmarks can be set as reference points to check the quality of registration between functional and structural images (Krieg et al. 2017). Another criterion to evaluate the fusion quality is to verify that cortical stimulation points are typically situated on gyri. Precise image fusion is crucial to visualize functional data during surgery (Krieg et al. 2012).

4.2.5 Fiber Tracking

In DTI-based fiber tracking, nTMS data can be defined as seed points. In clinical routine, a deterministic tracking algorithm is utilized to generate subcortical neural fiber networks (Krieg et al. 2012, 2017). Although deterministic tracking algorithms may not be the most advanced, they help to reduce false-positive fibers, especially in the vicinity of brain tumors, which are often accompanied by severe edema (Smith et al. 2020).

Currently, the subcortical localization of functions is restricted mainly to motor and language function in clinical practice. The motor function consists of neural connectivity networks, and the primary motor network, including the CST, as the main motor output pathway of this network (Isa 2012; Moreno-Lopez et al. 2016). Most studies focus on the CST to identify motor function (Krieg et al. 2014; Perez and Cohen 2009). Other neural fibers supporting motor functions, for example, frontostriatal tracts related to the supplementary motor area, are not routinely considered (Bozkurt et al. 2017). Mapping results regarding language function can be categorized based on different language error types and addressed separately in fiber tracking (Krieg et al. 2017; Picht et al. 2013). However, in clinical practice, tracking the comprehensive language network by aggregating all language error localization points is recommended for intraoperative applications.

Additionally, setting the FA threshold in fiber tracking is a critical consideration. There are various methods for defining the FA thresholds (Krieg 2017; Krieg et al. 2017; Kunimatsu et al. 2004) as there is ongoing controversy on the interpretation of the FA (Leow et al. 2009). However, considering individual variations, all the obtained tracts should be manually verified.

Moreover, establishing a standardized visualization of nTMS data and fiber tracts can facilitate data interpretation and consensus among medical professionals, making additional textual explanations redundant (Ille and Krieg 2021; Diehl et al. 2022). It involves determining the image display format and utilizing various colors to code-specific information. We present a color-coding standard established at our institution, which defines standardized color codes for each data type to promote consistent interpretation among various team members (Ille and Krieg 2021) (Fig. 4.2):

1. *Green*: cortical nTMS-based motor areas
2. *Gold*: CST based on nTMS-derived DTI-FT
3. *Purple*: nrTMS language areas (mother tongue, first acquired language) and corresponding language tracts based on nrTMS-derived DTI-FT
4. *Pink*: nrTMS language areas (second acquired language) and corresponding language tracts based on nrTMS-derived DTI-FT
5. *Red*: nrTMS language areas (third acquired language) and corresponding language tracts based on nrTMS-derived DTI-FT
6. *Blue*: nrTMS areas involved in arithmetic processing and corresponding subcortical tracts based on nrTMS-derived DTI-FT
7. *Flesh-color*: Lesion or tumor

Those imaging data will then be saved in the server system. Therefore, in the subsequent neurosurgical navigation based on the structural MR images, we can simultaneously visualize the cortical distribution and subcortical tracts of eloquent brain functions (Fig. 4.3).

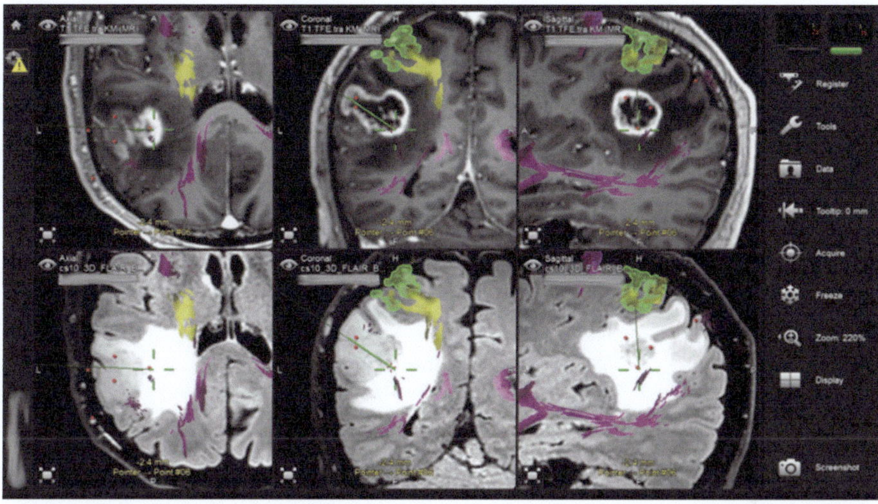

Fig. 4.3 Mapping results are visualized during surgery to localize the cortical and subcortical regions related to eloquent brain functions

4.3 Clinical Workflow

Establishing a clinical workflow is crucial to integrating functional data within your hospital or department. A customized workflow covering the entire process should be acquired, from data collection to creating clinically applicable reports or images. Its primary purpose is to delineate responsibilities and facilitate coordination among various departments or team members. It should be used to increase the efficiency and reliability of the clinical process and be continuously improved during daily practice. Figure 4.4 illustrates the workflow established in our hospital based on the single working steps applied.

4.3.1 Indication of nTMS-Mapping and Registration for the Mapping Test

Indications for nTMS examinations are based on the anatomical location of the lesion, planned surgical approach, and the patient's neurological status (Ottenhausen et al. 2015; Picht et al. 2013, 2016). Primarily, MRI data is used to assess the anatomical location of the lesion. In certain situations, however, such as patients at the ICU, computer tomographic imaging can also be applied for cortical mapping (Schramm et al. 2020). Indications for nTMS motor mapping are mostly lesions near motor-eloquent cortical areas or the internal capsule and the CST (Sollmann et al. 2021; Haddad et al. 2020), with or without motor deficits. Similarly, nrTMS language mapping is performed for lesions within or in close vicinity to

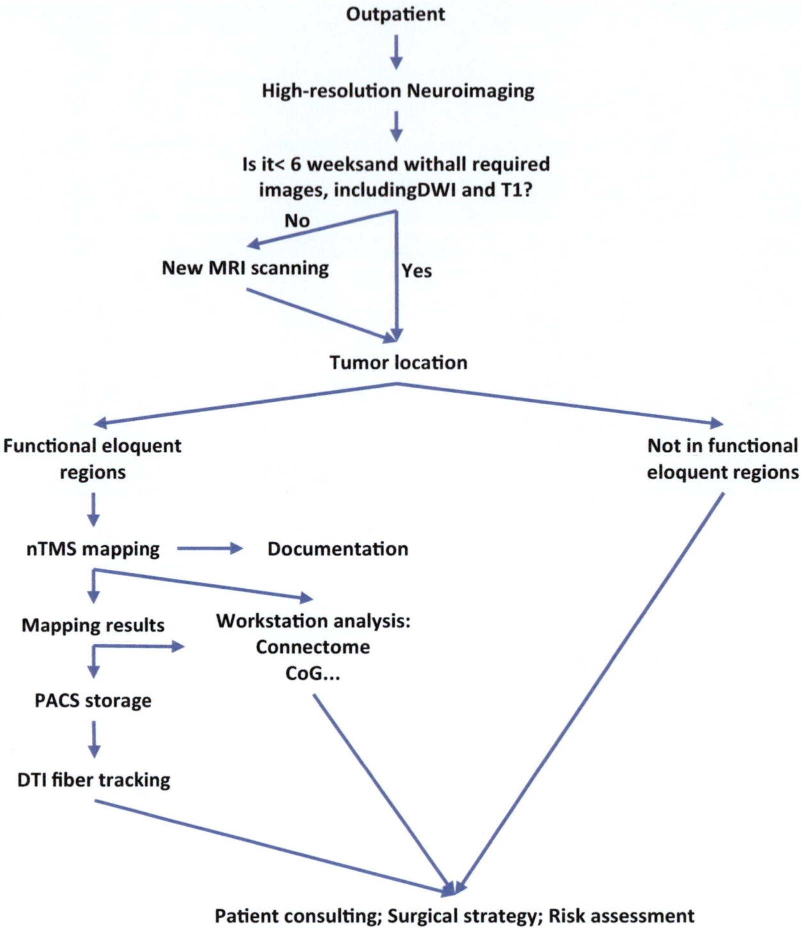

Fig. 4.4 Clinical workflow: the integration of nTMS-based functional data into the clinical workflow. The primary objective of this entire process is to support clinical decision-making by providing multimodal diagnostic imaging information

language-related cortical regions, and language eloquent fiber tracts, including the superior longitudinal fasciculus (SLF), arcuate fasciculus (AF), inferior fronto-occipital fasciculus (IFOF), and uncinate fasciculus (UF) (Picht et al. 2013; Benjamin et al. 2017). However, the distribution of cerebral function in cortical regions shows great interindividual variability involving neuronal plasticity at a functional and structural level. Those cortical regions and subcortical connections compose the networks for different cerebral functions (Shams et al. 2023; Bassett and Sporns 2017; Sporns 2011). So, it is reasonable to apply nTMS mapping to a majority of patients with intracranial lesions preoperatively to estimate the risk of postoperative functional deficits, especially regarding language function (Ottenhausen et al. 2015; Picht 2014; Lefaucheur and Picht 2016). Additionally, clinical manifestations, like aphasia or paresis, even transient or accompanied by seizures, indicate the necessity for preoperative mapping.

Once the decision to perform functional mapping is made, the clinical physician can submit a request for an nTMS examination appointment in the HIS while also confirming the availability of the patient's MRI. The nTMS technician is then assigned to schedule the examination based on the planned date of surgery and the patient's condition. To optimize scheduling these appointments, a specific calendar (Fig. 4.5) should be accessible for all team members to book available slots for future patients, and see the scheduled nTMS sessions.

4.3.2 Preparation for Functional Mapping

By comprehensively analyzing the patient's symptoms, neurological status, and imaging, neurosurgeons assess to what extent the patient's brain function is affected by the tumor and evaluate the risk of surgery-related neurological impairments. This includes the evaluation of whether brain mapping for relevant functions is needed. This can also be based on specific demands, such as localizing the specific regions responsible for functional performances or assessing the functional integrity of brain regions. The patient's medical records should be checked to rule out contraindications to the mapping procedure.

Adequate preparation prior to the mapping procedure is crucial to ensure reliable mapping results. The process involves a step-by-step approach: checking data quality and availability, evaluating the patient's clinical records, and obtaining the

Fig. 4.5 Calendar implemented for the management of appointments for nTMS examinations for patients

patient's consent for the examination. Quality control includes verifying data integrity, checking for artifacts or inconsistencies, and assessing scanning quality to identify and address any issues that could compromise the feasibility and accuracy of the nTMS mapping procedure. The first choice of imaging for neuronavigation is contrast-enhanced T1-weighted images. Next, the registration landmarks are identified, and stimulation targets and the assumed functional regions are predefined, such as the handknob area for the determination of the resting motor threshold and cortical parcellation system (CPS) targets for language mappings.

Like any other medical examination, informed consent before nTMS functional mapping is warranted. The treating physician should clearly explain the procedure to the patient, emphasizing its necessity and significance in their treatment, including potential risks and their likelihood. If the patient has the legal capacity to consent to the procedure, they should personally sign the consent form. Otherwise, the legal representative needs to consent to the procedure.

Emergency plans for potential complications should also be incorporated into the workflow. This includes the management of acute situations, such as seizures. These protocols ensure the safe conduct of clinical operations, particularly when newly establishing this examination in hospitals (Fig. 4.6).

4.3.3 Offline Functional Data Analysis and Storage

For data analysis, it is crucial to establish a collaboration between technical and research staff. The assessment of motor function data primarily involves postprocessing electromyographic (EMG) recordings (Krieg et al. 2017). Analysts must manually review changes in excitatory potentials corresponding to stimulation when the baseline signal is unstable. Merely excluding these EMG data sets with poor baseline performance would result in omitting motor localization information

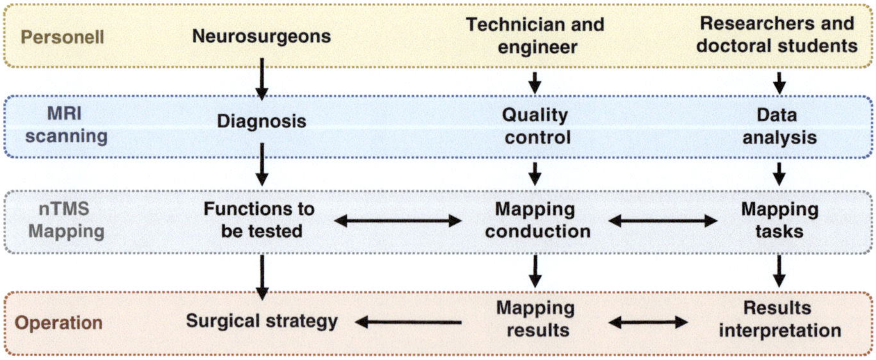

Fig. 4.6 Coordination of efforts and task allocation among neurosurgeons, technical staff, and researchers. Throughout the nTMS mapping process and the conceptualization of the surgical procedure, these three groups should collaborate to ensure clinical, technical, and research expertise to optimize patient care and treatment strategies

for the corresponding muscles. For different tasks, specific software, such as NBS Speech Analyzer (Nexstim, Helsinki, Finland) is specifically designed for language function tasks, in which the task performance during stimulation can be compared over the baseline to identify and categorize language errors (Krieg et al. 2017). Analysts should undergo specific training to identify and distinguish various language errors to guarantee a high standard of data analysis and reduce interrater variability. Therefore, close cooperation with linguistics and language therapists in the team is mandatory to build a structured workflow and to have assistance for demanding cases, such as patients who stutter (Sonsterud et al. 2019; Roux et al. 2003).

In the offline analysis, researchers determine which stimulation points are true positive nTMS stimulation sites. This processed data is then exported from the nTMS system in DICOM format. All exported image coordinate systems are registered to the coordinate system of the original MRI sequences and used for mapping, with potential errors of approximately 1–2 mm occurring during registration. nTMS data is exported in three DICOM files: (1) nTMS-positive points and registration points (nasion, crus spiral); (2) negative points and registration points during the mapping process; and (3) datasets containing the stimulation points projected onto the original MRI sequence.

The exported DICOM data sets are uploaded to the PACS system (Mildenberger et al. 2002; Bidgood Jr and Horii 1992).

4.3.4 Application of Functional Data in Neurosurgical Navigation System

The application of functional data from nTMS in neurosurgical navigation systems is aimed at visualizing functional brain areas during tumor resection procedures, allowing for effective tumor removal while preserving the patient's functionality (Rizzo et al. 2014). Currently, most navigation systems are directly integrated with PACS systems. For example, Elements (Brainlab AG, Germany; Version 3.2) can directly import offline analysis results from the server into the navigation system, avoiding additional data transfer-related concerns. Furthermore, the outpatient integrated Brainlab Net®, Brainlab BUZZ®, and Medtronic StealthViz® can provide enhanced data visualization for patient consultations.

It is crucial to note that maintaining consistency in color coding between the intraoperative navigation system and the preoperative fiber tracking system is essential. Different software may use varying color code systems, which can result in deviations regarding data visualization. Therefore, when selecting the neuronavigation system that suits your clinical practice, it is essential to ensure its applicability.

It is important to note, that the application of preoperative nTMS results during surgery under certain circumstances, for example, patients being unable to cooperate for intraoperative mapping, is very helpful for indicating the functional regions. However, if and to what extent this technology can replace intraoperative DCS still requires more research to verify, and the latter remains currently the gold standard for investigating brain functions. In exposed regions corresponding to the craniotomy, localization results from nTMS can expedite the identification of functional sites in the brain through DCS. This enables a reduction of the duration of intraoperative awake testing for functional assessment, thereby enhancing the safety of the surgical procedure.

4.4 Postoperative Application of Functional Data

In clinical routine, a postoperative MRI scan is performed to confirm the extent of resection. In case of surgery-related motor deficits related to perioperative and postoperative ischemia, a corresponding nTMS assessment can be performed. Depending on these findings, estimations of clinical outcome can be made, and tailored therapies such as nrTMS over a time period of 7 days can then be applied on the non-operated hemisphere to facilitate functional reorganization and neurological rehabilitation (Ille et al. 2021).

4.5 Training for the Application of Functional Data

Collecting and analyzing functional data for clinical application requires coordination among various departments, adequate staffing, equipment, and resource preparation. The operation and analysis of nTMS have been demonstrated to reach a stable technical proficiency level after practical training, typically involving a learning curve of approximately 12 cases (Sollmann et al. 2016). Therefore, when newly establishing nTMS, it is essential to provide specific training courses to ensure that sufficiently skilled technicians are available for the specific applications of nTMS, thus improving efficiency and reliability of the examination.

In the department of radiology, setting up and adjusting a data import interface requires coordination with appropriate services and equipment suppliers. Updating data transmission protocols to establish new data storage channels also involves the participation of the hospital's information technology department. This process involves addressing numerous data security concerns, and due to varying legal regulations in different countries, consultations with the legal department are essential before implementing new protocols.

Neurosurgeons also need to learn and adapt to the benefits and challenges associated with the availability of functional data. On the one hand, this data should also be carefully checked for plausibility and accuracy before applying it. On the other hand, the benefits regarding preoperative assessments, planning of the surgical procedure and the intraoperative use—combined with neuronavigation on DCS have to be illustrated and clear workflows have to be established to utilize the full potential.

4.6 Further Potential Clinical Applications

nTMS allows for non-invasive and highly accurate identification of eloquent brain areas and the underlying fiber tracts. This allows functional data to be applied to broad spectrum of clinical applications, reaching beyond the intraoperative use. The application of nTMS has demonstrated potential in enhancing radiotherapy planning and treatment. A study by Kallioniemi et al. (2017) showed that integrating nTMS data with DTI in pre-radiotherapy planning could significantly reduce the maximum dose to motor areas and the corticospinal tract by 6–54%, potentially minimizing radiation-induced damage to these critical regions (Kallioniemi et al. 2017). Furthermore, the role of nTMS motor mapping in adjuvant radiotherapy planning for patients with high-grade gliomas, as well as patients with supratentorial brain metastases has further validated the value of nTMS in accurately targeting tumors while preserving vital brain functions (Schwendner et al. 2018; Diehl et al. 2019).

Furthermore, it has emerged as a significant advancement in the treatment of major depressive disorder, particularly for patients who have not responded to traditional pharmacotherapy. Studies have shown that nTMS, especially when applied to the left prefrontal cortex, can effectively alleviate symptoms of depression (Janicak et al. 2013; Perera et al. 2016). Solvason et al. found that acute nTMS treatment improved functional status and quality of life in patients with major depression, with these benefits persisting over long-term follow-up (Solvason et al. 2014). Similarly, Perera et al. reported substantial evidence supporting the efficacy and safety of daily left-hemispheric prefrontal nTMS for treating acute phases of depression in treatment-resistant patients (Perera et al. 2016). These findings are corroborated by Janicak et al., who observed significant improvements in patient-reported quality of life and functional status outcomes following nTMS treatment of major depressive disorder (Janicak et al. 2013).

These advancements highlight the growing intersection between neurostimulation techniques like nTMS and the precision-focused field of individualized therapy, opening new avenues for patient-centered and function-based treatments.

4.7 Conclusion

In essence, the integration of nTMS-based functional data into neurosurgical practice offers a beneficiary fusion of medicine and engineering. This innovative approach, with its promise of improving patient prognosis, requires a meticulously structured workflow encompassing various crucial elements.

Thoughtful design of data-related workflows is paramount within neurosurgical departments to harness the collective expertise of medical professionals, engineers, and researchers. This synergy fosters innovation and guarantees that technology reaches its maximum potential. By establishing a tailored integration, the full potential of nTMS-based functional data to enhance patient prognosis and advance neurosurgery can be harnessed, enabling personalized and high-quality patient treatments.

References

Abbas R, Singh Y. PACS implementation challenges in a public healthcare institution: a South African vendor perspective. Healthcare Inf Res. 2019;25(4):324–31. https://doi.org/10.4258/hir.2019.25.4.324.

Abler B, Walter H, Wunderlich A, Grothe J, Schonfeldt-Lecuona C, Spitzer M, et al. Side effects of transcranial magnetic stimulation biased task performance in a cognitive neuroscience study. Brain Topogr. 2005;17(4):193–6. https://doi.org/10.1007/s10548-005-6028-y.

Bagarinao E, Watanabe H, Maesawa S, Mori D, Hara K, Kawabata K, et al. An unbiased data-driven age-related structural brain parcellation for the identification of intrinsic brain volume changes over the adult lifespan. Neuroimage. 2018;169:134–44. https://doi.org/10.1016/j.neuroimage.2017.12.014.

Bassett DS, Sporns O. Network neuroscience. Nat Neurosci. 2017;20(3):353–64. https://doi.org/10.1038/nn.4502.

Benjamin CF, Walshaw PD, Hale K, Gaillard WD, Baxter LC, Berl MM, et al. Presurgical language fMRI: mapping of six critical regions. Hum Brain Mapp. 2017;38(8):4239–55. https://doi.org/10.1002/hbm.23661.

Bidgood WD Jr, Horii SC. Introduction to the ACR-NEMA DICOM standard. Radiographics. 1992;12(2):345–55. https://doi.org/10.1148/radiographics.12.2.1561424.

Bozkurt B, Yagmurlu K, Middlebrooks EH, Cayci Z, Cevik OM, Karadag A, et al. Fiber connections of the supplementary motor area revisited: methodology of fiber dissection, DTI, and three dimensional documentation. J Vis Exp. 2017;123:55681. https://doi.org/10.3791/55681.

Catana C, Drzezga A, Heiss WD, Rosen BR. PET/MRI for neurologic applications. J Nucl Med. 2012;53(12):1916–25. https://doi.org/10.2967/jnumed.112.105346.

Caverzasi E, Hervey-Jumper SL, Jordan KM, Lobach IV, Li J, Panara V, et al. Identifying preoperative language tracts and predicting postoperative functional recovery using HARDI q-ball fiber tractography in patients with gliomas. J Neurosurg. 2016;125(1):33–45. https://doi.org/10.3171/2015.6.JNS142203.

Cheng S, Xu G, Zhou J, Qu Y, Li Z, He Z, et al. A multimodal meta-analysis of structural and functional changes in the brain of tinnitus. Front Hum Neurosci. 2020;14:28. https://doi.org/10.3389/fnhum.2020.00028.

Darmani G, Bergmann TO, Zipser C, Baur D, Muller-Dahlhaus F, Ziemann U. Effects of antiepileptic drugs on cortical excitability in humans: a TMS-EMG and TMS-EEG study. Hum Brain Mapp. 2019;40(4):1276–89. https://doi.org/10.1002/hbm.24448.

Diehl CD, Schwendner MJ, Sollmann N, Oechsner M, Meyer B, Combs SE, et al. Application of presurgical navigated transcranial magnetic stimulation motor mapping for adjuvant radiotherapy planning in patients with high-grade gliomas. Radiother Oncol. 2019;138:30–7. https://doi.org/10.1016/j.radonc.2019.04.029.

Diehl CD, Rosenkranz E, Schwendner M, Misslbeck M, Sollmann N, Ille S, et al. Dose reduction to motor structures in adjuvant fractionated stereotactic radiotherapy of brain metastases: nTMS-derived DTI-based motor Fiber tracking in treatment planning. Cancers (Basel). 2022;15(1):282. https://doi.org/10.3390/cancers15010282.

Frey D, Schilt S, Strack V, Zdunczyk A, Rosler J, Niraula B, et al. Navigated transcranial magnetic stimulation improves the treatment outcome in patients with brain tumors in motor eloquent locations. Neuro Oncol. 2014;16(10):1365–72. https://doi.org/10.1093/neuonc/nou110.

Haddad AF, Young JS, Berger MS, Tarapore PE. Preoperative applications of navigated transcranial magnetic stimulation. Front Neurol. 2020;11:628903. https://doi.org/10.3389/fneur.2020.628903.

Hindel R. The impact of image storage organization on the effectiveness of PACS. J Digit Imaging. 1990;3(4):205–10. https://doi.org/10.1007/BF03168116.

Ille S, Krieg SM. Functional mapping for glioma surgery, part 1: preoperative mapping tools. Neurosurg Clin N Am. 2021;32(1):65–74. https://doi.org/10.1016/j.nec.2020.08.004.

Ille S, Kulchytska N, Sollmann N, Wittig R, Beurskens E, Butenschoen VM, et al. Hemispheric language dominance measured by repetitive navigated transcranial magnetic stimulation and postoperative course of language function in brain tumor patients. Neuropsychologia. 2016;91:50–60. https://doi.org/10.1016/j.neuropsychologia.2016.07.025.

Ille S, Kelm A, Schroeder A, Albers LE, Negwer C, Butenschoen VM, et al. Navigated repetitive transcranial magnetic stimulation improves the outcome of postsurgical paresis in glioma patients – a randomized, double-blinded trial. Brain Stimul. 2021;14(4):780–7. https://doi.org/10.1016/j.brs.2021.04.026.

Isa T. The corticospinal tract and its role in motor control. Brain Nerve. 2012;64(11):1331–9.

Janicak PG, Dunner DL, Aaronson ST, Carpenter LL, Boyadjis TA, Brock DG, et al. Transcranial magnetic stimulation (TMS) for major depression: a multisite, naturalistic, observational study of quality of life outcome measures in clinical practice. CNS Spectr. 2013;18(6):322–32. https://doi.org/10.1017/S1092852913000357.

Jeltema HR, Ohlerth AK, de Wit A, Wagemakers M, Rofes A, Bastiaanse R, et al. Comparing navigated transcranial magnetic stimulation mapping and "gold standard" direct cortical stimulation mapping in neurosurgery: a systematic review. Neurosurg Rev. 2021;44(4):1903–20. https://doi.org/10.1007/s10143-020-01397-x.

Kallioniemi E, Palmgren J-E, Fraunberg M, Könönen M, Vanninen R, Julkunen P. P004 Application of TMS and DTI in pre-radiotherapy planning and the effect of radiation on motor function: a pilot study with two patients. Clin Neurophysiol. 2017;128(3):e12–e3.

Kalyanpur A, Singh J, Bedi R. Practical issues in picture archiving and communication system and networking. Indian J Radiol Imaging. 2010;20(1):2–5. https://doi.org/10.4103/0971-3026.59743.

Kaspar M, Liman L, Ertl M, Fette G, Seidlmayer LK, Schreiber L, et al. Unlocking the PACS DICOM domain for its use in clinical research data warehouses. J Digit Imaging. 2020;33(4):1016–25. https://doi.org/10.1007/s10278-020-00334-0.

Khoo MC. Linking engineering and medicine: fostering collaboration skills in interdisciplinary teams. IEEE Pulse. 2012;3(4):27–9. https://doi.org/10.1109/MPUL.2012.2196832.

Krieg SM. Navigated transcranial magnetic stimulation in neurosurgery. Cham: Springer; 2017.

Krieg SM, Buchmann NH, Gempt J, Shiban E, Meyer B, Ringel F. Diffusion tensor imaging fiber tracking using navigated brain stimulation--a feasibility study. Acta Neurochir (Wien). 2012;154(3):555–63. https://doi.org/10.1007/s00701-011-1255-3.

Krieg SM, Sabih J, Bulubasova L, Obermueller T, Negwer C, Janssen I, et al. Preoperative motor mapping by navigated transcranial magnetic brain stimulation improves outcome for motor eloquent lesions. Neuro Oncol. 2014;16(9):1274–82. https://doi.org/10.1093/neuonc/nou007.

Krieg SM, Lioumis P, Makela JP, Wilenius J, Karhu J, Hannula H, et al. Protocol for motor and language mapping by navigated TMS in patients and healthy volunteers; workshop report. Acta Neurochir (Wien). 2017;159(7):1187–95. https://doi.org/10.1007/s00701-017-3187-z.

Kunimatsu A, Aoki S, Masutani Y, Abe O, Hayashi N, Mori H, et al. The optimal trackability threshold of fractional anisotropy for diffusion tensor tractography of the corticospinal tract. Magn Reson Med Sci. 2004;3(1):11–7. https://doi.org/10.2463/mrms.3.11.

Lefaucheur JP, Picht T. The value of preoperative functional cortical mapping using navigated TMS. Neurophysiol Clin. 2016;46(2):125–33. https://doi.org/10.1016/j.neucli.2016.05.001.

Lennox-Chhugani N. Inter-disciplinary work in the context of integrated care – a theoretical and methodological framework. Int J Integr Care. 2023;23(2):29. https://doi.org/10.5334/ijic.7544.

Leow AD, Zhan L, Zhu S, Hageman N, Chiang M-C, Barysheva M, et al. White matter integrity measured by fractional anisotropy correlates poorly with actual individual fiber anisotropy. In:

2009 IEEE international symposium on biomedical imaging: from nano to macro. IEEE; 2009. p. 622–5.

Makela T, Vitikainen AM, Laakso A, Makela JP. Integrating nTMS data into a radiology picture archiving system. J Digit Imaging. 2015;28(4):428–32. https://doi.org/10.1007/s10278-015-9768-6.

Markello RD, Hansen JY, Liu ZQ, Bazinet V, Shafiei G, Suarez LE, et al. Neuromaps: structural and functional interpretation of brain maps. Nat Methods. 2022;19(11):1472–9. https://doi.org/10.1038/s41592-022-01625-w.

Mildenberger P, Eichelberg M, Martin E. Introduction to the DICOM standard. Eur Radiol. 2002;12(4):920–7. https://doi.org/10.1007/s003300101100.

Moreno-Lopez Y, Olivares-Moreno R, Cordero-Erausquin M, Rojas-Piloni G. Sensorimotor integration by corticospinal system. Front Neuroanat. 2016;10:24. https://doi.org/10.3389/fnana.2016.00024.

Ottenhausen M, Krieg SM, Meyer B, Ringel F. Functional preoperative and intraoperative mapping and monitoring: increasing safety and efficacy in glioma surgery. Neurosurg Focus. 2015;38(1):E3. https://doi.org/10.3171/2014.10.FOCUS14611.

Perera T, George MS, Grammer G, Janicak PG, Pascual-Leone A, Wirecki TS. The Clinical TMS Society consensus review and treatment recommendations for TMS therapy for major depressive disorder. Brain Stimul. 2016;9(3):336–46. https://doi.org/10.1016/j.brs.2016.03.010.

Perez MA, Cohen LG. The corticospinal system and transcranial magnetic stimulation in stroke. Top Stroke Rehabil. 2009;16(4):254–69. https://doi.org/10.1310/tsr1604-254.

Picht T. Current and potential utility of transcranial magnetic stimulation in the diagnostics before brain tumor surgery. CNS Oncol. 2014;3(4):299–310. https://doi.org/10.2217/cns.14.25.

Picht T, Krieg SM, Sollmann N, Rosler J, Niraula B, Neuvonen T, et al. A comparison of language mapping by preoperative navigated transcranial magnetic stimulation and direct cortical stimulation during awake surgery. Neurosurgery. 2013;72(5):808–19. https://doi.org/10.1227/NEU.0b013e3182889e01.

Picht T, Schilt S, Frey D, Vajkoczy P, Kufeld M. Integration of navigated brain stimulation data into radiosurgical planning: potential benefits and dangers. Acta Neurochir (Wien). 2014;156(6):1125–33. https://doi.org/10.1007/s00701-014-2079-8.

Picht T, Frey D, Thieme S, Kliesch S, Vajkoczy P. Presurgical navigated TMS motor cortex mapping improves outcome in glioblastoma surgery: a controlled observational study. J Neuro-Oncol. 2016;126(3):535–43. https://doi.org/10.1007/s11060-015-1993-9.

Pillai JJ. The evolution of clinical functional imaging during the past 2 decades and its current impact on neurosurgical planning. AJNR Am J Neuroradiol. 2010;31(2):219–25. https://doi.org/10.3174/ajnr.A1845.

Raffa G, Spiriev T, Zoia C, Aldea CC, Bartek J Jr, Bauer M, et al. The use of advanced technology for preoperative planning in cranial surgery – a survey by the EANS Young Neurosurgeons Committee. Brain Spine. 2023;3:102665. https://doi.org/10.1016/j.bas.2023.102665.

Rizzo V, Terranova C, Conti A, Germano A, Alafaci C, Raffa G, et al. Preoperative functional mapping for rolandic brain tumor surgery. Neurosci Lett. 2014;583:136–41. https://doi.org/10.1016/j.neulet.2014.09.017.

Roberson GH, Shieh YY. Radiology information systems, picture archiving and communication systems, teleradiology--overview and design criteria. J Digit Imaging. 1998;11(4 Suppl 2):2–7. https://doi.org/10.1007/BF03168169.

Rosenstock T, Giampiccolo D, Schneider H, Runge SJ, Bahrend I, Vajkoczy P, et al. Specific DTI seeding and diffusivity-analysis improve the quality and prognostic value of TMS-based deterministic DTI of the pyramidal tract. Neuroimage Clin. 2017;16:276–85. https://doi.org/10.1016/j.nicl.2017.08.010.

Rossi S, Hallett M, Rossini PM, Pascual-Leone A. Safety, ethical considerations, and application guidelines for the use of transcranial magnetic stimulation in clinical practice and research. Clin Neurophysiol. 2009;120(12):2008–39. https://doi.org/10.1016/j.clinph.2009.08.016.

Roux F-E, Boulanouar K, Lotterie J-A, Mejdoubi M, LeSage JP, Berry I. Language functional magnetic resonance imaging in preoperative assessment of language areas: correlation with direct cortical stimulation. Neurosurgery. 2003;52(6):1335–47.

Schramm S, Haddad AF, Chyall L, Krieg SM, Sollmann N, Tarapore PE. Navigated TMS in the ICU: introducing motor mapping to the critical care setting. Brain Sci. 2020;10(12):1005. https://doi.org/10.3390/brainsci10121005.

Schwendner MJ, Sollmann N, Diehl CD, Oechsner M, Meyer B, Krieg SM, et al. The role of navigated transcranial magnetic stimulation motor mapping in adjuvant radiotherapy planning in patients with supratentorial brain metastases. Front Oncol. 2018;8:424. https://doi.org/10.3389/fonc.2018.00424.

Shams B, Reisch K, Vajkoczy P, Lippert C, Picht T, Fekonja LS. Improved prediction of glioma-related aphasia by diffusion MRI metrics, machine learning, and automated fiber bundle segmentation. Hum Brain Mapp. 2023;44(12):4480–97. https://doi.org/10.1002/hbm.26393.

Silvanto J, Muggleton NG. New light through old windows: moving beyond the "virtual lesion" approach to transcranial magnetic stimulation. Neuroimage. 2008;39(2):549–52. https://doi.org/10.1016/j.neuroimage.2007.09.008.

Sirota-Cohen C, Rosipko B, Forsberg D, Sunshine JL. Implementation and benefits of a vendor-neutral archive and enterprise-imaging management system in an integrated delivery network. J Digit Imaging. 2019;32(2):211–20. https://doi.org/10.1007/s10278-018-0142-3.

Smith SM, Vidaurre D, Beckmann CF, Glasser MF, Jenkinson M, Miller KL, et al. Functional connectomics from resting-state fMRI. Trends Cogn Sci. 2013;17(12):666–82. https://doi.org/10.1016/j.tics.2013.09.016.

Smith RE, Calamante F, Connelly A. Mapping connectomes with diffusion MRI: deterministic or probabilistic tractography? Magn Reson Med. 2020;83(3):787–90. https://doi.org/10.1002/mrm.27916.

Sollmann N, Goblirsch-Kolb MF, Ille S, Butenschoen VM, Boeckh-Behrens T, Meyer B, et al. Comparison between electric-field-navigated and line-navigated TMS for cortical motor mapping in patients with brain tumors. Acta Neurochir (Wien). 2016;158(12):2277–89. https://doi.org/10.1007/s00701-016-2970-6.

Sollmann N, Krieg SM, Saisanen L, Julkunen P. Mapping of motor function with neuronavigated transcranial magnetic stimulation: a review on clinical application in brain tumors and methods for ensuring feasible accuracy. Brain Sci. 2021;11(7):897. https://doi.org/10.3390/brainsci11070897.

Solvason HB, Husain M, Fitzgerald PB, Rosenquist P, McCall WV, Kimball J, et al. Improvement in quality of life with left prefrontal transcranial magnetic stimulation in patients with pharmacoresistant major depression: acute and six month outcomes. Brain Stimul. 2014;7(2):219–25. https://doi.org/10.1016/j.brs.2013.10.008.

Sonsterud H, Kirmess M, Howells K, Ward D, Feragen KB, Halvorsen MS. The working alliance in stuttering treatment: a neglected variable? Int J Lang Commun Disord. 2019;54(4):606–19. https://doi.org/10.1111/1460-6984.12465.

Sporns O. The human connectome: a complex network. Ann N Y Acad Sci. 2011;1224:109–25. https://doi.org/10.1111/j.1749-6632.2010.05888.x.

Strickland NH. PACS (picture archiving and communication systems): filmless radiology. Arch Dis Child. 2000;83(1):82–6. https://doi.org/10.1136/adc.83.1.82.

Subbiah V. The next generation of evidence-based medicine. Nat Med. 2023;29(1):49–58. https://doi.org/10.1038/s41591-022-02160-z.

Sun W, Wu Q, Gao L, Zheng Z, Xiang H, Yang K, et al. Advancements in transcranial magnetic stimulation research and the path to precision. Neuropsychiatr Dis Treat. 2023;19:1841–51. https://doi.org/10.2147/NDT.S414782.

Taşgetiren S, Süzme NÖ, Özkan N, Yuran AF. Cooperation between health and engineering fields in biomedical research areas. In: 2015 Medical technologies national conference (TIPTEKNO); 2015. p. 1–4.

van de Ven VG, Formisano E, Prvulovic D, Roeder CH, Linden DE. Functional connectivity as revealed by spatial independent component analysis of fMRI measurements during rest. Hum Brain Mapp. 2004;22(3):165–78. https://doi.org/10.1002/hbm.20022.

Walton D, Spencer DC, Nevitt SJ, Michael BD. Transcranial magnetic stimulation for the treatment of epilepsy. Cochrane Database Syst Rev. 2021;4(4):CD011025. https://doi.org/10.1002/14651858.CD011025.pub3.

Zhang W, Ille S, Schwendner M, Wiestler B, Meyer B, Krieg SM. Tracking motor and language eloquent white matter pathways with intraoperative fiber tracking versus preoperative tractography adjusted by intraoperative MRI-based elastic fusion. J Neurosurg. 2022:1–10. https://doi.org/10.3171/2021.12.JNS212106.

Safety Issues for the Use of TMS

5

Riccardo Di Iorio, Benedetta Maria Angeloni, and Paolo Maria Rossini

5.1 Introduction

Since its introduction in 1985 (Barker et al.), transcranial magnetic stimulation (TMS) has experienced a significant surge in popularity within the international scientific community (Rossini et al. 1985a, b, 1987). Used for studying cognition, brain–behavior relationships, and the pathophysiology of neurologic and psychiatric disorders, TMS has seen advancements in coils, stimulus paradigms (such as patterned repetitive TMS), and integration with technologies like neuronavigation, electroencephalography (EEG), positron emission tomography (PET), and functional magnetic resonance imaging (fMRI). These developments have enhanced the accuracy and clinical value of research and clinical studies, expanding investigations into non-motor areas of the brain and testing therapeutic applications.

Over the last decade, numerous studies and clinical trials have demonstrated potential therapeutic uses of TMS. Various guidelines published in the literature cover theoretical and physiological aspects (Rossini et al. 2015), methodology (Groppa et al. 2012), and therapeutic applications (Lefaucheur et al. 2014, 2020). The increasing applications of TMS have prompted the scientific and medical community to evaluate the safety record of research studies and clinical applications, particularly repetitive TMS (rTMS). The first safety precautions and practice recommendations were established by the consensus conference held at the National Institutes of Health in June 1996. These recommendations were first summarized by

R. Di Iorio (✉)
Neurology Unit, Fondazione Policlinico A. Gemelli IRCCS, Rome, Italy

B. M. Angeloni
Department of Neuroscience, Università Cattolica del Sacro Cuore, Rome, Italy

P. M. Rossini
Department of Neuroscience and Neurorehabilitation, IRCCS San Raffaele, Rome, Italy
e-mail: paolomaria.rossini@policlinicogemelli.it

S. M. Krieg, T. Picht (eds.), *Navigated Transcranial Magnetic Stimulation in Neurosurgery*, https://doi.org/10.1007/978-3-031-97155-6_5

Wassermann et al. (1998) and subsequently adopted by *The International Federation of Clinical Neurophysiology [IFCN]* (Hallett et al. 1999). Ethical considerations were also addressed in the early phase (Green et al. 1997) and revised over the following years by experts in the field (Wolpe 2002; Mashour et al. 2005; Illes et al. 2006; Steven and Pascual-Leone 2006).

In 2008, on behalf of IFCN, a group of global experts—including neurologists, neurophysiologists, psychiatrists, experimental psychologists, cognitive neuroscientists, physicists, engineers, representatives of TMS equipment manufacturers, and representatives from various world regulatory agencies—convened to revise safety material since 1996. A consensus on safety, ethical issues, and recommendations for TMS use in research and clinical settings was reached, leading to the publication of updated guidelines in the late 2009 on *Clinical Neurophysiology* [i.e., the official Journal of IFCN] (Rossi et al. 2009). These guidelines covered risk and safety of conventional and emerging TMS protocols, stimulation parameter limits, monitoring, team expertise, and ethical considerations.

In 2018, a new consensus conference, promoted and supported by IFCN, took place in Siena (Italy) laying the foundations for the third article on safety of use of rTMS in clinical practice and research setting (Rossi et al. 2021). In this updated version, only emerging and new issues have been covered in detail, leaving still valid the 2009 recommendations regarding the description of conventional or patterned TMS protocols, the screening of subjects/patients, the need of neurophysiological monitoring for new protocols, the utilization of reference thresholds of stimulation, the managing of seizures, and the list of minor side effects.

The updated 2021 guidelines have addressed, in detail, the following new aspects: safety issues of newly developed stimulation devices and pulse configurations, outlining the responsibilities of device makers; novel TMS applications, including neuroimaging, imaging-guided, and robot-guided TMS; TMS interleaved with transcranial electrical stimulation (TES); safety during paired associative stimulation (PAS) interventions; the risks associated with using TMS to induce therapeutic seizures (magnetic seizure therapy); an update on the induction of seizures; new operational guidelines for safety in planning future trials, outlining minimal training requirements for operators; ethical considerations related to neuroenhancement.

This chapter provides a brief summary of essential safety aspects for using TMS in a diagnostic setting, highlighting some new safety issues emerging in 2021 guidelines.

5.2 Contraindications and Precautions

The 2009 safety study group report presented comprehensive guidelines on the safety aspects of TMS in both healthy individuals and patients. Examining numerous studies conducted prior to these guidelines, the authors highlighted key considerations with widespread consensus, which remain today widely accepted and still valid, considering that 2021 guidelines have not made any substantial changes except for some minor suggestions.

Notably, the unique absolute contraindication for TMS/rTMS is currently the presence of metallic hardware near the discharging coil, posing a risk of dysfunction in devices like cochlear implants, internal pulse generators (IPG), or medication pumps. However, several conditions have been identified as potentially increasing the risk of the most serious adverse TMS-related event, i.e., epileptic seizures. These conditions include specific stimulation protocols (especially novel paradigms or high-frequency protocols exceeding 2009 safety limits [see the appropriate section 5.3]), certain patient conditions (e.g., epilepsy history, brain lesions, drug interactions reducing seizure threshold), and factors like sleep deprivation or alcoholism. About advised caution in the application of TMS in persons taking medications known to lower seizure threshold suggested in 2009, update 2021 guidelines no longer support this recommendation, due to currently available data showing low seizure rate. However, documentation of concurrent intake of drugs and other potentially seizure threshold lowering factors during TMS application and systematic capture/reporting of side effect data are recommended to further inform the field. Other instances related to implanted brain electrodes [see the appropriate section 5.6.1], pregnancy and specific neurodegenerative or psychiatric diseases are also associated with increased or uncertain risk. Single-pulse (sp-TMS) or paired-pulse TMS (pp-TMS) or conventional low-frequency (LF-rTMS) or high-frequency rTMS (HF-rTMS) protocol including none of the previous conditions and using parameters of stimulation within the 2009 safety limits are to be considered "free-risk."

Summarizing, except for cases involving implanted devices, all other conditions do not represent rigorous contraindications to the use of the TMS. However, a risk analysis should be carefully evaluated before initiating any TMS study, especially—as suggested in 2021 guidelines—whenever there is a change in the TMS equipment or method of use, including hardware and software configuration, dose selection, environment, or subject population that result in potential new risks compared to TMS devices and methods with established safety records. A concise safety checklist, incorporating factors such as seizure history, brain diseases, medications affecting seizure risk, implanted devices, and pregnancy, is recommended for screening patients before TMS investigations. To this purpose, an updated version of the questionnaire for the screening of patients before TMS investigations published by Rossi et al. (2011) is here reproduced in full, given its great utility in clinical and research practice.

Screening 13-Item Questionnaire for rTMS Candidates

1. Do you have epilepsy or have you ever had a convulsion or a seizure?
2. Have you ever had a fainting spell or syncope? If yes, please describe on which occasion(s)?
3. Have you ever had a head trauma that was diagnosed as a concussion or was associated with loss of consciousness?

4. Do you have any hearing problems or ringing in your ears?
5. Do you have cochlear implants?
6. Are you pregnant or is there any chance that you might be?
7. Do you have metal in the brain, skull, or elsewhere in your body (e.g., splinters, fragments, clips)? If so, specify the type of metal.
8. Do you have an implanted neurostimulator (e.g., deep brain stimulation [DBS], epidural/subdural, vagus nerve stimulation VNS)?
9. Do you have a cardiac pacemaker or intracardiac lines?
10. Do you have a medication infusion device?
11. Are you taking any medications? (please list).
12. Did you ever undergo TMS in the past? If so, were there any problems?
13. Did you ever undergo MRI in the past? If so, were there any problems?

5.3 Considerations on Dosing TMS

As reported in the latest guidelines, stimulation dose safety refers to the impact of the TMS magnetic field on individuals exposed to it, including the subject, operator, bystander, or fetus. The TMS dose is defined by device parameters affecting the magnetic field and induced E-field in the body (Peterchev et al. 2012). TMS accessories and individual differences, such as seizure threshold or drug interactions, can influence the safety of stimulation. The safety may vary based on the targeted cortical region, E-field shape, and the presence of implants. Acceptable risk criteria differ for subjects, pregnant women, patients, and operators. While equipment design and device instructions may provide some limitations, the safety of stimulation dose depends largely on risk management by the operator.

In 1998, Wassermann established the initial safety limits for the four key parameters defining rTMS trains: intensity, frequency, train duration, and inter-train interval. By determining the maximum safe durations of single rTMS trains across various frequencies and intensities (refer to Table 3 in Wassermann 1998), these limits have proven highly effective in preventing seizures, the spread of excitation, or after-discharges of electromyography (EMG) activity, both in normal subjects and in patients with neurological and psychiatric conditions. It is worth noting that these guidelines were originally based on a relatively limited sample of normal subjects and focused exclusively on conventional rTMS. The 2009 safety consensus group revisited these safety limits, refining the acceptable intensity range for stimulation in research and clinical contexts from 100–220% to 90–130% of resting motor threshold (MT) using a figure-of-8 coil (see Table 4 in Rossi et al. 2009). As these previous safety tables were overall effective in preventing seizure occurrence both in healthy volunteers and patient populations, the last safety consensus group has decided not to provide a formal update of the previous safety tables, but instead to propose "operational guidelines" here reported (Rossi et al. 2021): "in all clinical trials and scientific studies that use conventional rTMS protocols, the Principal Investigator (PI) has to (i) balance the overall risk/benefit ratio of the proposed intervention; (ii) use neurophysiological monitoring (i.e., emergence of motor twitches during stimulation) as a warning for increased cortical excitation, in case the combination of parameters of stimulation exceeds the 2009 safety guidelines; (iii) reconsider the protocol of the trial if a seizure occurs under these circumstances; and (iv) alert the scientific community through dedicated scientific journals about the new possibly unsafe combinations of parameters."

Regarding dose, the updated guidelines in 2021 emphasize the importance of adhering to safety protocols when using conventional TMS devices and paradigms. For novel devices, paradigms, or diverse subject populations, an extra layer of risk analysis is advised due to potential variations in the impact on the body compared to conventional techniques. The risk assessment of the stimulation dose of new TMS devices or protocols involves several steps. Initial considerations should involve theoretical aspects related to the induced magnetic field, computational models estimating the induced electric field, and an understanding of how pulse train parameters influence relevant neuronal activation. The evaluation of potential risks includes but is not limited to effects on the brain (such as seizures or cognitive processing), impact on implanted objects (like cardiac pacemakers, brain implants, hearing aids, surgical clips), and risks to operators. Given that the most significant established dose-related risk of TMS is the induction of seizures, factors affecting seizure threshold should be carefully considered. This encompasses changes in the electric field distribution, alterations in temporal pulse repetition rates, patterns, or number of pulses, variations in pulse waveforms, and the selection and individualization of intensity. In cases where the risk analysis identifies new potential risks, additional safety studies, and vigilant safety monitoring during research or clinical studies become essential.

It should be remarked here that the suggested safety parameters mentioned above primarily stem from rTMS applied to the motor cortex. Most studies that address the individualization of stimulation intensity refer to the classical definition of RMT, defined as the minimum intensity required to elicit a muscle response of at least 50 μV with a 50% probability in a fully relaxed muscle, even when stimulating brain regions beyond the motor cortex (Rossini et al. 1994, 2015). For safe rTMS applications targeting cortical areas outside the motor cortex, a reasonable starting point is considering that the threshold for inducing after-discharges is lowest in the motor cortex compared to other cortical areas when stimulated electrically (Penfield and Jasper 1954). Nevertheless, comprehensive safety tables for rTMS application outside the motor cortex are currently unavailable and require further research. The phosphene threshold (PT), representing the minimal intensity needed to induce a phosphene in the contralateral visual hemifield (Marg and Rudiak 1994), may be more appropriate for individualizing stimulation intensity when targeting visual areas. However, determining PT is challenging in about half of the subjects, and the exact relationship between motor threshold (MT), PT, and the threshold for TMS activation of other non-motor areas remains unknown.

In the last decade, the rapid and large development of rTMS for the possible treatment of several neurological and psychiatric diseases (see Lefaucheur et al. 2014, 2020) has raised newer safety issues. In fact, the repeated application of rTMS in a great variety of clinical settings introduced additional dosing parameters that describe the cumulative exposure to rTMS, i.e., total pulses per session, sessions per day, days per week, weeks per acute course, and maintenance frequency. Furthermore, several studies explored the (sequential or simultaneous) combination of different protocols of stimulation, as well as new advances in terms of technology (new forms of coils and magnetic field geometry, neuro-navigation, integration with EEG and neuroimaging). Despite the fact that neither seizure occurrence nor other adverse events (AEs) emerged consistently in recent years—indicating that whatever the protocol of intervention, the technique can be considered basically safe—and that it is unrealistic to categorize all these variables into new tables, future safety investigations about these aspects are needed.

5.4 Adverse Events

5.4.1 Seizures

Epileptic seizures, defined as "transient occurrence of signs/symptoms because of abnormal excessive or synchronous neuronal activity in the brain" (Fisher et al. 2005), are the most serious AE related to TMS. When TMS protocols are applied, seizures can be seen due to the induction of excitatory or inhibitory changes in neural activity and/or connectivity. Nowadays, only a small number of seizures are reported in the literature during TMS, mostly prior to the definition of first 1998 safety limits: it is therefore possible to assert that is a rather rare AE.

Epileptic AEs have been described to varying degrees across all TMS protocols and with all coils, but rTMS has been the method considered most at risk of such a complication based on assumption and theoretical risk (which would increase in proportion to increasing stimulation frequency). Wassermann hypothesized that in addition to an immediate effect, rTMS could induce seizures by the so-called kindling effect, as an after-effect of modulation of cortical excitability (Wasserman et al., 1998). However, to date, no cases of kindling in humans have been described. Recently, Lerner et al. (2019) conducted a survey to quantify the occurrence of seizure and other AEs during TMS. The survey was conducted requesting information on numbers of TMS sessions performed and numbers of seizures between 2012 and 2016: 24 seizures were reported with an estimated rate of 0.08 of 1000 sessions (or less than 1 seizure per 60,000 sessions). In details, eight seizures were reported with rTMS at a frequency >1 Hz, seven occurred in individuals with risk factors and one in an individual without identifiable risks. Thirteen seizures were reported with cumulative sp-TMS and pp-TMS and three seizures with LF-rTMS; out of those 16 subjects, only 3 of them had no risk factor. In summary, out of the 24 seizures reported, 19 of these occurred in subjects with elevated risk and only 4 occurred when TMS was delivered within published guidelines to subjects without recognized risk factors; furthermore, the study implied that HF-rTMS delivered within published guidelines to individuals without known risk factors was no more likely to cause seizures than LF-rTMS and sp-TMS/pp-TMS.

In another paper, Chou et al. (2020) carried out a systematic literature search including 36 studies since the introduction of TMS that used the 1998 and 2009 guidelines, identifying a total of 41 TMS-induced seizures, mostly generalized (22 of 41, 54%), of which 13 were in healthy people and 28 in patients with clinical conditions, representing a per-session seizure rate of 0.0017%. In a more recent study by Caulfield et al. (2022) evaluating the safety profile of accelerated TMS (aTMS)—i.e., by applying two or more stimulation sessions within a day—that considered 85 studies from 2001 to 2022 with a total of 1543 participants, only a single seizure was reported in an 18-year-old patient: this result represents a per-session seizure rate of 0.0023%, similar to the 0.0017% per-session seizure rate of once daily rTMS.

Considering these data, as stated in the latest 2021 guidelines, the risk of TMS in inducing a seizure is definitely low, even in patient populations taking drugs acting on the central nervous system, at least with the use of traditional stimulation parameters and focal coils for which large data sets are available. However, although encouraging given the small number of epileptic adverse event, seizures can arise in any subject and using any TMS protocol. For this reason, the clinician should be aware that it is important to rule out potential predisposing conditions for the occurrence of seizures, such as through informal pre-treatment interview and through use of questionnaires designed to identify risk factors (Giustiniani et al. 2022).

5.4.2 Syncope

Vasovagal syncope is a relatively common AE of minimally invasive and/or discomfortable procedures, which may occur more often than epileptic seizures during TMS testing and treatment. Indeed, in the survey by Lerner et al. (2019), the most common AEs were syncope or presyncope (17% among 174 respondents). Some clinical features such as loss of consciousness and the possible manifestation of convulsive syncope could place syncope in differential diagnosis with seizures. Given the frequency of this complication, clinicians should consider the presence of premonitory symptoms (such as fainting sensation, distress, tachycardia, sweating, nausea), and a rapid recover of full consciousness after the event, more likely related to syncope than to seizures. It is important to remember that patients who develop syncope under TMS often have experienced similar episodes in the past. In managing the patient with presyncope/syncope, it is advisable to lay the subject down and place them in a safety position, replenishing fluids when possible and providing reassurance to mitigate stressful emotional states.

5.4.3 Minor Side Effects

Minor AEs during TMS include headache, tingling sensation, temporary acute tones hearing loss, and local mild discomfort. Other rare complaints include worsening of psychotic symptoms and other psychiatric symptoms such as anxiety, irritability, insomnia, joint pain, and tinnitus. However, most of subjects do not complain about discomfort, especially when they are familiarized with TMS procedures. Among the mild AEs, headache is certainly the most common reported overall in studies on TMS. In the study by Lerner et al. (2019), the incidence of post-TMS headache is reported to be around 16%, with a similar incidence to post-treatment seizures (17%). To the current knowledge, there is no literature available about patient's pre-TMS treatment characteristic (e.g., systematic data about personal history of headache). Anyway, posttreatment headache is generally described as transient and mild, often associated with application-site discomfort and mild neck tension, with a tendency toward spontaneous resolution.

5.5 Safety Issues About New Methodological and Technical Advancements

5.5.1 Patterned Repetitive TMS

Since its inception (Huang et al. 2005.), the utilization of theta burst stimulation (TBS) has progressively expanded in recent years, as its after-effects closely resemble those induced by conventional rTMS interventions but can be achieved faster through brief bursts of high-frequency stimuli. This characteristic raises the theoretical concern that TBS may pose a heightened risk of seizures compared to other rTMS protocols due to its delivery of high-frequency bursts.

The 2008 consensus meeting was hindered in providing safety guidelines for TBS due to the absence of specific literature and recommendations regarding maximum stimulation duration or intensity. At that time, a lone study addressing TBS safety in 24 healthy subjects (Grossheinrich et al. 2009) reported no serious AEs, except for lipotimic-like reactions in three subjects. Subsequently, Oberman et al. (2011) conducted a meta-analysis of 67 studies spanning from May 2004 to December 2009, involving 1040 subjects/patients with diverse neurological or psychiatric diagnoses. Most AEs associated with TBS were mild, occurring in 5% of subjects, primarily manifesting as headache and neck pain. The crude risk for mild AEs was estimated at 1.1%, and the risk of seizures was 0.02%, with only one documented seizure in a healthy control subject during continuous TBS (cTBS) (Obermann and Pascual-Leone 2009). This comprehensive review demonstrated that the most reported AEs during TBS were comparable to those observed in conventional rTMS interventions, both in terms of severity and overall incidence.

After these first evidence, subsequent studies explored the safety and tolerability of TBS in diverse populations, including children (Krishnan et al. 2015; Hong et al. 2015), patients with major depression (Bakker et al. 2015), patients with schizophrenia (Tikka et al. 2017), and patients with psychiatric disorders (Rachid 2017), suggesting that TBS could be considered a generally safe and effective technique. Indeed, only one additional seizure has been documented over the years (Lenoir et al. 2018) using the parameters originally described by Huang et al. (50 Hz bursts of 3 pulses repeated at 5 Hz; stimulus intensity of 80% AMT) (Huang et al. 2005).

Therefore, based on these data, it can be concluded that TBS is safe if applied with the reference parameters described in the latest guidelines, which do not differ substantially from the originals (Rossi et al. 2021; table S5 of Supplementary materials). However, it is crucial to emphasize that certain aspects of TBS safety, even in normal subjects, necessitate further evaluation. This includes considerations such as the total number of pulses, intervals between repeated TBS sessions, stimulation intensity, and establishing a safe combination of these factors.

Furthermore, about patterned rTMS, latest guidelines also have proposed safety parameters for Quadripulse stimulation (QPS) (see table S3 of Supplementary materials of Rossi et al. 2021), a technique invented to induce robust LTP/LTD effects in human brain using monophasic magnetic/electric pulses (Hamada and Ugawa 2010), which could be considered safe at least in normal subjects.

5.5.2 Combined TMS and TES

In recent investigations, TMS has been paired with other modalities of transcranial stimulation, incorporating the simultaneous application of mild electric currents to the scalp. This approach, involving transcranial direct current stimulation (tDCS), transcranial alternating current stimulation (tACS), and transcranial random noise stimulation (tRNS), aims to explore a potential "priming" effect or meta-plasticity effect on cortical excitability before the subsequent or during the concomitant application of TMS (Karabanov et al. 2015; Muller-Dahlhaus and Ziemann 2015). Theoretically, the interplay between transcranial electric stimulation (TES) and TMS raises concerns about an elevated risk of altering brain excitability, potentially exposing subjects to a higher risk of seizure or other AEs (Rossi et al. 2009).

Although safety-specific assessments were not carried out in most cases, studies in which TES was integrated with TMS (single-pulse, paired-pulse, repetitive and/or patterned) have reported no AEs during or after the combined interventions, except for a stronger scalp pain in one pilot study in patients with major depression (Loo et al. 2009).

In summary, current evidence does not suggest that the combination of tDCS/tACS/tRNS and rTMS/TMS used up to now poses safety concerns or is associated with any significant burden (Rossi et al. 2021).

5.5.3 PAS Protocols

Paired associative stimulation (PAS) protocols represent an experimental method to study neural plasticity in humans based on spike-timing-dependent plasticity (STDP) rules established in animal models. Originally developed by applying a single electrical stimulus to a peripheral nerve before a TMS pulse to the contralateral primary motor cortex (M1)—as originally described by Mariorenzi et al. (1991) and thereafter standardized according to the Stefan's protocol (Stefan et al. 2000; Wolters et al. 2005)—more recently PAS protocols have evolved with the introduction of repeated paired-coil focal TMS over different cortical areas to modify the activity of corticocortical networks. By using different interstimulus intervals and several cortical targets, novel corticocortical PAS (cc-PAS) protocols have been developed to investigate STDP within different interhemispheric cortical networks, being able to induce bidirectional modulation of cortical plasticity in the conditioned target area (Koganemaru et al. 2009; Rizzo et al. 2009; Arai et al. 2011; Buch et al. 2011; Koch et al. 2013; Veniero et al. 2013; Johnen et al. 2015; Ribolsi et al. 2017; Di Lorenzo et al. 2018; Fiori et al. 2018; Nord 2019; Zibman et al. 2019). The intensity, frequency, and duration of stimuli vary across studies, and the effects of PAS have been investigated in different cognitive domains and sensory systems. Additionally, some studies have extended PAS protocols to evaluate the effects of afferent volleys to M1 driven by subcortical structures (Lu et al. 2012; Udupa et al. 2016; Ni et al. 2018) and to explore the effects on motor axon activation and spinal reflexes at the spinal cord level (Taylor and Martin 2009; Leukel et al. 2012).

In summary, the mentioned studies have been designed to investigate physiological mechanisms of corticocortical plasticity in humans, not involving repeated sessions over several days to achieve clinical effects. Given that no significant AEs have been reported to date in both healthy and pathological conditions, latest guidelines raise no specific safety concern for studies of this nature (Rossi et al. 2021).

5.6 Safety Aspects of TMS Applied in Specific Patient Populations

5.6.1 TMS in Patients with Implanted Brain Electrodes

Several TMS studies have investigated patients with implanted stimulating/recording electrodes in both the central and peripheral nervous systems. These studies predominantly utilized sp-TMS or pp-TMS, occasionally employing rTMS. From the initial study by Kofler et al. (1991), various TMS investigations were performed involving implanted patient with five types of electrodes: (Arai et al. 2011) deep brain stimulation (DBS) electrodes (Pierantozzi et al. 2002; Bakker et al. 2015) epidural electrodes (implanted over the cerebral cortex or spinal cord); (Barker et al. 1987) subdural cortical electrodes; (Bender Pape et al. 2020) peripheral or cranial nerve stimulating electrodes (e.g., vagus nerve (VN) electrodes); and (Brasil-Neto et al. 2004) cardiac pacemakers. Some studies occurred shortly after implantation, with externalized electrode leads before the connection to a subcutaneous stimulus generator. In other instances, studies were conducted with leads connected to implanted stimulators.

Based on ex vivo and in vivo studies, TMS can be safely applied to patients with cardiac pacemakers, VNS systems, and spinal cord stimulator when the TMS coil is not in proximity (<10 cm) to electronic components such as the internal pulse generator (IPG) system located in the neck or chest. Furthermore, TMS can also be conducted safely in patients with implanted electrodes in the central and peripheral nervous systems that are not connected to an IPG, taking caution to minimize the currents induced in any connections to external stimulators or amplifiers. Importantly, TMS in subjects with cochlear implants should be avoided due to potential unsafe interactions between the TMS pulse and the implant.

Concerning potential risk in DBS implanted patients, applying rTMS may induce significant voltages in subcutaneous leads in the scalp. This could lead to unexpected electric currents in the electrode contacts used for DBS, whether the internal pulse generator (IPG) is turned "on" or "off." The situation may worsen due to coiling of the electrode lead into several loops near the insertion point in the skull. Deng et al. (2011) demonstrated ex vivo that TMS-induced voltages and currents in DBS electrodes, with the IPG set in various modes, can reach up to 100 V and 83 mA, respectively. These values are an order of magnitude higher than normal DBS pulses and may result in tissue damage. Therefore, when such systems are implanted, lead loops should be avoided if possible or wound with each turn circling in opposite direction (e.g., one turn clockwise and the next turn counterclockwise) to minimize electromagnetic induction.

However, additional safety studies should be conducted to evaluate the magnitude of the voltages and currents induced in implanted stimulation systems and to obtain detailed information on the safe distance between the TMS coil and the implanted stimulator, as well as the impact of coil shape and angulation. For this reason, TMS should only be administered in patients with implanted stimulators (especially DBS) if there are scientific or medical justifications, following a predefined experimental protocol and settings, and with approval from the Institutional Review Board or the Ethical Committee.

5.6.2 Navigated TMS for Preoperative Mapping in Neurosurgical Patients

In recent years, navigation systems have been developed to enhance the precision of TMS coil placement on specific targets in the brain by integrating anatomical or functional brain imaging (Ruohonen and Karhu 2010). These systems, particularly those dedicated to MRI-based neuronavigated TMS (nTMS) practice, have demonstrated reliability in determining cortical target locations and improving the consistency of coil placement across multiple sessions (Lefaucheur 2010).

The primary application of nTMS is pre-surgical motor and language mapping for patients with brain lesions in presumed eloquent and peri-eloquent regions. Multiple studies have shown that nTMS correlates well with intraoperative direct cortical stimulation (DCS) and is likely more accurate than fMRI and MEG (Krings et al. 2001; Picht et al. 2009; Krieg et al. 2012; Tarapore et al. 2012). Therefore, this technique has become a significant step in the standard preoperative workflow for neurosurgical patients (Krieg et al. 2017), even because evidence suggests that presurgical nTMS improves patient outcomes, leading to a greater extent of resection (Frey et al. 2014; Krieg et al. 2014) and to a fewer long-term neurological deficits compared to those without nTMS mapping (Krieg et al. 2014, 2015; Sollmann et al. 2015).

Given that neurosurgical patients are often at greater risk of neurological complication (including seizure), with the increasing use of nTMS in neurosurgery, it has been necessary to evaluate its safety and tolerability in this patient population. The 2009 safety group did not specifically discuss the risks associated with performing TMS in neurosurgical patients with brain lesions due to lack of published data at the time of meeting date. To address this issue, Tarapore et al. (2016) investigated the safety and tolerability of nTMS in a large, multi-center cohort of neurosurgical patients. The study included 733 patients undergoing functional mapping with single pulse and repetitive nTMS. Results from prospectively collected data across three institutions and two countries demonstrated minimal risk associated with presurgical motor and language mapping using nTMS. Notably, no AEs, including seizures, were observed over several hundred mapping procedures in patients with intracranial lesions. This is particularly significant as the study focused exclusively on a neurosurgical population, which typically carries a higher risk of seizures. Despite half of the patients having a history of seizure due to their lesions, neither single pulse nor repetitive nTMS sequences triggered seizures in any patient. More recently, Schiavao et al. (2022) performed a systematic review to assess the potential of nTMS to predict the occurrence of motor deficits and the capabilities of the nTMS technique in performing preoperative/intraoperative mapping of the motor cortex. Of the 38 included articles, 13 mentioned adverse event analysis, with 11 mentioning that any AE occurred, one reporting the occurrence of minor headaches or small scalp pinch (Neville et al. 2020) and one reporting that 7 of 12 patients had complained about strong facial contractions (Freigang et al. 2020). From the results above, latest guidelines have stated that preoperative motor and language mapping with navigated TMS is safe to perform in neurosurgical patients, even in patients with poorly controlled seizures.

5.6.3 rTMS in Patients with Epilepsy

rTMS is emerging as a potential therapeutic approach for suppressing seizures in patients with pharmacologically intractable epilepsy. Preliminary various studies, including open-label investigations and randomized controlled trials (Menkes and Gruenthal 2000; Daniele et al. 2003; Brasil-Neto et al. 2004; Fregni et al. 2005, 2006; Misawa et al. 2005), have shown a reduction in seizure frequency and epileptic discharges following rTMS applications, especially low-frequency stimulation paradigms. However, conflicting results in some well-designed, multicenter trials (Cantello et al. 2007), highlight the need for multiparametric studies with strict inclusion criteria to ensure data consistency and reproducibility of effects.

The safety of rTMS in epilepsy patients is a critical concern due to the risk of inducing seizures and other AEs. As mentioned before, existing research has reported instances of TMS-induced seizures in both patients with epilepsy and healthy volunteers. Nevertheless, in large-scale trials focused on LF-rTMS to reduce seizure frequency, no TMS-linked seizures were observed among 152 epilepsy patients receiving weekly applications (Theodore 2003; Tergau et al. 2003; Fregni et al. 2006; Cantello et al. 2007; Joo et al. 2007; Santiago-Rodríguez et al. 2008). A first comprehensive review on the topic (Bae et al. 2007), which included 30 studies published from 1990 to 2007, suggested a low overall risk of seizures (1.4%) among 280 subjects undergoing rTMS, albeit potentially mitigated by the use of antiepileptic drugs.

Later, Pereira et al. (2016) conducted a systematic review on rTMS in epilepsy, updating the data from January 1990 to August 2015. Out of 46 publications, 16 were new studies published after the previous safety review of 2007. The crude per subject seizure risk was estimated at 2.9%, with 12 seizures reported out of 410 subjects, among which only one seizure—moreover occurring during HF-rTMS—was considered atypical in terms of the clinical characteristics. Despite a slightly higher crude seizure risk compared to a 2007 review, confidence intervals overlapped, indicating a small risk of seizure induction in epilepsy patients undergoing rTMS. Furthermore, it was also found that among the studies that reported the presence or absence of AEs, the latter were observed in 18.3% of cases; in most cases (85%) these AES were mild, therefore suggesting that the risk of complications in patients with epilepsy undergoing rTMS is similar to that of rTMS applied to other conditions and to healthy subjects.

In a very recent article (Tavakoli and Heidarpanah 2023), reviewing original studies published from January 2000 to October 2021, authors examined the efficacy and side effects of rTMS with a specific look at its application in epilepsy. Apart from a resulting controversial effects of rTMS on reducing epileptic seizures, regarding AEs, this review showed that in the non-review studies that investigated the presence or absence of AEs (15 articles, 93.75%), either no reported AEs (6 articles, 40%) or the reported AEs were mild and tolerable (7 articles, 46.67%); however, severe AEs were reported in some cases (2 articles, 13.33%), in line with other previous analyses (Bender Pape et al. 2020; Walton et al. 2021).

In conclusions, the comparable safety profiles of rTMS in individuals with and without epilepsy suggest the potential for the technique as a safe therapy for seizure suppression. While the overall risk for patients with epilepsy seems low, additional caution and rigorous monitoring are advised for those with heightened risk factors. In these cases, it is recommended to fully endorse safety guidelines, which include electroencephalographic and electromyographic monitoring, along with video recording when available, to be able to analyze seizure characteristics in detail during the session.

5.6.4 MST in Patients with Epilepsy

Magnetic seizure therapy (MST) is a treatment involving the deliberate induction of seizures using TMS under anesthesia for conditions like depression or other serious psychiatric disorders (Lisanby et al. 2001, 2003). The method is considered safer than electroconvulsive therapy (ECT) by allowing more control over stimulation sites, thus saving by stimulation brain regions related to the AEs of ECT (Lisanby 2002). MST requires a multidisciplinary team and is performed under general anesthesia, involving EEG and motor seizure assessment; pre-procedure evaluation including specific medical tests like those for ECT is also necessary.

In contrast to conventional rTMS, for whose application parameters are selected to minimize seizure risk, optimal MST dosing involves the selection of parameters more efficient in inducing seizure. MST is commonly administered at 100% of maximal stimulator output, utilizing a frequency range of 25–100 Hz in a single train lasting up to 10 seconds. Notably, even at maximum intensity, research employing realistic head modeling to assess the electric field induced in the brain during MST indicates significantly lower and more targeted effects compared to ECT (Deng et al. 2009, 2011, 2013; Lee et al. 2014, 2017). Interestingly, studies have also revealed that the optimal frequency for inducing seizures with MST is lower than the frequencies typically employed in clinical ECT settings, specifically ranging from 18 to 25 Hz (Peterchev et al. 2015a, b). These differences in terms of frequency and intensity between MST and ECT procedures are believed to influence their respective efficacy and to contribute to superior side effect profile of MST. Indeed, clinical trials with MST demonstrate an excellent safety profile if compared to ECT with minimal-or-no detectable short–/long-term cognitive side effects, lower risk of cardiovascular and anesthetic complications, and absence of other serious potential complications such as death, cerebral herniation, or prolonged seizure (Lisanby et al. 2003; White et al. 2006; Kayser et al. 2011; Fitzgerald et al. 2013). However, to date, the optimal MST dosing for therapeutic benefits with minimal side effects is yet unknown and future research are needed to explore dose–response relationships, individualized intensity, and other factors influencing MST outcomes.

5.7 Conclusions

TMS is a now validated method for clinical and research purposes in various diseases and conditions that has been gaining popularity in recent years. However, as with any technology, it is crucial to ensure that it is being used in an ethical and safe manner. In addition to technical issues, the safety guidelines of 2009 and 2021 also address these concerns and provide essential recommendations on the ethical, legal, and regulatory aspects of TMS. These recommendations result crucial in ensuring that TMS is being administered by qualified professionals, in appropriate settings, and that emergency situations are properly managed.

However, there is a new ethical concern regarding the potential for TMS to enhance brain activity in an abnormal way, which is called neuroenhancement. This involves enhancing core brain functions beyond natural training, like pharmacological neuroenhancement using substances or devices. TMS is proposed for neuroenhancement based on theories like balance effects, entrainment theory, stochastic resonance, and the net zero-sum framework. Various studies claim cognitive improvement in specific functions following TMS sessions targeting distinct brain areas. This "brain doping" raises ethical and social concerns that require future research and safety considerations. For this reason, official approval for TMS use is crucial, with some protocols and devices being approved in the last decade for therapeutic purposes in specific countries. Endorsement by scientific societies for diseases with demonstrated efficacy should be the next step in globally regulating the clinical use of TMS.

References

Arai N, Müller-Dahlhaus F, Murakami T, Bliem B, Lu MK, Ugawa Y, Ziemann U. Statedependent and timing-dependent bidirectional associative plasticity in the human SMA-M1 network. J Neurosci. 2011;31:15376–83.

Bakker N, Shahab S, Giacobbe P, Blumberger DM, Daskalakis ZJ, Kennedy SH, Downar J. rTMS of the dorsomedial prefrontal cortex for major depression: safety, tolerability, effectiveness, and outcome predictors for 10 Hz versus intermittent theta-burst stimulation. Brain Stimul. 2015;8(2):208–15.

Barker AT, Freeston IL, Jalinous R, Jarrat JA. Magnetic stimulation of the human brain and peripheral nervous system: an introduction and the results of an initial clinical evaluation. Neurosurgery. 1987;20:100–9.

Bae EH, Schrader LM, Machii K, Alonso MA, Riviello Jr JJ, Pascual-Leone A, et al. Safety and tolerability of repetitive transcranial magnetic stimulation in patients with epilepsy: a review of the literature.Epilepsy Behav. 2007;10(4):521-8.

Bender Pape T, Kletzel SL, Aaronson AL, Guernon A, Carbone C, Chaudhry N, et al. Safety considerations for the use of transcranial magnetic stimulation as treatment for coma recovery in people with severe traumatic brain injury. J Head Trauma Rehabil. 2020;35(6):430–8.

Brasil-Neto JP, De Araújo DP, Teixeira WA, Araújo VP, Boechat-Barros R. Experimental therapy of epilepsy with transcranial magnetic stimulation: lack of additional benefit with prolonged treatment. Arq Neuropsiquiatr. 2004;62(1):21–5.

Buch ER, Johnen VM, Nelissen N, O'Shea J, Rushworth MF. Noninvasive associative plasticity induction in a corticocortical pathway of the human brain. J Neurosci. 2011;31:17669–79.

Cantello R, Rossi S, Varrasi C, Ulivelli M, Civardi C, Bartalini S, et al. Slow repetitive TMS for drug-resistant epilepsy: clinical and EEG findings of a placebo-controlled trial. Epilepsia. 2007;48(2):366–74.

Caulfield KA, Fleischmann HH, George MS, McTeague LM. A transdiagnostic review of safety, efficacy, and parameter space in accelerated transcranial magnetic stimulation. J Psychiatr Res. 2022;152:384–96.

Chou Y, That VT, Chen AY, Sundman M, Huang Y. TMS-induced seizure cases stratified by population, stimulation protocol, and stimulation site – a systematic literature search. Clin Neurophysiol. 2020;131(5):1019–20.

Daniele O, Brighina F, Piazza A, Giglia G, Scalia S, Fierro B. Low-frequency transcranial magnetic stimulation in patients with cortical dysplasia. J Neurol. 2003;250(6):761–2.

Deng ZD, Lisanby SH, Peterchev AV. Effect of anatomical variability on neural stimulation strength and focality in electroconvulsive therapy (ECT) and magnetic seizure therapy (MST). Conf Proc IEEE Eng Med Biol Soc. 2009;2009:682–8.

Deng ZD, Lisanby SH, Peterchev AV. Electric field strength and focality in electroconvulsive therapy and magnetic seizure therapy: a finite element simulation study. J Neural Eng. 2011;8:016007.

Deng ZD, Lisanby SH, Peterchev AV. Electric field depth-focality tradeoff in transcranial magnetic stimulation: simulation comparison of 50 coil designs. Brain Stimul. 2013;6:1–13.

Di Lorenzo F, Ponzo V, Motta C, Bonnì S, Picazio S, Caltagirone C, Koch G. Impaired spike timing dependent cortico-cortical plasticity in Alzheimer's disease patients. J Alzheimers Dis Preprint. 2018;66:1–9.
Fiori F, Chiappini E, Avenanti A. Enhanced action performance following TMS manipulation of associative plasticity in ventral premotor-motor pathway. NeuroImage. 2018;183:847–58.
Fisher RS, van Emde BW, Blume W, Elger C, Genton P, Lee P, Engel J Jr. Response: definitions proposed by the International League Against Epilepsy (ILAE) and the International Bureau for Epilepsy (IBE). Epilepsia. 2005;46:1701–2.
Fitzgerald PB, Hoy KE, Herring SE, Clinton AM, Downey G, Daskalakis ZJ. Pilot study of the clinical and cognitive effects of high-frequency magnetic seizure therapy in major depressive disorder. Depress Anxiety. 2013;30:129–36.
Fregni F, Thome-Souza S, Bermpohl F, Marcolin MA, Herzog A, Pascual-Leone A, et al. Antiepileptic effects of repetitive transcranial magnetic stimulation in patients with cortical malformations: an EEG and clinical study. Stereotact Funct Neurosurg. 2005;83(2–3):57–62.
Fregni F, Otachi PTM, DoValle A, Boggio PS, Thut G, Rigonatti SP, et al. A randomized clinical trial of repetitive transcranial magnetic stimulation in patients with refractory epilepsy. Ann Neurol. 2006;60(4):447–55.
Freigang S, Fresnoza S, Ali KM, Zaar K, Jehna M, Reishofer G, et al. Impact of priming on effectiveness of TMS in detecting language-eloquent brain areas in tumor patients. J Neurol Surg a Cent Eur Neurosurg. 2020;81(02):111–29.
Frey D, Schilt S, Strack V, Zdunczyk A, Rosler J, Niraula B, et al. Navigated transcranial magnetic stimulation improves the treatment outcome in patients with brain tumors in motor eloquent locations. Neuro-Oncology. 2014;16:1365–72.
Giustiniani A, Vallesi A, Oliveri M, Tarantino V, Ambrosini E, Bortoletto M, et al. A questionnaire to collect unintended effects of transcranial magnetic stimulation: a consensus based approach. Clin Neurophysiol. 2022;141:101–8.
Green RM, Pascual-Leone A, Wassermann EM. Ethical guidelines for rTMS research. IRB. 1997;2:1–7.
Groppa S, Oliviero A, Eisen A, Quartarone A, Cohen LG, Mall V, et al. A practical guide to diagnostic transcranial magnetic stimulation: report of an IFCN committee. Clin Neurophysiol. 2012;123:858–82.
Grossheinrich N, Rau A, Pogarell O, Hennig-Fast K, Reinl M, Karch S, et al. Theta burst stimulation of the prefrontal cortex: safety and impact on cognition, mood, and resting electroencephalogram. Biol Psychiatry. 2009;65:778–84.
Hallett M, Wassermann EM, Pascual-Leone A, Valls-Solé J, Deuschl G, Eisen A. Repetitive transcranial magnetic stimulation. Recommendations for the practice of clinical neurophysiology: guidelines of the international federation of clinical neurophysiology. Electroencephalogr Clin Neurophysiol (2nd ed.). 1999;(Suppl. 52):105–13.
Hamada M, Ugawa Y. Quadripulse stimulation – a new patterned rTMS. Restor Neurol Neurosci. 2010;28:419–24.
Hong YH, Wu SW, Pedapati EV, Horn PS, Huddleston DA, Laue CS, Gilbert DL. Safety and tolerability of theta burst stimulation vs. single and paired pulse transcranial magnetic stimulation: a comparative study of 165 pediatric subjects. Front Hum Neurosci. 2015;9:29.
Huang YZ, Edwards MJ, Rounis E, Bhatia KP, Rothwell JC. Theta burst stimulation of the human motor cortex. Neuron. 2005;45:201–6.
Illes J, Gallo M, Kirschen MP. An ethics perspective on transcranial magnetic stimulation (TMS) and human neuromodulation. Behav Neurol. 2006;7:149–57.
Johnen VM, Neubert FX, Buch ER, Verhagen L, O'Reilly JX, Mars RB, Rushworth MF. Causal manipulation of functional connectivity in a specific neural pathway during behaviour and at rest. Elife. 2015;4:04585.
Joo EY, Han SJ, Chung SH, Cho JW, Seo DW, Hong SB. Antiepileptic effects of low-frequency repetitive transcranial magnetic stimulation by different stimulation durations and locations. Clin Neurophysiol. 2007;118(3):702–8.

Karabanov AN, Ziemann U, Hamada M, George MS, Quartarone A, Classen J, et al. Consensus paper: probing homeostatic plasticity of human cortex with non-invasive transcranial brain stimulation. Brain Stim. 2015;8:993–1006.

Kayser S, Bewernick BH, Grubert C, Hadrysiewicz BL, Axmacher N, Schlaepfer TE. Antidepressant effects, of magnetic seizure therapy and electroconvulsive therapy, in treatment-resistant depression. J Psychiatr Res. 2011;45:569–76.

Koch G, Ponzo V, Di Lorenzo F, Caltagirone C, Veniero D. Hebbian and anti-Hebbian spike-timing-dependent plasticity of human cortico-cortical connections. J Neurosci. 2013;33:9725–33.

Kofler M, Leis AA, Sherwood AM, Delapasse JS, Halter JA. Safety of transcranial magnetic stimulation in patients with abdominally implanted electronic devices. Lancet. 1991;338:1275–6.

Koganemaru S, Mima T, Nakatsuka M, Ueki Y, Fukuyama H, Domen K. Human motor associative plasticity induced by paired bihemispheric stimulation. J Physiol. 2009;587:4629–44.

Krieg SM, Shiban E, Buchmann N, Gempt J, Foerschler A, Meyer B, et al. Utility of presurgical navigated transcranial magnetic brain stimulation for the resection of tumors in eloquent motor areas. J Neurosurg. 2012;116:994–1001.

Krieg SM, Sabih J, Bulubasova L, Obermueller T, Negwer C, Janssen I, et al. Preoperative motor mapping by navigated transcranial magnetic brain stimulation improves outcome for motor eloquent lesions. Neuro-Oncology. 2014;16:1274–82.

Krieg SM, Sollmann N, Obermueller T, Sabih J, Bulubas L, Negwer C, et al. Changing the clinical course of glioma patients by preoperative motor mapping with navigated transcranial magnetic brain stimulation. BMC Cancer. 2015;15:231.

Krieg SM, Lioumis P, Mäkelä JP, Wilenius J, Karhu J, Hannula H, et al. Protocol for motor and language mapping by navigated TMS in patients and healthy volunteers; workshop report. Acta Neurochir. 2017;159(7):1187–95.

Krings T, Foltys H, Reinges MH, Kemeny S, Rohde V, Spetzger U, et al. Navigated transcranial magnetic stimulation for presurgical planning–correlation with functional MRI. Minim Invasive Neurosurg. 2001;44:234–9.

Krishnan C, Santos L, Peterson MD, Ehinger M. Safety of noninvasive brain stimulation in children and adolescents. Brain Stimul. 2015;8(1):76–87.

Lee WH, Lisanby SH, Laine AF, Peterchev AV. Stimulation strength and focality of electroconvulsive therapy and magnetic seizure therapy in a realistic head model. Conf Proc IEEE Eng Med Biol Soc. 2014;2014:410–3.

Lee WH, Lisanby SH, Laine AF, Peterchev AV. Minimum electric field exposure for seizure induction with electroconvulsive therapy and magnetic seizure therapy. Neuropsychopharmacology. 2017;42:1192–200.

Lefaucheur JP. Why image-guided navigation becomes essential in the practice of transcranial magnetic stimulation. Neurophysiol Clin. 2010;40:1–5.

Lefaucheur JP, André-Obadia N, Antal A, Ayache SS, Baeken C, Benninger DH, et al. Evidence-based guidelines on the therapeutic use of repetitive transcranial magnetic stimulation (rTMS). Clin Neurophysiol. 2014;125:2150–206.

Lefaucheur JP, Aleman A, Baeken C, Benninger DH, Brunelin J, Di Lazzaro V, et al. Evidence-based guidelines on the therapeutic use of repetitive transcranial magnetic stimulation (rTMS): an update (2014–2018). Clin Neurophysiol. 2020;131(2):474–528.

Lenoir C, Algoet M, Mouraux A. Deep continuous theta burst stimulation of the operculo-insular cortex selectively affects adelta-fiber heat pain. J Physiol. 2018;596:4767–87.

Lerner AJ, Wassermann EM, Tamir DI. Seizures from transcranial magnetic stimulation 2012–2016: results of a survey of active laboratories and clinics. Clin Neurophysiol. 2019;176:139–48.

Leukel C, Taube W, Beck S, Schubert M. Pathway-specific plasticity in the human spinal cord. Eur J Neurosci. 2012;35:1622–9.

Lisanby SH. Update on magnetic seizure therapy: a novel form of convulsive therapy. J ECT. 2002;18:182–8.

Lisanby SH, Luber B, Finck AD, Schroeder C, Sackeim HA. Deliberate seizure induction with repetitive transcranial magnetic stimulation in nonhuman primates. Arch Gen Psychiatry. 2001;58:199–200.

Lisanby SH, Luber B, Schlaepfer TE, Sackeim HA. Safety and feasibility of magnetic seizure therapy (MST) in major depression: randomized within-subject comparison with electroconvulsive therapy. Neuropsychopharmacology. 2003;28:1852–65.

Loo C, Martin D, Pigot M, Arul-Anandam P, Mitchell P, Sachdev P. Transcranial direct current stimulation priming of therapeutic repetitive transcranial magnetic stimulation: a pilot study. JECT. 2009;25:256–60.

Lu MK, Tsai CH, Ziemann U. Cerebellum to motor cortex paired associative stimulation induces bidirectional STDP-like plasticity in human motor cortex. Front Hum Neurosci. 2012;6:260.

Marg E, Rudiak D. Phosphenes induced by magnetic stimulation over the occipital brain: description and probable site of stimulation. Optom Vis Sci. 1994;71:301–11.

Mariorenzi R, Zarola F, Caramia MD, Paradiso C, Rossini PM. Non-invasive evaluation of central motor tract excitability changes following peripheral nerve stimulation in healthy humans. Electroencephalogr Clin Neurophysiol. 1991;81(2):90–101.

Mashour GA, Walker EE, Martuza RL. Psychosurgery: past, present, and future. Brain Res Brain Res Rev. 2005;48:409–19.

Menkes DL, Gruenthal M. Slow-frequency repetitive transcranial magnetic stimulation in a patient with focal cortical dysplasia. Epilepsia. 2000;41(2):240–2.

Misawa S, Kuwabara S, Shibuya K, Mamada K, Hattori T. Low-frequency transcranial magnetic stimulation for epilepsia partialis continua due to cortical dysplasia. J Neurol Sci. 2005;234(1–2):37–9.

Muller-Dahlhaus F, Ziemann U. Metaplasticity in human cortex. Neuroscientist. 2015;21:185–202.

Neville IS, Dos Santos AG, Almeida CC, Hayashi CY, Fontoura Solla DJ, Galhardoni R, et al. Evaluation of changes in preoperative cortical excitability by navigated transcranial magnetic stimulation in patients with brain tumor. Front Neurol. 2020;11:582262.

Ni Z, Kim SJ, Phielipp N, Ghosh S, Udupa K, Gunraj CA, Lee DJ. Pallidal deep brain stimulation modulates cortical excitability and plasticity. Ann Neurol. 2018;83:352–62.

Nord C. The effect of frontoparietal paired associative stimulation on decisionmaking and working memory. Cortex. 2019;117:266.

Oberman L, Edwards D, Eldaief M, Pascual-Leone A. Safety of theta burst transcranial magnetic stimulation: a systematic review of the literature. J Clin Neurophysiol. 2011;28(1):67–74.

Obermann LM, Pascual-Leone A. Report of seizure induced by continuous theta burst stimulation. Brain Stim. 2009;2(4):246–7.

Penfield W, Jasper H. Epilepsy and the functional anatomy of the human brain. Boston: Little, Brown; 1954.

Pereira LS, Müller VT, da Mota GM, Rotenberg A, Fregni F. Safety of repetitive transcranial magnetic stimulation in patients with epilepsy: a systematic review. Epilepsy Behav. 2016;57(Pt A):167–76.

Peterchev AV, Wagner TA, Miranda PC, Nitsche MA, Paulus W, Lisanby SH. Fundamentals of transcranial electric and magnetic stimulation dose: definition, selection, and reporting practices. Brain Stimul. 2012;5:435–53.

Peterchev AV, Deng Z-D, Goetz SM. Advances in transcranial magnetic stimulation technology. In: Reti IM, editor. Brain stimulation: methodologies and interventions. Hoboken: Wiley Blackwell; 2015a. p. 165–89.

Peterchev AV, Sikes-Keilp C, Rosa MA, Lisanby SH. Re-evaluating the electroconvulsive therapy stimulus: frequency and directionality. Biol Psychiatry. 2015b;77:23S.

Picht T, Mularski S, Kuehn B, Vajkoczy P, Kombos T, Suess O. Navigated transcranial magnetic stimulation for preoperative functional diagnostics in brain tumor surgery. Neurosurgery. 2009;65:93–8.

Pierantozzi M, Palmieri MG, Mazzone P, Marciani MG, Rossini PM, Stefani A, et al. Deep brain stimulation of both subthalamic nucleus and internal globus pallidus restores intracortical inhibition in Parkinson's disease paralleling apomorphine effects: a paired magnetic stimulation study. Clin Neurophysiol. 2002;113(1):108–13.

Rachid F. Safety and efficacy of theta-burst stimulation in the treatment of psychiatric disorders: a review of the literature. J Nerv Ment Dis. 2017;205:823–39.

Rizzo V, Aricò I, Mastroeni C, Morgante F, Liotta G, Girlanda P, et al. Dopamine agonists restore cortical plasticity in patients with idiopathic restless legs syndrome. Mov Disord. 2009;24:710–5.

Ribolsi M, Lisi G, Ponzo V, Siracusano A, Caltagirone C, Niolu C, et al. Left hemispheric breakdown of LTP-like cortico-cortical plasticity in schizophrenic patients. Clin Neurophysiol. 2017;128(10):2037-2042.

Rossi S, Hallett M, Rossini PM, Pascual-Leone A. Safety of TMS consensus group. Safety, ethical considerations, and application guidelines for the use of transcranial magnetic stimulation in clinical practice and research. Clin Neurophysiol. 2009;120:2008–39.

Rossi S, Hallett M, Rossini PM, Pascual-Leone A. Screening questionnaire before TMS: An update. Clin Neurophysiol. 2011;122(8):1686.

Rossi S, Antal A, Bestmann S, Bikson M, Brewer C, Brockmöller J, et al. Safety and recommendations for TMS use in healthy subjects and patient populations, with updates on training, ethical and regulatory issues: expert guidelines. Clin Neurophysiol. 2021;132:269–306.

Rossini PM, Di Stefano E, Stanzione P. Nerve impulse propagation along central and peripheral fast conducting motor and sensory pathways in man. Electroencephalogr Clin Neurophysiol. 1985a;60(4):320–34.

Rossini PM, Marciani MG, Caramia M, Roma V, Zarola F. Nervous propagation along 'central' motor pathways in intact man: characteristics of motor responses to 'bifocal' and 'unifocal' spine and scalp non-invasive stimulation. Electroencephalogr Clin Neurophysiol. 1985b;61(4):272–86.

Rossini PM, Caramia MD, Zarola F. Mechanisms of nervous propagation along central motor pathways: noninvasive evaluation in healthy subjects and in patients with neurological disease. Neurosurgery. 1987;20(1):183–91.

Rossini PM, Barker AT, Berardelli A, Caramia MD, Caruso G, Cracco RQ, et al. Non-invasive electrical and magnetic stimulation of the brain, spinal cord and roots: basic principles and procedures for routine clinical application. Report of an IFCN committee. Electroencephalogr Clin Neurophysiol. 1994;91:79–92.

Rossini PM, Burke D, Chen R, Cohen LG, Daskalakis Z, Di Iorio R, et al. Non-invasive electrical and magnetic stimulation of the brain, spinal cord, roots and peripheral nerves: basic principles and procedures for routine clinical and research application. An updated report from an I.F.C.N. Committee. Clin Neurophysiol. 2015;126(6):1071–107.

Ruohonen J, Karhu J. Navigated transcranial magnetic stimulation. Neurophysiol Clin. 2010;40:7–17.

Santiago-Rodríguez E, Cárdenas-Morales L, Harmony T, Fernández-Bouzas A, Porras-Kattz E, Hernández A. Repetitive transcranial magnetic stimulation decreases the number of seizures in patients with focal neocortical epilepsy. Seizure. 2008;17(8):677–83.

Schiavao LJV, Neville Ribeiro I, Yukie Hayashi C, Gadelha Figueiredo E, Russowsky Brunoni A, Jacobsen Teixeira M, et al. Assessing the capabilities of transcranial magnetic stimulation (TMS) to aid in the removal of brain tumors affecting the motor cortex: a systematic review. Neuropsychiatr Dis Treat. 2022;18:1219–35.

Sollmann N, Ille S, Hauck T, Maurer S, Negwer C, Zimmer C, et al. The impact of preoperative language mapping by repetitive navigated transcranial magnetic stimulation on the clinical course of brain tumor patients. BMC Cancer. 2015;15:261.

Stefan K, Kunesch E, Cohen LG, Benecke R, Classen J. Induction of plasticity in the human motor cortex by paired associative stimulation. Brain. 2000;123:572–84.

Steven MS, Pascual-Leone A. Transcranial magnetic stimulation and the human brain: an ethical evaluation. In: Illes J, editor. 21st century neuroethics: defining the issues in research, practice and policy. Oxford: Oxford University Press; 2006. p. 201–11.

Tarapore PE, Tate MC, Findlay AM, Honma SM, Mizuiri D, Berger MS, et al. Preoperative multimodal motor mapping: a comparison of magnetoencephalography imaging, navigated transcranial magnetic stimulation, and direct cortical stimulation. J Neurosurg. 2012;117:354–62.

Tarapore PE, Picht T, Bulubas L, Shin Y, Kulchytska N, Meyer B, et al. Safety and tolerability of navigated TMS for preoperative mapping in neurosurgical patients. Clin Neurophysiol. 2016;127(3):1895–900.

Tavakoli H, Heidarpanah A. Literature review of the efficacy of repetitive transcranial magnetic stimulation on epilepsy. Iran J Child Neurol. 2023;17(1):9–28.

Taylor JL, Martin PG. Voluntary motor output is altered by spike-timing-dependent changes in the human corticospinal pathway. J Neurosci. 2009;29:11708–16.

Tergau F, Neumann D, Rosenow F, Nitsche MA, Paulus W, Steinhoff B. Can epilepsies be improved by repetitive transcranial magnetic stimulation? Interim analysis of a controlled study. Suppl Clin Neurophysiol. 2003;56:400–5.

Theodore WH. Transcranial magnetic stimulation in epilepsy. Epilepsy Curr. 2003;3(6):191–7.

Tikka SK, Nizamie SH, Venkatesh Babu GM, Aggarwal N, Das AK, Goyal N. Safety and efficacy of adjunctive theta burst repetitive transcranial magnetic stimulation to right inferior parietal lobule in schizophrenia patients with first-rank symptoms: a pilot. Exploratory Study J ECT. 2017;33(1):43–51.

Udupa K, Bahl N, Ni Z, Gunraj C, Mazzella F, Moro E, Chen R. Cortical plasticity induction by pairing subthalamic nucleus deep-brain stimulation and primary motor cortical transcranial magnetic stimulation in Parkinson's disease. J Neurosci. 2016;36:396–404.

Veniero D, Ponzo V, Koch G. Paired associative stimulation enforces the communication between interconnected areas. J Neurosci. 2013;33:13773–83.

Walton D, Spencer DC, Nevitt SJ, Michael BD. Transcranial magnetic stimulation for the treatment of epilepsy. Cochrane Database Syst Rev. 2021;4

Wassermann EM. Risk and safety of repetitive transcranial magnetic stimulation: report and suggested guidelines from the international workshop on the safety of repetitive transcranial magnetic stimulation, June 5–7, 1996. Electroencephalogr Clin Neurophysiol. 1998;108:1–16.

White PF, Amos Q, Zhang Y, Stool L, Husain MM, Thornton L, et al. Anesthetic considerations for magnetic seizure therapy: a novel therapy for severe depression. Eur PMC. 2006;103:76–80.

Wolpe PR. Treatment, enhancement, and the ethics of neurotherapeutics. Brain Cogn. 2002;50:387–95.

Wolters A, Schmidt A, Schramm A, Zeller D, Naumann M, Kunesch E. Timing-dependent plasticity in human primary somatosensory cortex. J Physiol Lond. 2005;565:1039–52.

Zibman S, Daniel E, Alyagon U, Etkin A, Zangen A. Interhemispheric cortico-cortical paired associative stimulation of the prefrontal cortex jointly modulates frontal asymmetry and emotional reactivity. Brain Stimul. 2019;12:139–47.

Part II

nTMS Motor Mapping

nTMS Motor Mapping: Basic Principles and Clinical Use

6

Matthew T. Muir and Sujit S. Prabhu

6.1 Rationale of Functional Mapping

Gliomas constitute a group of intrinsic brain neoplasms for which no curative therapies currently exist. Cytoreductive surgery remains the cornerstone of glioma treatment, and while surgical resection alone is not sufficient to halt tumor progression, the importance of maximizing the extent of resection (EOR) has been increasingly recognized for low-grade gliomas (LGG) as well as high-grade gliomas (HGG) (Li et al. 2016; Lacroix et al. 2001; Sanai et al. 2011; Jakola et al. 2012, Brown et al. 2016). However, preservation of neurological function postoperatively is equally important because it influences not only quality of life but also survival (McGirt et al. 2009). Many tools are available at neurosurgeons' disposal today to help maximize the EOR. The addition of only one modality, such as diffusion tensor imaging (DTI) of the corticospinal tract (CST) preoperatively or intraoperative DES and mapping, can improve the rates of complete resection by 30–40% while preserving neurological function (De Witt Hamer et al. 2012; Wu et al. 2007). Preoperative mapping tools like nTMS can also help to identify eloquent cortical areas and subcortical pathways and better prepare the surgeon to achieve a maximal, safe tumor resection.

Tumors within or close to eloquent motor areas, particularly the precentral gyrus, always involve a challenge between the EOR and the preservation of motor function. It is well known that gross total resection (GTR) not only improves quality of life but also overall and progression free survival in patients with glioma (Li et al.

M. T. Muir
Department of Neurosurgery, Division of Surgery, The University of Texas MD Anderson, Houston, TX, USA
e-mail: mmuir@mdanderson.org

S. S. Prabhu (✉)
Cancer Center, Houston, TX, USA
e-mail: sprabhu@mdanderson.org

S. M. Krieg, T. Picht (eds.), *Navigated Transcranial Magnetic Stimulation in Neurosurgery*, https://doi.org/10.1007/978-3-031-97155-6_6

2016; Lacroix et al. 2001). For this reason, although advances in diagnostic imaging have enabled the tumor margins to be more clearly defined, anatomic criteria alone are not always reliable in accurately localizing areas of motor function. Together, anatomical and functional information about the cortical and subcortical areas at risk is crucial for avoiding focal neurological deficits during tumor surgery (De Witt Hamer et al. 2012). Intraoperative direct electrical stimulation (DES) remains the gold standard for motor mapping; however, noninvasive preoperative methods are becoming increasingly accurate and useful to guide surgery, especially for presurgical planning. Such information can not only be used to plan the EOR but also the least invasive trajectory to the tumor, thereby effectively reducing the associated risks of operating in eloquent areas.

6.2 nTMS Motor Mapping in Neurosurgery

6.2.1 General Considerations

The Food and Drug Administration (FDA) in the United States approved image-guided nTMS (Ref. K091457, November 23, 2009) as a validated device for presurgical functional mapping of the motor cortex. In contrast to other presurgical mapping techniques like functional MRI (fMRI) and magnetoencephalography (MEG), which identify cortical areas activated during task execution, nTMS enables the surgeon to determine the eloquent areas that are required for a given function (also *see* Chap. 2). In recent years, nTMS has taken a more prominent place in the field of presurgical functional mapping, as a noninvasive approach by mimicking DES by probing and stimulating the cortex transcranially. The integrity of the CST can hence be measured using this technique. Upon coregistration of the patient's MRI scan with predetermined landmarks, a cortical electrical current is induced by applying a transcranial magnetic field to a specific area of the cortex under real-time navigation (also *see* Chap. 1) (Picht et al. 2016). This allows the direct activation of the cortex below the area of stimulation, resulting in a corresponding motor evoked potential (MEP).

In addition to the accuracy compared to DES, the use of nTMS preoperatively has been correlated with improved patient outcomes, resulting in 16–17% greater rates of GTR and longer progression-free survival of patients with LGG of 22.4 vs. 15.4 months in the control group (Krieg et al. 2014; Picht et al. 2016). Finally, low cost and relative ease of use contribute to the increasing use of this modality in preoperative planning.

6.2.2 Methodology and Execution

The full protocol of motor mapping by nTMS is outlined in Chap. 1. The parameters recorded during the nTMS procedure include peeling depth (peeling depth from the scalp surface represents the depth of cortical penetration of the magnetic field), latency, and the amplitude of the rMT. The peeling depth is set on a case-by-case basis to best reveal the cortical anatomy and its relation with the tumor. Surface electrodes are attached to the subject's extremity muscles. The following muscles are currently recommended for EMG recording in a standard setup, which can then be changed depending on the tumor's location:

Upper extremity:

- Abductor pollicis brevis (APB)
- First dorsal interosseous muscle (FDI)
- Abductor digiti minimi (ADM)
- Flexor carpi radialis muscle (FCR)
- Extensor carpi radialis muscle (ECR)
- Biceps brachii muscle (BIC)

Lower extremity:

- Tibialis anterior (TA)
- Plantar toe flexor muscle
- Abductor hallucis muscle
- Vastus medialis muscle
- Extensor hallucis longus muscle

Face:

- OrO
- Mentalis muscle (MEN)
- Inferior longitudinal muscle of the tongue

When performing studies on nTMS motor mapping that include intraoperative DES data, it is important to use the same muscles for EMG recording for nTMS and DES to maintain consistent measurements of stimulation response between preoperative and intraoperative modalities. The most important concern during motor mapping with nTMS is consistency. Proper technique and careful analysis of the hotspots will ensure reliable results.

6.2.3 Analysis

The nTMS technique of cortical mapping is performed by stimulating multiple points on the scalp of the patient corresponding to the possible cortical motor regions and by concomitantly recording MEPs for each stimulation point. The stimulation intensity is usually maintained just above the rMT to limit the neuronal excitation zone and to obtain the greatest functional anatomical accuracy (also *see* Chap. 1). To facilitate interpretation, the motor response to each stimulation spot is usually visualized with a color code corresponding to the value of the MEP amplitude. The latency for each MEP has also to be documented and considered during analysis. The resulting nTMS maps generally appear as colored markers corresponding to all of the stimulation spots placed on the individual patient's MRI.

6.2.4 Validation

For nTMS, the patients' cortical representation areas for individual muscles were mapped noninvasively by nTMS before the surgery and compared to locations obtained invasively during surgery by DES of the cortex (Picht et al. 2011). In the study of Picht et al. (2011), the locations where the largest MEPs were elicited from the target muscles (= "hotspots") were mapped with nTMS and DES. It is important to note that nTMS hotspots were not given to the operating surgeon for the DES guidance to guarantee a real-life clinical situation with DES. The operating surgeon and surgery planning team were only informed about the location of motor function around the tumor; they were not part of the team doing the nTMS mapping. After the mappings, DES hotspot coordinates were imported to the original nTMS coordinate system and displayed in the nTMS 3D navigation view to determine the overlap of the DES and nTMS hotspots. The nTMS and DES hotspots were located in the same gyrus in all patients. The mean ± standard error of mean (SEM) distance between the nTMS and DES hotspots was 7.8 ± 1.2 mm for APB (n = 15) and 7.1 ± 0.9 mm for the TA muscle (n = 8). DES mapping was performed under routine operation room conditions under time pressure and clinical demands; therefore, occasionally only a small number of DES stimulations were performed. In such cases, the distance between nTMS and DES hotspots was increased substantially (r = −0.86 for APB). After excluding the cases with fewer than 15 DES APB responses, the mean ± SEM distance between the hotspots was only 4.7 ± 1.1 mm for APB (n = 8).

Subsequent studies in patients with tumor like that of Forster et al. have compared nTMS and fMRI with the intraoperative DES (Forster et al. 2011). The authors determined average nTMS-to-DES and fMRI-to-DES distances by comparing distances between DES coordinates and nTMS coordinates and the centroid of fMRI activation coordinates for each muscle. They reported significantly smaller distances (mean ± standard deviation; SD) from nTMS to DES (10.5 ± 5.7 mm) than those from fMRI to DES (15.0 ± 7.6 mm; $p < 0.05$). In their conclusion, the clinical team notes that nTMS is "more precise than fMRI when correlated with intraoperative DES" and nTMS "anticipates information usually only enabled by direct cortical stimulation." Krieg et al. (2012) compared nTMS and DES by measuring the borders between positive and negative stimulation points on axial slices by using recalibrated screenshots and Brainlab iPlan Net Cranial 3.0.1 (Krieg et al. 2012). They reported a mean distance (mean ± SD) between the borders for nTMS versus DES of 4.4 ± 3.4 mm (range 1.9–9.2 mm). In addition, the investigators evaluated the difference between borders delineating M1 according to BOLD data on fMRI studies and nTMS. With the 9.8 ± 8.5 mm distance (mean ± SD) between nTMS and fMRI (range 5.3–39.7 mm) for the upper extremity and 14.7 ± 12.4 mm (range 8.4–33.5 mm) for the lower extremity, they concluded that nTMS correlates well with intraoperative DES, whereas nTMS and fMRI differed significantly from each other. Finally, Takahashi et al. (2013) reviewed multiple articles on the correlation of nTMS with DES in patients with Rolandic brain tumors. They reported a mean distance between both methods of 6.18 mm (Takahashi et al. 2013). This value nicely corresponds to the calculated accuracy of the used system (also *see* Chap. 1, Table 1.1).

Recent studies have also investigated the clinical utility of cortical-subcortical motor network reconstructions based on TMS and DTI tractography. One study found that TMS points without connecting white matter tracts identified at a particular fractional anisotropic threshold are safely resectable while TMS points with connecting tracts cause permanent motor deficits when surgically damaged (Muir et al. 2022a). Another demonstrated that TMS produces corticospinal reconstructions that more accurately correspond to postoperative neurological status than corticospinal reconstructions generated by traditional anatomical seeding with the precentral gyrus (Muir et al. 2022b). This study found that the precentral gyrus is frequently safely resectable due to neuroplastic reorganization of cortical motor sites. TMS seeded tractography has also been compared to fMRI seeded tractography. This analysis showed that while TMS can capture intraparenchymal cortical neuroplastic redistribution, fMRI BOLD signal consistently localizes to the central sulcal vasculature regardless of cortical reorganization (Muir et al. 2023).

Studies have also demonstrated the impact of TMS and DTI tractography on surgical outcomes for patients with motor eloquent tumors. These studies have reported modification of the initial surgical strategy in 25–70% of cases, leading to quicker and more extensive resections and favorable long-term outcomes related to EOR in patients who have had preoperative nTMS mapping, compared to the control group, in which no nTMS mapping was performed (Krieg et al. 2012; Picht et al. 2012, 2013, 2016; Rizzo et al. 2014). A recent study found that rates of gross total resection, overall survival, and progression-free survival are all significantly improved in patients who had preoperative TMS seeded tractography versus a matched control group (Lucas et al. 2022) (Fig. 6.1).

6.3 Clinical Application

The cortical areas subserving motor function, i.e., not resectable, versus nonfunctional areas, i.e., resectable, are easily determined upon visual inspection of these maps. These functional data merged with the anatomical MRI scans can be transferred to the neuronavigation system used in the operating room to guide surgery and DES, in case the latter is indicated (also see Chap. 3).

As a result of good patient compliance, especially for motor mapping, nTMS has been utilized in longitudinal studies to assess cortical plasticity in patients harboring tumors near the perirolandic area. In one such study, Conway et al. described the utility of assessing cortical plasticity in both HGGs and LGGs using a mathematical model design (Conway et al. 2017).

The anatomical and functional data provided by preoperative nTMS have shown a very good correlation with those provided by DES performed intraoperatively, with a difference in the location of the motor hotspot that remains less than 15 mm between the two techniques, varying between 4 and 8 mm on average in several studies (Krieg et al. 2012; Paiva et al. 2012; Picht et al. 2009, 2011). Moreover,

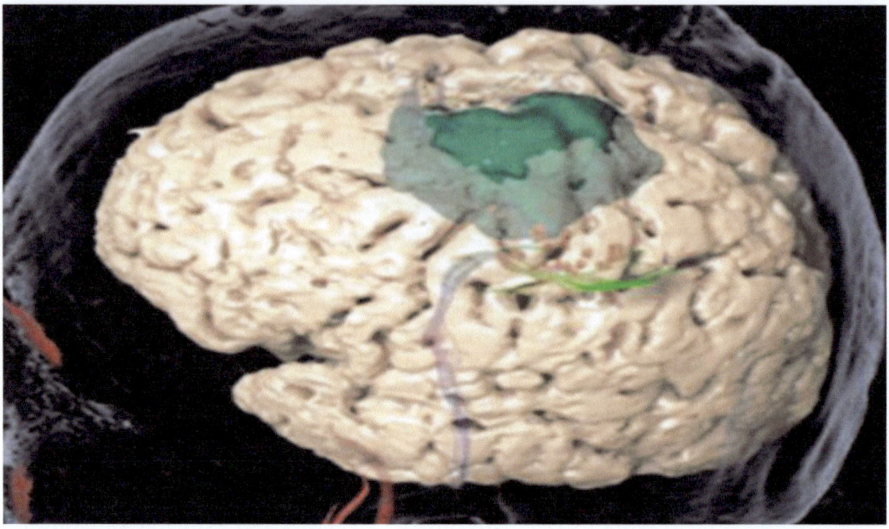

Fig. 6.1 Peritumoral functional anatomy. The nTMS data (red) is used to visualize corticospinal fibers (blue) in a glioma (green) infiltrating the motor cortex. This clearly improves the identification of the proper peritumoral functional anatomy and also helps our patients to understand the objectives and risks of surgery

compared to fMRI, which is widely utilized for identifying the motor cortex, nTMS has better spatial and temporal resolution and provides an immediate result, without recourse to complex postprocessing analysis (also *see* Chap. 2). While fMRI relies on neurovascular coupling, nTMS establishes a clearer causal relationship between the stimulation point and observed response (e.g., a MEP). The fMRI signal also depends on the local tumor milieu, including blood flow, edema, and mass effect. While fMRI requires active participation of the individual since it is task based, nTMS is more of a passive modality and presents fewer restrictions in clinical practice; this is even truer in children. Tarapore et al. also showed a good correlation among the motor map information provided by nTMS, MEG, and DES (Tarapore et al. 2012). The limitations for the use of MEG include its cost and the time involved in data processing.

Intraoperative mapping of the lower extremity is challenging for anatomical reasons, as the foot/leg motor homunculus is in close proximity to the superior sagittal sinus along the interhemispheric fissure, making its localization difficult to confirm by intraoperative DES. Also, there is no sulcus between the foot/leg motor homunculus and the supplementary motor area (SMA), making the distinction between the two areas difficult to determine on routine MRI (Fisicaro et al. 2015). Therefore, mapping in the preoperative setting is important for planning surgical intervention.

A case report showed that TMS seeded tractography has potential to not only optimize patient selection and surgical planning but also inform important intraoperative surgical decisions. The patient had a glioblastoma in the paracentral lobule that did not appear amenable to gross total resection based on anatomical criteria.

While TMS points spanned over the tumor, white matter tracts only connected to the TMS points most distant from the tumor. The surgical strategy was changed to pursue a gross total resection which was safely achieved (Muir et al. 2022c).

6.4 Conclusion

In summary, nTMS for cortical mapping of motor areas is a very robust tool to identify and preserve motor function. The information obtained can be applied to both surgical preplanning and intraoperative orientation to minimize neurological injury while maximizing the extent of resection for tumors within or adjacent to motor areas of the brain.

References

Brown TJ, Brennan MC, Li M, Church EW, Brandmeir NJ, Rakszawski KL, et al. Association of the Extent of Resection With Survival in Glioblastoma: A Systematic Review and Meta-analysis. JAMA Oncol. 2016;2(11):1460–9.

Conway N, Wildschuetz N, Moser T, Bulubas L, Sollmann N, Tanigawa N, et al. Cortical plasticity of motor-eloquent areas measured by navigated transcranial magnetic stimulation in patients with glioma. J Neurosurg. 2017;127(5):981–91.

De Witt Hamer PC, Robles SG, Zwinderman AH, Duffau H, Berger MS. Impact of intraoperative stimulation brain mapping on glioma surgery outcome: a meta-analysis. J Clin Oncol. 2012;30(20):2559–65.

Fisicaro RA, Jiao RX, Stathopoulos C, Petrovich Brennan NM, Peck KK, Holodny AI. Challenges in Identifying the Foot Motor Region in Patients with Brain Tumor on Routine MRI: Advantages of fMRI. AJNR Am J Neuroradiol. 2015;36(8):1488–93.

Forster MT, Hattingen E, Senft C, Gasser T, Seifert V, Szelényi A. Navigated transcranial magnetic stimulation and functional magnetic resonance imaging: advanced adjuncts in preoperative planning for central region tumors. Neurosurgery. 2011;68(5):1317–24; discussion 24–5.

Jakola AS, Myrmel KS, Kloster R, Torp SH, Lindal S, Unsgård G, et al. Comparison of a strategy favoring early surgical resection vs a strategy favoring watchful waiting in low-grade gliomas. Jama. 2012;308(18):1881–8.

Krieg SM, Shiban E, Buchmann N, Gempt J, Foerschler A, Meyer B, et al. Utility of presurgical navigated transcranial magnetic brain stimulation for the resection of tumors in eloquent motor areas. J Neurosurg. 2012;116(5):994–1001.

Krieg SM, Sabih J, Bulubasova L, Obermueller T, Negwer C, Janssen I, et al. Preoperative motor mapping by navigated transcranial magnetic brain stimulation improves outcome for motor eloquent lesions. Neuro Oncol. 2014;16(9):1274–82.

Lacroix M, Abi-Said D, Fourney DR, Gokaslan ZL, Shi W, DeMonte F, et al. A multivariate analysis of 416 patients with glioblastoma multiforme: prognosis, extent of resection, and survival. J Neurosurg. 2001;95(2):190–8.

Li YM, Suki D, Hess K, Sawaya R. The influence of maximum safe resection of glioblastoma on survival in 1229 patients: Can we do better than gross-total resection? J Neurosurg. 2016;124(4):977–88.

McGirt MJ, Mukherjee D, Chaichana KL, Than KD, Weingart JD, Quinones-Hinojosa A. Association of surgically acquired motor and language deficits on overall survival after resection of glioblastoma multiforme. Neurosurgery. 2009;65(3):463–9; discussion 9-70.

Muir M, Prinsloo S, Michener H, Traylor JI, Patel R, Gadot R, et al. TMS Seeded Diffusion Tensor Imaging Tractography Predicts Permanent Neurological Deficits. Cancers (Basel). 2022a;14(2).

Muir M, Prinsloo S, Michener H, Shetty A, de Almeida Bastos DC, Traylor J, et al. Transcranial magnetic stimulation (TMS) seeded tractography provides superior prediction of eloquence compared to anatomic seeded tractography. Neurooncol Adv. 2022b;4(1):vdac126.

Muir M, Prinsloo S, Traylor JI, Patel R, Ene C, Tummala S, et al. Transcranial magnetic stimulation tractography and the facilitation of gross total resection in a patient with a motor eloquent glioblastoma: illustrative case. J Neurosurg Case Lessons. 2022c;3(20).

Muir M, Gadot R, Prinsloo S, Michener H, Traylor J, Athukuri P, et al. Comparative study of preoperative functional imaging combined with tractography for prediction of iatrogenic motor deficits. J Neurosurg. 2023;139(1):65–72.

Paiva WS, Fonoff ET, Marcolin MA, Cabrera HN, Teixeira MJ. Cortical mapping with navigated transcranial magnetic stimulation in low-grade glioma surgery. Neuropsychiatr Dis Treat. 2012;8:197–201

Picht T, Mularski S, Kuehn B, Vajkoczy P, Kombos T, Suess O. Navigated transcranial magnetic stimulation for preoperative functional diagnostics in brain tumor surgery. Neurosurgery. 2009;65(6 Suppl):93–8; discussion 8–9.

Picht T, Schmidt S, Brandt S, Frey D, Hannula H, Neuvonen T, et al. Preoperative functional mapping for rolandic brain tumor surgery: comparison of navigated transcranial magnetic stimulation to direct cortical stimulation. Neurosurgery. 2011;69(3):581–8; discussion 8.

Picht T, Strack V, Schulz J, Zdunczyk A, Frey D, Schmidt S, et al. Assessing the functional status of the motor system in brain tumor patients using transcranial magnetic stimulation. Acta Neurochir (Wien). 2012;154(11):2075–81.

Picht T, Krieg SM, Sollmann N, Rösler J, Niraula B, Neuvonen T, et al. A comparison of language mapping by preoperative navigated transcranial magnetic stimulation and direct cortical stimulation during awake surgery. Neurosurgery. 2013;72(5):808–19.

Picht T, Frey D, Thieme S, Kliesch S, Vajkoczy P. Presurgical navigated TMS motor cortex mapping improves outcome in glioblastoma surgery: a controlled observational study. J Neurooncol. 2016;126(3):535–43.

Rizzo V, Terranova C, Conti A, Germanò A, Alafaci C, Raffa G, et al. Preoperative functional mapping for rolandic brain tumor surgery. Neurosci Lett. 2014;583:136–41.

Sanai N, Polley MY, McDermott MW, Parsa AT, Berger MS. An extent of resection threshold for newly diagnosed glioblastomas. J Neurosurg. 2011;115(1):3–8.

Tarapore PE, Tate MC, Findlay AM, Honma SM, Mizuiri D, Berger MS, et al. Preoperative multimodal motor mapping: a comparison of magnetoencephalography imaging, navigated transcranial magnetic stimulation, and direct cortical stimulation. J Neurosurg. 2012;117(2):354–62.

Takahashi S, Vajkoczy P, Picht T. Navigated transcranial magnetic stimulation for mapping the motor cortex in patients with rolandic brain tumors. Neurosurg Focus. 2013;34(4):E3.

Weiss Lucas C, Faymonville AM, Loução R, Schroeter C, Nettekoven C, Oros-Peusquens AM, et al. Surgery of Motor Eloquent Glioblastoma Guided by TMS-Informed Tractography: Driving Resection Completeness Towards Prolonged Survival. Front Oncol. 2022;12:874631.

Wu JS, Zhou LF, Tang WJ, Mao Y, Hu J, Song YY, et al. Clinical evaluation and follow-up outcome of diffusion tensor imaging-based functional neuronavigation: a prospective, controlled study in patients with gliomas involving pyramidal tracts. Neurosurgery. 2007;61(5):935–48; discussion 48-9

nTMS-Based DTI Fiber Tracking of Motor Pathways

7

Giovanni Raffa and Alfredo Conti

7.1 Introduction

Surgery of brain lesions located in or close to the motor pathway is a neurosurgical challenge because of the high risk of postoperative motor deficits causing a decline of postoperative quality of life. During surgery, intraoperative neurophysiological mapping and monitoring (IONM) is the gold standard technique for the identification and preservation of the motor pathway, thus improving patients' motor and oncological outcome (Bello et al. 2014; Berger and Ojemann 1992; Sanai and Berger 2010). However, nowadays navigated transcranial magnetic stimulation (nTMS) can provide a reliable preoperative neurophysiological mapping of the primary motor cortex (M1) that can be used to plan the best surgical strategy to achieve the maximal safe resection of the lesions, before entering the operative theater (Frey et al. 2014a; Picht et al. 2009; Takahashi et al. 2013; Krieg et al. 2014, 2012a, 2013; Raffa et al. 2019a, b, c). Indeed, nTMS motor mapping allows for the preoperative analysis of the spatial relationship between M1 and the lesion, as well as for the stratification of patients according to the risk of developing postoperative permanent motor deficits (Rosenstock et al. 2017a). It may also help surgeons to distinguish between true motor-eloquent lesions—which are very close or infiltrate M1—from false motor-eloquent lesions—which are at a safe distance from M1—thus also providing indications to the necessity of using IONM or not during surgery (Raffa et al. 2018).

G. Raffa (✉)
Division of Neurosurgery, BIOMORF Department, University of Messina, Messina, Italy
e-mail: giovanni.raffa@unime.it

A. Conti
IRCCS Istituto delle Sceinze Neurologiche di Bologna, Alma Mater Studiorium Università di Bologna, Bologna, Italy
e-mail: alfredo.conti2@unibo.it

S. M. Krieg, T. Picht (eds.), *Navigated Transcranial Magnetic Stimulation in Neurosurgery*, https://doi.org/10.1007/978-3-031-97155-6_7

However, as a matter of fact, postoperative motor deficit can derive from surgical damage to M1, but more often from injury to the descending motor pathway that is the corticospinal tract (CST). The preoperative identification of the CST is therefore mandatory to reliably predict the risk of damage to the motor pathway during the surgical resection of the lesion. That could be achieved performing preoperative CST fiber tracking (FT). This technique provides a time efficient, noninvasive, both qualitative and quantitative method to study brain connections (Catani et al. 2002; Catani and Thiebaut de Schotten 2008) that represents a valid in vivo alternative to classic fiber dissection studies in cadaver labs. In fact, although operator dependent, prone to several artifacts, with low resolution and persistent technical limitations, FT can be repeatedly applied in large populations and correlated with behavioral and other functional measures (Basser 1995; Goebell et al. 2006; Minati et al. 2007; Sarubbo et al. 2015; Hakulinen et al. 2012)

The combination of the nTMS neurophysiological preoperative mapping of M1 with the FT of the CST provides a reliable 3D reconstruction of the whole motor pathway (M1+CST) that can be used during the preoperative period to analyze the spatial relationship with lesions and stratify patients according to surgical risks, but also during surgery to guide IONM and surgical resection, while preserving the integrity of the pathway (Raffa et al. 2019b; Rosenstock et al. 2017a; Sollmann et al. 2018).

To date, several FT techniques are available for the reconstruction of the CST: they are based on different algorithms, being the *deterministic* and *probabilistic* ones the most diffused (Mandelli et al. 2014). While the *probabilistic* algorithm is the most used approach in neuroscience, the *deterministic* algorithm is probably the most diffused in clinical applications—including preoperative planning in neurosurgery—because of its simplicity and wide implementations into the neuronavigation systems used in the operating theater (Mandelli et al. 2014; Kupper et al. 2015). However, in the last few years also *probabilistic* approaches have been implemented into neurosurgical neuronavigation systems and begin to be more frequently used also in the clinical practice. There algorithms, despite being implemented by more complex software, seems to be more reliable because of their ability to solve some critical issue specific of the deterministic ones, such us the possibility to visualize crossing fibers in a single voxel (Mandelli et al. 2014; Kupper et al. 2015; Abhinav et al. 2014; Toselli et al. 2017). In fact, there is a significant number of publications in the last decades in which fiber streamlines extracted with *deterministic* FT and indicated as the CST, do not respect completely the classical course of this tract, thus making the data obtained potentially incorrect and not reproducible (Pujol et al. 2015; Kristo et al. 2013; Wakana et al. 2007 (Mandelli et al. 2014). Currently, FT using *probabilistic* algorithms—e.g., constrained spherical deconvolution (CSD)—seems to provide a more reliable reconstruction of the CST (Mandelli et al. 2014; Kupper et al. 2015; Sheng et al. 2021).

7.2 Rationale Behind nTMS-Based Fiber Tracking

Regardless the reconstruction algorithm used, the main drawback of FT is the choice of anatomical landmarks for the computation of the CST. Several papers reported the use of different seeding regions of interest (ROIs) for the FT of the CST, including M1, the internal capsule, the cerebral peduncle (Catani et al. 2002; Catani and Thiebaut de Schotten 2008; Kwon et al. 2011; Weiss et al. 2015; Niu et al. 2016; Holodny et al. 2005). The choice of different anatomical landmarks as seeding ROIs and/or ROI malpositioning makes FT, including that of the CST, an operator-dependent technique, thus reducing its reliability and reproducibility (Niu et al. 2016). Moreover, it has been widely demonstrated that anatomical landmarks do not necessarily correspond to the functional organization of neuronal networks, including the motor pathway. This is particularly evident in patients with brain tumor, in which the presence of the tumor itself can lead to a plastic reorganization of the motor network, especially at a cortical level, causing a mismatch between anatomical landmarks and functional pathways (Niu et al. 2016; Takahashi et al. 2012; Rizzo et al. 2021). Therefore, the FT of the CST usually does not provide any information about the functional organization of different fiber bundles composing the CST itself. Indeed, reconstructing the somatotopic organization of the corticospinal tract, distinguishing between fiber bundles from different cortical representation areas of different muscles, as well as the visualization of the most lateral segments of the CST remains a challenge for modern tractography methods (He et al. 2023). In particular, *deterministic* approaches usually provide the reconstruction of the "straight" component of the CST—which is usually originating from the cortical representation area of the lower limb muscles—while rarely detecting more lateral bundles originating from the upper limb M1 representation area, as well as the cortico-bulbar bundles originating from the M1 area corresponding to the face muscles, unless using appropriate cortical ROIs. Conversely, *probabilistic* tractography will be able to reproduce the most lateral portion of the CST—including cortico-bulbar bundles—more easily (Jenabi et al. 2015). However, regardless of the algorithm used, the functional value of the reconstruction is questionable because no verified functional information is used to perform the FT.

Such an issue can be overcome by combining the FT technique with functional neuroimaging, improving the reconstruction of the CST. nTMS represents the most accurate technique to provide a preoperative functional characterization of M1, being more reliable than fMRI and correlating with IOM findings (Takahashi et al. 2013; Krieg et al. 2012a; Picht et al. 2011a). The nTMS motor mapping, which provides the functional topographic organization of M1, can easily replace anatomical landmarks as seeding regions for FT. Such a strategy adds functional information to the FT process, thus resulting in the visualization of a more accurate reconstruction of the CST (Raffa et al. 2018). In recent years, several studies reported the feasibility and accuracy of the nTMS-based FT of the CST, also demonstrating that this strategy is more accurate than the standard FT performed using anatomical landmarks as seeding ROIs (Frey et al. 2014a; Rosenstock et al. 2017a; Raffa et al. 2018; Sollmann et al. 2018; Weiss et al. 2015; Rosenstock et al. 2017b;

Conti et al. 2014; Munnich et al. 2019). Moreover, nTMS-based FT improves the visualization of the CST in patients with brain tumor, in which anatomical distortion caused by mass effect and neuroplasticity phenomena induced by the tumor reduce the reliability of standard FT results (Weiss et al. 2015; Rosenstock et al. 2017b, 2021, 2022; Conti et al. 2014; Krieg et al. 2012b; Frey et al. 2012; Forster et al. 2015; Raffa et al. 2017; Belotti et al. 2021; Ivren et al. 2023). Eventually, the nTMS motor mapping provides the "somatotopic" organization of M1 that can be used as seeding ROI for the computation of different fiber bundles of the CST originating from the M1 areas where the motor representation of different muscles of the upper and the lower limb, as well of the face is located (Conti et al. 2014; Raffa et al. 2017).

Here we describe the technique for the "somatotopic" deterministic nTMS-based FT of the CST using as seeding ROIs the different areas of nTMS "somatotopic" M1 motor mapping. We report the accuracy of this technique in comparison to the standard FT technique performed using only anatomical landmarks as seeding ROIs, as verified by subcortical direct electrical stimulation (DES) during surgical resection in a series of motor-eloquent brain lesions. Finally, we also report the clinical impact of using this nTMS-based strategy by reporting the extent of resection and the incidence of permanent postoperative motor deficits in series of 228 patients.

7.3 Materials and Methods

7.3.1 Patients

We prospectively collected data of patients operated on for brain tumors located in or close to the motor pathway (M1 and/or CST) at the Department of Neurosurgery of the University of Messina, Italy, between January 2012 and August 2019. We included in the study adult patients affected by intrinsic motor-eloquent brain tumors located superficially and/or deeply within 10 mm from M1 and/or CST. Exclusion criteria were age less than 18 years, and any contraindication to undergo MRI scan, and/or nTMS mapping (pacemaker, cochlear implants, non-MRI compatible prosthesis, intractable epilepsy, etc.).

7.3.2 Preoperative MRI

All patients underwent preoperative brain MRI scan using a 3 Tesla scanner (Achieva 3T, Philips Medical Systems, The Netherlands). T1-weighted, gadolinium-enhanced multiplanar reconstruction (MPR) (FS = 3, TR/repetition time = 8.1, TE/echo time = 3.7); 3D FLAIR-VISTA/fluid attenuated inversion recovery volumetric isotropic T2w acquisition (FS = 3, TR = 8000, TE = 331.5/7); diffusion-weighted imaging (DWI with 32 directions or gradients, with 80 slices for each direction; FS = 3, TR = 2383.9, TE = 51.9) for FT computation were obtained.

7.3.3 nTMS Mapping of M1 and DTI-FT of the CST

Preoperative mapping of M1 was performed using the NBS system 4.3 (Nexstim Plc, Helsinki, Finland), using a well-established protocol previously reported (Rizzo et al. 2014).

The M1 mapping was performed with a stimulation output of 110% of the resting motor threshold (RMT), increasing the intensity to up to 130% of the RMT for the lower limb mapping. MEPs were recorded by using standard electromyography electrodes (Neuroline 720; Ambu, Ballerup, Denmark) from three channels, choosing at least one muscle for each body segment (face, arm, leg) to obtain the somatotopic organization of the M1. Muscles usually used were FDI, abductor pollicis brevis (APB), extensor carpi radialis (ECR), biceps (BIC), tibialis anterior (TA), mentalis (MEN), and orbicularis oris (OrO).

Only patients in whom we were able to simultaneously obtain the nTMS somatotopic motor map face, arm, and leg muscles were enrolled in the study. At the end of the procedure, the nTMS "somatotopic" map of M1 was exported in a DICOM format.

The DWI sequences were imported together with the nTMS map of M1 and with the anatomical reference MRI examination (contrast-enhanced T1-weighted or FLAIR sequences) into a StealthStation® S7 or S8 neuronavigation system (Medtronic Navigation, Coal Creek Circle Louisville, CO, USA). The FT computation was performed using a *deterministic* algorithm for diffusion tensor imaging (DTI) FT on the StealthStation® S7 or S8 navigation system using the StealthViz® software (Medtronic Navigation, Coal Creek Circle Louisville, CO, USA). A multiple-ROI technique was used for the somatotopic DTI FT. The first ROI was placed at the anterolateral portion of the ipsilateral cerebral peduncle (Fig. 7.1a). The nTMS map of M1 was selected as second ROI. It was divided into three different areas corresponding to the cortical motor representation of muscles of the three different body segments (face, arm, leg muscles), to track the CST with its somatotopic organization. Each area corresponding to a segmentation area including all the positive spots provided by the nTMS for each somatotopic area (face, arm, and leg) according to the different muscular response recorded during stimulation, was individually selected as second ROI. For each patient, we obtained at least three ROIs, corresponding, respectively, to the somatotopic representation of the contralateral face (OrO and/or MEN muscles), arm (FDI and/or APB and/or BIC muscles and/or ECR), and leg (TA muscle) (Fig. 7.1b). Then, we separately computed the DTI tractography of each functionally different CST fiber bundle using the anterolateral portion of the ipsilateral cerebral peduncle as first ROI and each somatotopically different cortical areas of M1 as second ROI (Fig. 7.1c).

The *fiber assignment by continuous tracking* (FACT) algorithm was used with the following parameters were set: fractional anisotropy (FA) threshold = 0.20; vector step length = 1 mm; minimum fiber length = 20 mm; seed density = 1.0; maximal directional change = 45 degrees. For the nTMS-based DTI-FT the maximal directional change was 45 degrees for arm and leg fibers and ranged 75–90 degrees for face fibers. In order to visualize face motor fibers, directional change was

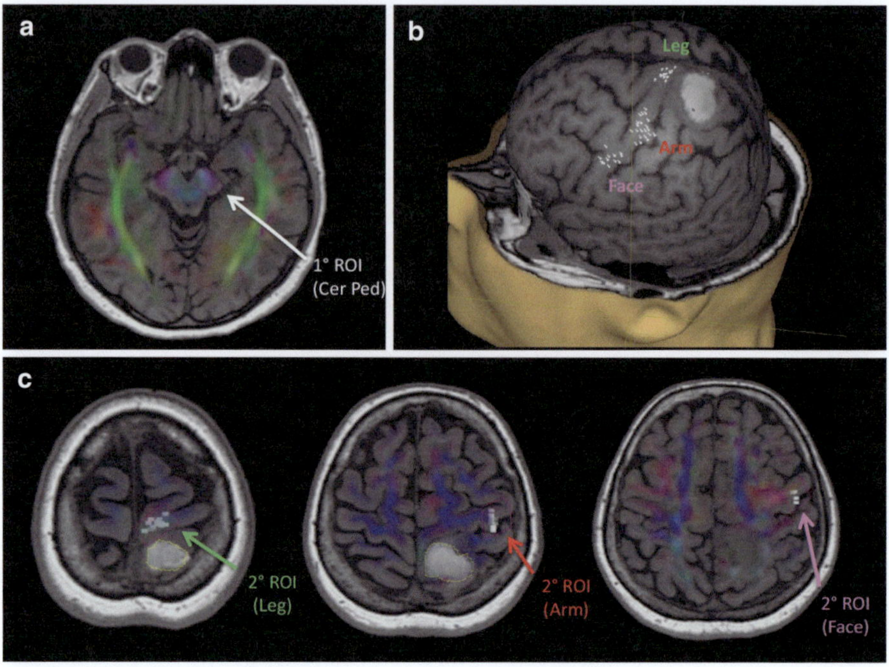

Fig. 7.1 nTMS-based DTI FT of the CST. nTMS-based DTI FT of the CST in a case of a post-rolandic lesion. (**a**) Directionally encoded color (DEC) map was co-registered with a reference MRI scan to select the first region of interest (ROI) corresponding to the anterolateral portion of the cerebral peduncle. (**b**) The somatotopic nTMS map of the motor cortex is divided into 3 ROIs, each corresponding to the cortical representation of leg, arm, and face muscles. (**c**) The 3 nTMS-based functional areas are separately used as second ROIs to compute 3 different components of the CST

increased progressively until the bundle could be identified. Both standard and nTMS-based DTI FT were performed by using the same computation parameters, except for angulation. In cases of excessive perilesional brain edema that hampers the DTI FT computation, the FA threshold values were progressively reduced in 0.02 steps until a sufficient number of fibers were visualized.

The simultaneous visualization of different fiber bundles obtained through the nTMS-based DTI FT represented the entire CST, reflecting its somatotopic organization (Fig. 7.2).

After surgery, standard DTI FT of the CST was performed by using a seeding ROI located in the anterolateral portion of the ipsilateral cerebral peduncle and a second ROI was placed over the entire area corresponding to M1 according to anatomical landmarks (i.e., omega sign).

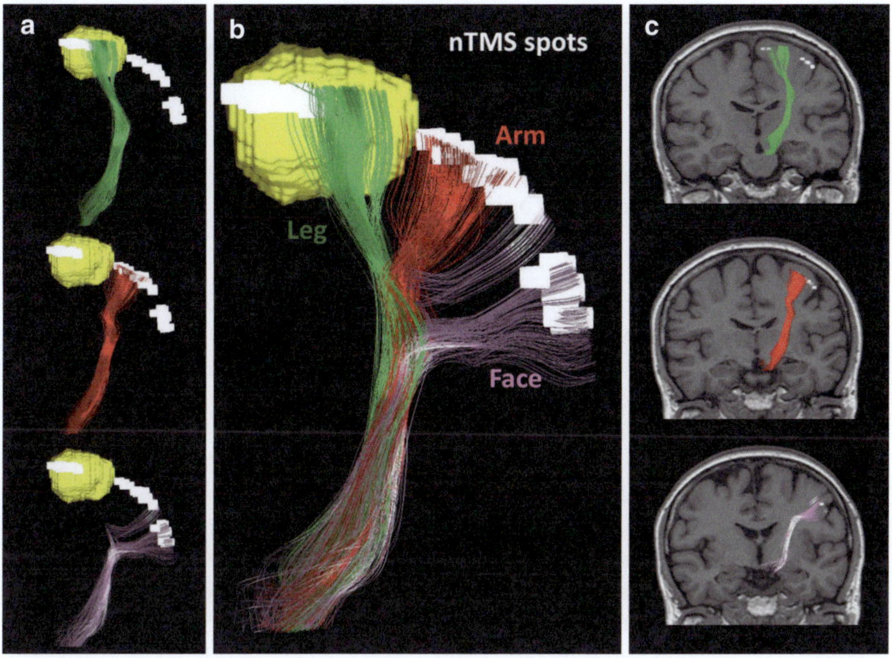

Fig. 7.2 3D reconstruction. nTMS-based somatotopic DTI-FT of the CST. (**a**) 3D reconstruction of each functionally different fiber bundles (leg = green, arm = red, face = pink) forming the CST with the corresponding nTMS map of M1 (white spots). (**b**) Coronal view of whole CST 3D somatotopic reconstruction. (**c**) Navigable tractography after co-registration with a reference anatomic MRI examination

7.3.4 Intraoperative Verification of the Accuracy of the nTMS-Based DTI FT of the CST

The nTMS-based DTI FT of the CST was used during surgery as a guide for subcortical stimulation in all patients. Subcortical DES was performed by monopolar stimulation through the NIM-Eclipse system (Medtronic, Minneapolis, MN, USA). From 2018 on, DES was performed using continuous dynamic mapping by integrating the mapping probe at the tip of an ultrasonic aspirator (Raabe et al. 2014; Seidel et al. 2020). In order to avoid damage to the CST, we continued with the lesion resection until obtaining a MEP at a stimulation intensity of 5 mA. In general, when a MEP was obtained at this stimulation intensity the tumor resection was discontinued. These stimulation points were stored into the neuronavigation system. We exported the coordinates of each subcortical DES point and reported them on the preoperative MRI scan co-registered with the CST computed using the two different FT techniques. The distance between the stimulated subcortical DES point and the CST was measured and stored together with the used subcortical DES intensity.

At this point, a correlation analysis between distance and subcortical DES stimulation intensity was performed to assess the reliability and eventual differences of the two DTI-FT strategies.

7.3.5 Impact of the nTMS-Based DTI FT on Surgery of Motor-Eloquent Lesions

The EOR was assessed on an early postoperative MRI scan in all cases. Lesion segmentation and EOR calculation were performed on T1-enhanced sequences or, for non-enhancing lesions, on FLAIR sequences using Horos Software© (free and open source code software program, distributed free of charge under the LGPL license at Horosproject.org and sponsored by Nimble Co LLC d/b/a Purview in Annapolis, MD USA). The EOR was defined as the difference between the preoperative and postoperative tumor volumes (ml) (Bloch et al. 2012): gross total resection (GTR) = no residual pathological tissue; subtotal resection (STR) = less than 10 ml of pathological tissue residue; partial resection (PR) $\geq$ 10 ml of tissue residue; biopsy $\geq$ 95% of tumor residue (Berger et al. 1994; Frey et al. 2014b).

The motor status was evaluated using the Medical Research Council (MRC) scale (Compston 2010) at the admission and after three months from surgery. Motor deficits recorded after three months from surgery were considered as permanent deficits.

7.4 Results

A total of 228 patients (134 males, 94 females, mean age 53.9 ± 13.9 years) were included in the study. Demographic and clinical characteristics of patients are reported in Table 7.1.

The nTMS-based FT allowed the visualization of the somatotopic organization of the CST in all cases (Fig. 7.3). We differentiated fiber bundles of the arm, leg, and face fibers, by using a color-coded technique (green for fibers from the leg M1 area, red for fibers from the arm area, and pink for fibers from the face area).

7.4.1 Intraoperative Analysis of Accuracy of nTMS-Based DTI FT of the CST

The mean distance between the last stimulation point and the CST as displayed by the nTMS-based DTI FT was 4.64 ± 0.78 mm (Fig. 7.4a). We observed a concordance between the type of recorded MEP and the functionally different CST fiber bundles in all cases. Indeed, we recorded specific MEPs (from arm, face, or leg muscles) that we expected according to the nTMS-based CST reconstruction (Fig. 7.4).

Table 7.1 Patients' demographics and outcome using nTMS-based preoperative mapping at the Division of Neurosurgery of the University of Messina (2012–2019)

Characteristics	nTMS motor mapping—Tumors close to motor pathway
No. of patients	228
Age in years (mean)	53.9 ± 13.9
Sex (%)	
Male	134 (58.8)
Female	94 (41.2)
Pathology (%)	
High grade gliomas	166 (72.8)
Low grade gliomas	40 (17.5)
Metastases	14 (6.2)
Others	8 (3.5)
Extent of resection (%)	
Gross total resection	154 (67.6)
Subtotal resection	55 (24.1)
Partial resection	14 (6.1)
Biopsy only	5 (2.2)
Post-op permanent deficits at 3 months from surgery (%)	17 (7.4)

7.4.2 Analysis of Correlation Between CST Distance and Subcortical DES Intensity

We retrieved a total of 366 stimulation points from the neuronavigation system. The mean distance from the CST was 9.8 ± 3.3 mm for the nTMS-based technique. Statistical analysis showed a positive correlation between the CST distance and the subcortical DES intensity ($R = 0.97$; $p < 0.0001$). Regression analysis showed a linear correlation expressed by the equation $y = 0.9331*X + 1.247$. Fitting with the linear model was stronger as compared to standard DTI FT ($R^2 = 0.9432$ vs. $R^2 = 0.7186$) (Fig. 7.5).

7.4.3 Extent of Resection and Incidence of Permanent Postoperative Motor Deficits

The clinical impact of the nTMS-based strategy is reported in Table 7.1.

On postoperative MRI scan, we observed a GTR in 67.6% of cases, a STR in 24.1%, a PTR in 6.1%. A biopsy only was performed in 2.2% of cases.

A permanent new motor deficit was observed after three months from surgery in 7.4% of patients.

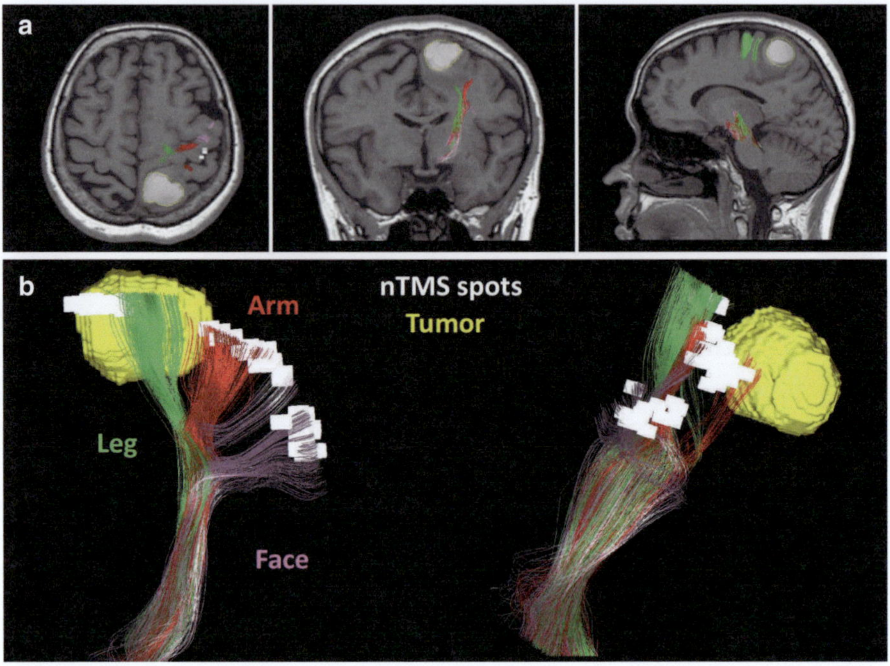

Fig. 7.3 Co-registration of the nTMS-based DTI FT of the CST. nTMS-based DTI FT in the case of a left post-rolandic lesion. (**a**) Co-registration of the nTMS-based DTI FT of the CST (leg fibers = green; arm fibers = red; pink fibers = face), with the reference T1-weighted anatomic examination. (**b**) 3D reconstruction of the CST and its spatial relationship with the tumor (yellow). The tumor was located just posteriorly to the leg fibers (green) of the CST

7.5 Discussion, Other Protocols, and Limitations

7.5.1 Discussion

The nTMS-based DTI-FT is a reliable technique that provides an increased preoperative awareness of spatial relationship between intrinsic brain lesions and the entire motor pathway, improving the evaluation of risks related to surgical treatment. The major advantage as compared to the standard anatomical-based DTI-FT is the reconstruction of the CST based on objective preoperative neurophysiological findings obtained through the nTMS mapping, and the ability to visualize its somatotopic organization.

Standard anatomical-based DTI FT is affected by several pitfalls, especially because it is an operator-dependent technique (Ivren et al. 2023). Final reconstruction can be influenced by several factors, including the choice of anatomical landmarks as seeding ROIs, the use of different reconstruction algorithms and software, and the presence of peritumoral edema (Weiss et al. 2015; Feigl et al. 2014; Duffau 2014; Lu et al. 2003).

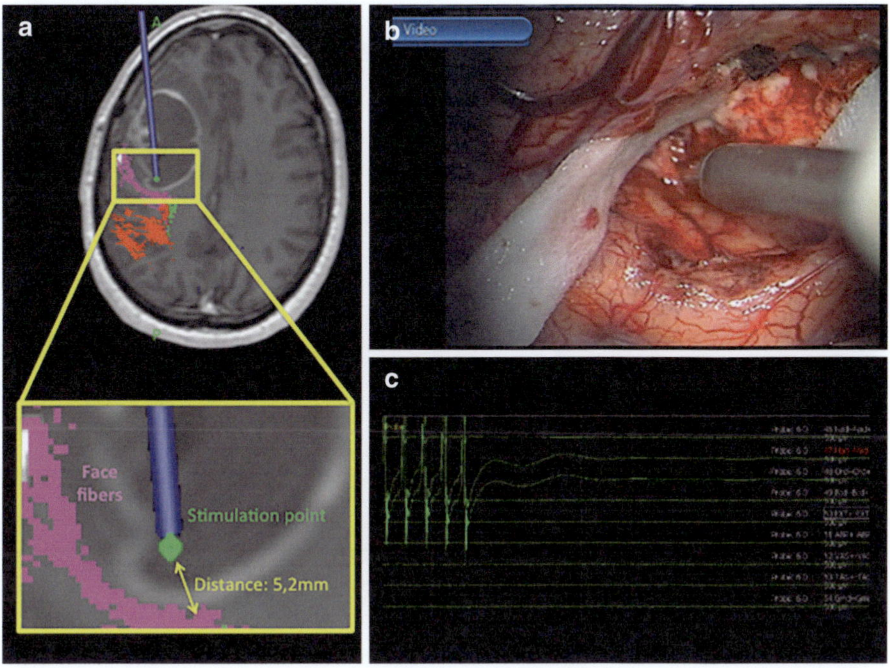

Fig. 7.4 Subcortical DES confirmed the spatial accuracy of the CST. Subcortical DES confirmed the spatial accuracy of the CST and its somatotopic organization, as delineated by nTMS-based DTI FT. (**a**) Snapshot from intraoperative navigation during removal of a left frontal glioblastoma (GBM). The somatotopic organization of the CST is represented by different colors of fibers (green = leg; red = arm; purple = face). The blue stylet represents the navigated probe used for subcortical DES. The green point at the end of the stick represents the subcortical DES point that evoked a MEP using a 6 mA stimulation intensity. (**b**) The corresponding microscope view during tumor resection. (**c**) Subcortical DES recording showing correspondence between the muscles activated by subcortical DES and the CST somatotopic organization, as displayed by the functional tractography. The response were evoked on the right masseter and orbicularis oris muscles and corresponded to the CST face fibers (purple). The distance between the stimulation point and the CST, as displayed by the neuronavigation system, was 5.2 mm

Anatomical landmarks for FT reconstruction of the CST are well defined in the literature (Catani et al. 2002; Catani and Thiebaut de Schotten 2008; Conturo et al. 1999). Nevertheless, it is well known that the presence of a space-occupying lesions, especially if slow-growing, can induce neuroplasticity phenomena or a simple derangement of anatomical structures that can make the placement of seeding ROIs much more difficult (Niu et al. 2016; Takahashi et al. 2012; Yen et al. 2009; Lehericy et al. 2000; Robles et al. 2008). The use of nTMS-based approach reduces this limitation, being based on objective neurophysiological muscular responses, thus considering an eventual reorganization of the motor cortex due to neuroplasticity induced by the lesion. This also increases the reproducibility of CST reconstructions as compared to the standard technique, increasing the agreement between operators.

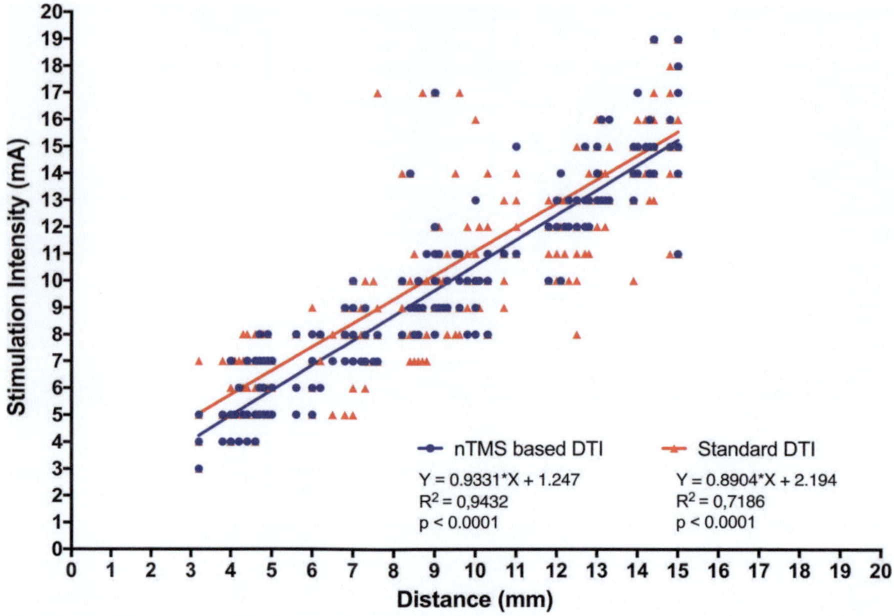

Fig. 7.5 Linear regression analysis showing the correlation between the distance from the CST. Linear regression analysis showing the correlation between the distance from the CST (mm) and the intensity of the subcortical DES (mA) at each stimulated point for both techniques. We found a linear positive correlation, as described by the equation $y = 0.9331x + 1.247$ for the nTMS-based technique (blue) and by the equation $y = 0.8904x + 2.194$ for the standard tractography (red). The correlation and the fitting with the linear model were stronger for the nTMS-based technique ($R = 0.97$ vs. 0.84; $R^2 = 0.9432$ vs. 0.7186). Note the different slopes and y intercepts of the 2 linear models, which suggest a different reliability between the two techniques

The use of functional nTMS motor mapping as seeding region for DTI FT has been described by several studies (Raffa et al. 2018, 2019b; Sollmann et al. 2018; Weiss et al. 2015, 2016; Rosenstock et al. 2017b; Conti et al. 2014; Krieg et al. 2012b; Frey et al. 2012; Forster et al. 2015; Belotti et al. 2021). An important advantage of using nTMS is the possibility to also map the cortical representation of face muscles (Weiss et al. 2013). This provides the somatotopic organization of M1 that can be successfully used to improve the visualization of the CST (Conti et al. 2014). Indeed, an important limitation of standard deterministic DTI FT is the low ability to visualize the most lateral CST fiber bundles that bend to connect to the cortical representation of face muscles (Burgel et al. 2009; Lee et al. 2015). Moreover, the visualization of these most lateral CST projections (directed to the hand and face M1 areas) still remains a challenge also for the most advanced tractography approaches, including those based on *probabilistic* algorithms (Mandelli et al. 2014; He et al. 2023). On the contrary, the use of the nTMS mapping areas on the most lateral portion of M1—those with the representation of hand and face muscles—as seeding ROIs for DTI-FT, allows for visualizing the most lateral components of the CST, provides its somatotopic organization, and increases the reliability of its computation. Indeed, in our series the nTMS-based DTI FT computation of the CST resulted to correlate well with DES findings, and to be more accurate as compared to the standard DTI-FT based on anatomical landmarks. We always observed a correspondence between the type of recorded MEP (recorded from arm, leg, or face muscles) and the type of the stimulated CST fiber bundle as it was visualized with its somatotopic characterization by the neuronavigation system. Moreover, the correlation between the distance from the nearest CST fiber bundle and the intensity of subcortical DES responses showed to be stronger for the nTMS-based DTI-FT as compared to the standard technique.

All these data suggest that the nTMS-based strategy is reliable and accurate, and can be successfully used for the preoperative planning for intrinsic brain lesions in or close to the motor pathway. It increases the awareness of the spatial relationship between the CST and the suspected motor-eloquent lesions, especially if deep-seated. In this latter case, its use could also modify surgical strategy and trajectory, thanks to an improved preoperative risk/benefit analysis of surgical treatment (Angileri et al. 2023).

In fact, the nTMS-based reconstruction of the CST can be also successfully used during surgery, as a further support to DES and lesion resection. The combination of CST somatotopic tractography and subcortical DES helps the surgeon to respect the planned operative strategy by orienting the subcortical stimulation. In the present series, tumor resection was performed layer-by-layer repeating the stimulation along with the resection of layers of tumor. The CST was accurately localized by nTMS-based DTI FT and MEPs were evoked by progressively decreasing current intensities along with tumor resection. Resection was then interrupted when the MEP was evoked by low-intensity stimulation (5 mA), indicating proximity of the CST. The mean distance from the CST was, at this point, 4.61 ± 1.32 mm. Analysis of the correlation between stimulation intensity and distance from the CST showed a strong linear positive correlation ($R = 0.97$; $R^2 = 0.9432$).

Five previous studies performed a quantitative analysis of the relationship between the distance form CST and the stimulation intensity, providing variable results (Nossek et al. 2011; Ostry et al. 2013; Maesawa et al. 2010; Ohue et al. 2012; Kamada et al. 2009). Among these studies, 4 found a linear correlation between the distance and the stimulation intensity (Nossek et al. 2011; Ostry et al. 2013; Maesawa et al. 2010; Ohue et al. 2012), whereas in one study a nonlinear relationship was found (Kamada et al. 2009). Furthermore, equations expressing such a correlation were remarkably different. Ohue et al. found a relation described by the equation $y = 0.972x + 0.120$ with a slope of 0.972 (close to 1) and a y intercept of 0.120 (close to 0) using postoperative tractography (Ohue et al. 2012). This is a quasi-ideal correlation with 1 mA corresponding to 1 mm from the CST. Our findings demonstrated a linear correlation with a good slope value (0.9331), but a relatively high y intercept value (1.247), meaning a closer distance for each mA. This can be explained by the brain shift due to the use of preoperative imaging. Nevertheless, the correlation and the fitting with the linear model was stronger for the nTMS-based DTI FT, indicating a significant higher accuracy as compared to standard DTI FT ($p = 0.02$).

The nTMS-based strategy for the preoperative neurophysiology-driven somatotopic computation of the CST, positively impacted also surgical strategy and patients' outcome. Several studies clearly confirmed the nTMS-based preoperative mapping and may improve the treatment of brain tumors located in the rolandic region (Frey et al. 2014a; Krieg et al. 2014; Raffa et al. 2018, 2019a, c, d; Krieg et al. 2016). More recent papers demonstrated that the nTMS-based DTI FT is useful for stratifying patients according to their risk of the occurrence of postoperative motor deficits (Rosenstock et al. 2017a, 2022; Sollmann et al. 2018). Such a stratification is based on the distance of the lesion form the nTMS-based reconstruction of the CST and can also predict the extent of resection of the lesion (Belotti et al. 2021). Moreover, recent studies demonstrated that the preoperative stratification using the preoperative nTMS-based DTI-FT is more accurate than the stratification based on anatomical landmarks only (Ivren et al. 2023)—and can also be used during surgery to guide DES, interpret ambiguous DES phenomena (such as irreversible MEP amplitude decrease ≤50%) and adjustment of the stimulation intensity (Rosenstock et al. 2021).

7.5.2 Other Protocols for nTMS-Based DTI FT

To date, six milestone papers have already reported different protocols for nTMS-based DTI FT of the CST (Weiss et al. 2015, 2016; Conti et al. 2014; Krieg et al. 2012b; Frey et al. 2012; Forster et al. 2015). Literature evidence suggests that the number of directions used for DTI acquisition can influence FT results. In particular, the higher is the number of directions, the more reliable is the reconstruction of tracts, being 6 the minimum number of needed gradients to compute the tensor (Yao et al. 2015; Lebel et al. 2012; Basser et al. 1994). In many of the published studies, nTMS-based DTI-FT has been computed by using ≥ 20 directions (Weiss et al. 2015, 2016; Conti et al. 2014; Frey et al. 2012; Forster et al. 2015), except in one earlier study (Krieg et al. 2012b). Reported DTI tracking parameters are quite variable, the following are the most common: vector step length = 1.6 mm (Weiss et al. 2015, 2016; Frey et al. 2012); minimum fiber length ≥100 mm (Weiss et al. 2015; Krieg et al. 2012b; Frey et al. 2012); angulation threshold = 30° (Weiss et al. 2015, 2016; Krieg et al. 2012b; Frey et al. 2012), even if it can be increased up to 90° to visualize the most lateral component of the CST (face fibers) (Conti et al. 2014). More attention has to be paid for choosing the best FA value to stop tracking: an attempt to standardize the selection of the FA stop value has been proposed by using a value corresponding to the 75% of the individual FA threshold (the lowest FA value at which a single CST fiber is visible) (Frey et al. 2012). Nevertheless, customized values need to be selected in each case according to peritumoral edema, balancing the necessity to identify the CST and the danger to visualize aberrant fibers (higher risk for low FA values).

Lastly, a two-seeding ROI technique has been used in all studies. The first ROI is commonly placed at the brainstem (anterior pons/cerebral peduncle), providing more accurate results than using the internal capsule (Weiss et al. 2015). The second

ROI is usually placed over an area including all the nTMS positive spots and converted to a volume representing the entire primary motor cortex (Krieg et al. 2012b; Frey et al. 2012). The somatotopic reconstruction of the CST can be obtained by drawing separate seeding areas corresponding to the somatotopic organization of the motor cortex (foot, hand, face areas) and by using each single area as second seeding ROI (Raffa et al. 2019a; Conti et al. 2014; Forster et al. 2015). Alternatively, the hotspots of different somatotopic areas can be enlarged to 2–3 mm and separately used as second seeding ROI for the somatotopic FT of the CST (Sollmann et al. 2018; Weiss et al. 2015, 2016; Krieg et al. 2012b).

More recent studies used the previously reported computation techniques for nTMS-based DTI FT with slight modifications of the protocol (Rosenstock et al. 2017a; Raffa et al. 2018; Sollmann et al. 2018; Belotti et al. 2021; Ivren et al. 2023). Interestingly, only one paper described the technique for nTMS-based FT of the CST using both a *deterministic* and a *probabilistic* algorithm (Munnich et al. 2019). That may be explained with the lack of implementation of probabilistic tractography software in the most diffused neuronavigation systems. However, it is plausible that novel neuronavigation models will implement this software, thus leading to the publication of novel technical strategies for the computation of the nTMS-based probabilistic FT. Table 7.2 summarizes most important nTMS-based FT protocols reported in the literature.

7.5.3 Limitations

Some DTI FT limitations cannot be avoided even if using the nTMS-based strategy. First of all, the presence of excessive perilesional edema seriously hampers an optimal computation of the water diffusion tensor, especially in the voxels nearest to the lesion (Lu et al. 2003; Yen et al. 2009). Despite the use of specific second ROIs consisting of the nTMS map represents a valid method to "force" software to compute fibers, the presence of edema dramatically reduces the values of anisotropy needed to obtain a reliable calculation of the tensor, hampering the visualization of reliable fibers. In this scenario, the only way to visualize CST fibers is to use low values of fractional anisotropy. This results in the visualization of aberrant fibers, reducing the accuracy of the DTI FT, regardless of the use of the standard or the nTMS-based strategy.

Moreover, in patients with severe preoperative motor impairment, nTMS could not elicit any motor response due to tumor infiltration of motor cortex/CST, and therefore it cannot be used as seeding region for the CST FT. Nevertheless, some reports have been published describing the possibility to obtain an nTMS mapping of the motor cortex even in hemiplegic patients (Picht et al. 2011b).

Lastly, the most important limitation of DTI FT is represented by the brain shift during surgery (Nimsky et al. 2005; Romano et al. 2011; Bozzao et al. 2010). This is unavoidable unless using intraoperative imaging and consists of an inward or outward displacement of the brain and therefore of the CST, especially at the final stages of surgery. Nevertheless, the intraoperative visualization of the DTI FT of the

Table 7.2 Different published protocols for the nTMS-based DTI FT of the CST

Study	DTI acquisition	DTI fiber tracking parameters				Seeding ROIs		Software
		Vector step length	Minimum fiber length	Angulation threshold	Fractional anisotropy (FA) value	1st ROI	2nd ROI	
Frey et al. (2012)	24 direction (*b* value = 1000)	1.6 mm	110 mm	30°	75% of FA threshold	Brainstem/ Internal Capsule	All nTMS positive spots (each enlarged to 6 mm)	BrainLab iPlan 2.0
Krieg et al. (2012a)	6 direction (*b* value = 800)	/	100 mm	30°	Variable (<0.20)	Brainstem (at the level of the tentorium)	All nTMS positive spots (each enlarged to 2 mm)	BrainLab iPlan 3.0.1
Conti et al. (2014)	32 direction (*b* value = 1000)	1 mm	20 mm	45° for leg and arm fibers; 75°–90° for face fibers	0.20	Brainstem (cerebral peduncle)	All nTMS positive spots (distinguishing between leg, arm and face spots)	Medtronic StealthViz—StealthStation S7
Weiss et al. (2015)	30 direction (*b* value = 800)	1.6 mm	1 mm (plausibility cut-off = 121.5 mm)	30°	100% FA threshold (plausibility cut-off = 0.105)	Brainstem (anterior pons)/Internal Capsule (posterior limb)	nTMS hotspots for foot, hand and tongue muscles (each enlarged to 2–3 mm)	BrainLab iPlan 3.0.0
Forster et al. (2015)	20 direction (*b* value = 700)	/	80 mm	/	0.20	Brainstem (ventral pons/ cerebral peduncle)	All nTMS positive spots (distinguishing between leg and hand spots)	BrainLab iPlan 3.0.3

Weiss et al. (2016)	30 direction (*b* value = 800)	1.6 mm	1 mm	30°	100% FA threshold	Brainstem (anterior pons)	nTMS hotspots for foot, hand and tongue muscles (each enlarged to 2–3 mm)	BrainLab iPlan 3.0.0
Rosenstock et al. (2017b)	24 direction (*b* value = 1000)	1.6 mm	110 mm	30°	75% of FA threshold	Brainstem (antero-lateral portion of the cerebral peduncle)	All nTMS positive spots (each enlarged to 3 mm)	BrainLab iPlan 2.0
Sollmann et al. (2018)	15 directions (*b* value = 800)	/	110 mm	/	50%, 75% and 100% FA threshold	Brainstem (at the level of the tentorium)	nTMS hotspots for foot, hand and tongue muscles (each enlarged to 2 mm)	BrainLab iPlan 3.0.1
Raffa et al. (2018)	32 direction (*b* value = 1000)	1 mm	110 mm	45° for leg and arm fibers; 75°–90° for face fibers	0.20	Brainstem (cerebral peduncle)	All nTMS positive spots (distinguishing between leg, arm and face spots)	Medtronic StealthViz—StealthStation S8

(continued)

Table 7.2 (continued)

Study	DTI acquisition	DTI fiber tracking parameters				Seeding ROIs		Software
		Vector step length	Minimum fiber length	Angulation threshold	Fractional anisotropy (FA) value	1st ROI	2nd ROI	
Munnich et al. (2019)	20 directions (*b* value = 700)	1 mm	3 mm	0.3	0.05	Posterior limb of the internal capsule	All nTMS-positive spots (distinguishing between leg and arm spots) enlarged to 1–2 mm	NeuroQLab—MeVisLab (Deterministic FT)
		0.5–7 mm[a]	1.5–6 mm[a]	0.2–0.65[a]	0.1–0.45[a]	Posterior limb of the internal capsule	2nd ROI: All nTMS positive spots (distinguishing between leg and arm spots) enlarged to 1–2 mm	NeuroQLab—MeVisLab (Probabilistic FT)
							3rd ROI: ventrolateral cerebral peduncle	
Belotti et al. (2021)	24 direction (*b* value = 1000)	1.6 mm	110 mm	30°	75% of FA threshold	Brainstem (antero-lateral portion of the cerebral peduncle)	All nTMS positive spots (each enlarged to 3 mm)	BrainLab iPlan 2.0
Ivren et al. (2023)	24 direction (*b* value = 1000)	1.6 mm	110 mm	30°	75% of FA threshold	Brainstem (ipsilateral pons)	All nTMS positive spots (each enlarged to 3 mm)	BrainLab iPlan 3.0

[a]Element out of this range

CST has to be considered exclusively as a visual feedback orienting subcortical DES that still remains the *gold standard* for guiding lesion resection. However, the stronger correlation with subcortical DES findings makes the nTMS-based DTI FT of the CST more accurate and therefore useful during surgery as compared to the standard DTI FT.

7.6 Conclusions

The nTMS-based DTI FT is more reliable and accurate for the reconstruction of the CST than standard anatomical-based tractography. The anatomical and functional details obtained through this somatotopic reconstruction provide a preoperative preliminary assessment of the spatial relationship between the lesion and parts of the motor fibers and improve the evaluation of risks for tumor resection. Moreover, the nTMS-based DTI FT of the CST can be successfully used during surgery as a guide for orienting DES and lesion resection. In particular, the somatotopic organization of the CST may be used for an improved neurophysiological exploration and to guide the extent of lesion resection according to an nTMS-based preoperative plan.

References

Abhinav K, Yeh FC, Pathak S, Suski V, Lacomis D, Friedlander RM, et al. Advanced diffusion MRI fiber tracking in neurosurgical and neurodegenerative disorders and neuroanatomical studies: a review. Biochim Biophys Acta. 2014;1842(11):2286–97.

Angileri FF, Raffa G, Curcio A, Granata F, Marzano G, Germano A. Minimally invasive surgery of deep-seated brain lesions using tubular retractors and navigated transcranial magnetic stimulation-based diffusion tensor imaging tractography guidance: the minefield paradigm. Oper Neurosurg (Hagerstown). 2023;24(6):656–64.

Basser PJ. Inferring microstructural features and the physiological state of tissues from diffusion-weighted images. NMR Biomed. 1995;8(7–8):333–44.

Basser PJ, Mattiello J, LeBihan D. MR diffusion tensor spectroscopy and imaging. Biophys J. 1994;66(1):259–67.

Bello L, Riva M, Fava E, Ferpozzi V, Castellano A, Raneri F, et al. Tailoring neurophysiological strategies with clinical context enhances resection and safety and expands indications in gliomas involving motor pathways. Neuro Oncol. 2014;16(8):1110–28.

Belotti F, Tuncer MS, Rosenstock T, Ivren M, Vajkoczy P, Picht T. Predicting the extent of resection of motor-eloquent gliomas based on TMS-guided fiber tracking. Brain Sci. 2021;11(11)

Berger MS, Ojemann GA. Intraoperative brain mapping techniques in neuro-oncology. Stereotact Funct Neurosurg. 1992;58(1-4):153–61.

Berger MS, Deliganis AV, Dobbins J, Keles GE. The effect of extent of resection on recurrence in patients with low grade cerebral hemisphere gliomas. Cancer. 1994;74(6):1784–91.

Bloch O, Han SJ, Cha S, Sun MZ, Aghi MK, McDermott MW, et al. Impact of extent of resection for recurrent glioblastoma on overall survival: clinical article. J Neurosurg. 2012;117(6):1032–8.

Bozzao A, Romano A, Angelini A, D'Andrea G, Calabria LF, Coppola V, et al. Identification of the pyramidal tract by neuronavigation based on intraoperative magnetic resonance tractography: correlation with subcortical stimulation. Eur Radiol. 2010;20(10):2475–81.

Burgel U, Madler B, Honey CR, Thron A, Gilsbach J, Coenen VA. Fiber tracking with distinct software tools results in a clear diversity in anatomical fiber tract portrayal. Cent Eur Neurosurg. 2009;70(1):27–35.

Catani M, Thiebaut de Schotten M. A diffusion tensor imaging tractography atlas for virtual in vivo dissections. Cortex. 2008;44(8):1105–32.

Catani M, Howard RJ, Pajevic S, Jones DK. Virtual in vivo interactive dissection of white matter fasciculi in the human brain. Neuroimage. 2002;17(1):77–94.

Compston A. Aids to the investigation of peripheral nerve injuries. Medical Research Council: Nerve Injuries Research Committee. His Majesty's Stationery Office: 1942; pp. 48 (iii) and 74 figures and 7 diagrams; with aids to the examination of the peripheral nervous system. By Michael O'Brien for the Guarantors of Brain. Saunders Elsevier: 2010; pp. [8] 64 and 94 Figures. Brain. 2010;133(10):2838–44.

Conti A, Raffa G, Granata F, Rizzo V, Germano A, Tomasello F. Navigated transcranial magnetic stimulation for "somatotopic" tractography of the corticospinal tract. Neurosurgery. 2014;10(Suppl 4):542–54. discussion 54

Conturo TE, Lori NF, Cull TS, Akbudak E, Snyder AZ, Shimony JS, et al. Tracking neuronal fiber pathways in the living human brain. Proc Natl Acad Sci U S A. 1999;96(18):10422–7.

Duffau H. The dangers of magnetic resonance imaging diffusion tensor tractography in brain surgery. World Neurosurg. 2014;81(1):56–8.

Feigl GC, Hiergeist W, Fellner C, Schebesch KM, Doenitz C, Finkenzeller T, et al. Magnetic resonance imaging diffusion tensor tractography: evaluation of anatomic accuracy of different fiber tracking software packages. World neurosurgery. 2014;81(1):144–50.

Forster MT, Hoecker AC, Kang JS, Quick J, Seifert V, Hattingen E, et al. Does navigated transcranial stimulation increase the accuracy of tractography? A prospective clinical trial based on intraoperative motor evoked potential monitoring during deep brain stimulation. Neurosurgery. 2015;76(6):766–76.

Frey D, Strack V, Wiener E, Jussen D, Vajkoczy P, Picht T. A new approach for corticospinal tract reconstruction based on navigated transcranial stimulation and standardized fractional anisotropy values. Neuroimage. 2012;62(3):1600–9.

Frey D, Schilt S, Strack V, Zdunczyk A, Rosler J, Niraula B, et al. Navigated transcranial magnetic stimulation improves the treatment outcome in patients with brain tumors in motor eloquent locations. Neuro-oncology. 2014a;16(10):1365–72.

Frey D, Schilt S, Strack V, Zdunczyk A, Rosler J, Niraula B, et al. Navigated transcranial magnetic stimulation improves the treatment outcome in patients with brain tumors in motor eloquent locations. Neuro Oncol. 2014b;

Goebell E, Fiehler J, Ding XQ, Paustenbach S, Nietz S, Heese O, et al. Disarrangement of fiber tracts and decline of neuronal density correlate in glioma patients – a combined diffusion tensor imaging and 1H-MR spectroscopy study. AJNR Am J Neuroradiol. 2006;27(7):1426–31.

Hakulinen U, Brander A, Ryymin P, Ohman J, Soimakallio S, Helminen M, et al. Repeatability and variation of region-of-interest methods using quantitative diffusion tensor MR imaging of the brain. BMC Med Imaging. 2012;12:30.

He J, Zhang F, Pan Y, Feng Y, Rushmore J, Torio E, et al. Reconstructing the somatotopic organization of the corticospinal tract remains a challenge for modern tractography methods. Hum Brain Mapp. 2023;44(17):6055–73.

Holodny AI, Watts R, Korneinko VN, Pronin IN, Zhukovskiy ME, Gor DM, et al. Diffusion tensor tractography of the motor white matter tracts in man: Current controversies and future directions. Ann N Y Acad Sci. 2005;1064:88–97.

Ivren M, Grittner U, Khakhar R, Belotti F, Schneider H, Poser P, et al. Comparison of anatomical-based vs. nTMS-based risk stratification model for predicting postoperative motor outcome and extent of resection in brain tumor surgery. Neuroimage Clin. 2023;38:103436.

Jenabi M, Peck KK, Young RJ, Brennan N, Holodny AI. Identification of the corticobulbar tracts of the tongue and face using deterministic and probabilistic DTI fiber tracking in patients with brain tumor. AJNR Am J Neuroradiol. 2015;36(11):2036–41.

Kamada K, Todo T, Ota T, Ino K, Masutani Y, Aoki S, et al. The motor-evoked potential threshold evaluated by tractography and electrical stimulation. J Neurosurg. 2009;111(4):785–95.

Krieg SM, Shiban E, Buchmann N, Gempt J, Foerschler A, Meyer B, et al. Utility of presurgical navigated transcranial magnetic brain stimulation for the resection of tumors in eloquent motor areas. J Neurosurg. 2012a;116(5):994–1001.

Krieg SM, Buchmann NH, Gempt J, Shiban E, Meyer B, Ringel F. Diffusion tensor imaging fiber tracking using navigated brain stimulation – a feasibility study. Acta Neurochir (Wien). 2012b;154(3):555–63.

Krieg SM, Shiban E, Buchmann N, Meyer B, Ringel F. Presurgical navigated transcranial magnetic brain stimulation for recurrent gliomas in motor eloquent areas. Clin Neurophysiol. 2013;124(3):522–7.

Krieg SM, Sabih J, Bulubasova L, Obermueller T, Negwer C, Janssen I, et al. Preoperative motor mapping by navigated transcranial magnetic brain stimulation improves outcome for motor eloquent lesions. Neuro Oncol. 2014;16(9):1274–82.

Krieg SM, Picht T, Sollmann N, Bahrend I, Ringel F, Nagarajan SS, et al. Resection of motor eloquent metastases aided by preoperative nTMS-based motor maps-comparison of two observational cohorts. Front Oncol. 2016;6:261.

Kristo G, Leemans A, de Gelder B, Raemaekers M, Rutten GJ, Ramsey N. Reliability of the corticospinal tract and arcuate fasciculus reconstructed with DTI-based tractography: implications for clinical practice. Eur Radiol. 2013;23(1):28–36. https://doi.org/10.1007/s00330-012-2589-9.

Kupper H, Groeschel S, Alber M, Klose U, Schuhmann MU, Wilke M. Comparison of different tractography algorithms and validation by intraoperative stimulation in a child with a brain tumor. Neuropediatrics. 2015;46(1):72–5.

Kwon HG, Hong JH, Jang SH. Anatomic location and somatotopic arrangement of the corticospinal tract at the cerebral peduncle in the human brain. AJNR Am J Neuroradiol. 2011;32(11):2116–9.

Lebel C, Benner T, Beaulieu C. Six is enough? Comparison of diffusion parameters measured using six or more diffusion-encoding gradient directions with deterministic tractography. Magn Reson Med. 2012;68(2):474–83.

Lee DH, Park JW, Park SH, Hong C. Have you ever seen the impact of crossing fiber in DTI?: demonstration of the corticospinal tract pathway. PLoS One. 2015;10(7):e0112045.

Lehericy S, Duffau H, Cornu P, Capelle L, Pidoux B, Carpentier A, et al. Correspondence between functional magnetic resonance imaging somatotopy and individual brain anatomy of the central region: comparison with intraoperative stimulation in patients with brain tumors. J Neurosurg. 2000;92(4):589–98.

Lu S, Ahn D, Johnson G, Cha S. Peritumoral diffusion tensor imaging of high-grade gliomas and metastatic brain tumors. AJNR Am J Neuroradiol. 2003;24(5):937–41.

Maesawa S, Fujii M, Nakahara N, Watanabe T, Wakabayashi T, Yoshida J. Intraoperative tractography and motor evoked potential (MEP) monitoring in surgery for gliomas around the corticospinal tract. World Neurosurg. 2010;74(1):153–61.

Mandelli ML, Berger MS, Bucci M, Berman JI, Amirbekian B, Henry RG. Quantifying accuracy and precision of diffusion MR tractography of the corticospinal tract in brain tumors. J Neurosurg. 2014;121(2):349–58.

Minati L, Grisoli M, Bruzzone MG. MR spectroscopy, functional MRI, and diffusion-tensor imaging in the aging brain: a conceptual review. J Geriatr Psychiatry Neurol. 2007;20(1):3–21.

Munnich T, Klein J, Hattingen E, Noack A, Herrmann E, Seifert V, et al. Tractography verified by intraoperative magnetic resonance imaging and subcortical stimulation during tumor resection near the corticospinal tract. Oper Neurosurg (Hagerstown). 2019;16(2):197–210.

Nimsky C, Ganslandt O, Hastreiter P, Wang R, Benner T, Sorensen AG, et al. Intraoperative diffusion-tensor MR imaging: shifting of white matter tracts during neurosurgical procedures – initial experience. Radiology. 2005;234(1):218–25.

Niu C, Liu X, Yang Y, Zhang K, Min Z, Wang M, et al. Assessing region of interest schemes for the corticospinal tract in patients with brain tumors. Medicine (Baltimore). 2016;95(12):e3189.

Nossek E, Korn A, Shahar T, Kanner AA, Yaffe H, Marcovici D, et al. Intraoperative mapping and monitoring of the corticospinal tracts with neurophysiological assessment and 3-dimensional ultrasonography-based navigation. Clinical article. J Neurosurg. 2011;114(3):738–46.
Ohue S, Kohno S, Inoue A, Yamashita D, Harada H, Kumon Y, et al. Accuracy of diffusion tensor magnetic resonance imaging-based tractography for surgery of gliomas near the pyramidal tract: a significant correlation between subcortical electrical stimulation and postoperative tractography. Neurosurgery. 2012;70(2):283–93. discussion 94
Ostry S, Belsan T, Otahal J, Benes V, Netuka D. Is intraoperative diffusion tensor imaging at 3.0T comparable to subcortical corticospinal tract mapping? Neurosurgery. 2013;73(5):797–807. discussion 6–7
Picht T, Mularski S, Kuehn B, Vajkoczy P, Kombos T, Suess O. Navigated transcranial magnetic stimulation for preoperative functional diagnostics in brain tumor surgery. Neurosurgery. 2009;65(6 Suppl):93–8. discussion 8–9
Picht T, Schmidt S, Brandt S, Frey D, Hannula H, Neuvonen T, et al. Preoperative functional mapping for rolandic brain tumor surgery: comparison of navigated transcranial magnetic stimulation to direct cortical stimulation. Neurosurgery. 2011a;69(3):581–8. discussion 8.
Picht T, Schmidt S, Woitzik J, Suess O. Navigated brain stimulation for preoperative cortical mapping in paretic patients: case report of a hemiplegic patient. Neurosurgery. 2011b;68(5):E1475–80. discussion E80
Pujol S, Wells W, Pierpaoli C, Brun C, Gee J, Cheng G, Vemuri B, Commowick O, Prima S, Stamm A, Goubran M, Khan A, Peters T, Neher P, Maier-Hein KH, Shi Y, Tristan-Vega A, Veni G, Whitaker R, Styner M, Westin CF, Gouttard S, Norton I, Chauvin L, Mamata H, Gerig G, Nabavi A, Golby A, Kikinis R. The DTI Challenge: Toward Standardized Evaluation of Diffusion Tensor Imaging Tractography for Neurosurgery. J Neuroimaging. 2015;25(6):875–82.
Raabe A, Beck J, Schucht P, Seidel K. Continuous dynamic mapping of the corticospinal tract during surgery of motor eloquent brain tumors: evaluation of a new method. J Neurosurg. 2014;120(5):1015–24.
Raffa G, Scibilia A, Germanò A, Conti A. nTMS-based DTI fiber tracking of motor pathways. In: Krieg MS, editor. Navigated transcranial magnetic stimulation in neurosurgery. Cham: Springer; 2017. p. 97–114.
Raffa G, Conti A, Scibilia A, Cardali SM, Esposito F, Angileri FF, et al. The impact of diffusion tensor imaging fiber tracking of the corticospinal tract based on navigated transcranial magnetic stimulation on surgery of motor-eloquent brain lesions. Neurosurgery. 2018;83(4):768–82.
Raffa G, Picht T, Angileri FF, Youssef M, Conti A, Esposito F, et al. Surgery of malignant motor-eloquent gliomas guided by sodium-fluorescein and navigated transcranial magnetic stimulation: a novel technique to increase the maximal safe resection. J Neurosurg Sci. 2019a;
Raffa G, Quattropani MC, Germano A. When imaging meets neurophysiology: the value of navigated transcranial magnetic stimulation for preoperative neurophysiological mapping prior to brain tumor surgery. Neurosurgical Focus. 2019b;47(6):E10.
Raffa G, Scibilia A, Conti A, Ricciardo G, Rizzo V, Morelli A, et al. The role of navigated transcranial magnetic stimulation for surgery of motor-eloquent brain tumors: a systematic review and metaanalysis. Clin Neurol Neurosurg. 2019c;
Raffa G, Scibilia A, Conti A, Cardali SM, Rizzo V, Terranova C, et al. Multimodal surgical treatment of high grade gliomas in the motor area: the impact of the combination of navigated transcranial magnetic stimulation and fluorescein-guided resection. World Neurosurg. 2019d;
Rizzo V, Terranova C, Conti A, Germano A, Alafaci C, Raffa G, et al. Preoperative functional mapping for rolandic brain tumor surgery. Neurosci Lett. 2014;583:136–41.
Rizzo V, Terranova C, Raffa G, Cardali SM, Angileri FF, Marzano G, et al. Cortical excitability and connectivity in patients with brain tumors. Front Neurol. 2021;12:673836.
Robles SG, Gatignol P, Lehericy S, Duffau H. Long-term brain plasticity allowing a multistage surgical approach to World Health Organization Grade II gliomas in eloquent areas. J Neurosurg. 2008;109(4):615–24.

Romano A, D'Andrea G, Calabria LF, Coppola V, Espagnet CR, Pierallini A, et al. Pre- and intraoperative tractographic evaluation of corticospinal tract shift. Neurosurgery. 2011;69(3):696–704. discussion 5

Rosenstock T, Grittner U, Acker G, Schwarzer V, Kulchytska N, Vajkoczy P, et al. Risk stratification in motor area-related glioma surgery based on navigated transcranial magnetic stimulation data. J Neurosurg. 2017a;126(4):1227–37.

Rosenstock T, Giampiccolo D, Schneider H, Runge SJ, Bahrend I, Vajkoczy P, et al. Specific DTI seeding and diffusivity-analysis improve the quality and prognostic value of TMS-based deterministic DTI of the pyramidal tract. Neuroimage Clin. 2017b;16:276–85.

Rosenstock T, Tuncer MS, Munch MR, Vajkoczy P, Picht T, Faust K. Preoperative nTMS and intraoperative neurophysiology – a comparative analysis in patients with motor-eloquent glioma. Front Oncol. 2021;11:676626.

Rosenstock T, Hani L, Grittner U, Schlinkmann N, Ivren M, Schneider H, et al. Bicentric validation of the navigated transcranial magnetic stimulation motor risk stratification model. J Neurosurg. 2022;136(4):1194–206.

Sanai N, Berger MS. Intraoperative stimulation techniques for functional pathway preservation and glioma resection. Neurosurg Focus. 2010;28(2):E1.

Sarubbo S, De Benedictis A, Merler S, Mandonnet E, Balbi S, Granieri E, et al. Towards a functional atlas of human white matter. Hum Brain Mapp. 2015;

Seidel K, Schucht P, Beck J, Raabe A. Continuous dynamic mapping to identify the corticospinal tract in motor eloquent brain tumors: an update. J Neurol Surg A Cent Eur Neurosurg. 2020;81(2):105–10.

Sheng Z, Yu J, Chen Z, Sun Y, Bu X, Wang M, et al. Constrained-spherical deconvolution tractography in the evaluation of the corticospinal tract in glioma surgery. Front Surg. 2021;8:646465.

Sollmann N, Wildschuetz N, Kelm A, Conway N, Moser T, Bulubas L, et al. Associations between clinical outcome and navigated transcranial magnetic stimulation characteristics in patients with motor-eloquent brain lesions: a combined navigated transcranial magnetic stimulation-diffusion tensor imaging fiber tracking approach. J Neurosurg. 2018;128(3):800–10.

Takahashi S, Jussen D, Vajkoczy P, Picht T. Plastic relocation of motor cortex in a patient with LGG (low grade glioma) confirmed by NBS (navigated brain stimulation). Acta Neurochir (Wien). 2012;154(11):2003–8. discussion 8.

Takahashi S, Vajkoczy P, Picht T. Navigated transcranial magnetic stimulation for mapping the motor cortex in patients with rolandic brain tumors. Neurosurg Focus. 2013;34(4):E3.

Toselli B, Tortora D, Severino M, Arnulfo G, Canessa A, Morana G, et al. Improvement in white matter tract reconstruction with constrained spherical deconvolution and track density mapping in low angular resolution data: a pediatric study and literature review. Front Pediatr. 2017;5:182.

Wakana S, Caprihan A, Panzenboeck MM, Fallon JH, Perry M, Gollub RL, Hua K, Zhang J, Jiang H, Dubey P, Blitz A, van Zijl P, Mori S. Reproducibility of quantitative tractography methods applied to cerebral white matter. Neuroimage. 2007;36(3):630–44.

Weiss C, Nettekoven C, Rehme AK, Neuschmelting V, Eisenbeis A, Goldbrunner R, et al. Mapping the hand, foot and face representations in the primary motor cortex – retest reliability of neuronavigated TMS versus functional MRI. Neuroimage. 2013;66:531–42.

Weiss C, Tursunova I, Neuschmelting V, Lockau H, Nettekoven C, Oros-Peusquens AM, et al. Improved nTMS- and DTI-derived CST tractography through anatomical ROI seeding on anterior pontine level compared to internal capsule. NeuroImage Clinical. 2015;7:424–37.

Weiss C, Tursunova I, Neuschmelting V, Nettekoven C, Oros-Peusquens A-M, Stoffels G, et al. Functional MRI vs. navigated TMS to optimize M1 seed volume delineation for DTI tractography. A prospective study in patients with brain tumours adjacent to the corticospinal tract. NeuroImage: Clinical. 2016;

Yao X, Yu T, Liang B, Xia T, Huang Q, Zhuang S. Effect of increasing diffusion gradient direction number on diffusion tensor imaging fiber tracking in the human brain. Korean J Radiol. 2015;16(2):410–8.

Yen PS, Teo BT, Chiu CH, Chen SC, Chiu TL, Su CF. White Matter tract involvement in brain tumors: a diffusion tensor imaging analysis. Surg Neurol. 2009;72(5):464–9. discussion 9

Risk Stratification and Clinical Decision-Making Support

8

Tizian Rosenstock

8.1 nTMS-Based Risk Stratification

Many studies have highlighted the advantages of the navigated transcranial magnetic stimulation (nTMS) motor mapping—such as precise functional localization of cortical motor-eloquent areas and improvement of the corticospinal tract (CST) tractography (Picht et al. 2011; Frey et al. 2012). In a first study of 100 patients with different brain tumor entities infiltrating or approximating the motor cortex, a correlation was found between the excitability level of the individual motor system and the postoperative motor outcome. In detail, the excitability level between the hemispheres was determined using the RMT_{ratio} ($RMT_{ratio} = RMT_{tumor}/RMT_{healthy}$) (Picht et al. 2012a). In a later study, an association between the motor outcome and the distance of nTMS-based fiber tracts and tumor (tumor-tract distance [TTD]) was investigated (Frey et al. 2012). This parameter demonstrated to be very sensitive, as only patients with a TTD <1 cm developed a new motor deficit. In order to improve the specificity of risk prediction, the integrity of the white matter, (represented by the fractional anisotropy [FA] value) was additionally analyzed. Even with the same TTD, an impaired fiber tract integrity (lower FA values) indicates the vulnerability of the motor system (Rosenstock et al. 2017a).

Based on an extensive cohort of prospectively analyzed patients, a multivariate risk stratification was developed to predict postoperative motor outcome for the short-term status on discharge day and the long-term status at 3 months postoperatively. In detail, the risk for the development of a new paresis or worsening of an

T. Rosenstock (✉)
Department of Neurosurgery, Charité–Universitätsmedizin Berlin, Corporate Member of Freie Universität, Berlin, Germany

Berlin Institute of Health at Charité – Universitätsmedizin Berlin, BIH Biomedical Innovation Academy, BIH Charité Digital Clinician, Scientist Program, Berlin, Germany
e-mail: tizian.rosenstock@charite.de

S. M. Krieg, T. Picht (eds.), *Navigated Transcranial Magnetic Stimulation in Neurosurgery*, https://doi.org/10.1007/978-3-031-97155-6_8

existing paresis can be estimated, using spatial and neurophysiological data from the nTMS motor mapping and spatial and structural data from the nTMS-based tractography. The identified risk factors differed in their specificity and were therefore analyzed with a regression tree analysis (Fig. 8.1).

In addition to risk prediction, the assessment also allows to identify the areas within the motor system that must be preserved and protected during surgery so that no new motor deficit is caused. These are, in particular, functional/highly eloquent areas of the motor cortex and the CST, which could not be delineated without the individual functional data of the nTMS. Thus, nTMS with nTMS-based tractography has also become established for determining resection strategy of motor-associated

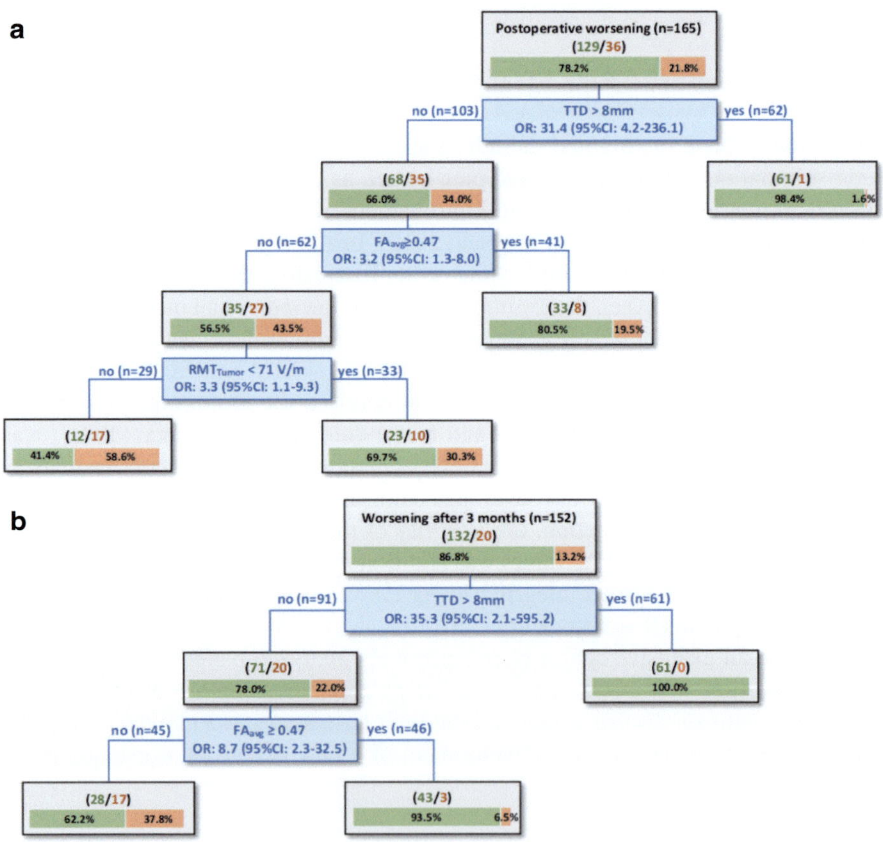

Fig. 8.1 nTMS-based risk stratification based on regression tree analysesThe following three parameters are used to predict the motor outcome at the day of discharge (**a**) and after 3 months (**b**): infiltration of the motor system (measured by tumor tract distance [TTD]), white matter integrity (measured by the average fractional anisotropy of the CST [FA_{avg}] and cortical motor excitability (measured by the resting motor threshold of the tumor hemisphere [RMT_{Tumor}]). The odds ratios (OR) indicate the risk ratio between the subgroups

lesions in order to optimize the approach and extent of resection without injuring eloquent areas (Picht et al. 2012b).

The nTMS-based risk stratification was calculated based on a prospective data set and validated/standardized with another prospective data set from two university neurosurgical hospitals (Rosenstock et al. 2017b, 2021). In those studies, motor-eloquent tumors with proximity to the motor cortex or CST (including larger insular gliomas) were included and operated under highest available standards including resection with fluorescence filter and application of intraoperative neurophysiological monitoring. A comparison to an alternative risk assessment that relies only on clinical data and structural magnetic resonance imaging (MRI) data, demonstrated the superiority of the nTMS-based risk stratification (Ivren et al. 2023). Patients can thus be given a precise preoperative counseling about the surgical risk. Young patients with a low risk profile and a non-contrast-enhancing tumor (suspected low-grade glioma) can be advised to undergo supramarginal resection. In such cases, the potential of the intraoperative neuromonitoring (IOM) is fully realized by lowering the subcortical stimulation intensity below 5 mA. A counter-example is a patient of advanced age with a suspected glioblastoma and high-risk profile, where subtotal resection is more likely to be considered while preserving functional integrity. In cases without increased risk for the motor outcome, IOM is typically not required and time and resources can be saved.

8.1.1 Infiltration of the Motor System

The spatial relationship of the motor-functional areas and the brain tumor represents the most important risk parameter. At the cortical level, tumors may infiltrate the functional (= nTMS positive) parts of the motor cortex. The tumor infiltration/proximity to the corticospinal tract is determined by the minimal distance between the tracts and tumor tissue. Exemplary cases are shown in Fig. 8.2.

To measure the tumor-tract distance (TTD), nTMS-based tractography is performed according to a standardized procedure that uses two regions of interest: (Picht et al. 2011) functional ROI based on nTMS mapping and (Frey et al. 2012) anatomic ROI in the anterior-inferior region of the pons. This algorithm has been shown to be reliable and robust even in the presence of a large tumor mass effect and significant peritumoral edema. In addition, nTMS-based tractography proved to be user-independent and can be easily performed even by less experienced examiners (Rosenstock et al. 2017a). A TTD of 8 mm has been evaluated as the cutoff value for safe resections. Thus, patients without tumorous infiltration of the motor cortex and with a TTD > 8 mm have a very low risk of 1% of experiencing a transient motor deficit and no risk of a long-term deficit due to direct injury of the motor system. In another study, a cut-off value of 11 mm was determined as the safety limit for the motor outcome (Sollmann et al. 2018). Slight differences in the technical implementation of tractography may lead to different threshold values. The exact individual risk profile is then determined by the additional consideration of the structural integrity of the CST and the neurophysiological motor excitability described later.

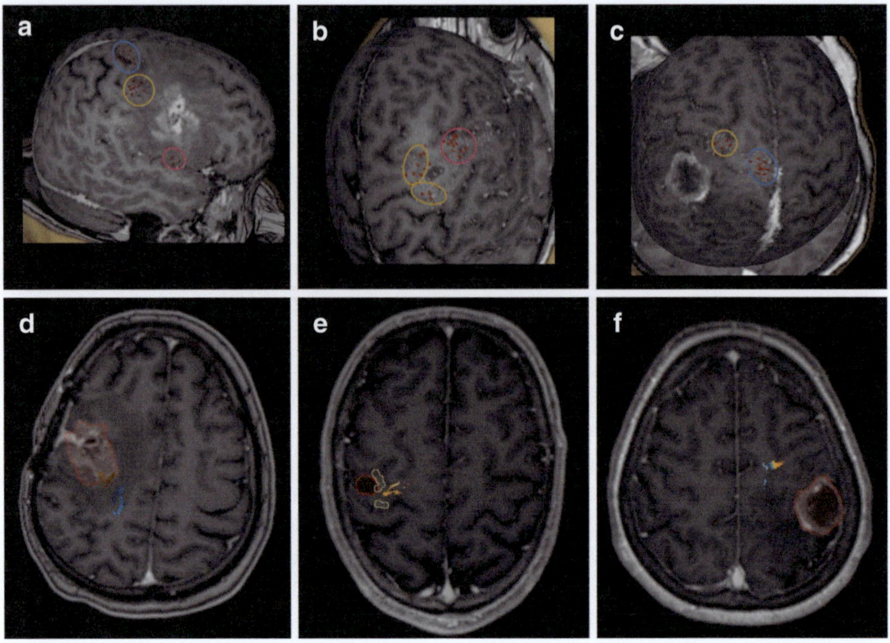

Fig. 8.2 Case examples of nTMS motor mapping and nTMS-based tractography for the upper extremity (yellow circle), lower extremity (blue circle), and for face muscles (pink circle). Gray spots indicate MEP-negative stimulation sites and colored spot MEP-positive stimulation sites
The first case is a 55-year-old woman with a recurrence of a glioblastoma and a mild hemiparesis British Medical Research Council (BMRC) 4/5. The cortical mapping shows a diffuse infiltration of the motor cortex (**a**) with an impaired excitability (RMT = 72 V/m). The nTMS-based tractography revealed a close relationship to the CST with a TTD of 0 mm (**d**) and an impaired white matter integrity (FA = 0.39). The risk for a transient deterioration was 58.6% and for permanent deterioration 30.3%. After counseling the patient, the decision was made not to perform another resection
Another case of a 51-year-old man with first diagnosis of a right precentral tumor is presented (**b**). He became symptomatic with a focal seizure and no neurological deficit was present. The cortical mapping revealed an infiltration of the motor cortex but without interfering with the nTMS-derived eloquent areas of the hand and face muscles. The cortical excitability showed an increased RMT of 79 V/m. Although a small TTD of 3 mm was measured (**e**), the diffusion integrity of the CST was maintained (FA = 0.51). The risk for a short-term motor deficit is 19.5% and for a long-term motor deficit 6.5%. Decision was made for surgery and a GTR could be performed without post-operative paresis
The third case is 57-year-old woman who presented with coordination disturbance of the right arm and leg. Imaging revealed a postcentral tumor without infiltration of the motor cortex (**c**) and with sufficient distance to the CST (20 mm) (**f**). With anticipated low risk of motor deterioration (short-term: 1.6%; long-term: 0%), complete tumor removal could be performed without inducing a new deficit

8.1.2 Integrity of the CST

In patients with a TTD ≤ 8 mm and/or infiltration of the motor cortex, the fiber tract integrity of the CST plays an important role for the risk assessment. This is represented by the mean fractional anisotropy (FA_{avg}) value of the CST which is calculated based on the diffusion tensor imaging (DTI) by the corresponding surgical planning tools. The fractional anisotropy value ranges between zero and one and indicates the degree/directionality of diffusion. A (theoretical) value of one means that all molecules would diffuse in the same direction and with a value of zero; on the contrary, all molecules would diffuse equally in all directions (isotropy). Several studies have shown that FA values are particularly impaired at the peritumoral level, where edema as well as cellular changes seem to play a role (Rosenstock et al. 2017a; Fekonja et al. 2020). With the help of a regression tree analysis (a method from the domain of machine learning), a cut-off value of 0.47 was calculated, which offers the best possible discrimination between increased and non-increased risk for the motor outcome. If the tract integrity is compromised (FA < 0.47), the risk for permanent deterioration of motor function increases to 16%. If the diffusion within the CST is intact (FA ≥ 0.47), the risk for a permanent deficit decreases to 7%. For the prediction of the short-term motor outcome (discharge day), the analysis of motor excitability is crucial.

8.1.3 Motor Excitability

The excitability of the motor system is represented by the resting motor threshold (RMT). This threshold indicates the minimum stimulation intensity of the coil at which a motor evoked potential (MEP) can be induced with a 50% probability. For the determination of the RMT, no difference is made whether the traditional Rossini-Rothwell method (five MEP positive stimulations out of ten total stimulations) or an adaptive threshold-hunting algorithm is used (Engelhardt et al. 2019). The RMT can be expressed in two different units: (Picht et al. 2011) percentage of the maximum output of the stimulation coil and (Frey et al. 2012) calculated e-field strength within the motor cortex. If the motor excitability was affected (RMT of tumor hemisphere ≥71 V/m), the risk for a transient deterioration of the motor status is 59%. If the motor excitability was not impaired (RMT < 71 V/m), the risk for transient deterioration is 30%.

Interestingly, patients with a preoperative motor deficit never showed an improvement in their paresis if the RMT ratio (affected/unaffected value) was >110%.

8.2 Impact on Clinical Decision-Making

Determining which surgical therapy is most appropriate for the individual patient is based on three key questions:

1. *Do we have a histopathological diagnosis?* In most cases not, so that there is almost always a surgical indication for sample acquisition.
2. *Can we improve the oncological prognosis by a supramarginal, gross total, or subtotal resection?* The answer to the question depends significantly on the histopathological entity/diagnosis, which is why in some departments a biopsy is often taken first and then a decision is made on definitive surgical therapy. However, for most entities, there is clear evidence that a gross total resection can optimize the neuro-oncologic outcome.
3. *Is there an increased risk of inducing a new motor deficit with the planned surgical strategy?* The nTMS-based risk stratification can be used to predict the risk for a new short-term and long-term motor deficit. Based on this evaluation, the patient can be counseled individually. It is important to discuss that not only the quality of life is significantly affected by a new paresis (Jakola et al. 2011). Studies have also shown that patients with new postoperative deficit are less likely to receive the necessary adjuvant therapy and thus have poorer survival rates (McGirt et al. 2009).

It becomes clear that an individual balance must be found for each patient in order to enable the greatest possible tumor resection without inducing a new deficit. Thereby, the detection of motor-eloquent areas impacts the surgical approach and the planned extent of resection (Picht et al. 2012b). A meta-analysis showed that preoperative nTMS motor mapping increased the average extent of resection while reducing the prevalence of postoperative motor deficits at the same time (Raffa et al. 2019). In addition to the cortical mapping, the nTMS-based tractography also became the focus of many studies. With its help, it is actually possible to visualize peritumoral tracts based on functional ROIs, which could not be visualized based on anatomical landmarks alone.

In another study, preoperative risk assessment was correlated with phenomena of the intraoperative neuromonitoring. Depending on the risk profile, the interpretation of ambiguous changes such as a transient MEP amplitude decrease or a permanent MEP amplitude decrease ≤50% is facilitated. In addition, it helps to select cases in which the subcortical stimulation intensity can be reduced to <5 mA without incurring an unacceptably high risk of morbidity.

Taking into account the previously discussed considerations, it has been suggested that optimal care for patients with motor-associated tumors can be achieved in facilities that are experienced with advanced preoperative mapping technologies such as nTMS (Ivren et al. 2023; Ille and Krieg 2021).

References

Engelhardt M, Schneider H, Gast T, Picht T. Estimation of the resting motor threshold (RMT) in transcranial magnetic stimulation using relative-frequency and threshold-hunting methods in brain tumor patients. Acta Neurochir. 2019;161(9):1845–51.

Fekonja LS, Wang Z, Aydogan DB, Roine T, Engelhardt M, Dreyer FR, et al. Detecting corticospinal tract impairment in tumor patients with fiber density and tensor-based metrics. Front Oncol. 2020;10:622358.

Frey D, Strack V, Wiener E, Jussen D, Vajkoczy P, Picht T. A new approach for corticospinal tract reconstruction based on navigated transcranial stimulation and standardized fractional anisotropy values. NeuroImage. 2012;62(3):1600–9.

Ille S, Krieg SM. Functional mapping for glioma surgery, part 1: preoperative mapping tools. Neurosurg Clin N Am. 2021;32(1):65–74.

Ivren M, Grittner U, Khakhar R, Belotti F, Schneider H, Poser P, et al. Comparison of anatomical-based vs. nTMS-based risk stratification model for predicting postoperative motor outcome and extent of resection in brain tumor surgery. NeuroImage Clin. 2023;38:103436.

Jakola AS, Gulati S, Weber C, Unsgard G, Solheim O. Postoperative deterioration in health related quality of life as predictor for survival in patients with glioblastoma: a prospective study. PLoS One. 2011;6(12):e28592.

McGirt MJ, Mukherjee D, Chaichana KL, Than KD, Weingart JD, Quinones-Hinojosa A. Association of surgically acquired motor and language deficits on overall survival after resection of glioblastoma multiforme. Neurosurgery. 2009;65(3):463–9. discussion 9–70

Picht T, Schmidt S, Brandt S, Frey D, Hannula H, Neuvonen T, et al. Preoperative functional mapping for rolandic brain tumor surgery: comparison of navigated transcranial magnetic stimulation to direct cortical stimulation. Neurosurgery. 2011;69(3):581–8. discussion 8.

Picht T, Strack V, Schulz J, Zdunczyk A, Frey D, Schmidt S, et al. Assessing the functional status of the motor system in brain tumor patients using transcranial magnetic stimulation. Acta Neurochir. 2012a;154(11):2075–81.

Picht T, Schulz J, Hanna M, Schmidt S, Suess O, Vajkoczy P. Assessment of the influence of navigated transcranial magnetic stimulation on surgical planning for tumors in or near the motor cortex. Neurosurgery. 2012b;70(5):1248–56. discussion 56–7

Raffa G, Scibilia A, Conti A, Ricciardo G, Rizzo V, Morelli A, et al. The role of navigated transcranial magnetic stimulation for surgery of motor-eloquent brain tumors: a systematic review and meta-analysis. Clin Neurol Neurosurg. 2019;180:7–17.

Rosenstock T, Giampiccolo D, Schneider H, Runge SJ, Bahrend I, Vajkoczy P, et al. Specific DTI seeding and diffusivity-analysis improve the quality and prognostic value of TMS-based deterministic DTI of the pyramidal tract. NeuroImage Clin. 2017a;16:276–85.

Rosenstock T, Grittner U, Acker G, Schwarzer V, Kulchytska N, Vajkoczy P, et al. Risk stratification in motor area-related glioma surgery based on navigated transcranial magnetic stimulation data. J Neurosurg. 2017b;126(4):1227–37.

Rosenstock T, Hani L, Grittner U, Schlinkmann N, Ivren M, Schneider H, et al. Bicentric validation of the navigated transcranial magnetic stimulation motor risk stratification model. J Neurosurg. 2021;136:1194–206.

Sollmann N, Wildschuetz N, Kelm A, Conway N, Moser T, Bulubas L, et al. Associations between clinical outcome and navigated transcranial magnetic stimulation characteristics in patients with motor-eloquent brain lesions: a combined navigated transcranial magnetic stimulation-diffusion tensor imaging fiber tracking approach. J Neurosurg. 2018;128(3):800–10.

Part III

nTMS Language Mapping

nTMS Language Mapping: Basic Principles and Clinical Use

9

Jyrki P. Mäkelä and Aki Laakso

9.1 Anatomy and Physiology of Language in Healthy Subjects and Patients

Mapping of areas processing language in the human brain was started in the early nineteenth century. The comprised model of language processing, still presented in some textbooks, includes the loosely defined frontal Broca and temporoparietal Wernicke areas in the left hemisphere, and their connections in the arcuate fascicle (for an elegant review, see Tremblay and Dick 2016). The model was created by data obtained from the effects of brain lesions in a limited number of patients, and the link between language disorders and focal brain lesions sometimes support, sometimes refute, these constructions. Theories based on hierarchical development of associative processes such as language by Hughlings Jackson, Goldstein, and Luria emphasized less localized presentations of language in the brain (Luria 1964).

Most aspects of language are difficult or impossible to study in animal models. Development of modern brain imaging techniques therefore provided a clear change in modeling of the language representations in the brain. The present models include a frontoparietal "dorsal stream" involved in mapping sound onto articulation-based representations, and a "ventral stream" in the temporal lobes, mapping sound onto meaning; particularly the ventral steam seems to have a clear representation both in the left and right hemisphere (Hickok and Pöppel 2007). Meta-analyses of modern

J. P. Mäkelä (✉)
BioMag Laboratory, HUS Diagnostic Center, University of Helsinki, and Helsinki University Hospital, Helsinki, Finland
e-mail: Jyrki.Makela@hus.fi

A. Laakso
Department of Neurosurgery, Helsinki University Hospital and Clinical Neurosciences, Helsinki, Finland
e-mail: aki.laakso@hus.fi

S. M. Krieg, T. Picht (eds.), *Navigated Transcranial Magnetic Stimulation in Neurosurgery*, https://doi.org/10.1007/978-3-031-97155-6_9

studies of lesion–symptom mapping (Na et al. 2022; Pial and Eikebloom 2023) support this dual pathway model encompassing widespread speech-related perisylvian areas.

Representations of language in the brain have also been studied when operating tumors in the vicinity of the presumed language eloquent areas. Under local anesthesia, the exposed cortex is stimulated directly via DES, and the effects of stimulation are tested on the patient's language. These studies (e.g., Ojemann et al. 1989) have revealed strong individual differences in cortical sites producing language disturbances. They pinpoint small, clearly delineated mosaic-like areas in traditional language-related cortical areas but also in the frontal lobe and middle parietal regions. Speech arrests can be observed in ventral prefrontal cortex, not in the typical Broca's area; anomia/aphasia can be induced from the Wernicke's area but also from the MFG and IFG and from the superior temporal gyrus (STG), and stimulation of right-hemisphere sites in homologous areas induce language disturbances as well (Tate et al. 2014). This individuality has been considered to explain a variability of language disturbances and their recovery after relatively similar lesions and emphasizes that language areas cannot be localized only on anatomical basis. This has led to the recent paradigm shift from the traditional localizationist view of language function in specific cortical regions toward a view of parallel, highly dynamic cortico-cortical and cortico-subcortical networks supporting speech and language function (Duffau et al. 2014). Plasticity of these speech representations has been well documented in gliomas (for a review, see, e.g., Nieberlein et al. 2023), and it is highly probable that active epileptic foci can induce functional plasticity as well (e.g., Mäkelä et al. 2013). The preservation of the functional integrity of the main language tracts may be even more important than preservation of cortical function (Duffay 2023). Nevertheless, even DES probing cortical and subcortical areas is not always sufficient in preserving the integrity of language in neurosurgical procedures: new postoperative chronic language disorders have been reported in 11% of previously asymptomatic patients despite DES during awake craniotomy (Ilmberger et al. 2008).

Knowing the cortical sites of language processing does not clarify how language processing is done. This question has been probed by studies of brain electrophysiology. Analysis of language-related evoked electroencephalography (EEG) and magnetoencephalography (MEG) responses has been used to estimate timing of language-related activations in the brain. A sequential processing model within 200–600 ms of language production has been comprised based on these studies (Indefrey 2011); it is supported by local field potential recordings in intracranial electrophysiology (Sahin et al. 2009). Nevertheless, it is highly probable that parallel processing of various language-related phenomena occurs as well. The neural access to word representations in the left perisylvian language-related cortex has been suggested to occur already 70 ms after language-related stimuli (Shtyrov and McGregor 2016).

Several studies emphasize the role of brain oscillatory activity in language processing. In signals induced by auditory language-type stimuli, 4 Hz MEG oscillations present syllables, whereas words formed from these syllables are coded with

2 Hz activity and sentences with 1 Hz activity; similar oscillatory activities can be recorded from subdural grids with the same stimuli from broad cortical areas including the temporal and frontal lobes in both hemispheres (Ding et al. 2016). Features of syllables may be determined by mechanical units of language apparatus defining natural oscillatory rhythms; natural mandibular cycles occur at about 4 Hz (Giraud et al. 2007). The oscillatory rhythms may also join processing of language features between different sensory modalities (Giraud and Poeppel 2012): when visual language input correctly predicts auditory language signals, a 3–4 Hz MEG activity develops over high-order language areas (Arnal et al. 2011). The importance of oscillatory brain activity in processing language is further emphasized by observations suggesting that electrocorticographic (ECoG) spectral mapping from subdural grid electrodes may identify some aspects of language processing better than stimulation of the grid electrodes (Cervenka et al. 2013).

Studies using rTMS to analyze language processing have found 4–8 Hz stimulus trains to be the most effective in disturbing language (Epstein et al. 1996; Devlin and Watkins 2007). In our preliminary tests, the nrTMS in this frequency range was found to be effective and well tolerated in search of language representations in healthy subjects and patients.

9.2 nrTMS Language Mapping in Healthy Subjects

nTMS induces currents within the cortex in an area that can be identified from the subject's or patient's own 3D brain MRI reconstruction (Ruohonen and Karhu 2010). The physiological changes induced by TMS and rTMS are qualitatively similar to those induced by DES. Consequently, the preoperative noninvasive nTMS is potentially well matched with the DES, the current gold standard of neurosurgical functional localization. The clinical experience from DES suggests that object naming is the most useful and sensitive experimental setup to map the language-related cortical areas intraoperatively (e.g., Petrovich Brennan et al. 2007). We developed a protocol for preoperative localization of language-related cortical areas by utilizing object naming, nrTMS, and videoed behavioral results (Lioumis et al. 2012). A commercial setup for the videoed language mapping procedure (NexSpeech™, Nexstim Ltd., Helsinki Finland) is in use in more than 60 neurosurgical centers around the world.

9.2.1 Methodology and Execution

Th methodology is described in detail in various publications (e.g., Krieg et al. 2017; Lioumis et al. 2023). Sets of pictures, normalized over visual and linguistic parameters (e.g., Brodeur et al. 2010), are displayed on a computer screen to the subject who is requested to name them as quickly and precisely as possible. The experiment starts with baseline sessions without nrTMS, followed by active nrTMS sessions. Both are video-recorded for offline analysis. Unfamiliar or incorrectly named images in the baseline session are removed from the image set, and only fluently named images are used during nrTMS. The stimulation is done with 5–10 Hz, 1–2 s nrTMS pulse trains delivered simultaneously or after the picture onset. The stimulus intensity over different brain regions is calibrated by the strength of the electric field induced within the cortex. The coil is hand-held and moved freely between the pulse trains. Usually, about 200–300 sites are stimulated within one hemisphere by moving the coil semirandomly between the trains of pulses, following a grid-like pattern so that the tested target sites cover systematically a wide frontotemporoparietal cortical area (for videos of the experiment, see Lioumis et al. 2012, 2023; Hernandez-Pavon et al. 2014), not limited to traditional Broca and Wernicke areas linked to speech processing. The baseline naming responses are compared with those recorded during nrTMS. The types of nrTMS-induced errors are classified; the corresponding nrTMS locations in parcellated cortex are marked as language-related and are tagged by the observed error type (for details, see Corina et al. 2010; Picht et al. 2013).

Sometimes the initially selected parameters do not produce language perturbations and require adjustments. Reduction of the interpicture interval to increase task difficulty, increase in nrTMS train frequency (from 5 to 7 Hz to 10 Hz), reduction of stimulation onset from 300 ms to 200–100 ms after the picture presentation for fast naming patients, and an increase in nrTMS intensity, if the patient tolerates it, are all possible. It is also useful to have different picture sets or tasks available to modify the cognitive demands of the task if needed, for example, in pediatric measurements (Narayana et al. 2021).

9.2.2 Results

The nrTMS mapping finds language-sensitive cortical sites most often in inferior frontal (IFG), superior temporal (STG), and supramarginal gyrus (SMG). A clear individual variability is observed in the sites where nrTMS induces language disturbance, paralleling individuality seen in DES results during awake surgery (Sanai et al. 2008; Corina et al. 2010). The methodology also provides an accurate monitoring and high-fidelity report of behavioral TMS experiments. It enables an unbiased offline analysis of stimulation effects as the stimulation sites can be blinded during data analysis. It is easier to recognize semantic and phonological paraphasia and performance errors from the video than during the measurement. Occasionally, the subjects smile or move their lips or jaw during anomia, indicating that no motor speech arrest is induced by nrTMS. Poststimulus consultation of the subject about the nrTMS-induced experiences and sensations is also feasible with joint viewing of the video recording. This is particularly useful in separating pain-induced changes (see below) from true effects of nrTMS. The spatial resolution of language nrTMS, evaluated from experiments in healthy subjects, suggests that language nrTMS is precise enough to evaluate language-related cortical areas on the gyral level (Sollmann et al. 2016). Naming in some healthy subjects is quite resistant to nrTMS.

The methodology provides ample opportunities for basic research as well. For example, sites interfering with naming are present also in the right hemisphere (Rösler et al. 2014; Sollmann et al. 2014). In electrocorticography recordings of patients speaking single syllables, consonants appear to be coded in the left hemisphere, vowels in both hemispheres and syllables in the right hemisphere, in line with theories of left hemisphere specialization for processing short-timescale linguistic units and right hemisphere processing of longer-duration units (Meier et al. 2023). Such fine-grained differences have not been tested with rnTMS.

Our anecdotal early observations indicated that the second rather than the primarily learned language is more easily disrupted by nrTMS. It appears that the cortical representations of the first and second languages differ somewhat (Tussis et al. 2017; Boerner et al. 2023), in line with previous DCS results (Lucas et al. 2004). The white matter pathways, however, are shared (Boerner et al. 2023). Moreover, cortical areas processing both first and second languages are differentially sensitive to nrTMS pulse timing: nrTMS induced significant language-specific delays of naming when applied over the left pIFG at 200 ms after the picture onset (Hämäläinen et al. 2018).

Types of naming errors may vary between repetitions of the stimulation of the same site (Fig. 9.1). Moreover, personal reports about the language difficulties induced by nrTMS to language-sensitive sites include reports of not remembering the word, in a manner of meeting a familiar face but not connecting it to the person's name, and breathing rhythm changes, which prevent naming. Modulations of attention during naming may produce errors similar to that induced by nrTMS. As attention enhances blood flow changes in language-related areas during word processing (Alho et al. 2003), its decrease may also decrease performance in, for example, naming tasks.

Targeting nrTMS to dorsal attention network disturbs semantic decisions (Capotosto et al. 2016). These features indicate that nrTMS affects language on several processing levels. As the active sites vary considerably between subjects, adequate intra- and interindividual statistics are useful for their analysis (cf. Hernandez-Pavon et al. 2014).

9.3 nrTMS Mapping in Patients

nrTMS language mapping combined with video recordings has been applied in preoperative functional localization in patients having tumors close to perisylvian language-related cortex. This approach was initially compared to DES during awake craniotomy (Picht et al. 2013; Tarapore et al. 2013; Jung et al. 2019), and the results were promising although not as systematic as those obtained in nTMS mapping of the motor cortex. Meta-analyses of multiple nrTMS–DES comparisons have provided general support for the initial findings (Indharty et al. 2023; Jeltema et al. 2021), and several reviews of the topic are available (e.g., Fang et al. 2019; Haddad et al. 2020; Natalizi et al. 2022). nrTMS language mapping has also found its way

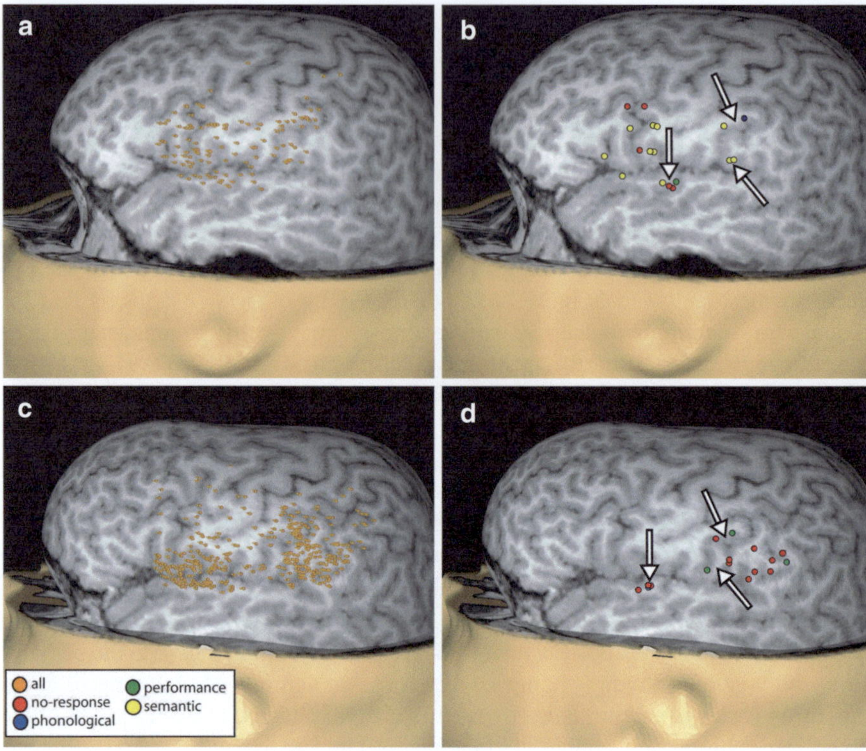

Fig. 9.1 Reproducibility of the nrTMS language mapping in a healthy subject. (**a**, **c**) illustrate all stimulated sites in two sessions, (**b**, **d**) illustrate the sites producing language disturbances in them. The arrows point to areas sensitive to nrTMS. Different error types are color-coded; all indicates all stimulated sites. Note that within the sensitive areas the error types vary between the two measurements. (Modified from Lioumis et al. (2012))

into description of clinical utility of TMS by International Federation of Clinical Neurophysiology (Vucic et al. 2023). The method is well tolerated in most adult subjects. In pediatric patients, mappings are slightly less successful but feasible and useful (see, e.g., Narayana et al. 2021; Lehtinen et al. 2018; Rosenstock et al. 2020). The youngest patient with successful preoperative language mapping in BioMag laboratory has been 6 years old.

9.3.1 Special Issues Related to Patients

The main indication for preoperative nrTMS language mapping is tumors located in "classic" language areas in the left perisylvian cortex, especially in the frontal operculum and temporoparietal region (Hervey-Jumper et al. 2015). Nevertheless, evidence about relevant speech and language function outside these areas, also in the nondominant hemisphere, is increasing (e.g., De Witt Hamer et al. 2012; Cogan et al. 2014); nrTMS is naturally applicable in patients with lesions in areas outside the left perisylvian cortex as well.

Maps and individual nrTMS-positive sites for language can be imported into the neuronavigation and hospital PACS via DICOM standard for data storage and flexible use from different hospital departments (Mäkelä et al. 2015a, b). The points can, for example, be transferred into the neuronavigation system of the neurosurgical operation room to add functionality to the anatomical visualization, or to be used as seeds for mapping the white matter tracts via DTI FT to be avoided during surgery (Fig. 9.2). The nrTMS experiment prepares the patient for the forthcoming awake craniotomy as stimulus presentation, stimulation, and experience of language impairment are comparable.

Moreover, the data enables consulting the patient preoperatively about the possible risks of extensive operation vs. remaining tumor tissue. Some patients feel empowered by the possibility to make such decisions about their treatment based on detailed visualization of the functional anatomy (e.g., Rosenstock et al. 2020).

No increased epileptic seizure rates have been associated with nrTMS language mapping (Tarapore et al. 2016; Braden et al. 2022), and it has been applied also in patients with medically intractable seizures in workup for epilepsy surgery without complications (Babajani-Feremi et al. 2016; Vitikainen et al. 2015; Lehtinen et al. 2018; Braden et al. 2022). In patients with epilepsy, seizures, with semiology consistent with their usual seizures, occurred during or after nrTMS in 6% of the patients. In DCS of a comparable cohort, seizures are induced in 35% of the studied patents (Braden et al. 2022). Speech rnTMS in patients with titanium implants from

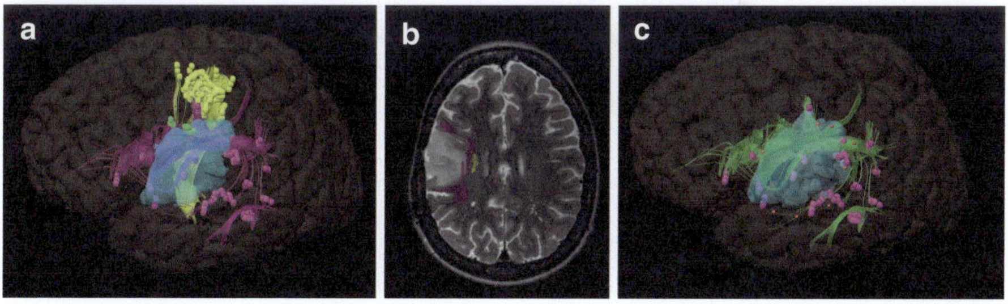

Fig. 9.2 Illustrative case. Data from a 48-year-old female patient with a recurrent oligoastrocytoma in the dominant left frontal operculum. Before awake surgery, she underwent nTMS for language and motor mapping. (**a**) Tumor outline in cyan, language nrTMS points in purple, face motor nTMS points in green, and hand motor nTMS points in yellow. Tractography was performed using nTMS points as seed ROIs: arcuate fascicle portion of the superior longitudinal fascicle in purple (generated with language nrTMS data), pyramidal tract in yellow (generated with motor nTMS data; additional ROI in the left side of the pons). (**b**) T2-weighted axial MRI of the same patient, demonstrating language nrTMS points and arcuate fascicle in purple, pyramidal tract in yellow. (**c**) Same patient with language nrTMS points in purple, arcuate fascicle in green. Smaller red points indicate the navigated intraoperative sites where DES elicited repeatable naming errors (anomia and paraphasia) using 3 mA and 50 Hz monopolar stimulation. (1.5 T MR images were processed using BrainLab iPlan Cranial 3.0 software)

previous surgery, ventriculoperitoneal shunts, vagus nerve stimulators, or dental or facial metal has been successfully done (Braden et al. 2022).

nrTMS speech mapping started with Finnish, German, and American patients and Nexstim nTMS stimulators. Nevertheless, other languages and devices have been tested as well. For example, Motomura et al. (2020) examined 61 Japanese patients with a Maxstim TMS device and custom-made electric field modeling. The method has been applied successfully also in Italian (Raffa et al. 2018) and Chinese (Han et al. 2021; Zhang et al. 2020) patients. This experience is in line with a recent DCS study, which found quite similar distribution in speech-sensitive sites in patients speaking English, French, and Chinese (Lu et al. 2021).

9.3.2 Validation

The mapping of language-related cortical areas by nrTMS was initially compared to DES during awake craniotomy in patients with brain tumors (Picht et al. 2013; Tarapore et al. 2013; Ille et al. 2015). nrTMS language mapping is sensitive but relatively nonspecific in detecting the sites producing language disturbance in DES. The first published comparison of the two methods in 20 patients found a sensitivity of 90%, specificity of 24%, positive predictive value (PPV) of 36%, and negative predictive value (NPV) of 84% for nrTMS prediction of DES within cortical parcellation areas. False-negative findings in comparison with DES were sparse and focused mainly on high parietal areas (Picht et al. 2013). Ille et al. (2015) found a sensitivity of 97%, specificity of 15%, PPV of 34%, and NPV of 91% for nrTMS with comparison to DES in 27 brain tumor patients. A study of 12 patients with brain tumors by nrTMS object naming compared the results with DES in five patients. nrTMS specificity was found to be 90%, sensitivity 98%, PPV 69%, and NPV 99% (Tarapore et al. 2013). These results may be due to limiting the distance between the compared DES and nrTMS sites to 1 cm instead of using the cortical parcellation system.

In six patients with epilepsy, the nrTMS predicted DES results with a sensitivity of 67%, specificity of 66% PPV of 24%, and NPV of 95% (Babajani-Feremi et al. 2016). Our experience on 20 patients with epilepsy surgery planning displayed similar values (Lehtinen et al. 2018). High NPV, indicating that nrTMS produces reliable "negative maps," has been a relatively consistent finding in nrTMS language mapping studies of brain tumor and epilepsy patients.

If DES is not applied, the sensitivity and specificity can be evaluated on the basis of clinical outcome. In series of 20 tumor patients, operated under general anesthesia with nrTMS-DTI guidance, speech worsened at 1-month evaluation in 6 out of 12 cases, whose tumors were preoperatively stratified as truly eloquent (PPV 50%). In eight patients, nrTMS-DTI indicated that the speech eloquence of the tumor, estimated by anatomical landmarks was not true. None of them had worsening of speech in 1-month testing, indicating a negative predictive value of 100% (Raffa et al. 2018).

9.3.3 Clinical Use of nrTMS Data

Cortical sites where repeated nrTMS could not disrupt language processing in an object-naming task appear to correspond rarely to language eloquent sites in intraoperative DES, suggesting a high correlation of "nrTMS language-negative" brain regions with DES (Krieg et al. 2014; Picht et al. 2013). Clusters of induced language disturbances, occurring within the same cortical area, indicate that the area is probably relevant for language processing. Isolated single deviations from the baseline performance indicate a need for further examination of the area by DES.

nrTMS language mapping is helpful for surgical planning as seed areas for nrTMS-based DTI FT for subcortical language tracts (Sollmann et al. 2015c; Raffa et al. 2016) (Fig. 9.2). nrTMS language mapping seems to work as a rescue strategy when awake mapping is not available (Ille et al. 2016b; Hendrix et al. 2017). These initial findings have largely been replicated in larger cohorts. Ille et al. (2021) prospectively enrolled 100 consecutive cases that were resected based on preoperative nrTMS language mapping (nrTMS group), and 47 cases via intraoperative DES mapping during awake craniotomy (awake group). Gross total resection (GTR) was achieved even more often in the nrTMS group than in the awake group. The authors state that the resection of language-eloquent gliomas purely based on nrTMS language mapping seems to be feasible and safe, and supported by similar functional and radiological outcomes compared to those of a cohort of patients who underwent DES-based glioma resection during awake craniotomy. Raffa et al. (2022) reported retrospectively collected data from 176 patients with rnTMS speech mapping and tractography who were not eligible for awake surgery and were operated under anesthesia. In 61 patients (35%), rnTMS mapping disclosed "false eloquence," a larger and safer distance between the tumor and the language network than hypothesized on the basis of the standard MRI scan. Permanent language deficits occurred in 7% of the patients. The percentage was 10% in true-eloquent tumors disclosed by preoperative nTMS mapping and 2% in false-eloquent tumors. GTR was obtained in 70% of patients with true-eloquent and in 100% of patients with false-eloquent lesions. The authors suggest that for patients not eligible for awake surgery, nTMS-based mapping and reconstruction of the language network is a valid alternative for performing tailored preoperative mapping of the network that may guide neurosurgeons to achieve personalized, safe, and effective resection of brain tumors in patients under general anesthesia.

It is naturally possible to make preoperative assessment of speech-related cortical areas also with extraoperative DCS, using implanted grid, strip, or stereoEEG electrodes in a manner applied in epilepsy surgery patients. Such recordings in tumor patients are, however, very resource-demanding and harassed by complications; some of them are severe (Fountas et al. 2022).

The efficacy of the nrTMS to map the language-related cortical areas varies considerably between individuals and experimental setups (Sollmann et al. 2013b). The specificity values reported in early studies (Picht et al. 2013; Tarapore et al. 2013; Ille et al. 2015) vary between 15% and 98% and suggest strong effects of slight variations in paradigms and interpretations of the results between the different laboratories using the methodology. A careful study of 20 tumor patients paid specific attention to map exposed cortex in detail with DCS (Bährend et al. 2020). The comparison of positive rnTMS and DCS overlaps revealed sensitivity of 35%, positive predictive value of 16%, specificity of 90%, and negative predictive value of 96%.

Some studies make speech rnTMS by activating predefined stimulation points according to the Cortical Parcellation System (Corina et al. 2005) or fixed grid points. The sites are tested in both hemispheres in a randomized order, and each stimulation point is targeted several times. This is naturally useful in standardizing the experimental setup. Nevertheless, use of fixed stimulus sites in defining cortical area sensitivity is relatively crude, as slight coli position and orientation changes may clearly alter the efficacy of the stimulation. In comparison with DCS, estimation of geometric distance (see, e.g., Tarapore et al. 2013; Lehtinen et al. 2018; Muir et al. 2022) may be a better solution, but no agreement is available about the exact measure to be applied.

It is also worth noticing that DES speech mapping techniques and stimuli are not standardized (e.g., Papazalas et al. 2022); this may induce variability to rnTMS-DES comparisons. Extraoperative DES sometimes shows low replicability across stimulation sessions. Thus, perfect accordance with DCS is not necessarily the optimal goal of a language mapping method (for references, see Lehtinen et al. 2018).

As the stimulation frequencies used in nrTMS language mapping are different from DES and the distance from the cortex is clearly longer, it is evident that a single nrTMS train does not give conclusive results about language sensitivity of any particular cortical site. Successful nrTMS language mapping is based on several nrTMS trains and meticulous observation and documentation of induced language changes, best done by video-based offline analysis (Lioumis et al. 2012). Final surgical decisions regarding the functional relevance of both cortical tissue and subcortical fiber tracts are made based on the intraoperative mapping and monitoring results (Bello et al. 2014; Soffietti et al. 2010).

Functional plasticity may be associated with the growth of the brain tumors. This plasticity may permit a multistage resection of tumors several years after the first surgery (Robles et al. 2008). In nrTMS studies, it appears that patients with left hemisphere tumors appear to have more right hemisphere sites interfering with language during nrTMS mapping than healthy control subjects (Rösler et al. 2014; Krieg et al. 2013). The possible functional significance of these changes is unclear at present. First reports on follow-up investigations by nrTMS showed plastic reshaping of cortical language areas after surgery (Kawashima et al. 2013). It has been suggested that surgical resection of brain gliomas may benefit from better understanding of dynamic interaction between the tumor and the reactional plasticity of the nervous system as a result of both tumor invasion and surgery (Duffay 2005). Language mapping by nrTMS may provide a tool to gather such understanding.

nrTMS induces more errors in patients with language disturbances than in those with fluent language (Rösler et al. 2014). If the error rates induced by nrTMS are clearly higher in the left than right hemisphere, the operation of the left hemisphere tumor harbors an increased risk of postoperative language disturbance (Ille et al.

2016a). nrTMS may thus be useful in defining hemispheric lateralization of language processes in patients. It is worth noticing, however, that language-sensitive sites are found in the right hemisphere of most investigated patients (Rösler et al. 2014; Sollmann et al. 2015b), even in those with left-dominant language representations found by other methods, including the gold standard WADA test. Previous experience on nonnavigated TMS in lateralizing language has been less than optimal (Epstein et al. 2000; for discussion, see Devlin and Watkins 2007). Encouraging results about the rnTMS-Wada comparison in terms of hemispheric speech lateralization have been published (Han et al. 2021).

The nrTMS language mapping has mainly been used for preoperative planning of tumor surgery. The sensitivity and specificity values of nrTMS language mapping to predict DES results in epilepsy surgery workup (Babajani-Feremi et al. 2016; Lehtinen et al. 2018; Zhang et al. 2020) are in line with those obtained in tumor surgery and suggest similar usefulness of nrTMS language mapping in planning epilepsy surgery as well.

The clinical usefulness of preoperative nrTMS language mapping is suggested by a study of two groups of 25 patients. Patients with an available nrTMS language mapping data had significantly smaller craniotomies and less postoperative aphasia in the first postoperative days than those without nrTMS maps. No difference in outcome, however, was observed in long-term follow-up of the two groups (Sollmann et al. 2015a).

The preoperative nrTMS mappings of language-related cortical areas are in routine clinical use in our hospital. nrTMS maps are incorporated into neuronavigation planning software and are used as ROIs for tractography (Fig. 9.2). During surgery, cortical mapping with DES is started at nTMS hotspots localized by neuronavigation. This usually facilitates rapid identification of positive DES sites and the required threshold current to elicit DES motor and language responses. In cases where eloquent areas are shifted from their presumed classical anatomical regions due to tumor- or epilepsy-induced plasticity, it is very comforting for the neurosurgeon to have this information already preoperatively, as it eliminates confusion and uncertainty during surgery. It is also worth noting that because the depth of the nrTMS effect is several millimeters, it may sometimes be unclear which of the adjacent gyri on opposite sides of a sulcus is actually responsible for the observed nrTMS response. Naturally, the fusion of nrTMS data with the intraoperative neuronavigation system has limitations due to brain shift after the skull opening and particularly during tissue resection.

The information provided by neuronavigation on the locations and directions of functionally crucial fiber tracts greatly facilitate intraoperative subcortical stimulation, which is started at the sites superficial to the tracts, using stimulation currents sufficient to elicit responses at safe distances. 1-mA current stimulates white matter motor tracts at the depth of approximately 1 mm (Kamada et al. 2009). We usually start locating the subcortical tracts with a current of 10 mA and estimate the distance to the tract on the basis of strength of the current required to elicit responses. The resection is halted latest when the threshold current reaches 3 mA to avoid permanent tract injury.

9.3.4 Comparison with fMRI and MEG

Other preoperative noninvasive methods for functional cortical mapping, such as fMRI or MEG, rely on different physiological changes than nrTMS and DES. fMRI detects blood oxygenation changes in the brain induced by stimuli or tasks, and MEG measures changes of brain's magnetic fields induced by sensory stimuli or motor activations. Combination of fMRI, MEG, and nrTMS may give a more complete picture about the pathophysiology and disease-related functional plasticity in patients with epilepsy (Mäkelä et al. 2013).

Language-related activations found by fMRI appear not to have a useful predictive value in estimating the results of DES in awake craniotomy (Giussani et al. 2010). nrTMS language mapping may be superior to fMRI in the vicinity of brain lesions since it is far less prone to artifacts based on increased vascularity and altered tissue oxygenation (Picht et al. 2013; Sollmann et al. 2013a; Ille et al. 2015; Muir et al. 2022). This may be particularly true for highly vascularized and high blood flow lesions, such as arteriovenous malformations, which may extend to eloquent regions (Juenger et al. 2009; Kronenburg et al. 2014). This may also indicate that the language networks in the brain as depicted by fMRI may contain both critical and participating areas (Mesulam 2000). These suggestions were supported by Ille et al. (2015), who described clearly lower sensitivity (40%) and NPV (79%) of language fMRI than nrTMS in the same patients (see above) in predicting the DES localization of the cortical language sites during awake surgery.

In healthy subjects, localization of the motor cortex is more straightforward with nTMS than MEG, and the found sites are more clustered in nTMS than MEG localizations (Pitkänen et al. 2019). Cortical excitatory modifications detected by nTMS may correlate only loosely by excitability changes estimated by MEG (Mäkelä et al. 2015a). These differences are probably reflected in localization of speech-related areas as well. Tarapore et al. (2013) compared localization results of nrTMS with those obtained by MEG beta band suppression during picture naming and verb generation, and with DES during awake craniotomy. MEG found 18 verb generation and 14 picture naming sites. Seven MEG language sites correlated with nrTMS sites in five patients and with DES in two patients; the result was clearly inferior to that in comparison with nrTMS and DES sites.

9.3.5 Mechanism of the rTMS Effect

Several mechanisms on how rTMS mediates its effects on the brain have been put forward. The most favored one suggests that rTMS induces a "virtual lesion" in the brain: no general agreement about exact mechanisms underlying the "lesion" is, however, available. rTMS may induce an aberrant inhibition/excitation state of the stimulated cortical target (Harris et al. 2008), introduce random activity that competes with the neural activity coding the signal (Walsh and Cowey 2000), or activate the neurons that have not been activated by target stimulus, thereby reducing signal-to-noise ratio and worsening stimulation detection (Silvanto and Muggleton 2008).

All TMS effects reflect complex interactions of the external and internal factors, and the state dependency of the TMS effects has been strongly emphasized (Hartwigsen and Silvanto 2023). rTMS effects (as well as those of DES; see Borchers et al. 2012) are not restricted to target regions but spread to adjacent and anatomically connected regions (Robertson et al. 2003; Walsh and Cowey 1998; Ilmoniemi et al. 1997; Valero-Cabre et al. 2005). Recently, a preprint described responses induced by dorsolateral prefrontal cortex TMS from electrodes in the anterior cingulate and insula in humans (Wang et al. 2022). It also appears that 1 Hz TMS has opposite impact on the functional connectivity of sensory and cognitive brain regions. TMS of the frontal cortex decreased local inhibition and disrupted feedforward and feedback connections estimated from resting-state fMRI recordings. Conversely, identical TMS increased local inhibition and enhanced forward signaling in the occipital cortex (Castrillon et al. 2020).

It is perfectly feasible that rTMS modulates the ongoing oscillatory electric brain activity and interferes with the normal brain processing. Rhythmic rTMS causes local entrainment of natural EEG oscillations (e.g., Thut et al. 2001; Veniero et al. 2015). Brief 20 Hz rTMS trains may interfere with suppression of alpha band oscillations associated with attention (Capotosto et al. 2016). The effects of rTMS depend on the state of the brain network and may vary with its variance despite precisely same stimulation parameters and site. This may contribute to variability of the effects of nrTMS at the same stimulation site (Fig. 9.1). nrTMS-EEG in studies of language-related oscillatory activity may reveal clues of its physiological effects. It is not clear what frequency ranges should be examined to this end. The most obvious choice is the theta frequency range induced by 4–8 Hz rTMS. Also 10 Hz inhibitory alpha bursts, known to be associated with the interference of memory tasks by rTMS (Capotosto et al. 2016), may be worth studying. Source localization of rTMS-EEG activations is, however, difficult (e.g., Litvak et al. 2007).

9.4 Open Questions

Language mapping protocols with nrTMS vary somewhat between centers performing the examinations. This is related to varying results, hampers comparison of the results across centers, and interferes with accepting the method in clinical use. An international workshop of experienced clinical nrTMS users prepared a recommendation of a joint protocol for further clinical studies (Krieg et al. 2017). Nevertheless, variability of the experimental setups still occurs.

In our experience, stimulation parameters may need to be adjusted according to individual properties of the subjects; the best sensitivity may be obtained with slightly different stimulation frequencies in different patients. Some patients name the images quite rapidly, and adjustment of the timing of the rTMS pulse train may be needed. Some patients with preexisting language difficulties may need long intervals between the picture presentation and rTMS train to be able to perform the task. These factors can usually be evaluated and tested during the baseline naming session. A quiet environment is useful to enable the full concentration on the naming task by the patient; it is difficult to separate rTMS-induced naming changes from those due to fluctuating attention to the task. Fatigue and boredom due to several repetitions of the task may also influence the task performance.

We have been able to induce language errors of some type in most patients with the nrTMS parameters described above. nrTMS trains with higher frequency may also induce language disturbances (Rogic et al. 2014; Nettenkoven et al. 2021). Pain and tetanization of face muscles may prevent high stimulation frequencies particularly when long nrTMS trains are delivered (Devlin and Watkins 2007).

The level of discomfort or pain experienced during mapping is a crucial factor in interpreting the language mapping results. rTMS-induced pain may limit the spatial extent of nrTMS language mapping, particularly when stimulating the orbital and polar inferior frontal gyrus (Krieg et al. 2016). A median visual analog scale (VAS) score of 4.5 (0 to 10; 10 = maximum pain intensity) for the maximum experienced pain has been reported, but the discomfort of nrTMS was not considered distressing by most patients (Tarapore et al. 2016). Search for optimal stimulus orientation (usually perpendicular to the fibers of the temporal muscle to minimize contraction) or reduction of stimulation intensity usually enables the completion of the mapping. Seventy percent of healthy children tolerated bilateral speech rnTMS mapping. Their mean VAS score was 3 and electric field strength was 115 V/m; the procedures to decrease pain described above were not reported (Greiner et al. 2023).

Small changes in stimulation site, orientation of the coil, and its tilt with respect to the head surface modify the strength of nTMS-induced MEPs (Schmidt et al. 2015). Similar sensitivity affects the nrTMS effects on object naming (Sollmann et al. 2015b). Stimulation frequency may also influence the type of the induced language disturbances (Hauck et al. 2015b). It is worth noting, however, that effects of sequential DES of the same site may vary considerably as well (Whitaker and Ojemann 1977; Lesser et al. 2008).

One source of the variability of nrTMS language mapping results may relate to the modeling of the induced cortical electric fields, which identifies the cortical target area. In the present nrTMS system accepted for clinical mapping of language-related cortex by the FDA, this is done by localized spheres. This type of modeling is fast to compute and is valid in brain areas with local sphericity, such as sensorimotor or parietal areas. It may, however, induce localization errors, for example, in nonspherical frontotemporal areas (Nummenmaa et al. 2013). As a result, source modeling errors and small changes in orientation may divert the actual created voltage to unexpected sites. Moreover, due to physical properties of the induced fields, nrTMS is not able to directly stimulate deep frontal or temporomesial structures.

The parameters of the nrTMS language mapping protocols are still not fully optimized for clinical use. For example, the false-negative findings in parietal regions during nrTMS language mapping are reported in nrTMS–DES comparisons. Object naming appears to be more sensitive to nrTMS-induced disturbances than action naming (Hernandez-Pavon et al. 2014; Hauck et al. 2015a), but use of action naming instead of object naming appears not to improve parietal sensitivity, contrary to suggestions based on DES (Lubrano et al. 2014). Several recent studies have applied action naming in rnTMS speech mapping (Ohlert et al. 2021; Bastianse et al. 2023; Ntemou et al. 2023a, b; for a review, see, e.g., Ntemou et al. 2023b). A mixed action and object naming stimulus patterns has been used for rnTMS speech mapping in seven patients, and the preoperative results aided the interpretation of postoperative language deficits (Ohlerth et al. 2022). A rnTMS setup for language comprehension has been described as well (Rejnö-Habte-Selassie et al. 2020). It is, however, unclear how the findings translate to clinical practice.

First tests of inducing language perturbations with nrTMS were done with a 300-ms delay between the picture onset an nrTMS to minimize the number of required nrTMS pulses and maximize the overlap of nrTMS and language processing, as the used coils had no cooling system and the coil heating limited the total number of deliverable pulses. Single-pulse nTMS to Broca's area delays naming significantly when it is delivered 225–300 ms after the picture onset, not before or after this time range (Wheat et al. 2013). In healthy subjects, varying the delay between images and nrTMS affects the type and distribution of the induced speech disturbances (Sollmann et al. 2017). DES typically aims at 0-ms delay between image presentation and stimulation although the stimulation is delivered manually, without automatized, precise time locking. nrTMS experiments using 0-ms delay produce language errors effectively and may increase sensitivity in the parietal areas as compared with DES (Krieg et al. 2014; Ille et al. 2015). Naturally, the comparability between DES and nrTMS should be increased by increasing similarity between the experimental setups. nrTMS may induce blinks, which may interfere with early image processing in the 0-ms paradigm (Corthout et al. 2003) and decrease specificity of the findings. Some recent studies, motivated by interest in studying linguistic phenomena, have suggested a 200–300 ms interval between the picture onset and rnTMS (Bastiaanse et al. 2023; Vasileiadi et al. 2023).

Suggested Improvements in Methodology

One problem of scoring the nrTMS-induced changes in object naming is the evaluation of delayed responses. Inducing a 50-ms delay of naming by rTMS of Broca's area is considered significant in basic research studies (e.g., Schuchmann et al. 2012), but such short delays are difficult to detect in visual analysis of language responses. An automated analysis of language-induced laryngeal vibrations by an accelerometer has been suggested useful for this purpose (Vitikainen et al. 2015). It is not sensitive to nrTMS-associated coil clicks or environmental sounds, for example, from the cooling system, and is thus preferable to recording vocalizations by a microphone. This may be of interest as hesitations were most often true positive findings in Lehtinen et al. (2018), possibly because these trials comprise adequate attention and effort but no interference with the motor output. Sound processing of microphone output may also provide useful behavioral data in association with nTMS speech experiments (Seynaeve et al. 2020). This method, however, has not yet been tested against DES recordings.

Brain tumors may be related to abnormal electric brain activity, and rTMS trains may induce such abnormal electric phenomena although overt seizures are not observed. The same possibility is naturally present in patients with frequent epileptic seizures as well. Such changes might induce speech disturbances although the stimulated sites were not related to speech processing. Systematic studies with TMS-EEG might be useful in detecting epileptiform phenomena (e.g., Vitikainen et al. 2013). The attachment of EEG electrodes requires time and prolongs distance between the TMS coil and cortex, requiring more stimulation output to achieve cortical activation. This may be problematic particularly in pediatric patients. No seizures were reported on nrTMS language mapping in a large international cohort of tumor patients (Tarapore et al. 2016), and in patients in epilepsy surgery workup, the observed rare seizures were consistent with their usual seizure semiology (Braden et al. 2022).

A better understanding of the effects of nrTMS on language networks is probably required to clarify the open questions and benefit its precise clinical use. Recently, speech rnTMS activations were used as a source for tractography for language graph-based connectome analysis in patients with gliomas. Aphasia induced by glioma appeared to deteriorate connectivity in the left hemisphere speech networks (Zhang et al. 2022) and strengthen the connectivity between hemispheres (Zhang et al. 2023). Resting-state fMRI has also been used in connectivity analysis of different types of rTMS-induced speech errors from the parietal cortex. The resting-state activity in sites where rnTMS-induced anomias and speech arrests from the parietal cortex was positively correlated with resting-state activity in the IFG, whereas activity in the sites producing phonological and semantic errors was not (Vasileiadi et al. 2023).

Naturally, the development of nTMS speech mapping requires experimentation for further development. For example, short bursts of high-frequency rnTMS may be better tolerated than the traditionally used 5–8 Hz stimulation (Nettenkoven et al. 2021). Integration of Brodmann maps with the anatomical MRI (Reijonen et al. 2021) to select stimulation targets may provide a new way to search for effective stimulation sites. Combining the nrTMS mappings with DTI FT already during the stimulation sessions may clarify the possible routes for the spreading effects. Imaging of the white matter tracts may be useful for optimizing nrTMS coil location and orientation for maximal stimulation of a predetermined axon bundle (Nummenmaa et al. 2014). Indeed, using the arcuate fascicle to inform selection of rnTMS sites seems to increase the occurrence of anomias (Reisch et al. 2022). On-line tractography of individual stimulation sites (Aydogan et al. 2022) may guide the selection of the TMS targets in a beneficial way. In the future, multilocus TMS (Nieminen et al. 2022) may speed up the selection of the stimulation sites most effective in producing speech disturbances. A multicoil TMS system, providing a possibility to incorporate also optically pumped magnetometers for a simultaneous MEG recording, has been introduced (Matsubara et al. 2023). These developments may produce new ways to probe the electrophysiological mechanisms inducing language problems by nrTMS, also in patients. Comparisons of these new experimental setups with DCS in patients are naturally required before their application in clinical use.

The two published meta-analyses recognize the lack of true double-blind comparisons in speech rnTMS–DCS studies but note that neuro-oncology is a particularly difficult area in research due to its ethical implications; this may complicate the buildup of blinded randomized studies.

9.5 Conclusions

Mapping language representations with nrTMS combined with video recordings for post hoc analysis appear to be a useful tool in basic research and particularly in preoperative planning of brain surgery. In our experience, it helps in surgical planning and eases evaluation of the need of intraoperative electrophysiology. It may also shorten the duration of awake craniotomies by making the search of positive sites and evaluation of the required simulation current in DES easier. It focuses on skull opening, resulting in smaller craniotomies. The language-negative sites are also valuable in planning surgery as they limit the area to be tested by DES during awake craniotomy. Although the first patient series using only nrTMS language mapping to localize the language representations have reported favorable results, its use without intraoperative awake mapping should be limited to patients having contraindications against awake surgery. DES during awake craniotomy is still the gold standard of functional localization in perisylvian neurosurgery. The liberal use for nrTMS language mapping in patients with tumors in the left perisylvian region and in the vicinity of the main language tracts is warranted. Particularly left-handed patients with right-sided tumors or patients with tumors outside the classical language areas who present with previous or transient clinical signs of language disruption probably benefit from preoperative nrTMS language mapping.

During the 7 years from the publication of the first version of this chapter, remarkable advances have been made in developing the nTMS speech mapping in patients. Particularly the use of the speech nTMS sites combined with tractography has developed, and this combination shows promise even in replacing DCS as a rescue strategy in some cases (Raffa et al. 2022; Ille et al. 2021). A multitude of suggestions improving the speech stimuli have appeared as well, but the clinical workhorse still seems to be the classical object naming task, and even the nrTMS stimulation parameters seem to have remained more or less the same. Time will tell if the new methodological developments will improve the specificity and positive predictive value of the methodology.

Acknowledgments The nTMS studies in BioMag laboratory have been supported by grants from SalWe Research program for Mind and Body (TEKES-the Finnish Funding Agency for Technology and Innovation grant 1104/10) and from the grants for project development by HUSLAB (MLE82TK005) and by HUS Medical Imaging Center (MLD81TK303 and MLD81TK304).

References

Alho K, Vorobyev V, Medvedev S, Pakhomov S, Roudas MS, Tervaniemi M, van Zuijen T, Näätänen R. Hemispheric lateralization of cerebral blood flow changes during selective listening to dichotically presented continuous speech. Cogn Brain Res. 2003;17:201–11.

Arnal LH, Wyart W, Giraud A-L. Transitions in neural oscillations reflect prediction errors generated in audiovisual speech. Nat Neurosci. 2011;14:797–801.

Babajani-Feremi A, Narayana S, Rezaie R, Choudhri AF, Fulton SP, Boop FA, et al. Language mapping using high gamma electrocorticography, fMRI, and TMS versus electrocortical stimulation. Clin Neurophysiol. 2016;127:1822–36.

Bährend I, Muench MR, Schneider H, Moshourab R, Dreyer FR, Vajkoczy P, Picht T, Faust K. Incidence and linguistic quality of speech errors: a comparison of preoperative transcranial magnetic stimulation and intraoperative direct cortex stimulation. J Neurosurg. 2020;134:1409–18. https://doi.org/10.3171/2020.3.JNS193085. PMID: 32470943

Baran Aydogan D, Souza VH, Matsuda RH, Lioumis P, Ilmoniemi RJ. Real-time tractography-assisted neuronavigation for TMS. bioRxiv preprint. 2022; https://doi.org/10.1101/2023.03.09.531565.

Bastiaanse R, Ohlerth A-K. Presurgical language mapping: what are we testing? J Pers Med. 2023;13:376. https://doi.org/10.3390/jpm13030376.

Bello L, Riva M, Fava E, Ferpozzi V, Castellano A, Raneri F, et al. Tailoring neurophysiological strategies with clinical context enhances resection and safety and expands indications in gliomas involving motor pathways. Neuro-Oncology. 2014;16:1110–28.

Boerner C, Schroeder A, Neyer B, Krieg SM, Ille S. Cortical location of language function may differ between languages while white matter pathways are similar in brain lesion patients. Brain Sci. 2023;13:1141. https://doi.org/10.3390/brainsci1308114.

Borchers S, Himmelbach M, Logothetis N, Karnath H-O. Direct electrical stimulation of human cortex-the gold standard for mapping brain functions? Nat Rev Neurosci. 2012;13:63–70.

Braden AA, Weatherspoon SE, Boardman T, Williard T, Adkins A, Gibbs SK, Wheless JW, Narayana S. Image-guided TMS is safe in a predominately pediatric clinical population. Clin Neurophysiol. 2022;137:193–206.

Brodeur MB, Dionne-Dostie E, Montreuil T, Lepage M. The bank of standardized stimuli (BOSS), a new set of 480 normative photos of objects to be used as visual stimuli in cognitive research. PLoS One. 2010;5:e10773.

Capotosto P, Baldassarre A, Sestieri C, Spadone S, Romani G-L, Corbetta M. Task and regions specific top-down modulation of alpha rhythms in parietal cortex. Cereb Cortex. 2016:1–8. https://doi.org/10.1093/cercor/bhw278.

Castrillon G, Sollmann N, Kurcyus K, Razi A, Krieg SM, Riedl V. The physiological effects of noninvasive brain stimulation fundamentally differ across the human cortex. Sci Adv. 2020;6:eaay2739.

Cervenka MC, Corines J, Boatman-Reich DF, Eloyan D, Sheng X, Franaszczuk PJ, Crone NE. Electrocorticographic functional mapping identifies human cortex critical for auditory and visual naming. NeuroImage. 2013;69:267–76.

Cogan GB, Thesen T, Carlson C, Doyle W, Devinsky O, Pesaran P. Sensory-motor transformations for speech occur bilaterally. Nature. 2014;507:94–8.

Corina DP, Loudermilk BC, Detwiler L, et al. Analysis of naming errors during cortical stimulation mapping: implications for models of language representation. Brain Lang. 2010;115:101–12.

Corina DP, Gibson EK, Martin R, Poliakov A, Brinkley J, Ojemann GA. Dissociation of action and object naming: evidence from cortical stimulation mapping. Hum Brain Mapp. 2005;24(1):1–10. https://doi.org/10.1002/hbm.20063. PMID: 15593268; PMCID: PMC6871733.

Corthout E, Hallett M, Cowey A. Interference with vision by TMS over the occipital pole: a fourth period. Neuroreport. 2003;14:651–5.

De Witt Hamer PC, Robles SG, Zwinderman AH, Duffau H, Berger MS. Impact of intraoperative stimulation brain mapping on glioma surgery outcome: a meta-analysis. J Clin Oncol. 2012;30:2559–65.

Devlin JT, Watkins KE. Stimulating language: insights from TMS. Brain. 2007;130:610–22.
Ding N, Melloni L, Zhang H, Tian X, Poeppel D. Cortical tracking of hierarchical linguistic structures in connected speech. Nat Neurosci. 2016;19:158–64.
Duffau H. Damaging a few millimeters of the deep white matter tracts during glioma surgery may result in a large-scale brain disconnection. J Neurosurg. 2023;11:1–4. https://doi.org/10.3171/2023.6.JNS231048. Epub ahead of print. PMID: 37874685
Duffay H. Lessons from brain mapping in surgery for low-grade glioma: insights into associations between tumour and brain plasticity. Lancet Neurol. 2005;4:476–86.
Duffay H, Moritz-Gasser S, Madonnet E. A re-examination of neural basis of language processing: proposal of a dynamic hodotopical model from data provided by brain stimulation mapping during picture naming. Brain Lang. 2014;131:1–10.
Epstein CM, Lah JJ, Meador K, et al. Optimum stimulus parameters for lateralized suppression of speech with magnetic brain stimulation. Neurology. 1996;47:1590–3.
Epstein CM, Woodard JL, Stringer AY, Bakay RAE, Henry TR, Pennell PB, Litt B. Repetitive transcranial magnetic stimulation does not replicate the Wada test. Neurology. 2000;55:1025–7.
Fang X, Liu M, Lu C, Zhao Y, Liu X. Current status and potential application of navigated transcranial magnetic stimulation in neurosurgery: a literature review. Chinese Neurosurg J. 2019;5:12.
Fountas KN, Brotis A, Paschalis T, Kapsalaki E. The role of extra-operative cortical stimulation and mapping in the surgical management of Intracranial Gliomas. Brain Sci. 2022;12:1434. https://doi.org/10.3390/brainsci12111434.
Giraud A-L, Poeppel D. Cortical oscillations and speech processing: emerging computational principles and operations. Nat Neurosci. 2012;15:511–7.
Giraud A-L, Kleinschmidt A, Poeppel D, Lund TE, Frackowiack RSJ, Laufs H. Endogenous cortical rhythms determine cerebral specialization for speech perception and production. Neuron. 2007;56:1127–34.
Giussani C, Roux FE, Ojemann J, Sganzerla EP, Pirillo D, Papagno C. Is preoperative functional magnetic resonance imaging reliable for language areas mapping in brain tumor surgery? Review of language functional magnetic resonance imaging and direct cortical stimulation correlation studies. Neurosurgery. 2010;66:113–20.
Greiner HM, Maue E, Horn PS, Vannest J, Vedala K, Leach JL, Tenney JR, Williamson B, Fujiwara H, Coghill RC, Mangano FC, Kadis DS. Tolerability of transcranial magnetic stimulation language mapping in children. Epilepsy Res. 2023;194:107183.
Haddad AF, Young JS, Berger MS, Tarapore PE. Preoperative applications of navigated transcranial magnetic stimulation. Front Neurol. 2020:11. https://doi.org/10.3389/fneur.2020.628903.
Hämäläinen S, Mäkelä N, Sairanen V, Lehtonen M, Kujala T, Leminen A. TMS uncovers details about sub-regional language-specific processing networks in early bilinguals. NeuroImage. 2018;171:209–21.
Han Y, Tong X, Wang X, Teng F, Deng Q, Zhou J, Guan Y, Yan Z, Chen L, Luan G, Wang M. A concordance study determining language dominance between navigated transcranial magnetic stimulation and the Wada test in patients with drug-resistant epilepsy. Epilepsy Behav. 2021;117:107711.
Harris JA, Clifford CW, Miniussi C. The functional effect of transcranial magnetic stimulation: signal suppression or neural noise generation? J Cogn Neurosci. 2008;20:734–40.
Hartwigsen G, Silvanto J. Noninvasive brain stimulation: multiple effects on cognition. Neuroscientist. 2023;29:639–53. https://doi.org/10.1177/10738584221113806.
Hauck T, Tanigawa N, Probst M, Wohlschlaeger A, Ille S, Sollmann N, Maurer S, Zimmer C, Meyer B, Ringel F, Krieg SM. Task type affects location of language-positive cortical regions by repetitive navigated transcranial magnetic stimulation mapping. PLoS One. 2015a;10:e0125298.
Hauck T, Tanigawa N, Probst M, Wohlschlaeger A, Ille S, Sollmann N, Maurer S, Zimmer C, Ringel F, Meyer B, Krieg SM. Stimulation frequency determines the distribution of language positive cortical regions during navigated transcranial magnetic brain stimulation. BMC Neurosci. 2015b;16:5.

Hendrix P, Senger S, Simgen A, Griessenauer CJ, Oertel J. Preoperative rTMS language mapping in speech-eloquent brain lesions resected under general anesthesia: a pair-matched cohort study. World Neurosurg 2017;100:425–33. https://doi.org/10.1016/j.wneu.2017.01.041.

Hernandez-Pavon JC, Mäkelä N, Lehtinen H, Lioumis P, Mäkelä JP. Effects of navigated TMS on object and action naming. Front Neurosci. 2014;8:Article 660, 1–9.

Hervey-Jumper SL, Li J, Lau D, Molinaro AM, Perry DW, Meng L, et al. Awake craniotomy to maximize glioma resection: methods and technical nuances over a 27-year period. J Neurosurg. 2015;123:325–39.

Hickok G, Poeppel D. The cortical organization of speech processing. Nat Rev Neurosci. 2007;8:393–402.

Ille S, Sollmann N, Hauck T, Maurer S, Tanigawa N, Obermueller T, et al. Impairment of preoperative language mapping by lesion location: a functional magnetic resonance imaging, navigated transcranial magnetic stimulation, and direct cortical stimulation study. J Neurosurg. 2015;123:314–24.

Ille S, Kulchytska N, Sollmann N, Wittig R, Beurskens E, Butenschoen VM, Ringel F, Vajkozy P, Meyer B, Picht T, Krieg SM. Hemispheric language dominance measured by repetitive navigated transcranial magnetic stimulation and postoperative course of language function in brain tumors. Neuropsychologia. 2016a;91:50–60.

Ille S, Sollmann N, Butenschoen VM, Meyer B, Ringel F, Krieg SM. Resection of highly language eloquent brain lesions based purely on rTMS language mapping without awake surgery. Acta Neurochir. 2016b;158:2265–75.

Ille S, Schroeder A, Albers L, Kelm A, Droese D, Meyer B, Krieg SM. Non-invasive mapping for effective preoperative guidance to approach highly language-eloquent Gliomas—a large scale comparative Cohort study using a new classification for language eloquence. Cancer. 2021;13:207. https://doi.org/10.3390/cancers13020207.

Ilmberger J, Ruge M, Kreth F-W, Briegel J, Reulen H-J, Tonn J-C. Intraoperative mapping of language functions: a longitudinal neurolinguistic analysis. J Neurosurg. 2008;109:583–92.

Ilmoniemi RJ, Virtanen J, Ruohonen J, Karhu J, Aronen HJ, Katila T. Neuronal responses to magnetic stimulation reveal cortical reactivity and connectivity. Neuroreport. 1997;8:3537–40.

Indefrey P. The spatial and temporal signatures of word production components: a critical update. Front Psychol. 2011; https://doi.org/10.3389/fpsyg.2011.00255.

Indharty RS, Japardi I, Sinta Irina R, Tandean S, Pangihutan Siahaan AM, Lumintang Loe M, Ivander A. Comparison of direct cortical stimulation and transcranial magnetic stimulation in brain tumor surgery: systematic review and meta analyses. J Neuro-Oncol. 2023;163:505–14.

Jeltema H-R, Ohlert AK, de Wit A, Wagemakers M, Rofes A, Bastiaanse R, Drost G. Comparing navigated transcranial stimulation mapping and "gold standard" direct cortical stimulation mapping in neurosurgery: a systematic review. Neurosurg Rev. 2021;44:1903–20.

Juenger H, Ressel V, Braun C, Ernemann U, Schuhmann M, Krägeloh-Mann I, Staudt M. Misleading functional magnetic resonance imaging mapping of the cortical hand representation in a 4-year-old boy with an arteriovenous malformation of the central region. J Neurosurg Pediatr. 2009;4:333–8.

Jung J, Lavrador J-P, Patel S, Giamouriadis S, Lam J, Bhangoo R, Ashkan K, Vergani F. First United Kingdom experience of navigated transcranial magnetic stimulation in preoperative mapping of brain tumors. World Neurosurg. 2019;122:e1578–87.

Kamada K, Todo T, Ota T, Ino K, Masutani Y, Aoki S, Takeuchi F, Kawai K, Saito N. The motor-evoked potential threshold evaluated by tractography and electrical stimulation. J Neurosurg. 2009;111:785–95.

Kawashima A, Krieg S, Faust K, Schneider H, Vajkozy P, Picht T. Plastic reshaping of cortical language areas evaluated by navigated transcranial magnetic stimulation in a surgical case of glioblastoma multiforme. Clin Neurol Neurosurg. 2013;115:2226–9.

Krieg SM, Sollmann N, Hauck T, Ille S, Foerschel A, Meyer B, Ringel F. Functional language shift to the right hemisphere in patients with language-eloquent brain tumors. PLoS One. 2013;8:e75403.

Krieg SM, Tarapore P, Picht T, Tanigawa N, Houde J, Sollmann N, Meyer B, Vajkoczy P, Berger MS, Ringel F, Nagarajan S. Optimal timing of pulse onset for language mapping with navigated repetitive transcranial magnetic stimulation. NeuroImage. 2014;100:219–36.

Krieg SM, Sollmann N, Tanigawa N, Foerschler A, Meyer B, Ringel F. Cortical distribution of speech and language errors investigated by visual object naming and navigated transcranial magnetic stimulation. Brain Struct Funct. 2016;221:2259–86.

Krieg SM, Lioumis P, Mäkelä JP, Wilenius J, Karhu J, Hannula H, Savolainen P, Weiss Lucas C, Seidel K, Laakso A, Islam M, Vaalto S, Lehtinen H, Vitikainen A-M, Tarapore PE, Picht T. Protocol for motor and language mapping by navigated TMS in patients and healthy volunteers; workshop report. Acta Neurochir. 2017;159:1187–95.

Kronenburg A, van Doormaal T, van Eijsden P, van der Zwan A, Leijten F, Han KS. Surgery for a giant arteriovenous malformation without motor deterioration: preoperative transcranial magnetic stimulation in a non-cooperative patient. J Neurosurg Pediatr. 2014;14:38–42.

Lehtinen H, Mäkelä JP, Mäkelä T, Lioumis P, Metsahonkala L, Hokkanen L, Wilenius J, Gaily E. Language mapping with navigated transcranial magnetic stimulation in pediatric and adult patients undergoing epilepsy surgery: comparison with extraoperative direct cortical stimulation. Epilepsia Open. 2018;3:224–35.

Lesser RP, Lee HW, Webber WRS, Prince B, Crone NE, Miglioretti DL. Short-term variations in response distribution to cortical stimulation. Brain. 2008;131:1528–39.

Lioumis P, Zhdanov A, Mäkelä N, Lehtinen H, Wilenius J, Neuvonen T, Hannula H, Deletis V, Picht T, Mäkelä JP. A novel approach for documenting naming errors induced by navigated transcranial magnetic stimulation. J Neurosci Meth. 2012;204:349–54.

Lioumis P, Autti S, Wilenius J, Vaalto S, Lehtinen H, Laakso A, Kirveskari E, Mäkelä JP, Liljeström M, Renvall H. Study design for navigated repetitive transcranial magnetic stimulation for speech cortical mapping. J Vis Exp. 2023. (193); https://doi.org/10.3791/64492.

Litvak V, Komssi S, Scherg M, Hoechstetter K, Classen J, Zaaror M, Pratt H, Kähkönen S. Artifact correction and source analysis of early electroencephalographic responses evoked by transcranial magnetic stimulation over primary motor cortex. NeuroImage. 2007;37:56–70.

Lu J, Zhao Z, Zhang J, Wu B, Zhu Y, Chang EF, Wu J, Duffau H, Berger MS. Functional maps of direct electrical stimulation-induced speech arrest and anomia: a multicenter retrospective study. Brain. 2021;144:2541–53.

Lubrano V, Filleron T, Démonet JF, Roux FE. Anatomical correlates for category-specific naming of objects and actions: a brain stimulation mapping study. Hum Brain Mapp. 2014;35:429–32.

Lucas TH, McKhann GM, Ojemann GA. Functional separation of languages in the bilingual brain: a comparison of electrical stimulation language mapping in 25 bilingual patients and 117 monolingual control patients. J Neurosurg. 2004;101:449–57.

Luria AR. Neuropsychology in the local diagnosis of brain damage. Cortex. 1964;1:3–18.

Mäkelä JP, Vitikainen A-M, Lioumis P, Paetau R, Ahtola E, Kuusela L, Valanne L, Blomstedt G, Gaily E. Functional plasticity of the motor cortical structures demonstrated by navigated TMS in two patients with epilepsy. Brain Stim. 2013;6:286–91.

Mäkelä JP, Lioumis P, Laaksonen K, Forss N, Tatlisumak T, Kaste M, Mustanoja S. Cortical excitability measured with nTMS and MEG during stroke recovery. Neural Plast. 2015a;2015:309546.

Mäkelä T, Vitikainen A-M, Laakso A, Mäkelä JP. Integrating nTMS data into a radiology picture archiving system. J Digit Imaging. 2015b;28:428–32.

Matsubara T, Daneshzand M, Navarro de Lara LI, Sundaram P, Hämäläinen M, Stufflebeam S, Paulson D, Pratt K, Mascarenas A, Knappe S, Okada Y. A 16-channel transcranial magnetic stimulation (TMS) system: its magnetic field and induced electric field patterns. Poster in Brain stimulation conference, 2023, Lisbon.

Meier A, Kuzdeba S, Jackson L, Daliri A, Tourville JA, Guenther FH, Greenlee JWH. Lateralization and time-course of cortical phonological representations during syllable production. Eneuro. 2023;0(10):ENEURO.0474-22.2023 1–18. https://doi.org/10.1523/ENEURO.0474-22.2023.

Mesulam MM. Principles of cognitive and behavioral neurology. New York: Oxford University Press; 2000. p. 1–2.

Motomura K, Takeuchi H, Nojima I, Aoki K, Chalise L, Iijima K, Wakabayashi T, Natsume A. Navigated repetitive transcranial magnetic stimulation as preoperative assessment in patients with brain tumors. Sci Rep. 2020;10:9044. https://doi.org/10.1038/s41598-020-65944-8.

Muir M, Patel R, Traylor J, de Almeida C, Bastos D, Prinsloo S, Liu H-L, Noll K, Wefel J, Tummala S, Kumar V, Prabhu S. Validation of non-invasive language mapping modalities for eloquent tumor resection: a pilot study. Front Neurosci. 2022;16:833073.

Na Y, Jung JY, Tench C, Auer DP, Pyun SB. Language systems from lesion-symptom mapping in aphasia: a meta-analysis of voxel-based lesion mapping studies. Neuroimage Clin. 2022;35:103038.

Narayana S, Gibbs SK, Fulton SP, McGregor AL, Mudigoudar B, Weatherspoon SE, Boop FA, Wheless JW. Clinical utility of transcranial magnetic stimulation (TMS) in the presurgical evaluation of motor, speech, and language functions in young children with refractory epilepsy or brain tumor: preliminary evidence. Front Neurol. 2021;12:650830. https://doi.org/10.3389/fneur.2021.650830.

Natalizi F, Piras F, Vecchio D, Spalletta G, Piras F. Preoperative navigated transcranial magnetic stimulation: new insight for brain tumor-related language mapping. J Pers Med. 2022;12:1589.

Nettekoven C, Pieczewski J, Neuschmelting V, Jonas K, Goldbrunner R, Grefkes C, Weiss LC. Improving the efficacy and reliability of rTMS language mapping by increasing the stimulation frequency. Hum Brain Mapp. 2021;42:5309–21. https://doi.org/10.1002/hbm.25619.

Nieberlein L, Rampp S, Gussew A, Prell J, Hartwigsen G. Reorganization and plasticity of the language network in patients with cerebral gliomas. NeuroImage Clin. 2023; https://doi.org/10.1016/j.nicl.2023.103326.

Nieminen JO, Sinisalo H, Souza VH, Malmi M, Yuryev M, Tervo AE, Stenroos M, Milardovich D, Korhonen JT, Koponen LM, Ilmoniemi RJ. Multi-locus transcranial magnetic stimulation system for electronically targeted brain stimulation. Brain Stim. 2022;15:116–24.

Ntemou E, Reisch K, Burchert F, Jonkers R, Picht T, Rofes A. Mapping action naming in patients with gliomas: the influence of transitivity. Neuroimage Rep. 2023a;3:100184.

Ntemou E, Svaldi C, Jonkers R, Picht T, Rofes A. Verb and sentence processing with TMS: a systematic review and meta-analysis. Cortex. 2023b;162:38–55.

Nummenmaa A, Stenroos M, Ilmoniemi RJ, Okada Y, Hämäläinen MS, Raij T. Comparison of spherical and realistically shaped boundary element head models for transcranial magnetic stimulation navigation. Clin Neurophysiol. 2013;124:1995–2007.

Nummenmaa A, Mc Nab JA, Savadijev P, Okada Y, Hämäläinen MS, Wang R, Wald LL, Pascual-Leone A, Wedeen JJ, Raij T. Targeting of white matter tracts with transcranial magnetic stimulation. Brain Stim. 2014;7:80–4.

Ohlerth A-K, Bastiaanse R, Negwer C, Sollmann N, Schramm S, Schröder A, Krieg SM. Benefit of action naming over object naming for visualization of subcortical language pathways in navigated transcranial magnetic stimulation-based diffusion tensor imaging-fiber tracking. Front Hum Neurosci. 2021;15:748274.

Ohlerth A-K, Bastiaanse R, Nickels L, Neu B, Zhang W, Ille S, Sollmann N, Krieg SM. Dual-task nTMS mapping to visualize the Cortico-subcortical language network and capture postoperative outcome—a patient series in neurosurgery. Front Oncol. 2022;11:788122. https://doi.org/10.3389/fonc.2021.788122.

Ojemann G, Ojemann J, Lettich E, Berger M. Cortical language localization in left, dominant hemisphere. An electrical stimulation mapping investigation in 117 patients. J Neurosurg. 1989;71:316–26.

Papatzalas C, Fountas K, Kapsalaki E, Papathanasiou I. The use of standardized intraoperative language tests in awake craniotomies: a scoping review. Neuropsychol Rev. 2022;32:20–50. https://doi.org/10.1007/s11065-021-09492-6.

Petrovich Brennan NM, Whalen S, de Morales BD, O'Shea JP, Norton IH, Golby AJ. Object naming is a more sensitive measure of speech localization than number counting: converging evidence from direct cortical stimulation and fMRI. NeuroImage. 2007;37:S100–8.

Pial V, Eikelboom D. Brain areas critical for picture naming: a systematic review and meta-analysis of lesion symptom mapping studies. Neurobiol Lang. 2023;4:280–96.

Picht T, Krieg SM, Sollmann N, Rösler J, Niraula B, Neuvonen T, Savolainen P, Mäkelä JP, Lioumis P, Deletis V, Meyer B, Vajkoczy P, Ringel F. A comparison of language mapping by presurgical navigated transcranial magnetic stimulation and direct cortical stimulation during awake surgery. Neurosurgery. 2013;72:808–19.

Pitkänen M, Yazawa S, Airaksinen K, Lioumis P, Nurminen J, Pekkonen E, Mäkelä JP. Localization of sensorimotor cortex using navigated transcranial magnetic stimulation and magnetoencephalography. Brain Topogr. 2019;32:873–81. https://doi.org/10.1007/s10548-019-00716-w.

Raffa G, Bährend I, Schneider H, Faust K, Germano A, Vajkozy P, Picht T. A novel technique for region and linguistic specific nTMS-based DTI fiber tracking of language pathways in brain tumor patients. Front Neurosci. 2016;10:art 552.

Raffa G, Quattropani MC, Scibilia S, Conti A, Angileri FF, Esposito F, Sindorio C, Cardali SM, Germanò A, Tomasello F. Surgery of language-eloquent tumors in patients not eligible for awake surgery: the impact of a protocol based on navigated transcranial magnetic stimulation on presurgical planning and language outcome, with evidence of tumor-induced intra-hemispheric plasticity. Clin Neurol Neurosurg. 2018;168:127–39.

Raffa G, Marzano G, Curcio A, Espahbodinea S, Germanò A, Angileri FF. Personalized surgery of brain tumors in language areas: the role of preoperative brain mapping in patients not eligible for awake surgery. Neurosurg Focus. 2022;53(6):E3.

Reijonen J, Könönen M, Tuunanen P, Määttä S, Julkunen P. Atlas-informed computational processing pipeline for individual targeting of brain areas for therapeutic navigated transcranial magnetic stimulation. Clin Neurophysiol. 2021;132:1612–21.

Reisch K, Böttcher F, Tuncer MS, Schneider H, Vajkoczy P, Picht T, Fekonja LS. Tractography-based navigated TMS language mapping protocol. Front Oncol. 2022;12:1008442. https://doi.org/10.3389/fonc.2022.1008442.

Rejnö-Habte Selassie G, Pegenius G, Karlsson T, Viggedal G, Thallböök T, Elam M. Cortical mapping of receptive language processing in children using navigated transcranial magnetic stimulation. Epilepsy Behav. 2020;103:106836.

Robertson ER, Theoret H, Pascual-Leone A. Studies in cognition: the problems solved and created by transcranial magnetic stimulation. J Cogn Neurosci. 2003;15:948–60.

Robles SG, Gatignol P, Lehericy S, Duffay H. Long term brain plasticity allowing multistage surgical approach to World Health Organization grade II gliomas in eloquent areas. J Neurosurg. 2008;109:615–24.

Rogic M, Deletis V, Fernandez-Conejero I. Inducing transient language disruptions by mapping of Broca's area with modified patterned repetitive transcranial magnetic stimulation protocol. J Neurosurg. 2014;120:1033–41.

Rosenstock T, Picht T, Schneider H, Vajkoczy P, Thomale U-W. Pediatric navigated transcranial magnetic stimulation motor and language mapping combined with diffusion tensor imaging tractography: clinical experience. J Neurosurg Pediatr. 2020;26:583–93.

Rösler J, Niraula B, Strack V, Zdunzyck A, Schilt S, Savolainen P, Lioumis P, Mäkelä J, Vajkozy P, Frey D, Picht T. Language mapping in healthy volunteers and brain tumor patients with a novel navigated TMS system: evidence of tumor-induced plasticity. Clin Neurophysiol. 2014;125:526–36.

Ruohonen J, Karhu J. Navigated transcranial magnetic stimulation. Neurophysiol Clin. 2010;40:7–17.

Sahin NT, Pinker S, Cash SS, Schomer D, Halgren E. Sequential processing of lexical, grammatical, and phonological information within Broca's area. Science. 2009;326:445–9.

Sanai N, Mirzadeh Z, Berger MS. Functional outcome after language mapping for glioma resection. N Engl J Med. 2008;358:18–27.

Schmidt S, Bathe-Peters R, Fleischmann R, Rönnefarth M, Scholz M, Brandt SA. Nonphysiological factors in navigated TMS studies; confounding covariates and valid intracortical estimates. Hum Brain Mapp. 2015;36:40–9.

Schuhmann T, Schiller NO, Goebel R, Sack AT. Speaking of which: dissecting neurocognitive network of language production in picture naming. Cer Cortex. 2012;22:701–9.

Seynaeve L, Van Baby D, Hamme H, Dupont P, van Paesschen V. Automated speech analysis to improve TMS-based language mapping: algorithm and proof of concept. Brain Stim. 2020;13:267–9.
Shtyrov J, McGregor LJ. Near-instant access to visually presented words in the human neocortex: neuromagnetic evidence. Sci Rep. 2016;6:26558. https://doi.org/10.1038/srep26558.
Silvanto J, Muggleton NG. New light through old windows: moving beyond the "virtual lesion" approach to transcranial magnetic stimulation. NeuroImage. 2008;39:549–52.
Soffietti R, Baumert BG, Bello L, von Deimling A, Duffau H, Frenay M, et al. Guidelines on management of low-grade gliomas: report of an EFNS-EANO Task Force. Eur J Neurol. 2010;17:1124–33.
Sollmann N, Picht T, Mäkelä JP, Meyer B, Ringel Krieg SM. Navigated TMS refutes negative left sided fMRI activation for language in a patient with a left fronto-opercular GBM: case report. J Neurosurg. 2013a;118:175–9.
Sollmann N, Hauck T, Hapfelmeier A, Meyer B, Ringel F, Krieg SM. Intra- and interobserver variability of language mapping by navigated transcranial magnetic brain stimulation. BMC Neurosci. 2013b;14:150.
Sollmann N, Tanigawa N, Ringel F, Zimmer K, Meyer B, Krieg SM. Language and its right-hemispheric distribution in healthy brains: an investigation by repetitive transcranial magnetic stimulation. NeuroImage. 2014;102:776–88.
Sollmann N, Ille S, Hauck T, Maurer S, Negwer C, Zimmer K, Ringel F, Meyer B, Krieg SM. The impact of preoperative language mapping by repetitive navigated transcranial magnetic stimulation on the clinical course of brain tumor patients. BMC Cancer. 2015a;15:261.
Sollmann N, Ille S, Obermueller T, Negwer C, Ringel F, Meyer B, Krieg SM. The impact of repetitive navigated transcranial magnetic stimulation coil positioning and stimulation parameters on human language function. Eur J Med Res. 2015b;20:47.
Sollmann N, Giglhuber K, Tussis L, Meyer B, Ringel F. Krieg SM nTMS-based DTI fiber tracking for language pathways correlates with language function and aphasia – a case report. Clin Neurol Neurosurg. 2015c;136:25–8.
Sollmann N, Hauck T, Tussis L, Ille S, Maurer S, Boekckh-Behrens T, Ringel F, Meyer B, Krieg SM. Results on the spatial resolution of repetitive transcranial magnetic stimulation for cortical language mapping during object naming in healthy subjects. BMC Neurosci. 2016;17:67.
Sollmann N, Ille S, Newger C, Boekckh-Behrens T, Ringel F, Meyer B, Krieg SM. Cortical time course of object naming investigated by repetitive navigated transcranial magnetic stimulation. Brain Imaging Behav. 2017;11:1192–206. https://doi.org/10.1007/s11682-016-9574-x.
Tarapore PE, Findlay AM, Honma SM, Mizuiri D, Houde JF, Berger MS, Nagarajan SS. Language mapping with navigated repetitive TMS: proof of technique and validation. NeuroImage. 2013;82:260–72.
Tarapore PE, Picht T, Bulubas L, Shin Y, Kulchytska N, Meyer B, et al. Safety and tolerability of navigated TMS for preoperative mapping in neurosurgical patients. Clin Neurophysiol. 2016;127:1895–900.
Tate M, Herbet G, Mortiz-Gassier S, Tate JE, Duffau H. Probabilistic map of critical functional regions of the human cerebral cortex: Broca area revisited. Brain. 2014;137:2773–82.
Thut G, Veniero D, Romei V, Miniussi C, Schuns P, Gross J. Ehythmic TMS causes local entrainment of natural oscillatory signatures. Curr Biol. 2001;21:1176–85.
Tremblay P, Dick AS. Broca and Wernicke are dead, or moving past the classic model of language neurobiology. Brain Lang. 2016;162:60–71.
Tussis L, Sollmann N, Boeckh-Behrens T, Meyer B, Krieg SM. Identifying cortical first and second language sites via navigated transcranial magnetic stimulation of the left hemisphere in bilinguals. Brain Lang. 2017;168:106–16. https://doi.org/10.1016/j.bandl.2017.01.011.
Valero-Cabre A, Payne BR, Rushmore J, Lomber SG, Pascual-Leone A. Impact of repetitive transcranial magnetic stimulation of the parietal cortex on metabolic activity: a 14C-2DG tracing study in the cat. Exp Brain Res. 2005;163:1–12.

Vasileiadi M, Schuler A-L, Woletz M, Linhardt D, Windischberger C, Tik M. Functional connectivity explains how neuronavigated TMS of posterior temporal subregions differentially affect language processing. Brain Stim. 2023;16:1062–71.
Veniero D, Vossen A, Gross J, Thut G. Lasting EEG/MEG aftereffects of rhythmic transcranial brain stimulation: level of control over oscillatory network activity. Front Cell Neurosci. 2015;9:Article 477. https://doi.org/10.3389/fncel.2015.00477.
Vitikainen A-M, Salli E, Lioumis P, Mäkelä JP, Metsähonkala L. Applicability of nTMS in locating the motor cortical representation areas in patients with epilepsy. Acta Neurochir. 2013;155:507–18.
Vitikainen A-M, Mäkelä E, Lioumis P, Jousmäki V, Mäkelä JP. Accelerometer-based automatic voice onset detection in rTMS. J Neurosci Meth. 2015;253:70–7.
Vucic S, et al. Clinical diagnostic utility of transcranial magnetic stimulation in neurological disorders. Updated report of an IFCN committee. Clin Neurophysiol. 2023;150:131–75.
Walsh V, Cowey A. Magnetic stimulation studies of visual cognition. Trends Cogn Sci. 1998;2:103–10.
Walsh V, Cowey A. Transcranial magnetic stimulation and cognitive neuroscience. Nat Rev Neurosci. 2000;1:73–80.
Wang JB, Bruss JE, Oya H, Uitermarkt BD, Trapp NT, Gander PE, Howard MA, Keller CJ, Boes AJ. Effects of transcranial magnetic stimulation on the human brain recorded with intracranial electrocorticography: first-in-human study. bioRxiv preprint. 2022; https://doi.org/10.1101/2022.01.18.476811.
Wheat KL, Cornelissen PL, Sack AT, Schuchmann T, Goebel R, Blomert L. Charting the functional relevance of Broca's area for visual word recognition and picture naming in Dutch using fMRI-guided TMS. Brain Lang. 2013;125:223–30.
Whitaker HA, Ojemann GA. Graded localization of naming from electrical stimulation mapping of left cerebral cortex. Nature. 1977;270:50–1.
Zhang X, Zhang G, Yu T, Xu C, Yan X, Ma K, Du W, Gao R, Li Y. Multitask preoperative language mapping in epilepsy surgery: a combination of navigated transcranial magnetic stimulation and extra-operative electrical cortical stimulation. J Clin Neurophysiol. 2020;79:259–65.
Zhang H, Ille S, Sogerer L, Schwendner M, Schröder A, Meyer B, Wiestler B, Krieg S. Elucidating the structural–functional connectome of language in glioma-induced aphasia using nTMS and DTI. Hum Brain Mapp. 2022;43:1836–49.
Zhang H, Tehlan K, Ille S, Schwendner M, Gong Z, Schroeder A, Meyer B, Krieg SM. Interhemispheric connections in the maintenance of language performance and prognosis prediction: fully connected layer-based deep learning model analysis. Neurosurg Focus. 2023;54:E6. https://doi.org/10.3171/2023.3.FOCUS2363. PMID: 37283401

Combining nTMS and Tractography for Language Mapping: An Integrated Paradigm for Neurosurgical Planning

10

Ann-Katrin Ohlerth, Jose Pedro Lavrador, Francesco Vergani, and Stephanie J. Forkel

10.1 Introduction

In the intricate tapestry of the human brain, the white matter tracts serve as high-speed communication networks, facilitating the complex interplay between different regions. Tractography, a remarkable neuroimaging technique, leverages the diffusion of water molecules to visualize and map these pathways, offering us a glimpse into the structural highways of neural communication. By tracking the movement of water along axons, tractography provides us with the unique ability to delineate the course of white matter tracts within the living brain, a feat unattainable by traditional imaging methods.

As we navigate deeper into neurosurgery, the role of tractography becomes particularly pivotal when focusing on function of language—a fundamentally human domain. Language tractography reveals the architecture underlying our verbal and written communication and is a critical guide in neurosurgical planning and intervention. The nuanced understanding of language pathways helps avoid functional

A.-K. Ohlerth
Neurobiology of Language Department, Max Planck Institute for Psycholinguistics, Nijmegen, The Netherlands
e-mail: ann-katrin.ohlerth@mpi.nl

J. P. Lavrador · F. Vergani
Department of Neurosurgery, King's College Hospital NHS Foundation Trust, London, UK
e-mail: josepedro.lavrador@nhs.net; francesco.vergani@nhs.ne

S. J. Forkel (✉)
Donders Institute for Brain Cognition Behaviour, Radboud University, Nijmegen, The Netherlands

Brain Connectivity and Behaviour Laboratory, Sorbonne Universities, Paris, France

Centre for Neuroimaging Sciences, Department of Neuroimaging, Institute of Psychiatry, Psychology and Neuroscience, King's College London, London, UK

S. M. Krieg, T. Picht (eds.), *Navigated Transcranial Magnetic Stimulation in Neurosurgery*, https://doi.org/10.1007/978-3-031-97155-6_10

compromise during surgery, thereby preserving the essence of human interaction postoperatively.

Enter the domain of navigated transcranial magnetic stimulation (nTMS)—a technology that marries the anatomical insights gained from tractography with functional localization. nTMS for language mapping is a noninvasive, patient-specific approach allowing us to "navigate" the cortical surface precisely. By combining subcortical anatomical white matter mapping derived from tractography with surface functional probing of nTMS, neurosurgeons can tailor their approach to each patient's unique neural landscape.

In this chapter, we will discuss the foundational principles of tractography to the cutting-edge applications of nTMS in language mapping. We will explore how the synergy of these two technologies is revolutionizing preoperative planning and maximizing surgical outcomes while safeguarding the quintessential human capacity for language.

10.2 The Methods

10.2.1 Navigated Transcranial Magnetic Stimulation (nTMS)

Navigated transcranial magnetic stimulation (nTMS) has emerged as a critical instrument in delineating cortical regions implicated in functional processes, particularly in motor and language mapping. The efficacy of TMS has been significantly enhanced by integrating MRI-based navigation systems. These systems allow for the concurrent visualization of an individual's structural brain scan, providing a real-time guide to target and precisely navigate the stimulation across the cortical landscape. This advancement marks a transformative step from the earlier, less precise techniques, enabling a tailored and patient-specific approach to brain mapping and standing in contrast to earlier mapping techniques, which often involved somewhat indiscriminate stimulation of the cortex without imaging support, or relied on standardized brain models (Krings et al. 2001; Rotenberg et al. 2014; Ruohonen and Karhu 2010). The necessity for individualized imaging becomes especially critical in cases involving brain tumor patients, where significant anatomical changes occur due to the presence and growth of a tumor. This individualized approach ensures a more accurate and safe intervention, accounting for the unique anatomical variations present in each patient.

The setup for nTMS for language mapping involves a series of coordinated steps and equipment, designed to accurately stimulate specific cortical regions involved in language processing (Chap. 9). Here is an outline of the typical setup:

1. *Patient Preparation and MRI Scanning*: Before the nTMS procedure, the patient undergoes an MRI scan. This scan provides detailed images of the patient's brain structure, which are crucial for the navigation during the TMS procedure. The MRI data, commonly in DICOM format, is used to create a 3D model of the

patient's brain. Additional sequences can be simultaneously acquired for later processing steps, including diffusion-weighted imaging (DWI).

2. Integration of MRI Data with nTMS System: The MRI images are uploaded into the nTMS system. Advanced software integrates these images with the nTMS navigation system, allowing for real-time guidance and precise targeting of specific brain areas.
3. Positioning of the Patient: The patient is seated comfortably in a chair or lies on a bed. Their head is positioned to ensure optimal access for the TMS coil and patient comfort.
4. Aligning of MRI with Patient's Head: By means of neuronavigation including an infrared camera, several markers on the patient's head are recorded and aligned on screen with the 3D image of the patient's brain, based on the individual MRI.
5. Setting Up the nTMS Equipment: The nTMS machine, equipped with a stimulation coil, is brought near the patient. The coil is positioned over the patient's scalp, targeting specific predefined brain areas. The position and orientation of the coil are guided by the navigational software, visualized on screen.
6. Language Task Execution: The patient is typically asked to perform language tasks during the stimulation. These tasks may involve naming objects and/or actions, or reading. The task items are timely aligned with the stimulation targeting the cortex.
7. Stimulation and Mapping: The TMS coil delivers magnetic pulses to the targeted brain areas. The responses and effects of these stimulations on the patient's language capabilities are monitored and recorded. This helps to identify language processing areas by observing any disruptions or changes in the patient's ability to perform these tasks when different brain regions are stimulated.
8. Data Analysis and Interpretation: The data collected from the nTMS session is analyzed to identify the brain regions essential for language function. All areas that were found to be involved in language, i.e., showing disruptions in performing the designated task, are marked as a positive language map in the MRI.
9. Integration with Surgical Planning: The language map obtained from nTMS is burned into the DICOM that can further be integrated into the surgical planning process.

The nTMS setup for language mapping is depicted in Fig. 10.1.

Although nTMS provides a reasonably good understanding of the cortical regions crucial for language, it cannot currently map the subcortical language network. This limitation includes the inability to effectively reach subcortical gray and white matter (WM) structures. There is no dependable method for nTMS to target the subcortex directly. This is in contrast to direct electrical stimulation (DES) used intraoperatively, which can be applied directly to the exposed brain tissues. In the case of nTMS, the magnetic pulses need to traverse not only the scalp and skull but also various layers of potentially relevant cortical gray matter, as well as other cellular and white matter structures, before they can reach the deeper subcortical areas of interest (Chap. 1). As a result, this can give rise to a situation where a region of interest might be falsely identified as relevant (a false positive). This occurs when

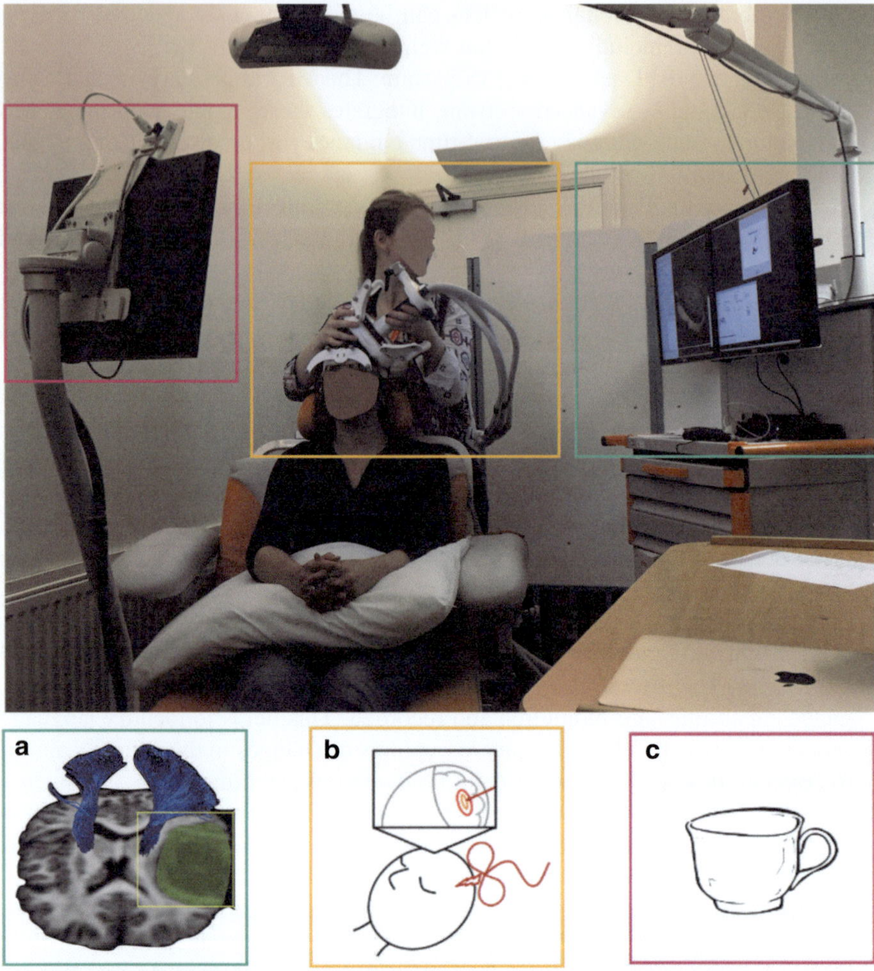

Fig. 10.1 The clinical setup for presurgical language mapping using tractography and TMS. (**a**) 3D model of patient's neuroanatomy based on pre-TMS MRI scanning (including tumor entity) to aid coil placement and thereby TMS pulse on cortex, with the possibility to visualize a priori segmented tractography results. (**b**) The experimenter places the stimulation coil on the head of the patient, guided by (**a**). (**c**) Patients view a screen where picture stimuli of a language task are presented. (Original artwork by Victor Xandri Antolin; © University of Groningen)

the observed effects are due to the stimulation and disturbance of lateral structures surrounding the intended target areas, rather than the target areas themselves.

10.2.2 Diffusion-Weighted Data Acquisition

In response to these challenges, significant strides have been made in reconstructing individual white matter (WM) bundles in the brain using diffusion-weighted imaging techniques. These methods track the diffusion pathways of water molecules, enabling the reconstruction of presumed white matter tracts where these molecules are present. Among the various algorithms available for reconstructing the brain's connective tissues, deterministic tractography, mainly based on diffusion tensor imaging, has become a standard and widely used approach in neurosurgical settings (Chap. 21).

We provide a brief overview of optimizing diffusion protocol for presurgical planning in Chap. 21 under Sect. 21.2.2. When preparing for the surgical treatment of language-related brain tumors, the implementation of an effective diffusion protocol for tractography is crucial. The protocol's nuances significantly influence the quality and utility of the resulting data for presurgical planning. The data is primed for processing using various algorithmic models with these acquisition parameters set. In the clinical setting, the FACT algorithm is the most common, which makes a single fiber orientation decision per voxel allowing to propagate from voxel to voxel following the primary orientation of the diffusion tensor in each voxel, hence diffusion tensor imaging (DTI). More advanced models would also account for the diffusion patterns in neighboring voxels and integrate this information.

10.2.2.1 Tractography Parameters and Algorithms

Once the data is acquired and preprocessed for tractography reconstructions (Chap. 21), one can reconstruct the whole connectome (i.e., all connections) or segment out individual connections of interest. The segmentation method choice depends on the preprocessing and the tracking algorithm used. Broadly speaking, two methods are used: (1) virtually segmenting a pathway from a whole brain connectome based on anatomical regions and (2) tracking individual pathways based on anatomical or functional seed and target regions. A concrete example would be delineating the language network by placing an anatomical region of interest (ROI) into a priori-defined brain areas (e.g., inferior frontal gyrus, IFG). This approach will visualize all streamlines passing through this region. For the second example, a seed ROI can be placed in the IFG to initiate repeated tracking from this part of the brain. Both these methods can be used with nTMS-positive and nTMS-negative points serving as ROIs.

10.2.2.2 Anatomical Placement of ROIs

Anatomical ROIs are a pivotal component in tractography, offering a tailored approach to mapping brain connections. They can be defined manually, typically preferred in clinical settings, or based on established brain atlases. ROIs are strategically placed in critical "bottleneck" areas, acting as essential passageways for white matter tracts (Catani et al. 2002; Hau et al. 2016).

Consider the inferior fronto-occipital fasciculus (IFOF), which spreads across the frontal and occipital lobes but narrows significantly as it traverses the external capsule. Placing an ROI in this narrow region effectively captures most of the IFOF's streamlines. However, for more nuanced dissection, such as differentiating the IFOF in the context of the ventral triade alongside the uncinate fasciculus and the inferior longitudinal fasciculus, employing three ROIs is advisable. Similarly, the arcuate fasciculus, connecting the frontal and temporal lobes, can be segmented by placing an ROI in the converging zone within the inferior parietal white matter (referenced in Catani et al. (2005)).

Not all connections, however, converge into bottlenecks. In such instances, employing multiple ROIs helps define connections as traversing between distinct brain regions, like the superior longitudinal fasciculi linking the frontal and parietal lobes (Thiebaut de Schotten et al. 2011). When multiple ROIs are used, each can act as a filter, including or excluding streamlines based on Boolean logic ("AND," "OR," "NOT").

Moreover, the function of an ROI can vary: a "waypoint" ROI captures or excludes streamlines passing through it, while an "endpoint" ROI focuses on streamlines that start or finish within it. This distinction is crucial for accurately mapping different types of fibers, such as short U-shaped fibers connecting adjacent gyri, where the "both ends" endpoint criterion is particularly effective. Utilizing these ROI rules and configurations, exploring almost every white matter connection in the brain becomes feasible, providing a comprehensive and precise understanding of its intricate network. Examples of this approach as applied to the language networks are provided in Fig. 10.2.

10.2.2.3 Functional ROIs

In contrast to anatomical ROI creation, functional ROI creation begins with identifying areas of functional involvement specific to the individual. These could be areas delineated using functional magnetic resonance imaging readouts (fMRI; for more information, see Kleiser et al. (2010)) or can be based on nTMS mapping results.

While fMRI is widely available in most medical institutions, its alignment with TMS-positive sites presents limitations, as the correlation is not always direct. Moreover, fMRI typically offers lower spatial resolution, which can be challenging when dealing with brain tumor patients (Giussani et al. 2010; Ille et al. 2015; Babajani-Feremi et al. 2016).

TMS, on the other hand, can be directly employed for functional-based mapping. If the ROIs originate from nTMS, they were typically conducted a few days before surgery for high-grade lesions or at a longer interval for lower-grade lesions. This mapping provides an up-to-date representation of the patient's functional cortical hubs for language.

In short, this involves targeting predefined cortical areas across most of the left hemisphere with nTMS pulses synchronously with a language task (for detailed

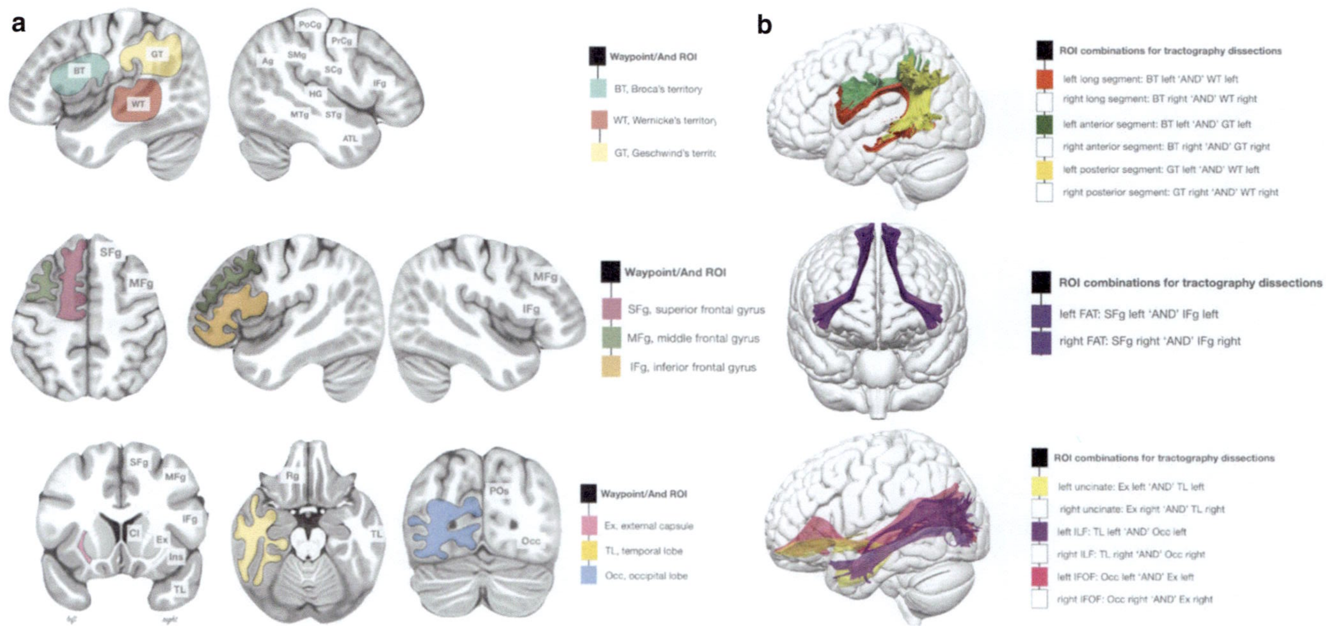

Fig. 10.2 Anatomical diagram of the human brain showing various regions and pathways. Panel (**a**) illustrates different brain territories: Broca's territory (BT), Wernicke's territory (WT), and Geschwind's territory (GT), with color-coded regions. Additional regions include the superior frontal gyrus (SFg), middle frontal gyrus (MFg), and inferior frontal gyrus (IFg). Panel (**b**) displays tractography dissections with color-coded pathways, including left and right segments of the arcuate fasciculus and frontal aslant tract (FAT). Labels indicate specific regions of interest (ROI) combinations for tractography

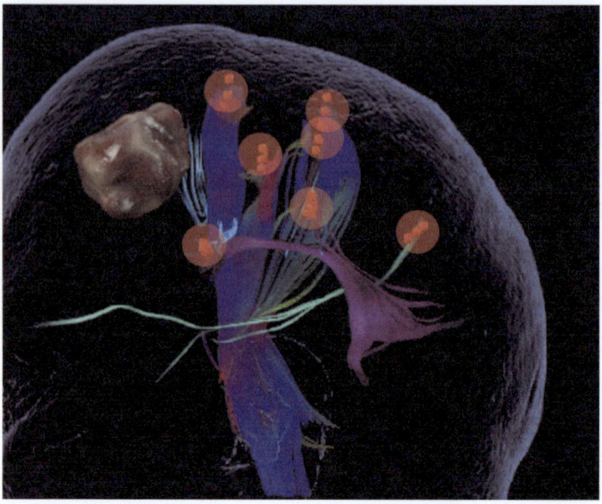

Fig. 10.3 DTI fiber tracking on the basis of nTMS-positive spots for language in relation to tumor mass (here in brown). All areas that, due to stimulation, resulted in a disruption of speech and language are marked as orange spheres of 1–2 mm. The spheres' diameter is increased to 5 mm to enlarge the region of interest (ROI) in order to mimic the size of focal stimulation and reach subcortical white matter. Fibers originating from these ROIs are then visualized and color-coded: CST (blue), arcuate (purple), MdLF (green), callosal fibers (light blue)

description, see Chap. 9, and for a commonly applied protocol (Krieg et al. 2017). Any stimulations that cause disruption in the patient's language output, such as difficulties in correctly and fluently naming a picture, are identified. These disruptions are marked and are later transformed into cortical spots (sphere of 1–2 mm^2), which are used as seed ROIs or objects within the imaging data. This is achieved by enlarging their size to a 5 mm radius to create an ROI of sufficient size to encompass white matter areas. Fibers originating from these cortical nTMS hubs can then be reconstructed and visualized in 3D. Finally, the data are coregistered and ideally placed onto the neuronavigation system in the OR, ensuring precise alignment and integration of functional data for surgical planning (see Fig. 10.3).

However, it is essential to note that this method primarily identifies functionally relevant components of neural networks rather than mapping the entire anatomical network. Studies such as Negwer and colleagues have demonstrated that this approach can effectively reveal white matter connections critical for cognitive functions, and their disruption could lead to significant cognitive deficits (Negwer et al. 2018).

10.2.3 How to Integrate TMS and Tractography for Language

Three main approaches have been described on the integration of TMS and tractography to preoperatively define the language network of the individual patient:

1. *TMS-informed tractography* (Sollmann et al. 2016a; Negwer et al. 2017a): In contrast to the anatomical ROI creation, with this approach, no a priori tracts are selected, but only those tracts that surpass, end, or start in the created nTMS sphere ROIs on the cortex are visualized (Sollmann et al. 2016a; Raffa et al. 2016, 2022; Silva et al. 2021; Ohlerth et al. 2022). While this certainly limits white matter presentation to functionally relevant tracts, cortical ROI placement entails drawbacks as well: it often produces aberrant fiber representations that do not align with anatomically nor functionally plausible structures and hence requires postprocessing of the tractogram before clinical use (Raffa et al. 2022; Silva et al. 2021). This is due to false positive responses with TMS for language.
2. *Tractography-informed TMS* (Reisch et al. 2022). In this approach, the subcortical anatomical information guides the TMS cortical mapping to decrease the false-positive rate. A TMS response cannot be considered if there is no subcortical tract whereby that stimulation can influence the subcortical network.
3. *Independently performed TMS and tractography* (Lucena et al. 2023; Hazem et al. 2021). In this approach, tractography and TMS are performed independently (using anatomical and functional ROIs), and the results are merged. This method tries to minimize one technique's false-positive and false-negative responses over the other. It aims for a probability approach given by the degree of overlap of the two techniques.

Different studies have compared the above methods and how they correlate with DES mapping in the OR. While some reports mention that TMS-informed tractography did not improve the intraoperative identification of positive sites for language when compared with TMS-only approach (Sollmann et al. 2016b), others support better outcomes with TMS-based tractography when compared with DES-based tractography when the quality of tract reconstruction is considered (Li et al. 2023). Conflicting results emerge when comparing TMS-based tractography versus anatomy-based tractography supporting either a superiority of one method or the other when comparing visualization of tracts involved in the subcortical language network (Silva et al. 2021; Negwer et al. 2017b). Therefore, a standardized approach is still lacking and open to discussion within the neurosurgical community. A combined tractogram, employing both subcortical hand-drawn ROIs and nTMS-based functional information, holds potential in visualizing specific pathways of the major language tracts more accurately (Silva et al. 2021). This avenue demands further attention and calls for prospective, large cohort investigations, as it can overcome both shortcomings of the single approaches, if done correctly.

10.3 The Applications

10.3.1 Functional Mapping of Language

Language is a highly intricate cognitive function, spanning two modalities (expressive and receptive) and encompassing several forms of communication (spoken;

written; gestures). It is broadly based on three subskills in all these modalities, entailing *meaning building* (conceptual and semantic processing, thereby selecting the correct meaning leading to lexical retrieval of words in the mind), *structure building* (morphological and syntactic processes of putting elements together to form more complex words and later sentences), and *sound building* (phonological encoding and finally phonetic-articulatory motor processes by assigning sounds to elements and initiating motor movements of speech articulators). Inversely, an equally complex chain of processes is in place during receptive language, namely comprehension, of dissecting strings of sound waves into elements, words, and sentences and correctly inferring meaning to the entire utterance.

As it is the ultimate goal in neurosurgical and neuro-oncological care to preserve as many of these subskills as possible, a line of research has been dedicated to how best to elicit and evaluate language skills during mapping techniques. In the search for designing and validating linguistic tasks covering the above-mentioned processes most comprehensively, the recent decades have brought about several available protocols of intraoperative tasks to be used with DES (for a review, see Papatzalas et al. (2022)). These language test batteries exist in several languages (such as the DULIP, the ECCO, the VAN-POP, and the MULTIMAP) and allow for task selection based on the patient's profile and preoperative status. Moreover, the respective tests are then used intraoperatively to identify positive areas relevant for language skills while approaching brain tumours (Alves et al. 2021; De Witte et al. 2015; Ohlerth et al. 2020; Gisbert-Muñoz et al. 2021; Dragoy et al. 2016).

Language mapping tasks for nTMS are not as vast compared to mapping with the gold standard DES. The highly restricted time frame of roughly 1 s of stimulation duration during nTMS compared to up to 4 s in DES precludes an adaptation of the extensive intraoperative DES tasks to the preoperative setting with nTMS. Faced with this limitation, most institutions at present opt for an expressive task (Natalizi et al. 2022; Jeltema et al. 2021) that prompts a short verbal output to a visual stimulus, is easily disrupted by nTMS interference, and can be thereby straightforward to evaluate.

This is most commonly done with the picture naming variant *object naming*. Depicting an object as an easy-to-decipher, black-and-white drawing triggers the patient or participant to recognize the concept and produce it as a noun. In addition to being the most established task and facilitating comparisons with earlier work, accurately naming an object can engage various stages of language function, ranging from conceptual retrieval to articulation. While a robust tool, this task alone does not do justice to the complexity of language in general (Bastiaanse and Ohlerth 2023). This shortcoming has been addressed recently by studies adding more depth to the production paradigm in the form of verb tasks and by venturing into possibilities of adding comprehension tasks: To incorporate a second, crucial word class of verbs to the paradigm and to trigger grammatical skills, so-called *action naming* can be employed under nTMS (Ohlerth et al. 2020). Eliciting verb retrieval and encoding, and later employing it in mapping cortical representation, reveals cortical hubs for verb knowledge that can then be considered for the preservation of function (Ohlerth et al. 2022, 2021a, b).

On top of that, utilizing *comprehension tasks* under nTMS could potentially not only aid in uncovering hubs for this modality but also enable the inclusion of individuals with highly impaired production of language: In these patients, picture naming, hence expressive language, might be impossible to administer, but a receptive task might render mapping viable after all: Hearing a noun label, matching it with visually presented options of object drawings and answering with a button press circumvents verbal response and enables mapping (Kram et al. 2023).

More data is needed to manifest the added benefit of the newly proposed tasks. Only case reports or small groups of patients have been presented piloting these approaches in brain tumor patients (Ohlerth et al. 2022; Kram et al. 2023). Nonetheless, evidence from other demographics (e.g., stroke) and methodologies (DES, fMRI, MEG) give rise to the assumption that a variety of tasks might be beneficial over single tasks (Hillis et al. 2002; Mätzig et al. 2009; Miceli et al. 1984; Zingeser and Berndt 1988; Rofes et al. 2017). Object naming remains the most used and validated variant for task choice in nTMS mapping (Natalizi et al. 2022).

10.3.1.1 Tractography and Prediction of Outcome

In neuro-oncology, tractography has become a valuable tool for predicting surgical outcomes, particularly in cases involving tumors located in brain regions crucial for language. The technique is used preoperatively to assess the extent to which language-related tracts, such as the inferior fronto-occipital fasciculus (IFOF), have been infiltrated or displaced by the tumor. When a brain mass, such as a glioma, exerts pressure on the surrounding tissue, it can cause significant deviation in the shape and location of adjacent white matter fibers from their typical anatomy. For instance, a tumor in the lateral frontal lobe may compress and shift a segment of the corticospinal tract (CST) toward the medial wall. This displacement, however, usually affects only the CST portion close to the tumor, altering its path when compared with a white matter atlas (Forkel et al. 2023). Crucially, these alterations are unique to each individual, making it impossible to predict the precise configuration of the white matter without employing tractography for each specific case. It is also crucial to consider the hemispheric asymmetries in the anatomy of white matter, which complicates using the contralesional hemisphere as a reference point (Satoer et al. 2025). This inherent variability between the two hemispheres means that relying on the unaffected side as a mirror image for surgical planning or tractography interpretation in the affected hemisphere may lead to inaccuracies. Understanding these asymmetries is vital for accurate mapping and surgical intervention strategies.

The indispensability of tractography in accounting for such patient-specific anatomical variations has cemented its role as a vital instrument in neurosurgical procedures. Its clinical utility has been extensively documented and advocated in numerous studies and publications, including those by (Cochereau et al. 2020; Dragoy et al. 2020; Duffau 2008; Kemerdere et al. 2016; Leclercq et al. 2010; Mirchandani et al. 2020; Sollmann et al. 2020; Teichmann et al. 2015; Thiebaut de Schotten et al. 2005). These works highlight the pivotal role of tractography in enhancing the precision and safety of neurosurgical interventions, particularly in cases where normal anatomical landmarks are distorted by pathological processes.

Furthermore, the preoperative involvement of specific tracts has been linked to a heightened risk of transient and permanent postoperative cognitive and linguistic deficits (Kinoshita et al. 2014; Castellano et al. 2017; Bello et al. 2010). These findings underscore the importance of detailed tract mapping in surgical planning to minimize the risk of functional impairment.

Tractography offers a nuanced means of evaluating surgical risks by measuring the lesion-to-tract distance (LTD), which is the proximity between a specific brain tract and a tumor. This metric has been instrumental in predicting language outcomes through anatomical ROIs (Tuncer et al. 2021) and functional ROIs derived from TMS data (Sollmann et al. 2020). In cases where preoperative infiltration of tracts is observed, specific LTD thresholds have been linked to language outcomes. For instance, an LTD of 16 mm or less for the arcuate fasciculus (AF), or 25 mm or less for other language tracts like the IFOF, uncinate fasciculus (UF), frontal aslant tract (FAT), and inferior longitudinal fasciculus (ILF) have been associated with an increased risk of poor language outcomes postsurgery (Sollmann et al. 2020). Additionally, Sollmann N. (2020) proposed safety cutoffs for language-sparing resections: LTDs of at least 8 mm for the AF and 11 mm for the SLF, ILF, UF, and IFOF.

However, it is absolutely crucial to acknowledge the inherent variability in tractography results based on the acquisition, preprocessing, and processing parameters (Chap. 21). Different tractography algorithms and parameters can significantly alter the estimates of tract dimensions, extensions, and proximity to the surgical target. This variation underscores a significant caveat in applying LTD measurements during surgery: the accuracy and reliability of the language tract reconstruction heavily depend on the specific algorithm and parameters used. Thus, careful consideration and standardization of these factors are essential for effectively utilizing LTD in surgical planning.

10.3.2 Surgical Anatomy of Language

Moving beyond the localizationist perspectives of language prevalent in the nineteenth and twentieth centuries, contemporary research increasingly supports the view that language relies on a widespread network involving multiple regions and white matter tracts (Thiebaut de Schotten and Forkel 2022). This perspective contrasts with the more focal and clearly defined networks associated with primary functions, such as those seen in the motor system. The language network is thought to be more expansive and includes diverse areas and tracts. Recent studies have also highlighted that variability in anatomy is associated with variability in cognitive profiles, including language functions (see Forkel et al. (2023) and Croxson et al. (2018) for more details).

This complexity and variability underscore the critical need for individualized mapping of the language network in the preoperative stages of neurosurgical interventions. This need is further emphasized by the fact that damage to white matter structures is often less reversible than damage to cortical hubs (as discussed in references Duffau (2014) and Trinh et al. (2013)). While cortical neuroplasticity allows for functional shifting to new cortical areas—a phenomenon observed over time in brains infiltrated by tumors (Rösler et al. 2014; Raffa et al. 2018)—white matter pathways cannot regrow over time. Therefore, a thorough understanding of their location and function is essential to prevent additional damage through disconnection. It makes precise white matter mapping indispensable in preserving language function during neurosurgical procedures.

Apart from more recent discoveries, like the middle longitudinal fascicle (MdLF), whose role is not yet understood and the corticospinal tract (CST) for motor speech articulation, the contemporary understanding of the language network encompasses the following key pathways:

The arcuate fasciculus is a key interlobar (frontoparietal-temporal) association tract connecting the inferior frontal gyrus, inferior parietal lobe (SMg, Ag), and posterior temporal lobe (STg, MTg). This fasciculus can be viewed as a singular frontotemporal pathway or as a structure comprising multiple branches (e.g., Catani et al. 2005). The nomenclature for its components varies across studies. Generally, there is agreement that the arcuate fasciculus includes additional segments that are more lateral compared to the primary frontotemporal pathway: a frontoparietal segment (sometimes called the horizontal or anterior component) and a parietal-temporal segment (often referred to as the vertical or posterior segment) (Catani et al. 2005; Forkel et al. 2022a; Frey et al. 2008; Giampiccolo and Duffau 2022; Kaplan et al. 2010). However, the exact endpoints of these components remain a topic of ongoing discussion and vary with different methodologies (Giampiccolo and Duffau 2022).

Functionally, the subsegments of the arcuate fasciculus (AF) primarily play a role in phonological processing and encoding, which involves the association of sounds with their respective meanings (Catani et al. 2005, 2007; Duffau et al. 2002; Mandonnet et al. 2007; Sierpowska et al. 2017). Additionally, the AF's involvement in semantic processing, which relates to the construction of meaning, and its role in syntactic processes, or the organization of sentence structure, have been increasingly recognized (Duffau et al. 2002; Friederici 2015; Papoutsi et al. 2011; Blecher et al. 2019; Flöel et al. 2009; Vidorreta et al. 2011; Anderson et al. 1999). This multifaceted functionality positions the AF as a critical network component supporting a variety of linguistic subskills, as highlighted in references Turken and Dronkers (2011) and Ivanova et al. (2021) (also see Vavassori et al. (2023) for a recent study in neurosurgery). Damage to the AF has a detrimental role to language, leading to deficits in expressive and receptive language abilities (Geva et al. 2011; Hillis et al. 2018; Hosomi et al. 2009; Kim and Jang 2013; Rosso et al. 2015; Forkel et al. 2014a).

Historically, and still prevalent in some literature, the terms "superior longitudinal fasciculus" (SLF) and "arcuate fasciculus" have been used interchangeably. While there is an overlap between the networks, such as between the SLF-III and the anterior segment of the arcuate fasciculus, other branches and segments are distinct. From anatomical and etymological standpoints, the superior longitudinal fasciculus should be defined as the fibers connecting frontal and parietal regions, aligning with its descriptor "superior and longitudinal" (Thiebaut de Schotten et al. 2011). In contrast, the arcuate fasciculus is appropriately characterized as the frontotemporal connection, with "arcuatus" meaning "arching" around the Sylvian fissure (Dulyan & Forkel, submitted).

The frontal aslant tract (FAT) represents an intralobar connection within the frontal lobe, linking the inferior frontal gyrus to the superior frontal gyrus (Thiebaut de Schotten et al. 2011; Vergani et al. 2014). The FAT can be effectively delineated using a method involving two ROIs. It is believed to play a crucial role in facilitating fluent speech and aiding in speech initiation (Vergani et al. 2014; Catani et al. 2012; Dick et al. 2019). Damage to the FAT has been associated with various speech impairments, including stuttering, speech arrests, and in less severe cases, a reduction in speech rate (Dragoy et al. 2020; Kemerdere et al. 2016; Vassal et al. 2014; Li et al. 2017). This highlights the FAT's significance in the neural network governing speech production and execution.

The uncinate fasciculus (UF) forms a crucial frontotemporal association connection and is recognized as a component of both the limbic network and the language system (Forkel et al. 2022a; Catani et al. 2013). The uncinate fasciculus converges in the ventral portion of the external capsule and extends into the orbitofrontal and anterior temporal lobes. The functional significance of the UF remains debated within the neurological community. The inconsistency in deficit patterns arising from damage or virtual lesions to the UF, induced through stimulation, further complicates our understanding. These deficits encompass a range of disturbances, from phonological and phonemic to semantic (Raffa et al. 2016; Mandonnet et al. 2009; Papagno et al. 2011, 2016). Such varied outcomes mirror the observations from neurological pathologies, suggesting a complex and multifaceted role of the UF in cognitive processing. These findings, collated from clinical interventions and pathological studies in patients with primary progressive aphasia (Mesulam et al. 2009; D'Anna et al. 2016), highlight the need for a more nuanced understanding of the UF's contribution to brain function.

The inferior longitudinal fasciculus (ILF) is an association tract linking the occipital and ventral temporal lobes. It comprises both long-range medial connections and shorter lateral ones, bridging early and late visual cortices and extending into the occipital and temporal lobes (Catani et al. 2003). The ILF plays a role in connecting the occipital visual areas with the amygdala and hippocampus and higher-order visual areas within the temporal lobe. The role of the Inferior longitudinal fasciculus (ILF) in cognitive functions related to literacy and visual object recognition is more clearly established and substantiated in the literature. Notably, damage to the ILF frequently results in consistent deficits in semantic processing, reading, and writing. However, these impairments are not always irreversible, suggesting potential for recovery or neuroplastic adaptation (Mandonnet et al. 2007; Ortibus et al. 2012; Sarubbo et al. 2015; Gil-Robles et al. 2013). Further expanding on this, research has indicated that the ILF is crucial in integrating visual information with linguistic processing, linking visual cortex areas to language-processing centers. This connection is vital for comprehending and interpreting written language and recognizing objects and faces, an essential aspect of visual cognition

(Mandonnet et al. 2007, 2009; Catani and Thiebaut de Schotten 2008; Martino and De Lucas 2014).

The inferior fronto-occipital fasciculus (IFOF) is a white matter tract that facilitates connections between the occipital and frontal cortices. It originates from the cuneus, lingual gyrus, posterior fusiform gyrus, and occipital pole in the posterior region. As it progresses from the occipital lobe into the temporal stem, its fibers converge at the level of the external/extreme capsule, just above the uncinate fasciculus. In the frontal lobe, these fibers form a thin sheet that extends toward the inferior frontal gyrus, medial orbitofrontal region, and frontal pole. The IFOF's complex history can be reviewed in references Forkel et al. (2014b), Schmahmann and Pandya (2007), and Türe et al. (1997). Its nomenclature, with the term "inferior" suggesting the existence of a "superior" fronto-occipital fasciculus, is a remnant of its historical interpretation. While this term persists in some atlases, such as the John Hopkins University (JHU) atlas (Mori et al. 2005), several studies using diverse methodologies have questioned the existence of a "superior" counterpart (see Forkel et al. (2014b), Liu et al. (2020), and Meola et al. (2015)).

The IFOF plays a key role in assigning meaning to visual inputs. Disruptions to the IFOF, particularly during surgical procedures, have frequently been observed to result in semantic deficits. This correlation has been extensively documented in various studies, including references Leclercq et al. (2010), Jarret et al. (2022), Duffau et al. (2005), Maldonado et al. (2011), Botha et al. (2015), Ivanova et al. (2016), Bello et al. (2007, 2008), and De Witt Hamer et al. (2011). These findings underscore the IFOF's significant role in the neural processing of semantic information, highlighting its importance in the cognitive network of language and visual processing.

The vertical occipital fasciculus (VOF) forms connections between the dorsolateral and ventrolateral visual cortex. This pathway includes the visual word form area (VWFA), located in the ventral occipitotemporal cortex, which specializes in the visual processing of words and reading (Dehaene et al. 2002; Forkel et al. 2022b; Wandell et al. 2012). This area is significant for its role in reading and word recognition. The earliest detailed account of the occipital lobe's white matter was provided by Heinrich Sachs in his atlas, "The White Matter of the Human Cerebrum. Part 1. The Occipital Lobe" (Forkel et al. 2015; Forkel 2015). Under the encouragement of his mentor Carl Wernicke, Sachs made significant contributions to our understanding of cerebral anatomy. The atlas was envisioned as part of a more extensive series exploring each tract's function and clinical relevance. With advancements in tractography and Klingler dissections, researchers have recently revisited and further elucidated the occipital white matter (Vergani et al. 2014; Bugain et al. 2021). To accurately identify the VOF, two regions of interest (ROIs) are placed—one in the ventral and another in the dorsal occipital lobe. These ROIs are outlined on axial slides to be perpendicular to the vertical fiber system. The dorsal ROI is posterior to the parieto-occipital sulcus (POS), while the ventral ROI is posterior to the occipital notch. The most lateral part of the occipital region, featuring the preoccipital (or temporo-occipital) notch, serves as a boundary between the inferior temporal gyrus and the ventral surface of the inferior occipital gyrus, sometimes continuing with

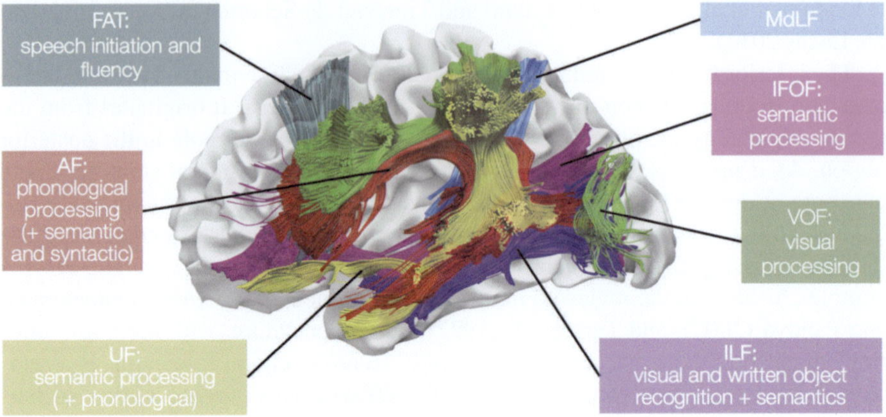

Fig. 10.4 Summary of language networks in the left hemisphere

the inferior temporal sulcus. The delineation of the occipital lobe's borders is based on landmarks between the POS and the notch, distinguishing the occipital lobe from the adjacent temporal and parietal lobes.

Figure 10.4 summarizes the language-related network in the left hemisphere (LH), outlining their presumed functions based on current research. This overview illustrates the specific contributions of tracts like the ILF and the interconnected nature of these pathways in supporting the broader network of language processing and cognitive functions in the brain.

The traditional viewpoint in neuroscience has long emphasized the dominance of the left hemisphere (LH) in language functions, leading clinical research to concentrate primarily on LH regions. However, this perspective is being reevaluated due to emerging evidence and clinical observations. Notably, in cases involving slow-growing tumor masses, neuroplasticity can lead to the right hemisphere (RH) assuming more significant roles in language processes, traditionally attributed to the LH. Recent studies are increasingly questioning the notion of the RH playing only a secondary role in language (Sarubbo et al. 2020; Vilasboas et al. 2017; Hartwigsen et al. 2010, 2013; Jung and Lambon Ralph 2016; Rice et al. 2015).

Consequently, contemporary clinical methodologies advocate for mapping cortical and subcortical white matter (WM) areas in the RH. This approach aims to preserve these areas from potential damage during surgical resection. Intraoperative language mappings have highlighted the importance of networks in the RH, emphasizing its involvement in various linguistic processes, both verbal and nonverbal, that are crucial to preserving (Vilasboas et al. 2017; Chang et al. 2011; Tate et al. 2014; Vassal et al. 2010; Herbet et al. 2017).

In recognition of the integral roles played by both hemispheres and encompassing all relevant lobes and tracts, the upcoming subsection will detail the methodologies for effectively capturing and visualizing individual language networks in a case report. This comprehensive approach highlights the importance of considering the entire cerebral landscape in the study and clinical treatment of language functions.

10.4 Illustrative Case

In this section, we delve into a complex and instructive case of a 51-year-old patient presenting with language difficulties due to a brain lesion. The case highlights the intersection of advanced neuroimaging, meticulous surgical planning, and the critical role of neuromonitoring in managing brain pathologies located in functionally significant areas. Through this patient's journey, we explore how integrated approaches in neurosurgery, combining preoperative tractography, incorporating preoperative nTMS data in the OR, and performing minimally invasive surgery, and minimally invasive surgical techniques can effectively address challenging neuro-oncological cases while preserving essential cognitive and motor functions.

A 51-year-old individual with a history of lung sarcoidosis came to the hospital, experiencing newly developed difficulties in finding words and hesitancy in speech. Diagnostic imaging revealed a lesion in the anterior limb of the internal capsule, characterized by solid and cystic components, exerting a mass effect in the area (as shown in Fig. 10.5). Following thorough discussions involving the patient and the neuro-oncology multidisciplinary team (MDT), it was decided to proceed with a maximal safe resection. This approach involved using a transsulcal minimally invasive parafascicular surgery (MIPS) technique, complemented by integrated neuromonitoring and mapping before and during the operation.

Preoperative tractography targeting specific tracts of interest was conducted using an anatomically guided, manual segmentation approach. This method involved careful delineation based on anatomical landmarks to accurately identify and visualize the relevant neural pathways:

1. For the corticospinal tract (CST), ROIs were precisely placed in the precentral gyrus and the ipsilateral cerebral peduncle to ensure accurate tract visualization.
2. For the fronto-aslant tract (FAT), ROIs were established in the superior frontal gyrus and the ipsilateral frontal operculum for detailed mapping.
3. In the case of the arcuate fasciculus (AF), ROIs were placed in the frontal operculum and extended to the posterior temporal lobe, encompassing areas toward the angular and supramarginal gyri for comprehensive tract visualization.
4. The inferior fronto-occipital fasciculus (IFOF) was mapped with an ROI positioned in the frontal lobe, lateral to the frontal horn, and another ROI in the ipsilateral stratum sagittal to capture its full trajectory.

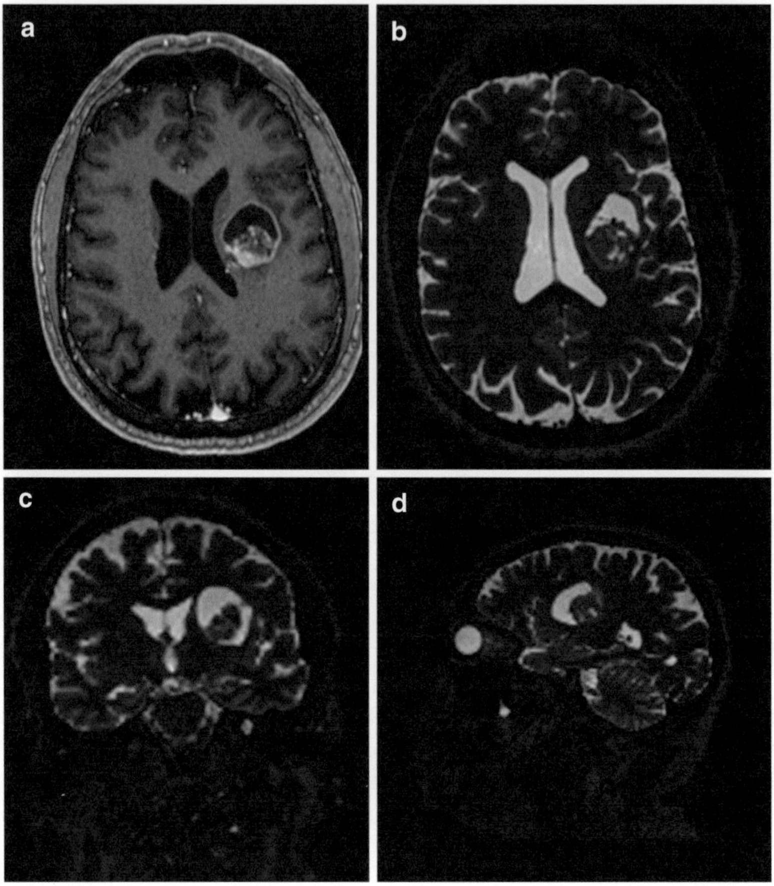

Fig. 10.5 Diagnostic structural image. Axial T1-weighted with gadolinium (**a**), T2-weighted axial (**b**), coronal (**c**), and sagittal (**d**) images showing an intrinsic lesion with mixed cystic-solid characteristics within the anterior limb of corona radiata (central core)

nTMS was carried out on the same side as the lesion (ipsilateral). For motor mapping of both upper and lower limbs, resting motor thresholds (RMT) for upper and lower limb were measured, with positive responses defined by an amplitude threshold of 50 μV (Chap. 6). Language mapping involved using repetitive nTMS at 7 Hz, set to 110% of RMT, utilizing the VAN-POP Test (Ohlerth et al. 2020), which is tailored for native English speakers. Positive language responses were detected at the conjunction of the AF and FAT in the pars opercularis during object naming, and at most anterior and posterior superior frontal gyrus terminations of the FAT during past tense action naming tasks, as shown in Fig. 10.6.

A three-dimensional model of the tumor was created and seamlessly integrated with both cortical and subcortical mapping data. Utilizing this combined model, a

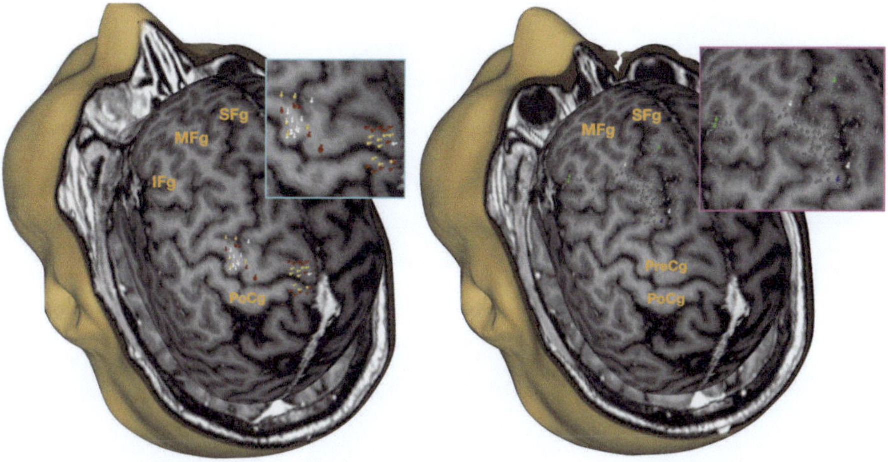

Fig. 10.6 nTMS motor (left) and language (right) mapping. The motor mapping is visually represented using a color-coded voltage scale: red dots indicate responses within the 50–500 μV range, yellow dots for 500–1000 μV, and white dots for responses exceeding 1000 μV. In the language mapping panel, different colors denote various response types: gray dots represent negative responses, green dots indicate performance errors, white dots signify no response, and blue dots are used to mark semantic errors. This color-coding provides a clear and concise visual representation of the motor and language areas as identified by nTMS

transsulcal approach for brain cannulation was planned through the superior frontal sulcus. This carefully chosen surgical corridor ran parallel to the AF and the CST, and navigated through the middle third of the FAT, deliberately avoiding areas where the FAT terminates that showed positive responses in nTMS language mapping. Additionally, the cortical region surrounding the selected cannulation site showed no language-related activity in the nTMS mapping, as indicated in Fig. 10.7.

An awake craniotomy was conducted, allowing for real-time intraoperative monitoring and mapping. The frontal operculum remained unexposed due to the minimally invasive parafascicular surgery approach targeting the superior frontal gyrus. Owing to the absence of positive language responses in the nTMS mapping and the lack of speech arrest, cortical mapping was not deemed necessary. During the procedure, brain cannulation took place concurrently with VAN-POP testing, which did not reveal any language deficits. Throughout the tumor resection, continuous language and motor mapping were employed. This process was halted once functional margins were identified, marked by reversible language disturbances such as hesitations (laterally) and semantic errors (inferiorly), as well as corticospinal tract responses at 4 mA with high-frequency stimulation. The details of this phase are depicted in Fig. 10.8.

Following the surgery, the patient successfully recovered without any language or motor impairments. Postoperative care included the continuation of oncological treatment (Fig. 10.9).

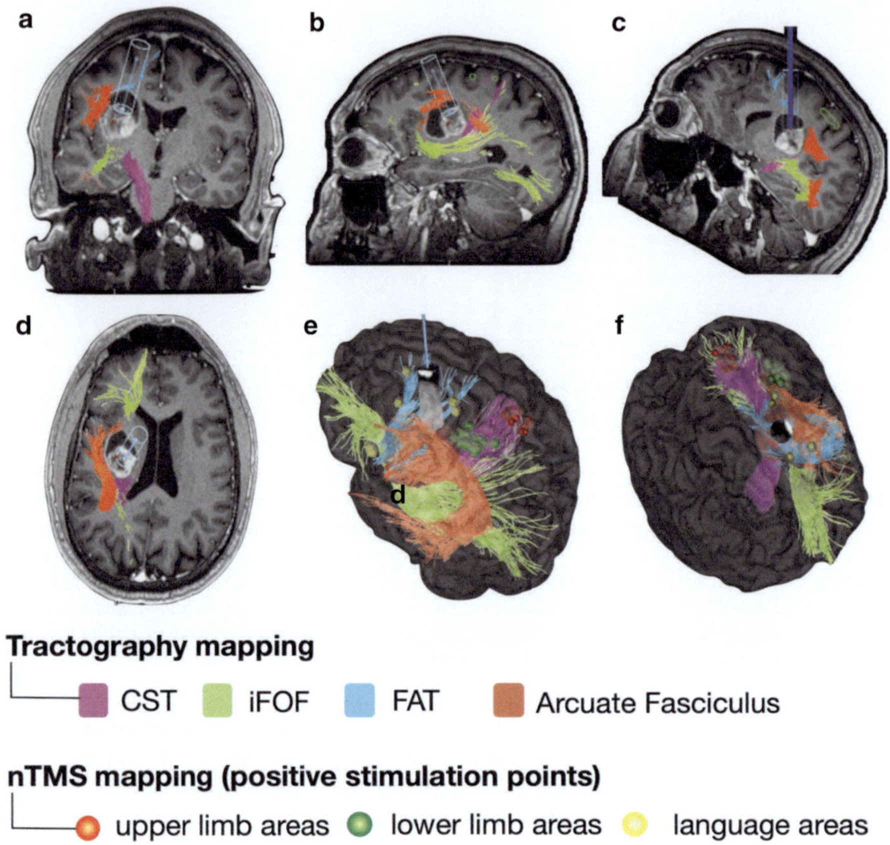

Fig. 10.7 Integrated preoperative cortical and subcortical mapping. This figure illustrates detailed motor and language mapping in various perspectives: coronal (**a**), sagittal (**b**), a trajectory view (**c**), axial (**d**), and a comprehensive 3D model (**e**, **f**)

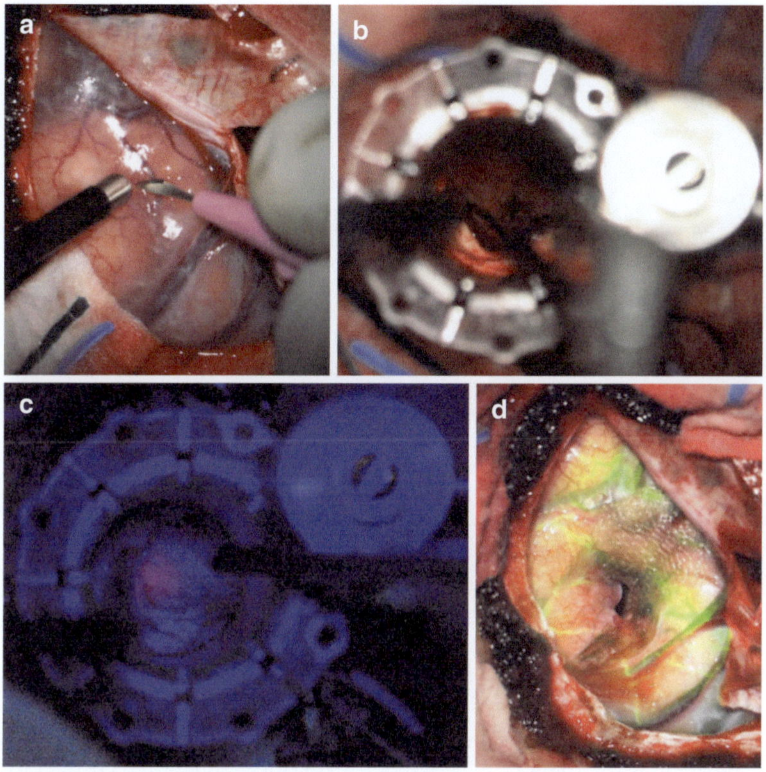

Fig. 10.8 Surgical procedure. It begins with a small craniotomy precisely centered over the superior frontal sulcus, followed by the opening of the overlying arachnoid membrane (panel **a**). The process then proceeds to brain cannulation and the resection of a tumor that tested positive for 5-aminolevulinic acid (5-ALA), as shown in panels (**b**) and (**c**). Finally, panel (**d**) illustrates the decannulation stage post-tumor resection, with use of indocyanine green to confirm the patency and integrity of the cortical vascular structures

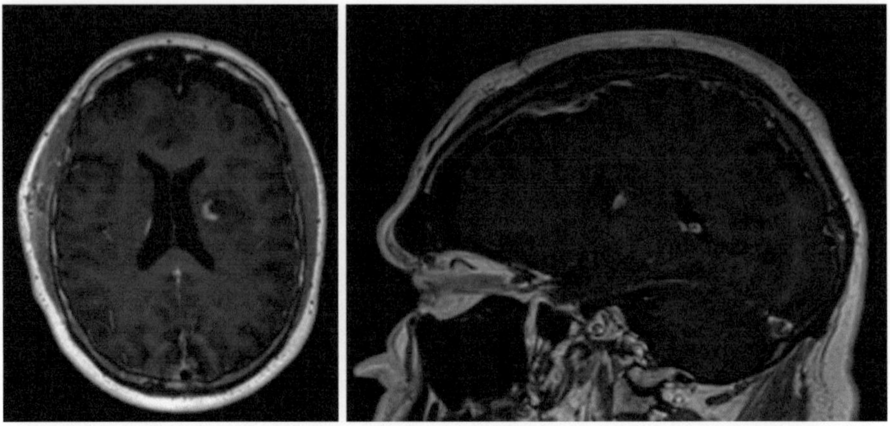

Fig. 10.9 Postoperative imaging. Axial (**a**) and sagittal (**b**) views captured after the tumor resection. These images confirm the absence of complications and show the presence of hemostatic material at the surgery site, indicating a successful procedural outcome

10.5 Conclusion

In conclusion, this chapter has highlighted the critical importance of tractography and navigated transcranial magnetic stimulation (nTMS) in neurosurgery, particularly for preserving language functions. Tractography provides invaluable insights into the intricate white matter tracts, while nTMS offers functional cortical localization. Their combined use in preoperative planning has revolutionized surgical approaches, maximizing outcomes and safeguarding vital human capabilities.

The presented case exemplifies the effectiveness of these techniques in managing complex brain lesions. Integrating advanced neuroimaging and neuromonitoring facilitated a tailored surgical intervention, successfully preserving the patient's cognitive and motor functions. This case underscores the transformative potential of combining anatomical and functional mapping in neurosurgery, paving the way for more personalized and precise treatments. As the field continues to evolve, such integrated approaches will undoubtedly play a pivotal role in enhancing patient care and outcomes in neurosurgery.

Acknowledgments We extend our heartfelt gratitude to the patients whose participation and cooperation were indispensable in the development of this chapter. Their willingness to engage in our studies and share their experiences has been invaluable in advancing our understanding of the brain's language system.

Funding This project received funding from the Donders Theme Leadership Award (SJF, LANGUAGE & COMMUNICATION) and the Donders Mohrmann Fellowship No. 2401515 (SJF, NEUROVARIABILITY).

References

Alves J, Cardoso M, Morgado M, De Witte E, Satoer D, Hall A, et al. Language assessment in awake brain surgery: the Portuguese adaptation of the Dutch linguistic intraoperative protocol (DuLIP). Clin Linguist Phon. 2021;35(12):1113–31.

Anderson JM, Gilmore R, Roper S, Crosson B, Bauer RM, Nadeau S, et al. Conduction aphasia and the arcuate fasciculus: a reexamination of the Wernicke–Geschwind model. Brain Lang. 1999;70(1):1–12.

Babajani-Feremi A, Narayana S, Rezaie R, Choudhri AF, Fulton SP, Boop FA, et al. Language mapping using high gamma electrocorticography, fMRI, and TMS versus electrocortical stimulation. Clin Neurophysiol. 2016;127(3):1822–36.

Bastiaanse R, Ohlerth AK. Presurgical language mapping: what are we testing? J Pers Med. 2023;13(3):376. Available from: https://doi.org/10.3390/jpm13030376.

Bello L, Gallucci M, Fava M, Carrabba G, Giussani C, Acerbi F, et al. Intraoperative subcortical language tract mapping guides surgical removal of gliomas involving speech areas. Neurosurgery. 2007;60(1):67.

Bello L, Gambini A, Castellano A, Carrabba G, Acerbi F, Fava E, et al. Motor and language DTI Fiber Tracking combined with intraoperative subcortical mapping for surgical removal of gliomas. NeuroImage. 2008;39(1):369–82.

Bello L, Castellano A, Fava E, Casaceli G, Riva M, Scotti G, et al. Intraoperative use of diffusion tensor imaging fiber tractography and subcortical mapping for resection of gliomas: technical considerations. Neurosurg Focus. 2010;28(2):E6.

Blecher T, Miron S, Schneider GG, Achiron A, Ben-Shachar M. Association between white matter microstructure and verbal fluency in patients with multiple sclerosis. Front Psychol. 2019;10:1607.

Botha H, Duffy JR, Whitwell JL, Strand EA, Machulda MM, Schwarz CG, et al. Classification and clinicoradiologic features of primary progressive aphasia (PPA) and apraxia of speech. Cortex. 2015;69:220–36.

Bugain M, Dimech Y, Torzhenskaya N, Thiebaut de Schotten M, Caspers S, Muscat R, et al. Occipital Intralobar fasciculi: a description, through tractography, of three forgotten tracts. Commun Biol. 2021;4(1):433.

Castellano A, Cirillo S, Bello L, Riva M, Falini A. Functional MRI for surgery of gliomas. Curr Treat Options Neurol. 2017;19(10):34.

Catani M, Thiebaut de Schotten M. A diffusion tensor imaging tractography atlas for virtual in vivo dissections. Cortex. 2008;44(8):1105–32.

Catani M, Howard RJ, Pajevic S, Jones DK. Virtual in vivo interactive dissection of white matter fasciculi in the human brain. NeuroImage. 2002;17(1):77–94.

Catani M, Jones DK, Donato R, Ffytche DH. Occipito-temporal connections in the human brain. Brain. 2003;126(9):2093–107.

Catani M, Jones DK, Ffytche DH. Perisylvian language networks of the human brain. Ann Neurol. 2005;57(1):8–16.

Catani M, Allin MPG, Husain M, Pugliese L, Mesulam MM, Murray RM, et al. Symmetries in human brain language pathways correlate with verbal recall. Proc Natl Acad Sci USA. 2007;104(43):17163–8.

Catani M, Dell'acqua F, Vergani F, Malik F, Hodge H, Roy P, et al. Short frontal lobe connections of the human brain. Cortex. 2012;48(2):273–91.

Catani M, Dell'acqua F, Thiebaut de Schotten M. A revised limbic system model for memory, emotion and behaviour. Neurosci Biobehav Rev. 2013;37(8):1724–37.

Chang EF, Wang DD, Perry DW, Barbaro NM, Berger MS. Homotopic organization of essential language sites in right and bilateral cerebral hemispheric dominance. J Neurosurg. 2011;114(4):893–902.

Cochereau J, Lemaitre AL, Wager M, Moritz-Gasser S, Duffau H, Herbet G. Network-behavior mapping of lasting executive impairments after low-grade glioma surgery. Brain Struct Funct. 2020;225(8):2415–29.

Croxson PL, Forkel SJ, Cerliani L, Thiebaut de Schotten M. Structural variability across the primate brain: a cross-species comparison. Cereb Cortex. 2018;28(11):3829–41.

D'Anna L, Mesulam MM, Thiebaut de Schotten M, Dell'Acqua F, Murphy D, Wieneke C, et al. Frontotemporal networks and behavioral symptoms in primary progressive aphasia. Neurology. 2016;86(15):1393–9.

De Witt Hamer PC, Moritz-Gasser S, Gatignol P, Duffau H. Is the human left middle longitudinal fascicle essential for language? A brain electrostimulation study. Hum Brain Mapp. 2011;32(6):962–73.

De Witte E, Satoer D, Robert E, Colle H, Verheyen S, Visch-Brink E, et al. The Dutch Linguistic Intraoperative Protocol: a valid linguistic approach to awake brain surgery. Brain Lang. 2015;140:35–48.

Dehaene S, Le Clec'H G, Poline JB, Le Bihan D, Cohen L. The visual word form area: a prelexical representation of visual words in the fusiform gyrus. Neuroreport. 2002;13(3):321–5.

Dick AS, Garic D, Graziano P, Tremblay P. The frontal aslant tract (FAT) and its role in speech, language and executive function. Cortex. 2019;111:148–63.

Dragoy O, National Research University Higher School of Economics, Chrabaszcz A, Tolkacheva V, Buklina S. Russian intraoperative naming test: a standardized tool to map noun and verb production during awake neurosurgeries. Russ J Cogn Sci. 2016;3(4):4–25.

Dragoy O, Zyryanov A, Bronov O, Gordeyeva E, Gronskaya N, Kryuchkova O, et al. Functional linguistic specificity of the left frontal aslant tract for spontaneous speech fluency: evidence from intraoperative language mapping. Brain Lang. 2020;208:104836.

Duffau H. The anatomo-functional connectivity of language revisited. New insights provided by electrostimulation and tractography. Neuropsychologia. 2008;46(4):927–34.

Duffau H. The huge plastic potential of adult brain and the role of connectomics: new insights provided by serial mappings in glioma surgery. Cortex. 2014;58:325–37.

Duffau H, Capelle L, Sichez N, Denvil D, Lopes M, Sichez JP, et al. Intraoperative mapping of the subcortical language pathways using direct stimulations. An anatomo-functional study. Brain. 2002;125(1):199–214.

Duffau H, Gatignol P, Mandonnet E, Peruzzi P, Tzourio-Mazoyer N, Capelle L. New insights into the anatomo-functional connectivity of the semantic system: a study using cortico-subcortical electrostimulations. Brain. 2005;128(Pt 4):797–810.

Flöel A, de Vries MH, Scholz J, Breitenstein C, Johansen-Berg H. White matter integrity in the vicinity of Broca's area predicts grammar learning success. NeuroImage. 2009;47(4):1974–81.

Forkel SJ. Heinrich Sachs (1863–1928). J Neurol. 2015;262(2):498–500.

Forkel SJ, Thiebaut de Schotten M, Dell'Acqua F, Kalra L, Murphy DGM, Williams SCR, et al. Anatomical predictors of aphasia recovery: a tractography study of bilateral perisylvian language networks. Brain. 2014a;137(Pt 7):2027–39.

Forkel SJ, Thiebaut de Schotten M, Kawadler JM, Dell'Acqua F, Danek A, Catani M. The anatomy of fronto-occipital connections from early blunt dissections to contemporary tractography. Cortex. 2014b;56:73–84.

Forkel SJ, Mahmood S, Vergani F, Catani M. The white matter of the human cerebrum: part I The occipital lobe by Heinrich Sachs. Cortex. 2015;62:182–202.

Forkel SJ, Friedrich P, Thiebaut de Schotten M, Howells H. White matter variability, cognition, and disorders: a systematic review. Brain Struct Funct. 2022a;227(2):529–44.

Forkel SJ, Labache L, Nachev P, Thiebaut de Schotten M, Hesling I. Stroke disconnectome decodes reading networks. Brain Struct Funct. 2022b;227(9):2897–908.

Forkel S, Bortolami C, Dulyan L, Barrett RL, Beyh A. Dissecting white matter pathways: a neuroanatomical approach [Internet]. 2023. Available from: https://doi.org/10.31234/osf.io/9xwgm.

Frey S, Campbell JSW, Pike GB, Petrides M. Dissociating the human language pathways with high angular resolution diffusion fiber tractography. J Neurosci. 2008;28(45):11435–44.

Friederici AD. White-matter pathways for speech and language processing. Handb Clin Neurol. 2015;129:177–86.

Geva S, Correia M, Warburton EA. Diffusion tensor imaging in the study of language and aphasia. Aphasiology. 2011;25(5):543–58.

Giampiccolo D, Duffau H. Controversy over the temporal cortical terminations of the left arcuate fasciculus: a reappraisal. Brain. 2022;145(4):1242–56.

Gil-Robles S, Carvallo A, Jimenez MDM, Gomez Caicoya A, Martinez R, Ruiz-Ocaña C, et al. Double dissociation between visual recognition and picture naming: a study of the visual language connectivity using tractography and brain stimulation. Neurosurgery. 2013;72(4):678–86.

Gisbert-Muñoz S, Quiñones I, Amoruso L, Timofeeva P, Geng S, Boudelaa S, et al. MULTIMAP: multilingual picture naming test for mapping eloquent areas during awake surgeries. Behav Res Methods. 2021;53(2):918–27.

Giussani C, Roux FE, Ojemann J, Sganzerla EP, Pirillo D, Papagno C. Is preoperative functional magnetic resonance imaging reliable for language areas mapping in brain tumor surgery? Review of language functional magnetic resonance imaging and direct cortical stimulation correlation studies. Neurosurgery. 2010;66(1):113.

Hartwigsen G, Baumgaertner A, Price CJ, Koehnke M, Ulmer S, Siebner HR. Phonological decisions require both the left and right supramarginal gyri. Proc Natl Acad Sci USA. 2010;107(38):16494–9.

Hartwigsen G, Saur D, Price CJ, Ulmer S, Baumgaertner A, Siebner HR. Perturbation of the left inferior frontal gyrus triggers adaptive plasticity in the right homologous area during speech production. Proc Natl Acad Sci USA. 2013;110(41):16402–7.

Hau J, Sarubbo S, Perchey G, Crivello F, Zago L, Mellet E, et al. Cortical terminations of the inferior fronto-occipital and uncinate fasciculi: anatomical stem-based virtual dissection. Front Neuroanat. 2016;10:58.

Hazem SR, Awan M, Lavrador JP, Patel S, Wren HM, Lucena O, et al. Middle frontal gyrus and area 55b: perioperative mapping and language outcomes. Front Neurol. 2021;12:646075.

Herbet G, Moritz-Gasser S, Duffau H. Direct evidence for the contributive role of the right inferior fronto-occipital fasciculus in non-verbal semantic cognition. Brain Struct Funct. 2017;222(4):1597–610.

Hillis AE, Tuffiash E, Caramazza A. Modality-specific deterioration in naming verbs in nonfluent primary progressive aphasia. J Cogn Neurosci. 2002;14(7):1099–108.

Hillis AE, Beh YY, Sebastian R, Breining B, Tippett DC, Wright A, et al. Predicting recovery in acute poststroke aphasia. Ann Neurol. 2018;83(3):612–22.

Hosomi A, Nagakane Y, Yamada K, Kuriyama N, Mizuno T, Nishimura T, et al. Assessment of arcuate fasciculus with diffusion-tensor tractography may predict the prognosis of aphasia in patients with left middle cerebral artery infarcts. Neuroradiology. 2009;51(9):549–55.

Ille S, Sollmann N, Hauck T, Maurer S, Tanigawa N, Obermueller T, et al. Impairment of preoperative language mapping by lesion location: a functional magnetic resonance imaging, navigated transcranial magnetic stimulation, and direct cortical stimulation study. J Neurosurg. 2015;123(2):314–24.

Ivanova MV, Isaev DY, Dragoy OV, Akinina YS, Petrushevskiy AG, Fedina ON, et al. Diffusion-tensor imaging of major white matter tracts and their role in language processing in aphasia. Cortex. 2016;85:165–81.

Ivanova MV, Zhong A, Turken A, Baldo JV, Dronkers NF. Functional contributions of the arcuate fasciculus to language processing. Front Hum Neurosci. 2021;15:672665.

Jarret J, Ferré P, Chedid G, Bedetti C, Bore A, Joanette Y, et al. Functional network and structural connections involved in picture naming. Brain Lang. 2022;231:105146.

Jeltema HR, Ohlerth AK, de Wit A, Wagemakers M, Rofes A, Bastiaanse R, et al. Comparing navigated transcranial magnetic stimulation mapping and "gold standard" direct cortical stimulation mapping in neurosurgery: a systematic review. Neurosurg Rev. 2021;44(4):1903–20.

Jung J, Lambon Ralph MA. Mapping the dynamic network interactions underpinning cognition: a cTBS-fMRI study of the flexible adaptive neural system for semantics. Cereb Cortex. 2016;26(8):3580–90.

Kaplan E, Naeser MA, Martin PI, Ho M, Wang Y, Baker E, et al. Horizontal portion of arcuate fasciculus fibers track to pars opercularis, not pars triangularis, in right and left hemispheres: a DTI study. NeuroImage. 2010;52(2):436–44.

Kemerdere R, de Champfleur NM, Deverdun J, Cochereau J, Moritz-Gasser S, Herbet G, et al. Role of the left frontal aslant tract in stuttering: a brain stimulation and tractographic study. J Neurol. 2016;263(1):157–67.

Kim SH, Jang SH. Prediction of aphasia outcome using diffusion tensor tractography for arcuate fasciculus in stroke. AJNR Am J Neuroradiol. 2013;34(4):785–90.

Kinoshita M, Nakada M, Okita H, Hamada JI, Hayashi Y. Predictive value of fractional anisotropy of the arcuate fasciculus for the functional recovery of language after brain tumor resection: a preliminary study. Clin Neurol Neurosurg. 2014;117:45–50.

Kleiser R, Staempfli P, Valavanis A, Boesiger P, Kollias S. Impact of fMRI-guided advanced DTI fiber tracking techniques on their clinical applications in patients with brain tumors. Neuroradiology. 2010;52(1):37–46.

Kram L, Ohlerth AK, Ille S, Meyer B, Krieg SM. CompreTAP: feasibility and reliability of a new language comprehension mapping task via preoperative navigated transcranial magnetic stimulation. Cortex. 2023;171:347–69.

Krieg SM, Lioumis P, Mäkelä JP, Wilenius J, Karhu J, Hannula H, et al. Protocol for motor and language mapping by navigated TMS in patients and healthy volunteers; workshop report. Acta Neurochir. 2017;159(7):1187–95.

Krings T, Chiappa KH, Foltys H, Reinges MH, Cosgrove GR, Thron A. Introducing navigated transcranial magnetic stimulation as a refined brain mapping methodology. Neurosurg Rev. 2001;24(4):171–9.

Leclercq D, Duffau H, Delmaire C, Capelle L, Gatignol P, Ducros M, et al. Comparison of diffusion tensor imaging tractography of language tracts and intraoperative subcortical stimulations. J Neurosurg. 2010;112(3):503–11.

Li M, Zhang Y, Song L, Huang R, Ding J, Fang Y, et al. Structural connectivity subserving verbal fluency revealed by lesion-behavior mapping in stroke patients. Neuropsychologia. 2017;101:85–96.

Li S, Mu Y, Rao Y, Sun C, Li X, Liu H, et al. Preoperative individual-target transcranial magnetic stimulation demonstrates an effect comparable to intraoperative direct electrical stimulation in language-eloquent glioma mapping and improves postsurgical outcome: a retrospective fiber-tracking and electromagnetic simulation study. Front Oncol [Internet]. 2023;13:1089787. Available from: https://www.frontiersin.org/articles/10.3389/fonc.2023.1089787.

Liu X, Kinoshita M, Shinohara H, Hori O, Ozaki N, Nakada M. Does the superior fronto-occipital fascicle exist in the human brain? Fiber dissection and brain functional mapping in 90 patients with gliomas. Neuroimage Clin. 2020;25:102192.

Lucena O, Lavrador JP, Irzan H, Semedo C, Borges P, Vergani F, et al. Assessing informative tract segmentation and nTMS for pre-operative planning. J Neurosci Methods. 2023;396:109933.

Maldonado IL, Moritz-Gasser S, de Champfleur NM, Bertram L, Moulinié G, Duffau H. Surgery for gliomas involving the left inferior parietal lobule: new insights into the functional anatomy provided by stimulation mapping in awake patients. J Neurosurg. 2011;115(4):770–9.

Mandonnet E, Nouet A, Gatignol P, Capelle L, Duffau H. Does the left inferior longitudinal fasciculus play a role in language? A brain stimulation study. Brain. 2007;130(Pt 3):623–9.

Mandonnet E, Gatignol P, Duffau H. Evidence for an occipito-temporal tract underlying visual recognition in picture naming. Clin Neurol Neurosurg. 2009;111(7):601–5.

Martino J, De Lucas EM. Subcortical anatomy of the lateral association fascicles of the brain: a review. Clin Anat. 2014;27(4):563–9.

Mätzig S, Druks J, Masterson J, Vigliocco G. Noun and verb differences in picture naming: past studies and new evidence. Cortex. 2009;45(6):738–58.

Meola A, Comert A, Yeh FC, Stefaneanu L, Fernandez-Miranda JC. The controversial existence of the human superior fronto-occipital fasciculus: connectome-based tractographic study with microdissection validation. Hum Brain Mapp. 2015;36(12):4964–71.

Mesulam M, Rogalski E, Wieneke C, Cobia D, Rademaker A, Thompson C, et al. Neurology of anomia in the semantic variant of primary progressive aphasia. Brain. 2009;132(Pt 9):2553–65.

Miceli G, Silveri MC, Villa G, Caramazza A. On the basis for the agrammatic's difficulty in producing main verbs. Cortex. 1984;20(2):207–20.

Mirchandani AS, Beyh A, Lavrador JP, Howells H, Dell'Acqua F, Vergani F. Altered corticospinal microstructure and motor cortex excitability in gliomas: an advanced tractography and transcranial magnetic stimulation study. J Neurosurg. 2020;134(5):1368–76.

Mori S, Wakana S, van Zijl PCM, Nagae-Poetscher LM. MRI atlas of human white matter. Amsterdam: Elsevier; 2005. 276 p.

Natalizi F, Piras F, Vecchio D, Spalletta G, Piras F. Preoperative navigated transcranial magnetic stimulation: new insight for brain tumor-related language mapping. J Pers Med [Internet]. 2022;12(10). Available from: https://doi.org/10.3390/jpm12101589.

Negwer C, Ille S, Hauck T, Sollmann N, Maurer S, Kirschke JS, et al. Visualization of subcortical language pathways by diffusion tensor imaging fiber tracking based on rTMS language mapping. Brain Imaging Behav. 2017a;11(3):899–914.

Negwer C, Sollmann N, Ille S, Hauck T, Maurer S, Kirschke JS, et al. Language pathway tracking: comparing nTMS-based DTI fiber tracking with a cubic ROIs-based protocol. J Neurosurg. 2017b;126(3):1006–14.

Negwer C, Beurskens E, Sollmann N, Maurer S, Ille S, Giglhuber K, et al. Loss of subcortical language pathways correlates with surgery-related aphasia in patients with brain tumor: an investigation via repetitive navigated transcranial magnetic stimulation—based diffusion tensor imaging fiber tracking. World Neurosurg. 2018;111:e806–18.

Ohlerth AK, Valentin A, Vergani F, Ashkan K, Bastiaanse R. The verb and noun test for perioperative testing (VAN-POP): standardized language tests for navigated transcranial magnetic stimulation and direct electrical stimulation. Acta Neurochir. 2020;162(2):397–406.

Ohlerth AK, Bastiaanse R, Negwer C, Sollmann N, Schramm S, Schröder A, et al. Bihemispheric navigated transcranial magnetic stimulation mapping for action naming compared to object naming in sentence context. Brain Sci [Internet]. 2021a;11(9). Available from: https://doi.org/10.3390/brainsci11091190.

Ohlerth AK, Bastiaanse R, Negwer C, Sollmann N, Schramm S, Schröder A, et al. Benefit of action naming over object naming for visualization of subcortical language pathways in navigated transcranial magnetic stimulation-based diffusion tensor imaging-fiber tracking. Front Hum Neurosci. 2021b;15:748274.

Ohlerth AK, Bastiaanse R, Nickels L, Neu B, Zhang W, Ille S, et al. Dual-task nTMS mapping to visualize the cortico-subcortical language network and capture postoperative outcome—a patient series in neurosurgery. Front Oncol [Internet]. 2022;11:788122. Available from: https://www.frontiersin.org/articles/10.3389/fonc.2021.788122.

Ortibus E, Verhoeven J, Sunaert S, Casteels I, de Cock P, Lagae L. Integrity of the inferior longitudinal fasciculus and impaired object recognition in children: a diffusion tensor imaging study. Dev Med Child Neurol. 2012;54(1):38–43.

Papagno C, Miracapillo C, Casarotti A, Romero Lauro LJ, Castellano A, Falini A, et al. What is the role of the uncinate fasciculus? Surgical removal and proper name retrieval. Brain. 2011;134(Pt 2):405–14.

Papagno C, Casarotti A, Comi A, Pisoni A, Lucchelli F, Bizzi A, et al. Long-term proper name anomia after removal of the uncinate fasciculus. Brain Struct Funct. 2016;221(1):687–94.

Papatzalas C, Fountas K, Kapsalaki E, Papathanasiou I. The use of standardized intraoperative language tests in awake craniotomies: a scoping review. Neuropsychol Rev. 2022;32(1):20–50.

Papoutsi M, Stamatakis EA, Griffiths J, Marslen-Wilson WD, Tyler LK. Is left fronto-temporal connectivity essential for syntax? Effective connectivity, tractography and performance in left-hemisphere damaged patients. NeuroImage. 2011;58(2):656–64.

Raffa G, Bährend I, Schneider H, Faust K, Germanò A, Vajkoczy P, et al. A novel technique for region and linguistic specific nTMS-based DTI Fiber tracking of language pathways in brain tumor patients. Front Neurosci. 2016;10:552.

Raffa G, Quattropani MC, Scibilia A, Conti A. Surgery of language-eloquent tumors in patients not eligible for awake surgery: the impact of a protocol based on navigated transcranial magnetic stimulation on …. Clin Neurol Neurosurg [Internet]. 2018. Available from: https://www.sciencedirect.com/science/article/pii/S0303846718301124.

Raffa G, Marzano G, Curcio A, Espahbodinea S, Germanò A, Angileri FF. Personalized surgery of brain tumors in language areas: the role of preoperative brain mapping in patients not eligible for awake surgery. Neurosurg Focus. 2022;53(6):E3.

Reisch K, Böttcher F, Tuncer MS, Schneider H, Vajkoczy P, Picht T, et al. Tractography-based navigated TMS language mapping protocol. Front Oncol. 2022;12:1008442.

Rice GE, Lambon Ralph MA, Hoffman P. The roles of left versus right anterior temporal lobes in conceptual knowledge: an ALE meta-analysis of 97 functional neuroimaging studies. Cereb Cortex. 2015;25(11):4374–91.

Rofes A, Spena G, Talacchi A, Santini B, Miozzo A, Miceli G. Mapping nouns and finite verbs in left hemisphere tumors: a direct electrical stimulation study. Neurocase. 2017;23(2):105–13.

Rösler J, Niraula B, Strack V, Zdunczyk A, Schilt S, Savolainen P, et al. Language mapping in healthy volunteers and brain tumor patients with a novel navigated TMS system: evidence of tumor-induced plasticity. Clin Neurophysiol. 2014;125(3):526–36.

Rosso C, Vargas P, Valabregue R, Arbizu C, Henry-Amar F, Leger A, et al. Aphasia severity in chronic stroke patients: a combined disconnection in the dorsal and ventral language pathways. Neurorehabil Neural Repair. 2015;29(3):287–95.

Rotenberg A, Horvath JC, Pascual-Leone A. The transcranial magnetic stimulation (TMS) device and foundational techniques. In: Rotenberg A, Horvath JC, Pascual-Leone A, editors. Transcranial magnetic stimulation. New York: Springer New York; 2014. p. 3–13.

Ruohonen J, Karhu J. Navigated transcranial magnetic stimulation. Neurophysiol Clin. 2010;40(1):7–17.

Sarubbo S, De Benedictis A, Merler S, Mandonnet E, Balbi S, Granieri E, et al. Towards a functional atlas of human white matter. Hum Brain Mapp. 2015;36(8):3117–36.

Sarubbo S, Tate M, De Benedictis A, Merler S, Moritz-Gasser S, Herbet G, et al. Mapping critical cortical hubs and white matter pathways by direct electrical stimulation: an original functional atlas of the human brain. NeuroImage. 2020;205:116237.

Satoer D, Dulyan L, Forkel S. Asymmetries in language-relevant brain tumors. In: Corballis P, Papagno C, editors. Handbook of clinical neurology – cerebral asymmetries. Amsterdam: Elsevier; 2025.

Schmahmann JD, Pandya DN. The complex history of the fronto-occipital fasciculus. J Hist Neurosci. 2007;16(4):362–77.

Sierpowska J, Gabarrós A, Fernandez-Coello A, Camins À, Castañer S, Juncadella M, et al. Words are not enough: nonword repetition as an indicator of arcuate fasciculus integrity during brain tumor resection. J Neurosurg. 2017;126(2):435–45.

Silva LL, Tuncer MS, Vajkoczy P, Picht T, Rosenstock T. Distinct approaches to language pathway tractography: comparison of anatomy-based, repetitive navigated transcranial magnetic stimulation (rTMS)–based, and rTMS-enhanced diffusion tensor imaging–fiber tracking. J Neurosurg. 2021;136(2):589–600.

Sollmann N, Negwer C, Ille S, Maurer S, Hauck T, Kirschke JS, et al. Feasibility of nTMS-based DTI fiber tracking of language pathways in neurosurgical patients using a fractional anisotropy threshold. J Neurosci Methods. 2016a;267:45–54.

Sollmann N, Kubitscheck A, Maurer S, Ille S, Hauck T, Kirschke JS, et al. Preoperative language mapping by repetitive navigated transcranial magnetic stimulation and diffusion tensor imaging fiber tracking and their comparison to intraoperative stimulation. Neuroradiology. 2016b;58(8):807–18.

Sollmann N, Zhang H, Fratini A, Wildschuetz N, Ille S, Schröder A, et al. Risk assessment by presurgical tractography using navigated TMS maps in patients with highly motor- or language-eloquent brain tumors. Cancers [Internet]. 2020;12(5). Available from: https://doi.org/10.3390/cancers12051264.

Tate MC, Herbet G, Moritz-Gasser S, Tate JE, Duffau H. Probabilistic map of critical functional regions of the human cerebral cortex: Broca's area revisited. Brain. 2014;137(Pt 10):2773–82.

Teichmann M, Rosso C, Martini JB, Bloch I, Brugières P, Duffau H, et al. A cortical-subcortical syntax pathway linking Broca's area and the striatum. Hum Brain Mapp. 2015;36(6):2270–83.

Thiebaut de Schotten M, Forkel SJ. The emergent properties of the connected brain. Science. 2022;378(6619):505–10.

Thiebaut de Schotten M, Urbanski M, Duffau H, Volle E, Lévy R, Dubois B, et al. Direct evidence for a parietal-frontal pathway subserving spatial awareness in humans. Science. 2005;309(5744):2226–8.

Thiebaut de Schotten M, Dell'Acqua F, Forkel S, Simmons A, Vergani F, Murphy DGM, et al. A lateralized brain network for visuospatial attention. Nat Neurosci. 2011;14(10):1245–6.

Trinh VT, Fahim DK, Shah K, Tummala S, McCutcheon IE, Sawaya R, et al. Subcortical injury is an independent predictor of worsening neurological deficits following awake craniotomy procedures. Neurosurgery. 2013;72(2):160–9.

Tuncer MS, Salvati LF, Grittner U, Hardt J, Schilling R, Bährend I, et al. Towards a tractography-based risk stratification model for language area associated gliomas. Neuroimage Clin. 2021;29:102541.

Türe U, Yaşargil MG, Pait TG. Is there a superior occipitofrontal fasciculus? A microsurgical anatomic study. Neurosurgery. 1997;40(6):1226–32.

Turken AU, Dronkers NF. The neural architecture of the language comprehension network: converging evidence from lesion and connectivity analyses. Front Syst Neurosci. 2011;5:1.

Vassal M, Le Bars E, Moritz-Gasser S, Menjot N, Duffau H. Crossed aphasia elicited by intraoperative cortical and subcortical stimulation in awake patients. J Neurosurg. 2010;113(6):1251–8.

Vassal F, Boutet C, Lemaire JJ, Nuti C. New insights into the functional significance of the frontal aslant tract: an anatomo–functional study using intraoperative electrical stimulations combined with diffusion tensor imaging-based fiber tracking. Br J Neurosurg. 2014;28(5):685–7.

Vavassori L, Venturini M, Zigiotto L, Annicchiarico L, Corsini F, Avesani P, et al. The arcuate fasciculus: combining structure and function into surgical considerations. Brain Behav. 2023 Aug;13(8):e3107.

Vergani F, Lacerda L, Martino J, Attems J, Morris C, Mitchell P, et al. White matter connections of the supplementary motor area in humans. J Neurol Neurosurg Psychiatry. 2014;85(12):1377–85.

Vidorreta JG, Garcia R, Moritz-Gasser S, Duffau H. Double dissociation between syntactic gender and picture naming processing: a brain stimulation mapping study. Hum Brain Mapp. 2011;32(3):331–40.

Vilasboas T, Herbet G, Duffau H. Challenging the myth of right nondominant hemisphere: lessons from corticosubcortical stimulation mapping in awake surgery and surgical implications. World Neurosurg. 2017;103:449–56.

Wandell BA, Rauschecker AM, Yeatman JD. Learning to see words. Annu Rev Psychol. 2012;63:31–53.

Zingeser LB, Berndt RS. Grammatical class and context effects in a case of pure anomia: implications for models of language production. Cogn Neuropsychol. 1988;5(4):473–516.

Risk Stratification by nTMS Language Mapping

11

Florian Ringel and Alice Dauth

11.1 Introduction

The resection of brain lesions within regions of highly functional brain is a major challenge in neurosurgery. While number one priority in cases of eloquently located lesions is the preservation of the patients' functional integrity, the achievement of a maximum safe resection needs to be fulfilled especially in oncological cases. Therefore, oncological aims as a complete resection of a primary brain tumor need to be balanced against the aim to preserve essential brain function. Previously, generalized functional anatomy was used to guide resections in or adjacent to functional areas. However, generalized functional anatomy is not reliable to preserve individual higher cognitive functions as functional topographic anatomy can be highly variable. The availability of intraoperative direct electrical stimulation (DES) and identification of individual functional anatomy is used in modern neurosurgery to achieve maximum resection with functional preservation (Hervey-Jumper et al. 2015; Ojemann and Whitaker 1978; Sanai et al. 2008; Gerritsen et al. 2022; Morshed et al. 2024; Semonche et al. 2024). In addition to many individual studies supporting the use of DES to improve patient functional outcome and to increase extent of resection, meta-analyses could clearly demonstrate that the use of intraoperative electrical cortical and subcortical mapping and monitoring of functional cortex and subcortical white matter tracts allows for a higher number of GTR while maintaining functional integrity of the patient (De Witt Hamer et al. 2012) (Peters et al. 2023). Prospective randomized trials are currently recruiting patients to further increase the evidence for the use of asleep or awake brain mapping and monitoring for eloquent tumor surgery (Gerritsen et al. 2020, 2021).

F. Ringel (✉) · A. Dauth
Department of Neurosurgery, University Medical Center,
Johannes Gutenberg-University Mainz, Mainz, Germany
e-mail: florian.ringel@unimedizin-mainz.de; alice.dauth@unimedizin-mainz.de

S. M. Krieg, T. Picht (eds.), *Navigated Transcranial Magnetic Stimulation in Neurosurgery*, https://doi.org/10.1007/978-3-031-97155-6_11

Presently, the use of intraoperative DES mapping is regarded as the gold standard to identify relevant functional cortex during resections of eloquently located lesions. However, in order to allow mapping and monitoring of higher cortical functions as language, awake surgery is mandatory. These awake craniotomies have been popularized and increased the safety of surgery in functional brain areas. But prior to the intraoperative identification of relevant functional brain structures, a preoperative identification of functional anatomy is demanded in order to evaluate the surgical risks and allow for preoperative risk stratification (Rammeloo et al. 2023). Factors such as resectability, planned EOR, the surgical approach, preoperative awareness of eloquent cortex at risk, and the identification of starting points for intraoperative stimulation could contribute to the decision for resective surgery, preoperative patient counseling, intraoperative safety, and the reduction of awake time. Thereby, brain tumors initially deemed nonresectable on a purely generalized anatomical basis can be individually re-evaluated by pre- and intraoperative functional mapping and changed to a resectable lesion (Krieg et al. 2013a; Southwell et al. 2018).

Furthermore, while awake surgery is feasible for the majority of patients, a subgroup of patients is not amendable for awake language mapping or fail intraoperative language mapping (Sanai et al. 2008; Milian et al. 2014; Nossek et al. 2013; Picht et al. 2013). For this smaller subgroup of patients, presurgical mapping of language areas would be even more helpful and might allow for safe asleep resections of otherwise unresectable lesions.

The most commonly used technique to provide preoperative insights in individual functional anatomy is fMRI. However, especially in the vicinity of intrinsic brain tumors, fMRI language mapping can be associated with false-negative results, thereby making fMRI language mapping unreliable for brain tumors located in language-eloquent brain areas (Giussani et al. 2010; Lakhani et al. 2023). As elaborated in previous chapters, nTMS allows for presurgical identification of language areas in a noninvasive manner and could serve as a valuable technique for preoperative risk stratification prior to awake surgery or allow safe asleep surgical resection of language-eloquent lesions in patients not amendable to awake surgery.

The feasibility of nTMS motor mapping has been evaluated nicely in a previous study revealing that in 27.4% of cases, presurgical nTMS motor mapping had an objective benefit and, in 54.8%, an impact on the surgical resection of motor-eloquent lesions (Picht et al. 2012). Many further studies clearly supported the benefits of presurgical nTMS motor mapping with nTMS-based DTI-FT for preoperative risk stratification resulting in improved extents of resection and patient outcome (Ivren et al. 2023; Krieg et al. 2014a, 2016; Moser et al. 2017; Raffa et al. 2019; Sollmann et al. 2017).

This chapter aims to summarize the potential influence of presurgical nTMS language mapping on risk stratification for lesions located in or adjacent to language-eloquent cortical areas. Therefore, the topics (i) identification of hemispheric language dominance, (ii) reliability (sensitivity and specificity) of nTMS language mapping, (iii) spatial resolution of nTMS language mapping, and (iv) evaluated benefits of nTMS language mapping are elaborated in the following regarding its benefit for presurgical risk stratification.

11.2 Identification of Hemispheric Language Dominance

The classical concept of cortical language representation localized language function to the dominant hemisphere, which is the left-sided in the majority of individuals. However, language localization to the nondominant hemisphere is not excluded. The classical method to assess hemispheric dominance prior to surgery is the so-called Wada test utilizing a selective intracarotid sodium amytal injection to a hemisphere. The sodium amytal injection, although considered as gold standard to identify hemispheric dominance, is invasive and therefore loosing relevance with the availability of noninvasive methods as nTMS to detect language localization. A comparison of nTMS to functional MRI and high gamma electrocorticography revealed a sensitivity of 67%, specificity of 66%, and an AUC of 0.68 for TMS to detect hemispheric language localization (Babajani-Feremi et al. 2016). Furthermore, in a comparison of intracarotid sodium amytal injection, fMRI, magnetoencephalography, and nTMS, a sensitivity of 74%, specificity of 72%, PPV of 78%, and NPV of 68% to identify the hemispheric dominance by nTMS were reported (Rezaie et al. 2020).

With respect to hemispheric language localization and dominance of language representation, several studies could identify cortical regions within the nondominant hemisphere participating in language function. Right hemispheric language function has been identified in healthy participants as well as in left-hemispheric stroke or tumor patients by a variety of techniques (Baum et al. 2012; Baumgaertner et al. 2013; Brennan and Pylkkanen 2012; Briganti et al. 2012; Devlin and Watkins 2007; Schuhmann et al. 2012; Thiel et al. 2005; Vigneau et al. 2011). The right-sided inferior frontal gyrus was shown to contribute to language function in a study using non-navigated TMS, which has a suboptimal spatial resolution in comparison to nTMS (Thiel et al. 2005, 2006). The technique of nTMS allows a superior spatial resolution, precise localization, and quantification of left- and right-sided language areas and, by the comparison of error frequencies, a calculation of an hemispheric dominance ratio (HDR) (Krieg et al. 2013b).

In a recent nTMS language mapping study, the right and left hemispheric distribution of language was assessed in healthy volunteers as well as in patients harboring lesions in left-sided language-eloquent cortical regions (Krieg et al. 2013b). In all healthy volunteers and tumor harboring patients, language errors were elucidated upon stimulation of the right hemisphere supporting a role of language processing of the right hemisphere. In order to assess hemispheric dominance, the frequency of left- and right-sided language errors was compared calculating an HDR. The HDR is calculated by dividing the ER of the left hemispheric through the corresponding right hemispheric brain region. This can be the whole hemisphere, a lobe or even a subgyrus. Thus, an HDR >1 means left-sided language dominance (according to nTMS), while an HDR <1 means right-sided language dominance.

While a left-sided hemispheric dominance was predominant throughout patients and volunteers, there was a significantly higher rate of right-sided language regions in patients with left-sided perisylvian lesions in comparison to healthy volunteers (Fig. 11.1). These results were supported by another study comparing healthy

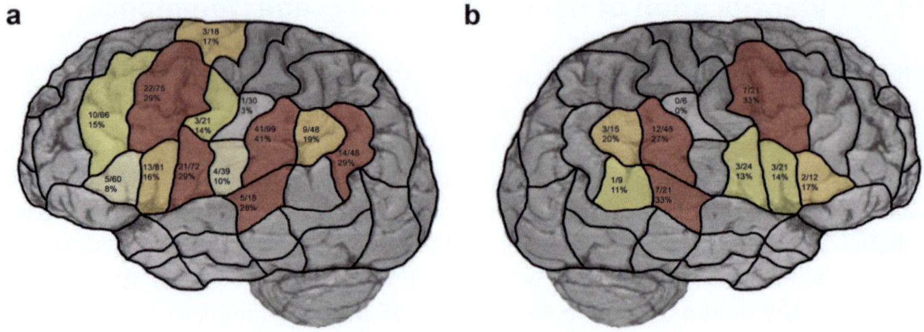

Fig. 11.1 Error rates of different brain regions. This brain template shows the ER of different brain regions (separated according to the CPS) for the left (**a**) and right (**b**) hemisphere in a patient with a left-sided angular gyrus anaplastic astrocytoma WHO grade III. The ERs show that nTMS was able to induce a considerable amount of naming errors within the right hemisphere

volunteers and patients with left hemispheric gliomas (Rosler et al. 2014). In healthy volunteers, language errors were almost exclusively produced by nTMS stimulation of the left hemisphere; in tumor-harboring patients, a higher frequency of errors in the right hemisphere occurred suggesting tumor-induced plasticity and language reorganization. These studies indicate a language shift toward the right hemisphere by brain plasticity induced by left-sided language-eloquent lesions reducing the left hemispheric dominance. In consequence, left-sided language regions might become less essential with an increasing dominance of the right hemisphere. But despite the presence of a language shift, it remains to be elucidated what extent of language shift to the right hemisphere allows sacrificing left-sided language areas during surgery while a sufficient compensation by the right hemisphere is secured. So far, these results were not validated regarding their significance by any further method to assess hemispheric dominance as Wada testing. Thereby, it remains to be elucidated whether the language shift to the right hemisphere as assessed by nTMS in the patient group translates into a clinically relevant right-hemispheric compensation of left-sided language function. As an approximation to this most important question, the hemispheric language dominance as measured by nTMS was correlated with the postoperative language outcome after resection of left-sided perisylvian lesions (Ille et al. 2016a). A significant difference of the hemispheric dominance for anterior language regions was found for patients with a new postoperative deficit compared to patients without a new postoperative deficit. This means patients with a shift of language function to the right hemisphere might be at lower risk for a postoperative new deficit after resection of language-eloquent regions by right hemispheric compensation. However, the overall number of new permanent language deficits in the study was at 4%, which lowers the strength of the conclusion.

Overall, the assessment of preoperative language dominance by nTMS might be a parameter to allow for preoperative risk assessment with respect to new language deficits. However, to further elucidate values of hemispheric dominance and the associated risk of surgery-induced deficits, studies including higher numbers of patients are necessary.

11.3 Reliability of nTMS Language Mapping Results

The most crucial point determining the feasibility of preoperative nTMS language mapping for risk stratification is the reliability of nTMS-identified language-positive or language-negative areas, i.e., the sensitivity, specificity, PPV, and NPV. To determine these values, presurgical language mapping results were compared to the present gold standard of language mapping, which is intraoperative DES during awake surgery (Fig. 11.2). Depending on the protocol of nTMS language mapping and the algorithm of analysis of copositive or conegative stimulation points with nTMS and DES, the sensitivity was 90%, the specificity 24–98%, the PPV 36–69%, and the NPV 84–99% (Picht et al. 2013; Tarapore et al. 2013). By an analysis of anterior language points surrounding Broca's area, only sensitivity and NPV were found to reach 100%. A further refinement of the protocol with regard to the timing of stimulation could even improve the NPV of nTMS language mapping (Krieg et al. 2014b). The low specificity and low PPV demonstrate that depending on the protocol of nTMS mapping used, a high number of false-positive responses might result. In

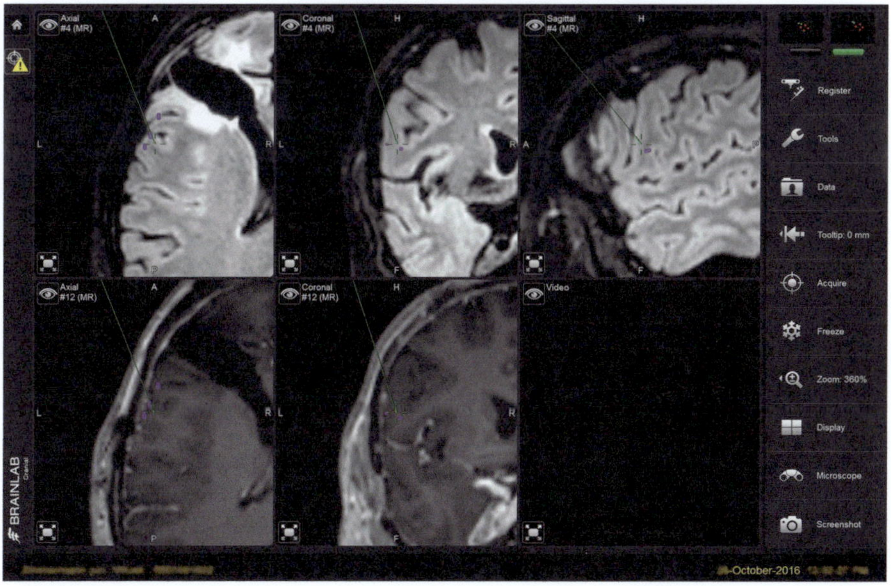

Fig. 11.2 Preoperative nTMS language mapping vs. intraoperative DES mapping during awake surgery. This screenshot shows the intraoperative neuronavigation during awake surgery of a 21 year old patient suffering from a recurrent opercular diffuse astrocytoma WHO grade II. The preoperatively nTMS-positive language areas (purple) correspond well with the intraoperative DES-positive spot identified during awake surgery

consequence, when mapping results would be used to guide the resection of a brain lesion, the false-positive points could result in a premature unnecessary termination of resection leaving potentially resectable tumor behind. However, in order to maintain patients´ functional integrity, the high NPV and low number of false-negative sites are important. This means the likelihood that a negatively nTMS-mapped point was positive during surgery was 1% and thereby the reliability of a negative point is very high. Therefore, a nTMS mapping-guided resection would have a very low risk of resection of language-relevant cortical tissue if tumor in negatively mapped tissue was resected, only. A resection based on nTMS language mapping, only, could result in unnecessary tumor remnants but associated with a very low risk of cortical functional damage.

Therefore, the use of presurgical nTMS language mapping based primarily on negative mapping results could potentially lead to a reliable estimation of resectability, corticotomy, and extent of resection of a language-eloquent lesion (Fig. 11.3). In cases where a lesion is surrounded by nTMS-negative points, even resection without intraoperative electrical stimulation might be justified in patients, which

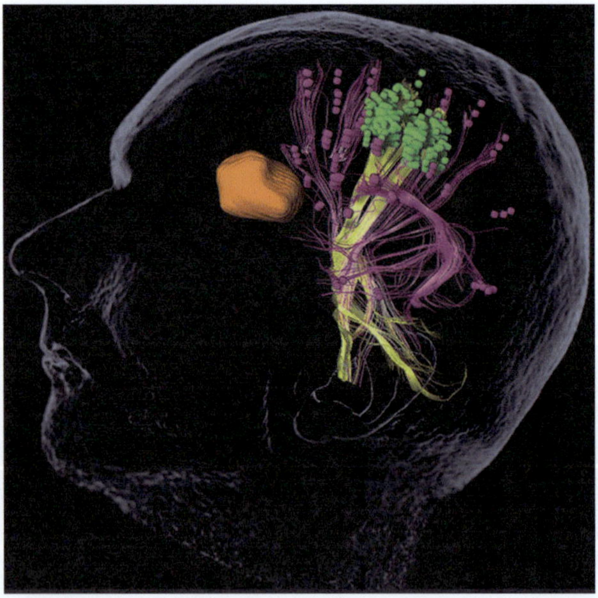

Fig. 11.3 nTMS mapping data to confirm asleep resectability. This patient suffering from an anaplastic astrocytoma WHO grade III within the triangular part of the IFG. Preoperative nTMS language mapping of the whole left hemisphere showed that the tumor does not affect language-involved cortical or subcortical structures. Since the whole hemisphere was mapped, the cortex above the tumor can be regarded as language-negative. Orange = tumor. Purple = language-positive cortical and subcortical areas according to preoperative nTMS language mapping and nTMS-based DTI FT of language pathways. Green = motor area as mapped by nTMS. Yellow = CST visualized via nTMS-based DTI FT

could not undergo awake surgery (Ille et al. 2016b). Due to the high NPV, the risk of a new postoperative deficit would be low.

Several further studies support the sensitivity, specificity, PPV, and NPV data reported so far. Several recent studies comparing nTMS and DES reported sensitivities of 35% and 83%, specificities of 90% and 67%, a PPV of 16% and 16%, and a NPV of 96% and 99%, thereby providing further information that a surgical resection can be guided by nTMS-negative areas (Bahrend et al. 2020; Muir et al. 2022).

11.4 Spatial Accuracy and Resolution of nTMS Language Mapping

In addition to the predictive reliability, spatial accuracy and spatial resolution are highly important with regard to the usability of nTMS for risk stratification. Spatial accuracy means how accurate the TMS pulse is delivered to the stimulation spot as projected on the 3D MRI dataset in the navigated system. The spatial accuracy is composed of several factors potentially contributing to inaccuracy, such as optical tracking of coil localization in the navigation system, inaccuracies of head tracking, the electric field computation model, and the registration to anatomical MRIs. The mean error of the real electric field hotspot to the virtual hotspot projected on the imaging dataset was calculated for a popular nTMS system to be 5.7 mm (Ruohonen and Karhu 2010). Furthermore, the spatial resolution, i.e., the minimum necessary distance of two language-relevant cortical areas that allows differentiation of these two spots is of importance. A recent volunteer study tried to evaluate the spatial resolution by the measurement of the minimum distance necessary to discriminate a point with a low ER with a point of high ER. A mean distance of 13.8 mm was described as spatial accuracy (Sollmann et al. 2016a). As the average width of a human brain cortical gyrus is 10–20 mm, a stimulation at the center of a gyrus should hit the stimulated gyrus according to the spatial accuracy, and upon stimulation of adjacent gyri, it should be possible to differentiate language function of these gyri according to the spatial resolution of nTMS language mapping. Therefore, the technique allows for identification of language function on a gyral level, which is slightly inferior to the differentiation of approximately 5 mm of DES during awake surgery. In conclusion, nTMS language mapping allows for identification of structures at risk on the spatial level of different gyri as depicted in the cortical parcellation system (CPS) by Corina (Corina et al. 2010).

11.5 Clinical Benefits and Outcome Prediction of nTMS Mapping Versus No Mapping

As nTMS for language can detect language-negative cortical areas with a high resolution and reliability, the question remains whether this does translate into a clinical benefit for the patient. Aspects which are highly relevant are indication for surgery, estimation of EOR, reduction of awake mapping time by knowledge of starting points for intraoperative electrical stimulation or even increased safety of asleep resections in patients where awake mapping is not feasible, and overall improved surgical outcome. Several of these aspects have been evaluated so far.

11.5.1 rnTMS-Based DTI-FT to Visualize Subcortical Language Pathways

While language function depends on the areas of cortical language representation which can be identified by nTMS, subcortical language tracts are likewise important for the correct execution of language function. DTI-FT allows for preoperative non-invasive reconstruction and identification of subcortical language pathways. Usually, generalized anatomical information is used as seed regions for DTI-FT of subcortical language pathways. However, use of the functional cortical language information from nTMS could potentially improve DTI-FT. Several studies assessed this aspect. Using different protocols for DTI-FT, Sollmann et al. could visualize nine language-related subcortical pathways with nTMS-based seeding (Sollmann et al. 2016b). A comparison to the classical anatomical cubic ROI-based approach to nTMS-based tracking of language pathways revealed superior and more reliable tracking result with the nTMS-based approach (Negwer et al. 2017a, b). In contrast, a comparison of different DT-FT approaches using anatomy-based and nTMS-based tracking strategies to visualize five principle language pathways could not reveal an advantage of the nTMS-based approach (Silva et al. 2022).

11.5.2 Risk Assessment and Reduction by Preoperative nTMS Language Mapping and nTMS-Based DTI-FT

Postoperative language outcome following resection of language-eloquent tumors clearly correlates with the integrity of cortical and subcortical language areas as assessed by nTMS (Ille et al. 2018; Kram et al. 2024). Therefore, prevention of resection of nTMS-identified language-eloquent cortical sites and subcortical pathways should improve patient outcome. Whether the surgeons´ knowledge of preoperative nTMS language data do make a difference in outcome for patients undergoing resection of language-eloquent lesions using awake language mapping has been already assessed (Sollmann et al. 2015). In a matched cohort analysis of 25 patients per group, the group in which presurgical language mapping results were available during surgery showed an improved language outcome and smaller craniotomies while the EOR, overall rate of perioperative complications and the duration of surgery did not differ. As an initial important step in planning of a resection of a language-eloquent tumor preoperative nTMS allows for identification true eloquence of a tumor location and potential disproval of eloquence. In a series of 20 patients with tumors in assumed language-eloquent location, presurgical nTMS language mapping combined with nTMS-based DTI-FT verified eloquence in 60% and disproved eloquence in 40% of patients. This approval or disproval of eloquent location allows for a risk/benefit estimation of a surgical resection and could guide the resections toward a more defensive or aggressive tumor reduction (Raffa et al. 2018).

The risk of new postoperative deficits following resection of language-eloquent tumors is furthermore predicted by the tumor or lesion to tract distance (LTD). An LTD of ≤16 mm to the arcuate fascicle and ≤ 25 mm to any other language-relevant fascicle was associated with at least temporary language deficits while LTD ≤8 mm to the AF and ≤ 11 mm to any other tracts was associated with permanent deficits (Kram et al. 2024; Sollmann et al. 2019, 2020).

In addition to the LTD, an increased interhemispheric connectivity of cortical language sites could be identified as a risk factor for new language deficits after surgical resections with a very high specificity of 70–93% (Ille et al. 2016a, 2022; Sollmann et al. 2016c; Zhang et al. 2023).

A recent nTMS-based connectome analysis of the whole brain could also reveal a predictability of surgery-related aphasia with an accuracy of 73.3% and a sensitivity of 78.3%.

By the determination of true eloquence, the LTD, and hemispheric connectivity, nTMS does contribute to the preoperative risk assessment for the resection of a lesion in assumed language-eloquent localization. Does this allow for resections based on nTMS language information only without intraoperative DES?

11.5.3 Resections of Language-Eloquent Lesions Based Purely on nTMS Language Mapping

In general, the individual patient's language distribution is assessed by DES during awake surgery. However, not all patients qualify for awake surgery for different reasons. Tumor resection based on preoperative nTMS language mapping is an alternative for this subgroup of patients. Several studies assessed the feasibility of asleep surgery for language-eloquent tumors based on preoperative nTMS language mapping and nTMS-based DTI-FT.

Four cases of patients with left-sided perisylvian lesions which have been resected on the basis of preoperative nTMS language data only because they were not testable during awake surgery have been reported (Ille et al. 2016b). None of the patients had a permanent new deficit after surgery. Hendrix et al. reported on 20 patients resected asleep, based on nTMS data which were matched to 20 asleep resections without nTMS data. They could show a more favorable language outcome for the mapped patient cohort in comparison to the nonmapped cohort (Hendrix et al. 2017). Raffa et al. reported that the visualization of cortical language areas, subcortical tracts, and interhemispheric connectivity allowed for safe surgical resections in 20 patients (Raffa et al. 2018). In a larger comparison of 47 patients with awake DES versus 100 asleep patients with nTMS language information, no difference in functional outcome was found, while gross total resection rate was slightly higher in awake patients (87%) versus asleep patients (72%) (Ille et al. 2021). A classification of the degree of eloquence revealed that awake DES-operated patients had a higher degree of eloquent location. This allows for the conclusion that presurgical nTMS language mapping and DTI-FT do allow for safe asleep resections in a relevant subgroup of patients with lower degrees of eloquence while highly eloquent lesion location seems to demand awake surgery to achieve high extents of resection in combination with good functional outcome.

11.6 Conclusion

In conclusion, nTMS-based preoperative language mapping does allow for an objective risk stratification. nTMS allows for presurgical assessment of the eloquence of a perisylvian lesion in the dominant hemisphere. The hemispheric dominance ratio, interhemispheric connectivity, and the lesion to subcortical language tract distance allow for a stratification of the resection risk. In less critical localizations, even a resection purely based on nTMS language mapping and nTMS-based fiber tracking in an asleep patient is feasible without an increased risk for resection-related language deficits. Therefore, nTMS is an increasingly valuable tool in the treatment of language-eloquent lesions.

References

Babajani-Feremi A, Narayana S, Rezaie R, et al. Language mapping using high gamma electrocorticography, fMRI, and TMS versus electrocortical stimulation. Clin Neurophysiol. 2016;127(3):1822–36. https://doi.org/10.1016/j.clinph.2015.11.017.

Bahrend I, Muench MR, Schneider H, et al. Incidence and linguistic quality of speech errors: a comparison of preoperative transcranial magnetic stimulation and intraoperative direct cortex stimulation. J Neurosurg. 2020;134(5):1409–18. https://doi.org/10.3171/2020.3.JNS193085.

Baum SH, Martin RC, Hamilton AC, Beauchamp MS. Multisensory speech perception without the left superior temporal sulcus. NeuroImage. 2012;62(3):1825–32. https://doi.org/10.1016/j.neuroimage.2012.05.034.

Baumgaertner A, Hartwigsen G, Roman SH. Right-hemispheric processing of non-linguistic word features: implications for mapping language recovery after stroke. Hum Brain Mapp. 2013;34(6):1293–305. https://doi.org/10.1002/hbm.21512.

Brennan J, Pylkkanen L. The time-course and spatial distribution of brain activity associated with sentence processing. NeuroImage. 2012;60(2):1139–48. https://doi.org/10.1016/j.neuroimage.2012.01.030.

Briganti C, Sestieri C, Mattei PA, et al. Reorganization of functional connectivity of the language network in patients with brain gliomas. AJNR Am J Neuroradiol. 2012;33(10):1983–90. https://doi.org/10.3174/ajnr.A3064.

Corina DP, Loudermilk BC, Detwiler L, Martin RF, Brinkley JF, Ojemann G. Analysis of naming errors during cortical stimulation mapping: implications for models of language representation. Brain Lang. 2010;115(2):101–12. https://doi.org/10.1016/j.bandl.2010.04.001.

De Witt Hamer PC, Robles SG, Zwinderman AH, Duffau H, Berger MS. Impact of intraoperative stimulation brain mapping on glioma surgery outcome: a meta-analysis. J Clin Oncol Off J Am Soc Clin Oncol. 2012;30(20):2559–65. https://doi.org/10.1200/JCO.2011.38.4818.

Devlin JT, Watkins KE. Stimulating language: insights from TMS. Brain J Neurol. 2007;130(Pt 3):610–22. https://doi.org/10.1093/brain/awl331.

Gerritsen JKW, Klimek M, Dirven CMF, et al. The SAFE-trial: safe surgery for glioblastoma multiforme: awake craniotomy versus surgery under general anesthesia. Study protocol for a multicenter prospective randomized controlled trial. Contemp Clin Trials. 2020;88:105876. https://doi.org/10.1016/j.cct.2019.105876.

Gerritsen JKW, Dirven CMF, De Vleeschouwer S, et al. The PROGRAM study: awake mapping versus asleep mapping versus no mapping for high-grade glioma resections: study protocol for an international multicenter prospective three-arm cohort study. BMJ Open. 2021;11(7):e047306. https://doi.org/10.1136/bmjopen-2020-047306.

Gerritsen JKW, Zwarthoed RH, Kilgallon JL, et al. Effect of awake craniotomy in glioblastoma in eloquent areas (GLIOMAP): a propensity score-matched analysis of an international, multicentre, cohort study. Lancet Oncol. 2022;23(6):802–17. https://doi.org/10.1016/S1470-2045(22)00213-3.

Giussani C, Roux FE, Ojemann J, Sganzerla EP, Pirillo D, Papagno C. Is preoperative functional magnetic resonance imaging reliable for language areas mapping in brain tumor surgery? Review of language functional magnetic resonance imaging and direct cortical stimulation correlation studies. Neurosurgery. 2010;66(1):113–20. https://doi.org/10.1227/01.NEU.0000360392.15450.C9.

Hendrix P, Senger S, Simgen A, Griessenauer CJ, Oertel J. Preoperative rTMS language mapping in speech-eloquent brain lesions resected under general anesthesia: a pair-matched cohort study. World Neurosurg. 2017;100:425–33. https://doi.org/10.1016/j.wneu.2017.01.041.

Hervey-Jumper SL, Li J, Lau D, et al. Awake craniotomy to maximize glioma resection: methods and technical nuances over a 27-year period. J Neurosurg. 2015;123(2):325–39. https://doi.org/10.3171/2014.10.JNS141520.

Ille S, Kulchytska N, Sollmann N, et al. Hemispheric language dominance measured by repetitive navigated transcranial magnetic stimulation and postoperative course of language function in brain tumor patients. Neuropsychologia. 2016a;91:50–60. https://doi.org/10.1016/j.neuropsychologia.2016.07.025.

Ille S, Sollmann N, Butenschoen VM, Meyer B, Ringel F, Krieg SM. Resection of highly language-eloquent brain lesions based purely on rTMS language mapping without awake surgery. Acta Neurochir. 2016b;158(12):2265–75. https://doi.org/10.1007/s00701-016-2968-0.

Ille S, Engel L, Kelm A, Meyer B, Krieg SM. Language-eloquent white matter pathway Tractography and the course of language function in glioma patients. Front Oncol. 2018;8:572. https://doi.org/10.3389/fonc.2018.00572.

Ille S, Schroeder A, Albers L, et al. Non-invasive mapping for effective preoperative guidance to approach highly language-eloquent gliomas-a large scale comparative cohort study using a new classification for language eloquence. Cancers (Basel). 2021;13(2):207. https://doi.org/10.3390/cancers13020207.

Ille S, Zhang H, Sogerer L, et al. Preoperative function-specific connectome analysis predicts surgery-related aphasia after glioma resection. Hum Brain Mapp. 2022;43(18):5408–20. https://doi.org/10.1002/hbm.26014.

Ivren M, Grittner U, Khakhar R, et al. Comparison of anatomical-based vs. nTMS-based risk stratification model for predicting postoperative motor outcome and extent of resection in brain tumor surgery. Neuroimage Clin. 2023;38:103436. https://doi.org/10.1016/j.nicl.2023.103436.

Kram L, Schroeder A, Meyer B, Krieg SM, Ille S. Function-guided differences of arcuate fascicle and inferior fronto-occipital fascicle tractography as diagnostic indicators for surgical risk stratification. Brain Struct Funct. 2024;229:2219. https://doi.org/10.1007/s00429-024-02787-3.

Krieg SM, Schnurbus L, Shiban E, et al. Surgery of highly eloquent gliomas primarily assessed as non-resectable: risks and benefits in a cohort study. BMC Cancer. 2013a;13:51. https://doi.org/10.1186/1471-2407-13-51.

Krieg SM, Sollmann N, Hauck T, et al. Functional language shift to the right hemisphere in patients with language-eloquent brain tumors. PLoS One. 2013b;8(9):e75403. https://doi.org/10.1371/journal.pone.0075403.

Krieg SM, Sabih J, Bulubasova L, et al. Preoperative motor mapping by navigated transcranial magnetic brain stimulation improves outcome for motor eloquent lesions. Neuro-Oncology. 2014a;16(9):1274–82. https://doi.org/10.1093/neuonc/nou007.

Krieg SM, Tarapore PE, Picht T, et al. Optimal timing of pulse onset for language mapping with navigated repetitive transcranial magnetic stimulation. NeuroImage. 2014b;100:219–36. https://doi.org/10.1016/j.neuroimage.2014.06.016.

Krieg SM, Picht T, Sollmann N, et al. Resection of motor eloquent metastases aided by preoperative nTMS-based motor maps-comparison of two observational cohorts. Front Oncol. 2016;6:261. https://doi.org/10.3389/fonc.2016.00261.

Lakhani DA, Sabsevitz DS, Chaichana KL, Quinones-Hinojosa A, Middlebrooks EH. Current state of functional MRI in the Presurgical planning of brain tumors. Radiol Imaging Cancer. 2023;5(6):e230078. https://doi.org/10.1148/rycan.230078.

Milian M, Tatagiba M, Feigl GC. Patient response to awake craniotomy – a summary overview. Acta Neurochir. 2014;156(6):1063–70. https://doi.org/10.1007/s00701-014-2038-4.

Morshed RA, Cummins DD, Clark JP, et al. Asleep triple-modality motor mapping for perirolandic gliomas: an update on outcomes. J Neurosurg. 2024;140(4):1029–37. https://doi.org/10.3171/2023.8.JNS231036.

Moser T, Bulubas L, Sabih J, et al. Resection of navigated transcranial magnetic stimulation-positive Prerolandic motor areas causes permanent impairment of motor function. Neurosurgery. 2017;81(1):99–110. https://doi.org/10.1093/neuros/nyw169.

Muir M, Patel R, Traylor J, et al. Validation of non-invasive language mapping modalities for eloquent tumor resection: a pilot study. Front Neurosci. 2022;16:833073. https://doi.org/10.3389/fnins.2022.833073.

Negwer C, Ille S, Hauck T, et al. Visualization of subcortical language pathways by diffusion tensor imaging fiber tracking based on rTMS language mapping. Brain Imaging Behav. 2017a;11(3):899–914. https://doi.org/10.1007/s11682-016-9563-0.

Negwer C, Sollmann N, Ille S, et al. Language pathway tracking: comparing nTMS-based DTI fiber tracking with a cubic ROIs-based protocol. J Neurosurg. 2017b;126(3):1006–14. https://doi.org/10.3171/2016.2.JNS152382.

Nossek E, Matot I, Shahar T, et al. Failed awake craniotomy: a retrospective analysis in 424 patients undergoing craniotomy for brain tumor. J Neurosurg. 2013;118(2):243–9. https://doi.org/10.3171/2012.10.JNS12511.

Ojemann GA, Whitaker HA. Language localization and variability. Brain Lang. 1978;6(2):239–60. (PM:728789)

Peters DR, Halimi F, Ozduman K, et al. Resection of the contrast-enhancing tumor in diffuse gliomas bordering eloquent areas using electrophysiology and 5-ALA fluorescence: evaluation of resection rates and neurological outcome-a systematic review and meta-analysis. Neurosurg Rev. 2023;46(1):185. https://doi.org/10.1007/s10143-023-02064-7.

Picht T, Schulz J, Hanna M, Schmidt S, Suess O, Vajkoczy P. Assessment of the influence of navigated transcranial magnetic stimulation on surgical planning for tumors in or near the motor cortex. Neurosurgery. 2012;70(5):1248–56.; discussion 1256-7. https://doi.org/10.1227/NEU.0b013e318243881e.

Picht T, Krieg SM, Sollmann N, et al. A comparison of language mapping by preoperative navigated transcranial magnetic stimulation and direct cortical stimulation during awake surgery. Neurosurgery. 2013;72(5):808–19. https://doi.org/10.1227/NEU.0b013e3182889e01.

Raffa G, Quattropani MC, Scibilia A, et al. Surgery of language-eloquent tumors in patients not eligible for awake surgery: the impact of a protocol based on navigated transcranial magnetic stimulation on presurgical planning and language outcome, with evidence of tumor-induced intra-hemispheric plasticity. Clin Neurol Neurosurg. 2018;168:127–39. https://doi.org/10.1016/j.clineuro.2018.03.009.

Raffa G, Scibilia A, Conti A, et al. The role of navigated transcranial magnetic stimulation for surgery of motor-eloquent brain tumors: a systematic review and meta-analysis. Clin Neurol Neurosurg. 2019;180:7–17. https://doi.org/10.1016/j.clineuro.2019.03.003.

Rammeloo E, Schouten JW, Krikour K, et al. Preoperative assessment of eloquence in neurosurgery: a systematic review. J Neuro-Oncol. 2023;165(3):413–30. https://doi.org/10.1007/s11060-023-04509-x.
Rezaie R, Schiller KK, Embury L, Boop FA, Wheless JW, Narayana S. The clinical utility of transcranial magnetic stimulation in determining hemispheric dominance for language: a magnetoencephalography comparison study. J Clin Neurophysiol. 2020;37(2):90–103. https://doi.org/10.1097/WNP.0000000000000499.
Rosler J, Niraula B, Strack V, et al. Language mapping in healthy volunteers and brain tumor patients with a novel navigated TMS system: evidence of tumor-induced plasticity. Clin Neurophysiol. 2014;125(3):526–36. https://doi.org/10.1016/j.clinph.2013.08.015.
Ruohonen J, Karhu J. Navigated transcranial magnetic stimulation. Neurophysiologie clinique = Clin Neurophysiol. 2010;40(1):7–17. https://doi.org/10.1016/j.neucli.2010.01.006.
Sanai N, Mirzadeh Z, Berger MS. Functional outcome after language mapping for glioma resection. N Engl J Med. 2008;358(1):18–27. https://doi.org/10.1056/NEJMoa067819.
Schuhmann T, Schiller NO, Goebel R, Sack AT. Speaking of which: dissecting the neurocognitive network of language production in picture naming. Cereb Cortex. 2012;22(3):701–9. https://doi.org/10.1093/cercor/bhr155.
Semonche A, Lee A, Negussie MB, et al. The association between task complexity and cortical language mapping accuracy. Neurosurgery. 2024;95:1126. https://doi.org/10.1227/neu.0000000000002981.
Silva LL, Tuncer MS, Vajkoczy P, Picht T, Rosenstock T. Distinct approaches to language pathway tractography: comparison of anatomy-based, repetitive navigated transcranial magnetic stimulation (rTMS)-based, and rTMS-enhanced diffusion tensor imaging-fiber tracking. J Neurosurg. 2022;136(2):589–600. https://doi.org/10.3171/2020.12.JNS204028.
Sollmann N, Ille S, Hauck T, et al. The impact of preoperative language mapping by repetitive navigated transcranial magnetic stimulation on the clinical course of brain tumor patients. BMC Cancer. 2015;15:261. https://doi.org/10.1186/s12885-015-1299-5.
Sollmann N, Hauck T, Tussis L, et al. Results on the spatial resolution of repetitive transcranial magnetic stimulation for cortical language mapping during object naming in healthy subjects. BMC Neurosci. 2016a;17(1):67. https://doi.org/10.1186/s12868-016-0305-4.
Sollmann N, Negwer C, Ille S, et al. Feasibility of nTMS-based DTI fiber tracking of language pathways in neurosurgical patients using a fractional anisotropy threshold. J Neurosci Methods. 2016b;267:45–54. https://doi.org/10.1016/j.jneumeth.2016.04.002.
Sollmann N, Ille S, Tussis L, et al. Correlating subcortical interhemispheric connectivity and cortical hemispheric dominance in brain tumor patients: a repetitive navigated transcranial magnetic stimulation study. Clin Neurol Neurosurg. 2016c;141:56–64. https://doi.org/10.1016/j.clineuro.2015.12.010.
Sollmann N, Bulubas L, Tanigawa N, Zimmer C, Meyer B, Krieg SM. The variability of motor evoked potential latencies in neurosurgical motor mapping by preoperative navigated transcranial magnetic stimulation. BMC Neurosci. 2017;18(1):5. https://doi.org/10.1186/s12868-016-0321-4.
Sollmann N, Fratini A, Zhang H, Zimmer C, Meyer B, Krieg SM. Associations between clinical outcome and tractography based on navigated transcranial magnetic stimulation in patients with language-eloquent brain lesions. J Neurosurg. 2019;132(4):1033–42. https://doi.org/10.3171/2018.12.JNS182988.
Sollmann N, Zhang H, Fratini A, et al. Risk assessment by Presurgical Tractography using navigated TMS maps in patients with highly motor- or language-eloquent brain tumors. Cancers (Basel). 2020;12(5):1264. https://doi.org/10.3390/cancers12051264.
Southwell DG, Birk HS, Han SJ, Li J, Sall JW, Berger MS. Resection of gliomas deemed inoperable by neurosurgeons based on preoperative imaging studies. J Neurosurg. 2018;129(3):567–75. https://doi.org/10.3171/2017.5.JNS17166.
Tarapore PE, Findlay AM, Honma SM, et al. Language mapping with navigated repetitive TMS: proof of technique and validation. NeuroImage. 2013;82:260–72. https://doi.org/10.1016/j.neuroimage.2013.05.018.

Thiel A, Habedank B, Winhuisen L, et al. Essential language function of the right hemisphere in brain tumor patients. Ann Neurol. 2005;57(1):128–31. https://doi.org/10.1002/ana.20342.
Thiel A, Habedank B, Herholz K, et al. From the left to the right: how the brain compensates progressive loss of language function. Brain Lang. 2006;98(1):57–65. https://doi.org/10.1016/j.bandl.2006.01.007.
Vigneau M, Beaucousin V, Herve PY, et al. What is right-hemisphere contribution to phonological, lexico-semantic, and sentence processing? Insights from a meta-analysis. NeuroImage. 2011;54(1):577–93. https://doi.org/10.1016/j.neuroimage.2010.07.036.
Zhang H, Tehlan K, Ille S, et al. Interhemispheric connections in the maintenance of language performance and prognosis prediction: fully connected layer-based deep learning model analysis. Neurosurg Focus. 2023;54(6):E6. https://doi.org/10.3171/2023.3.FOCUS2363.

Part IV

Special Aspects

Mapping of Further Brain Functions

12

Sebastian Ille

12.1 Introduction

Today, all specialized neuro-oncological centers use DES to locate individual functional areas of the brain. At least for the mapping of language function, patients have to be awake during surgery (Ojemann and Whitaker 1978; Ojemann et al. 1989; Haglund et al. 1994; Sanai et al. 2008; De Witt Hamer et al. 2012).

For decades, preserving the patient's motor and language functions had priority over other essential brain functions. Meanwhile, specialized centers also map further essential brain functions during awake surgery, such as working memory, arithmetic processing, visuospatial functions, judgment, recognition of facial emotions, or even playing instruments and singing (Thiebaut de Schotten et al. 2005; Duffau et al. 2002; Brandling-Bennett et al. 2012; Giussani et al. 2010a; Plaza et al. 2008; Roux et al. 2009a).

Noninvasive mapping techniques have also examined some of the above-mentioned brain functions apart from motor and language functions. Arithmetic processing, for instance, was already mapped by fMRI and TMS (Cohen et al. 2000; Rusconi et al. 2005). Although replacing intraoperative awake mapping with preoperative noninvasive mapping should not be the aim in neuro-oncology, noninvasive techniques have advantages: they can provide us with information before considering the indication of surgery at all. Moreover, they can also be performed on healthy subjects to gain information about brain functions for basic research. This becomes important, especially for brain functions, which are not standardly mapped during awake surgery.

S. Ille (✉)
Department of Neurosurgery, Heidelberg University Hospital, Heidelberg, Germany
e-mail: Sebastian.Ille@med.uni-heidelberg.de

S. M. Krieg, T. Picht (eds.), *Navigated Transcranial Magnetic Stimulation in Neurosurgery*, https://doi.org/10.1007/978-3-031-97155-6_12

Moreover, for patients, preoperative noninvasive mapping is performed in a more pleasant and relaxing environment compared to intraoperative mapping. Thus, such preoperative mapping can prepare the patient for the awake procedure, enable the neuropsychologist to better select the appropriate tasks, allow the surgeon to tailor his surgical approach before entering the OR, and serve as a backup option if awake surgery fails. Moreover, as it was observed for language mapping by nrTMS, the preoperative data allows for more targeted intraoperative DES mapping (Picht et al. 2013; Tarapore et al. 2013; Sollmann et al. 2015b). Regarding the preoperative mapping technique, fMRI has shown to be less reliable in tumor patients, especially in the vicinity of lesions (McGraw et al. 2001; Ille et al. 2015; Giussani et al. 2010b). With this in mind, particularly nrTMS language mapping has shown a good correlation compared to results obtained by DES during awake surgery as the gold standard technique, at least in terms of negative mapping (Picht et al. 2013; Tarapore et al. 2013).

The neurosurgical application of TMS significantly increased after the introduction of nTMS. However, since many studies concerning the mapping of higher cortical functions by TMS have already been performed before the age of nTMS, this chapter outlines stimulation protocols and test setups of both approaches: nTMS and non-navigated TMS.

12.2 Basic Principles of Mapping Further Brain Functions

As repetitively described for language mapping by rTMS, the mapping of further brain functions is most often conducted by stimulating with repetitive pulses. At the same time, the patient or healthy subject performs a particular task.

Since this chapter discusses navigated and non-navigated protocols, we must differentiate between navigated rTMS (nrTMS) and non-navigated rTMS (=rTMS).

In contrast to the application of single pulses, repetitive pulses are mainly used for the examination of higher cortical brain functions to induce a "virtual lesion" (Tables 12.1, 12.2, 12.3, and 12.4) (Pascual-Leone et al. 1991). However, the underlying mechanisms of the "virtual lesion" model are not yet precise (Miniussi et al. 2010). Probably, it is a combined effect of the suppression of neural signals (Harris et al. 2008) and the induction of random neural activity (Ruzzoli et al. 2010; Walsh and Cowey 2000) within the underlying cortical region, which also depends on anatomofunctional characteristics of the tissue (Miniussi et al. 2010; Siebner et al. 2009). The most often used stimulation frequency is 10 Hz/5 pulses (Tables 12.1, 12.2, 12.3, and 12.4).

As for language mapping, the mapping of further brain functions starts in many published protocols with the determination of the rMT via rough motor mapping, as described in earlier chapters. The stimulation intensity of the mapping can then be related to the rMT. In most cases, rTMS or nrTMS mappings are performed with a stimulation intensity of 100% rMT. In contrast, a large part of research groups does not relate the mapping intensity to the individual subject's rMT. Still, it stimulates all participants of a study with a similar intensity as defined by % of maximum stimulator output, which is, unfortunately, impossible to transfer directly to other

Table 12.1 Arithmetic processing

Study	Stimulation	Frequency (Hz)	Pulses	PTI (ms)	ISI (ms)	DT (ms)	Intensity	Task	Subjects	Cases
Fresnoza et al. (2020)	fMRI-guided nrTMS	1	900	–	–	3000	110% active MT	Subtraction and multiplication	Healthy volunteers	16
Ille et al. (2018)	nrTMS	5	10	0	3000	700	100% rMT	Four basic arithmetic operations	Patients with parietal tumors	26
Maurer et al. (2016)	nrTMS	5	10	0	3000	700	100% rMT	Four basic arithmetic operations	Healthy volunteers	20
Andres et al. (2011)	fMRI-guided nrTMS	10	4	100	5000	150	65% stimulator output	Subtraction, multiplication	Healthy volunteers	10
Cohen Kadosh et al. (2007)	fMRI-guided nrTMS	10	3	220	6000	1000	60% stimulator output	Numerical and physical size comparison	Healthy and dyscalculic volunteers	5 healthy/5 dyscalculic
Rusconi et al. (2005)	nrTMS	10	5	0	2500	65	60% stimulator output	Parity- and magnitude-matching	Healthy volunteers	8
Sandrini et al. (2004)	rTMS	15	4	0	Reaction times	Reaction times	110% rMT	Number comparison	Healthy volunteers	9
Gobel et al. (2001)	rTMS	10	5	Before stimulus	Reaction times	Reaction times	105% rMT	Number classification	Healthy volunteers	6

The table shows a choice of studies on the mapping of arithmetic processing by TMS. Garcia-Sanz and colleagues performed a systematic review on different TMS protocols for the examination of numerical cognition (Garcia-Sanz et al. 2022)

Table 12.2 Visuospatial attention

Study	Stimulation	Frequency (Hz)	Pulses	PTI (ms)	ISI (ms)	DT (ms)	Intensity	Task	Subjects	Cases
Guidali et al. (2023)	nrTMS	2	100	Before task	10	150	120% rMT	Landmark + line bisection	Healthy volunteers	25
Capotosto et al. (2023)	nrTMS	20	3	0	4000	500	100% rMT	Words + cue stimulus	Healthy volunteers	18
Middag-van Spanje et al. (2022)	rTMS/ cTBS	50	600	<5 min	1400–1800	200	100% rMT	Lateralized-attention network test	Healthy volunteers	34
Schwab et al. (2021)	nrTMS	1	1200	Before task	n.s.	n.s.	110% rMT	Virtual reality task	Healthy volunteers	7
Ahrens et al. (2019)	nrTMS	Double pulse	2	175/75	n.s.	10	100% rMT	Visuospatial attention paradigm	Healthy volunteers	22
Giglhuber et al. (2018)	nrTMS	5	10	0	3000	50	100% rMT	Grayscale	Healthy volunteers	10
Heinen et al. (2017)	rTMS/ cTBS	5/50	600	Before task	420	300	80% aMT	Cued visuospatial attention shifting paradigm	Healthy volunteers	16
Wang et al. (2016)	nrTMS	1	Continuous	0	Reaction times	Reaction times	100% rMT	Visuospatial attention shifting	Healthy volunteers	16
Wu et al. (2016)	nrTMS	1	Continuous	−300,000 + task	1500	500	100% rMT	Detection	Healthy volunteers	16
Giglhuber et al. (2016)	nrTMS	5	10	0	3000	50	100% rMT	Line bisection	Healthy volunteers	10
Bagattini et al. (2015)	rTMS	1	1800	n.s.	n.s.	n.s.	90% rMT	Line bisection + detection	Healthy volunteers	14

Studer et al. (2014)	rTMS	5	450	Before task	n.s.	350	40% stimulator output	Roulette betting	Healthy volunteers	28
Salatino et al. (2014)	TMS	Single pulse	1	150	≥4000	50	115% rMT	Line bisection	Healthy volunteers	8
Mahayana et al. (2014b)	rTMS	10	5	0	Reaction times	200	60% stimulator output	Line bisection	Healthy volunteers	15
Plow et al. (2014)	nrTMS	1	900,000	Before task	n.s.	3000	75% stimulator output	Visual tracking	Healthy volunteers	10
Mahayana et al. (2014a)	nrTMS	10	5	0	Reaction times	Individual threshold	60% stimulator output	Visual search	Healthy volunteers	24
Ricci et al. (2012)	TMS	Single pulse	1	150	n.s.	50	115% rMT	Line bisection	Healthy volunteers	3
Sauseng et al. (2011)	rTMS	1	900	n.s.	2000–3000	83	110% rMT	Detection	Healthy volunteers	12
Heinen et al. (2011)	nrTMS	11	3	90, 180, 270	n.s.	270	120% rMT	Detection	Healthy volunteers	12
Blankenburg et al. (2010)	rTMS	10 Hz	5	0	2430	570	75% stimulator output	Detection	Healthy volunteers	8
de Graaf et al. (2009)	nrTMS	Triple pulse	3	Four conditions	6000, 7000, 8000	300	120% rMT	Clock	Healthy volunteers	13
Oliver et al. (2009)	rTMS	10 Hz	5	−100 to 400	n.s.	Individual	100% rMT	Gap detection	Healthy volunteers	9
Van Ettinger-Veenstra et al. (2009)	fMRI-guided nrTMS	Triple pulse	3	−60, −30, 0	n.s.	120	120% rMT	Cue/ postponed saccade	Healthy volunteers	10

(continued)

Table 12.2 (continued)

Study	Stimulation	Frequency (Hz)	Pulses	PTI (ms)	ISI (ms)	DT (ms)	Intensity	Task	Subjects	Cases
Cattaneo et al. (2009)	nrTMS	Triple pulse	3	−500	n.s.	200	65% stimulator output	Line bisection	Healthy volunteers	9
Nyffeler et al. (2008)	rTMS	30	801	n.s.	n.s.	5500	90% rMT	Visual exploration	Healthy volunteers	12
Sack et al. (2007)	rTMS	13.3	5	0	2000	300	100% stimulator output	Clock	Healthy volunteers	8
Neggers et al. (2007)	fMRI-guided nrTMS	Triple pulse	3	60, 90, 120	n.s.	120	110% stimulator output	Cue/ postponed saccade	Healthy volunteers	8
Muggleton et al. (2006)	rTMS	10	5	0	Reaction times	60–260	65% stimulator output	Detection	Healthy volunteers	10
Meister et al. (2006)	TMS	Single pulse	1	150/250	3250	40	60% stimulator output	Detection	Healthy volunteers	14
Dambeck et al. (2006)	TMS	Single pulse	1	150/250	3210	40	60% stimulator output	Detection	Healthy volunteers	10
Hung et al. (2005)	nrTMS	10	5	0	n.s.	160	60% stimulator output	Partial−/ color-report control	Healthy volunteers	9
Thut et al. (2005)	rTMS	1	25 min	n.s.	n.s.	40	80% stimulator output	Detection	Healthy volunteers	10
Mevorach et al. (2005)	rTMS	1	600	n.s.	n.s.	80	90% rMT	Letters and shapes	Healthy volunteers	22

Koch et al. (2005)	rTMS	Single/ paired pulse	1/2	100/150	1, 3, 5, 10	40	130% rMT	Detection	Healthy volunteers	9
Kim et al. (2005)	rTMS	10	1000	Blocks	2180	180	80% rMT	Line bisection	Healthy volunteers	20
Ellison et al. (2004)	nrTMS	4/10	2/5	0	4000	Reaction times	65% stimulator output	Line bisection / visual search	Healthy volunteers	5
Chambers et al. (2004)	nTMS	Single pulse	1	30–360	n.s.	100	63% stimulator output	Detection	Healthy volunteers	3
Muri et al. (2002)	rTMS	Double pulse	2	120, 270, 520	3000–5000	120	80% stimulator output	Symbols	Healthy volunteers	10
Bjoertomt et al. (2002)	rTMS	n.s.	n.s.	0	Reaction times	200	65% stimulator output	Line bisection	Healthy volunteers	6
Hilgetag et al. (2001)	rTMS	1	600	≤ 300,000	2250	40	90% rMT	Detection	Healthy volunteers	7
Fierro et al. (2001)	TMS	Single pulse	1	150, 225, 300	≥30,000	50	115% rMT	Line bisection	Healthy volunteers	10
Fierro et al. (2000)	rTMS	25	10	0	≥30,000	50	115% rMT	Line bisection	Healthy volunteers	11
Pascual-Leone et al. (1994)	rTMS	25	5	0	100	n.s.	115% rMT	Detection	Healthy volunteers	6

The table gives an overview of studies on the mapping of visuospatial attention by TMS (n.s. = not specified; negative PTI means TMS onset before stimulus onset)

Table 12.3 Face processing

Study	Stimulation	Frequency (Hz)	Pulses	PTI (ms)	ISI (ms)	DT (ms)	Intensity	Task	Subjects	Cases
Handwerker et al. (2020)	rTMS/ cTMS	50	900	2.5–5 min	n.s.	n.s.	80% aMT	Video clips of faces	Healthy volunteers	17
Eick et al. (2020)	rTMS	n.s.	n.s.	−100, 0, 100	2000	750	65% stimulator output	Grayscale portraits	Healthy volunteers	45
Maurer et al. (2017)	nrTMS	10	5	0	3000	700	100% rMT	Famous faces	Healthy volunteers	20
Ferrari et al. (2016)	nrTMS	10	3	−200	n.s.	n.s.	60% stimulator output	Trustworthyness of faces	Healthy volunteers	12
Zachariou et al. (2017)	fMRI-guided nrTMS	10	5	0	4600–5800	500	70% stimulator output	Same–different face detection	Healthy volunteers	20
Solomon-Harris et al. (2016)	fMRI-guided nrTMS	1	1200	n.s.	200	800	60% stimulator output	Face identity and butterflies	Healthy volunteers	13
Gamond and Cattaneo (2016)	nrTMS	10	3	0	2700	300	60% stimulator output	Emotion recognition	Healthy volunteers	20
Bona et al. (2015)	fMRI-guided nrTMS	10	3	0	Response time	75, 500	40% stimulator output	Symmetry detection, facial features	Healthy volunteers	14
Pitcher (2014)	fMRI-guided nrTMS	10 / paired-pulse	5 / 2	0/20–60, 60–100, 100–140, 130–170, 170–210	1500	250	60% stimulator output	Face identity and expression	Healthy volunteers	14

Renzi et al. (2013)	nrTMS	20	3	100	2000	200	60% stimulator output	*Jane* faces	Healthy volunteers	16
Mattavelli et al. (2013)	nrTMS	Single pulse	1	100	1200—1400	700	62 ± 3% stimulator output	Face identity and expression	Healthy volunteers	11
Rochas et al. (2013)	nrTMS	10	5	0	± 4400	50	80% rMT	Facial emotion recognition	Healthy volunteers	20
Pitcher et al. (2011)	fMRI-guided nrTMS	10	5	0	1500	250	60% stimulator output	Face orientation	Healthy volunteers	10
Mattavelli et al. (2011)	nTMS	Single pulse	1	Immediately before	n.s.	Until response	65% stimulator output	Emotion-related word and face	Healthy volunteers	20
Dzhelyova et al. (2011)	nrTMS	10	5	0	3500–4250	500	65% stimulator output	Sex and trustworthyness	Healthy volunteers	12
Kadosh et al. (2011)	nrTMS	10/ paired-pulse	5	0/130–170, 170–210, 210–250, 250–290, 290–330	Response time	Until response	60% stimulator output	Emotion recognition	Healthy volunteers	8
Pitcher et al. (2009)	rTMS	10	5	0	1500	500	60% stimulator output	Face, body, and object	Healthy volunteers	16

(continued)

Table 12.3 (continued)

Study	Stimulation	Frequency (Hz)	Pulses	PTI (ms)	ISI (ms)	DT (ms)	Intensity	Task	Subjects	Cases
Pitcher et al. (2008)	rTMS	10/ paired-pulse	5/2	0/20–60, 60–100, 100–140, 130–170, 170–210, 210–250	1500	250	60% stimulator output	Identity and expression discrimination	Healthy volunteers	28
Pitcher et al. (2007)	rTMS	10/ paired-pulse	5/2	0/20–60, 60–100, 100–140, 130–170, 170–210, 210–250	1500	250	60% stimulator output	Face part discrimination	Healthy volunteers	25
Pourtois et al. (2004)	TMS	Single pulse	1	100/200	3250	100	110% rMT	Facial emotion recognition	Healthy volunteers	12

The table gives an overview of studies on the mapping of face processing by TMS (n.s. = not specified; negative PTI means TMS onset before stimulus onset)

Table 12.4 Categorization

Study	Stimulation	Frequency (Hz)	Pulses	PTI (ms)	ISI (ms)	DT (ms)	Intensity	Task	Subjects	Cases
Maurer et al. (2021)	nrTMS	10	5	0	3000	700	100% rMT	Living/nonliving	Healthy volunteers	20
Jacquet and Avenanti (2015)	nTMS	Single pulse	1	0	1000	1500	110% rMT	Goal–/ grip-recognition	Healthy volunteers	27
Passeri et al. (2015)	rTMS	10	5	0	1200	190	100% rMT	Verbal category membership	Healthy volunteers	18
Cattaneo et al. (2010)	nTMS	Single pulse	1	0	n.s.	Until response	65% stimulator output	Tool and animal words	Healthy volunteers	12
Fuggetta et al. (2009)	rTMS	10	5	−750	2450	1500	60–65% stimulator output	Picture–word verification	Healthy volunteers	9

The table gives an overview of studies on the categorization mapping by TMS (n.s. = not specified; negative PTI means TMS onset before stimulus onset) stimulators and coils (Tables 12.1, 12.2, 12.3, and 12.4). However, the intensity must be adapted to the patient's comfort on the one hand but also to the possibility of inducing specific errors on the other hand in all cases.

The use of nrTMS is strongly recommended if available. Neuronavigation enables us to better locate the stimulation sites and to analyze the mappings more precisely. Most researchers only stimulate single targets based on the results of prior studies. Of course, this is reasonable; however, this practice limits the knowledge gained from further eloquent cortical sites. Up to now, we only partially understand the underlying processes of the following neuropsychological brain functions. Hence, it is recommended that cortical sites be stimulated further to examine these complex functions more accurately.

12.3 Arithmetic Processing

As examined by multiple neuroimaging studies using fMRI or PET as well as lesion-based studies, arithmetic processing was assumed to be located within the inferior parietal lobe of the dominant hemisphere (Burbaud et al. 1999; Cohen et al. 2000; Cowell et al. 2000; Dehaene et al. 1996; Hayashi et al. 2000; Lee 2000; Martins et al. 1999; Mayer et al. 1999; Zago et al. 2001). Based on the knowledge of these studies, Whalen and later Duffau were the first to ask patients with a left-sided tumor within the parietal lobe to perform arithmetic processing tasks during bipolar DES in the setting of an awake craniotomy (Whalen et al. 1997; Duffau et al. 2002). The results of Whalen's 1997 and Duffau's 2002 reports have also been reproduced and extended by DES in the last decade (Roux et al. 2003, 2009b Kurimoto et al. 2006; Maldonado et al. 2011; Pu et al. 2011; Yu et al. 2011). Most importantly, Della Puppa et al. were even able to show the involvement of the right-sided parietal lobe in arithmetic processing by DES as well as the mapping of subcortical fiber tracts involved in arithmetic processing by subcortical DES (Della Puppa et al. 2013, 2015a, b).

Gobel et al. applied rTMS to the parietal lobe to show its involvement in number representation (Gobel et al. 2001). In 2004, Sandrini et al. started to examine number processing using rTMS (Sandrini et al. 2004). At this time, they used a simple number comparison task. They concluded that they could slow down the subject's number processing by applying rTMS to the left inferior parietal lobe but not when using it to the right. Similarly, another group applied rTMS to the left and right parietal lobe while subjects performed an additional task. They could find significantly longer reaction times during stimulations over the left but not over the right hemisphere (Gobel et al. 2006b). Also, Rusconi et al. were able to disrupt number processing when applying rTMS to the left angular gyrus in a study in which they tried to reproduce Gerstmann's syndrome (Rusconi et al. 2005). In 2007, it was shown that virtual dyscalculia is also inducible by fMRI-guided nrTMS over the right parietal lobe of healthy volunteers (Cohen Kadosh et al. 2007). Andres et al. also performed a study using a two-step approach: first, they identified cortical regions within the parietal lobes involved in subtraction and multiplication by fMRI. As a second step, they stimulated these regions by nrTMS to induce a virtual lesion. Their results showed arithmetic processing function in the left and right intraparietal sulcus (Andres et al. 2011). Fresnoza and colleagues also used fMRI-guided nrTMS to examine arithmetic operations. They found that left angular gyrus stimulation was detrimental to the retrieval and online calculation of solutions for multiplication problems and the retrieval (but not online calculation) of the solutions to subtraction problems. In contrast, the left horizontal segment of the intraparietal sulcus stimulation had no detrimental effect on both operations, regardless of strategy (Fresnoza et al. 2020).

Another nrTMS study on the cortical mapping of arithmetic processing was published in 2016 (Maurer et al. 2016). Maurer et al. performed nrTMS mappings of both hemispheres in 20 right-handed healthy subjects using a stimulation intensity of 100% rMT and a stimulation frequency of 5 Hz and 10 pulses. During nrTMS stimulations, healthy subjects performed a task with simple arithmetic operations consisting of addition, subtraction, multiplication, and division. The core result of this study is the feasibility of detecting cortical arithmetic processing function in healthy subjects by nrTMS, mainly to differentiate the location of arithmetic processing between predefined cortical subareas. Despite these promising results showing the feasibility of inducing arithmetic processing errors by nrTMS in both hemispheres, they also found the highest ER for all errors in all subjects within the right-sided ventral precentral gyrus (vPrG). With this in mind, the problem of differentiating between actual arithmetic processing errors and the impairment of language processing or dysarthria induced by nrTMS must be seen as a limitation or potential pitfall despite most of the positive sites found in this study being in good accordance with current literature (Maurer et al. 2016). However, the study showed the feasibility of mapping arithmetic processing by nrTMS and enabled to approve the results by patient studies which the same group subsequently did: The study aimed to correlate the results of mapping arithmetic processing with functional patient outcomes and included 26 patients with parietal brain tumors. Due to preoperative impairment of arithmetic processing, the mapping was impossible in eight patients (31%). Fifty-two predefined cortical sites were stimulated by nrTMS while patients performed a calculation task. Pre- and postoperatively, patients underwent a standardized number-processing and calculation test (NPCT). The tumor resection was blinded to nrTMS results, and the change in NPCT performance was correlated to resected arithmetic processing-positive sites as identified by nrTMS. The resection of arithmetic processing-positive sites associated with a worsening of the postoperative NPCT results in 12 cases. In three cases, no arithmetic processing-positive sites were resected, and the postoperative NPCT result was similar to or better than preoperatively. Also, in three cases, the postoperative NPCT result was better than preoperatively, although arithmetic processing-positive sites were resected. Although this study only presented a low number of cases, it was shown for the first time that nrTMS could be a valuable tool for preoperative mapping of arithmetic processing in patients with brain tumors (Ille et al. 2018). Additional studies and a meta-analysis including TMS data to examine numerical cognition in patients have been conducted (Faye et al. 2019).

Meanwhile, the subcortical representation and correlating white matter pathways of the arithmetic processing network are being examined. After prior studies on the subcortical representation of this brain function (van Eimeren et al. 2008, 2010; Peters and De Smedt 2018; Matejko et al. 2013), Zhang and colleagues showed a strong correlation between the volume of tracts in the nrTMS-based connectome and patients' performance in arithmetic processing tasks. Anatomically, the anterior cingulum, the frontal aslant tract, and the frontal superior longitudinal fasciculus could be identified as part of this network in correlation with nrTMS error rates (data under review).

To summarize, Table 12.1 gives an overview of publications for mapping arithmetic processing by TMS. Up to now, TMS for the mapping of arithmetic processing has most often been used in healthy volunteers and by the application of different numbers of arithmetic processing tasks (Table 12.1). For its application in neurosurgery to find eloquent regions preoperatively, the use of a task combining the four basic and easy arithmetic operations seems to be highly effective, such as 9 + 1; 5–2; 2 × 7; 12:4 (Maurer et al. 2016).

12.4 Visuospatial Attention

Neglect-like symptoms and visuospatial deficits can be observed in patients who underwent resections of parietal lobe tumors (Russell et al. 2005; Hommet et al. 2004). Sanai et al. analyzed 119 cases of parietal tumor resection. They found not otherwise specified visual deficits in 9.2% (6.7% permanent deficits) and parietal lobe symptoms such as right-left confusion, finger agnosia, sensory extinction, or astereognosis in 8.4% (2.5% permanent deficits) of cases (Sanai et al. 2012). With this in mind, the feasibility of mapping visuospatial attention by DES during awake surgery has already been shown (Bartolomeo et al. 2007; Thiebaut de Schotten et al. 2005).

Visuospatial attention and its processing is based on a complex network including cortical and subcortical levels (Corbetta et al. 2005; Heilman 1980; Kinsbourne 1977; Lunven et al. 2015; Umarova et al. 2014; Suchan et al. 2014; Duecker and Sack 2014; Sack 2010). Most importantly, neglect or neglect-like distortions significantly influence the postoperative functional outcome and the patient's QOL (Jehkonen et al. 2000, 2006; Katz et al. 1999).

Using a number bisection task, Gobel et al. were able to evoke neglect-like symptoms by applying rTMS to the right-sided parietal lobe of healthy subjects. For this task, subjects have to name the midpoint of a numerical interval without the necessity of arithmetic processing (Gobel et al. 2006a). Researchers often use the line bisection or detection tasks to map visuospatial attention (Fig. 12.1, Table 12.2). The line bisection task has also been used intraoperatively (Roux et al. 2011). The task is designed as a horizontal line bisected by a vertical landmark. The bisection can be symmetrical with an equal length of the left and right parts or asymmetrical with a longer left or right segment. Subjects are instructed to bisect the horizontal lines and to name the longer or the shorter segment (Fig. 12.1) (Giglhuber et al. 2016).

The feasibility of inhibiting visuospatial orientation by TMS has even been shown in animals by Valero-Cabre et al. In this sham-controlled study, they stimulated the

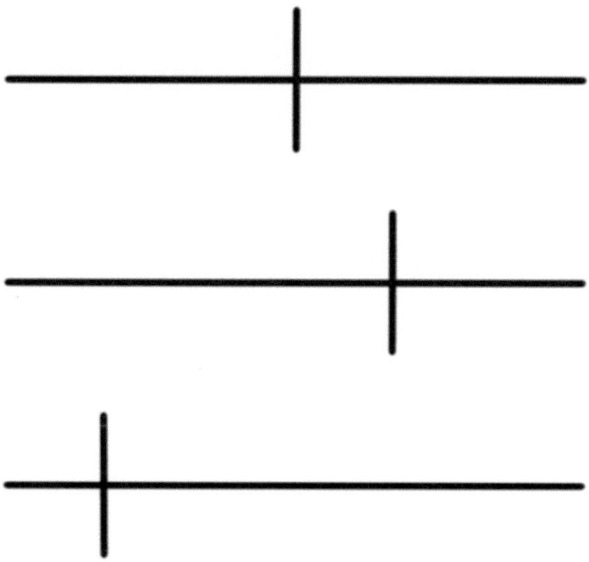

Fig. 12.1 Line bisection task. The figure shows the line bisection task. The task is designed as a horizontal line bisected by a vertical landmark. The bisection can either be symmetrical with an equal length of the left and right part or asymmetrically with a longer left or right segment. Subjects are instructed to bisect the horizontal lines and to name the longer or the shorter segment (Giglhuber et al. 2016)

parietal lobe of cats with a frequency of 1 Hz for 20 min. Interestingly, they also showed that the induced effects lasted about 20 min (Valero-Cabre et al. 2006). By the measurement of eye movements during the application of theta burst rTMS over the right parietal cortex, Nyffeler et al. were able to induce visual neglect-like effects in healthy subjects (Nyffeler et al. 2008). Koch et al. also used theta burst rTMS, however, with another intention: they stimulated the left-sided parietal lobe in patients suffering from hemispatial neglect due to right-hemispheric stroke. By inhibiting the hyperexcitability of the left hemisphere with rTMS, they accelerated the recovery from hemispatial neglect after 2 weeks of stimulations (Koch et al. 2012). Despite only being done in three healthy subjects, Ricci et al. were able to prove these promising results even for single-pulse TMS by combined TMS-fMRI sessions. They evoked neglect-like behavior by stimulating the right-sided parietal lobe. Afterward they showed decreased neuronal activity within frontoparietal areas according to the results of lesion-based studies (Ricci et al. 2012). Using a line bisection task, visual neglect like symptoms were observed during the stimulation with 5 Hz and 10 pulses over 52 predefined cortical sites of both hemispheres (Fig. 12.1).

Regarding the line bisection task, the authors found significantly more rightward errors during right-sided stimulation and more leftward errors during the stimulation of the left hemisphere (Giglhuber et al. 2016). This study also shows that it is crucial to select the right task, particularly for mapping visuospatial attention and the induction of neglect-like effects (Bonato 2012; Coello et al. 2013). Regarding former studies, the line bisection task has proven beneficial for mapping visuospatial attention (Fig. 12.1, Table 12.2). Another well-working task for mapping visuospatial attention is the grayscale task. This task shows different pictures with pairs of horizontal grayscales (Fig. 12.2). Subjects are asked to say which of the two grayscales appears darker. After

Fig. 12.2 Grayscale task. The figure shows an example of the grayscale task. This task shows different pictures with pairs of horizontal grayscales. Subjects are asked to respond by saying which of the two grayscales appears darker. The task can also be used to map visuospatial attention

baseline testing, the subjects present a known phenomenon called pseudo-neglect to the left side. In one nrTMS study, the right hemisphere showed a higher overall ER than the left hemisphere. Additionally, leftward errors were elicited by stimulations to the superior frontal gyrus (SFG) and again posterior parietal areas. In contrast, rightward errors were evoked due to stimulations of the inferior frontal gyrus (IFG) and the temporoparietal junction (Giglhuber et al. 2018).

Table 12.2, summarizing an overview of studies performed to examine visuospatial function using TMS, shows that most researchers initially used the line bisection task. Moreover, repetitive pulses with a frequency of 5–10 Hz and a pulse onset simultaneously to the presentation of the stimulus cover the process of visuospatial attention. Most importantly, the DT of the stimulus presentation must be short. The use of a DT of 50 ms increases the sensitivity crucially. However, more recent studies applied new tasks and modified stimulation paradigms (Ahrens et al. 2019; Guidali et al. 2023; Capotosto et al. 2023; Middag-van Spanje et al. 2022; Schwab et al. 2021; Jin et al. 2021; Stengel et al. 2021; Di Gregorio et al. 2021; Schintu et al. 2021; Gießing et al. 2020; Ciricugno et al. 2020; Salatino et al. 2019; Zuanazzi and Cattaneo 2017; Heinen et al. 2017). Likewise, studies have now been carried out not only on healthy volunteers but also on patients, primarily with neurodegenerative patients such as Parkinson's disease (Ruwaili et al. 2023; Bhat et al. 2023a, b; Tang et al. 2023; Bahun et al. 2022; Di Gregorio et al. 2021; Rawji et al. 2020). With this in mind, studies on neurosurgical patients, especially the preoperative examination of visuospatial attention and the postoperative follow-up of this highly eloquent function for the patient's quality of life, are still missing.

12.5 Face Processing

The mapping of the cortical representation of face processing seems like a very experimental approach from the first view. However, face perception also includes judging emotion, identity, and trustworthiness (Atkinson and Adolphs 2011). When this brain function is disturbed by stroke or other kinds of brain lesions, it means a decrease in QOL of patients but also their social environment and thereby justifies its mapping and preservation (Barton 2014; Busigny et al. 2014; Hier et al. 1983; Rapcsak et al. 2001; Young et al. 1993; Gainotti and Marra 2011). The underlying cortical locations of face processing and its subcortical network are complex. The cognitive process consists of two parts: the visual perception of faces and the matching to the face memory (Rapcsak 2003).

Moreover, face processing is bilaterally located with a predominance of the right hemisphere (Gainotti and Marra 2011; Landis et al. 1986). Surgical cases and intraoperative mapping have also demonstrated this bilaterality of face processing and its complexity (Corrivetti et al. 2016; Giussani et al. 2010a). Moreover, an anatomo-functional study examined the underlying subcortical structures on the one hand and showed the feasibility of mapping face processing during awake surgeries on the other hand (De Benedictis et al. 2014).

Concerning noninvasive modalities, face processing has already been described and visualized by fMRI (Hadjikhani and de Gelder 2002; Keenan et al. 2000; Gauthier et al. 2000; Druzgal and D'Esposito 2003; Fruhholz et al. 2011; Kitada et al. 2009). A meta-analysis combining fMRI data of healthy subjects showed involvement in face processing within the occipital, temporoparietal, and prefrontal cortex, the limbic system, and the cerebellum as well as within the according subcortical areas (Fusar-Poli et al. 2009).

Face processing has also been examined by rTMS and reported in various articles (Table 12.3). Pitcher et al. could reproduce findings, which fMRI had already discussed. Using rTMS, they localized the right-sided inferior occipital gyrus (occipital face area) as an early stage of the face-processing stream (Gauthier et al. 2000; Pitcher et al. 2007). A subsequent study of the same group confirmed these results. Moreover, they showed that applying rTMS over the right-sided lateral occipital area impairs the discrimination of objects but not of faces. The same results were found when stimulating the right-sided extrastriate body area (Pitcher et al. 2009). In 2008, the Pitcher group had already described the involvement of the right somatosensory area in face processing as examined by rTMS.

Similarly, this proves the contribution of nonvisual areas to expression processing (Pitcher et al. 2008). The latter also replicates the results of another rTMS study (Pourtois et al. 2004). Furthermore, nrTMS could prove the contribution of the occipital face area and the superior temporal sulcus in judging the gender and trustworthiness of faces (Dzhelyova et al. 2011). Another well-experienced group in nrTMS showed the involvement of the right-sided occipital face area in the integrative processing of facial identity and expression (Kadosh et al. 2011).

Maurer et al. performed a study on the mapping of face processing using nrTMS. Healthy subjects were asked to name 80 portraits of famous persons during the stimulation of predefined cortical sites of the whole hemisphere (Maurer et al. 2017). They differentiated between language errors and errors regarding the wrong identification of persons. Although the results did not show statistically significant differences, the locations were in good accordance with the current literature (Barton 2014; Yang et al. 2014; Gomez et al. 2015). Moreover, they found a more critical role of the right-sided frontal lobe in face processing than previously expected. The latter also correlates with earlier results of rTMS and PET studies (Campanella et al. 2001; Renzi et al. 2013). However, as Maurer et al. also conclude, particularly the mapping of face processing reveals the limitations of nrTMS: parts of the functional network of face processing, such as the limbic system, are not reachable for standard TMS since they are located subcortically (Leonard et al. 1985; Gothard et al. 2007).

According to multiple studies' experiences, applying emotion and identity recognition tasks based on pictures of standardized datasets seems to be most effective (Table 12.3). Since the course of face processing is more complex, DT has to be longer, for example, 700 ms. Again, repetitive pulses with a frequency of 5–10 Hz and simultaneous pulse and stimulus onset are recommended for a reliable inhibition of face processing. Still, as shown by the overview of Table 12.3, containing initial and more recent studies on the examination of face recognition (Ambrus et al. 2017a, b, 2019; Bona et al. 2018; Cattaneo et al. 2020; Eick et al. 2020; Ferrari et al. 2017, 2018; Guo et al. 2020; Handwerker et al. 2020; Janssens et al. 2020; Jeschke et al. 2023; Mattavelli et al. 2019; Paracampo et al. 2017; Pitcher et al. 2020; Rossion 2022; Sliwinska et al. 2020; Sliwinska and Pitcher 2018; Zachariou et al. 2017; Ziegler et al. 2018), stimulation paradigms, localizations, and tasks still differ in a wide range for the complex mapping of this higher brain function.

12.6 Categorizing

Neuropsychological literature discusses different theories to explain the basis of the humans' ability to categorize (Caramazza and Shelton 1998; Caramazza and Mahon 2003; Humphreys and Forde 2001; Martin et al. 1996; Warrington and Shallice 1984). Furthermore, it has been discussed that natural/living and artificial/nonliving domains of the underlying semantic knowledge are represented in separate subsystems and anatomical locations (Paz-Caballero et al. 2006; Devlin et al. 1998; Martin et al. 1996; Perani et al. 1995). The subsystems of these theories have already been visualized by fMRI, PET, and event-related potentials (ERPs) as measured by EEG or event-related fields (ERFs) as measured by MEG (Chao et al. 1999; Moore and Price 1999; Grafton et al. 1997; Damasio et al. 1996; Martin et al. 1996; Perani et al. 1995; Paz-Caballero et al. 2006; Kiefer 2001, 2005; Sim and Kiefer 2005; Dehaene 1995).

The latter technique has also been combined with TMS to show the functional representation of living and nonliving domains across both hemispheres. In 2009, Fuggetta et al. were able to impair the categorization of artificial/nonliving items by disrupting Wernicke's area with rTMS. They thereby supported the theory that semantic knowledge is associated with different conceptual domains based on a network of segregated systems of functionally connected cortical areas (Fuggetta et al. 2009). Passeri et al. also applied rTMS to Wernicke's area and its right-sided homolog to investigate both hemispheres' semantic categorization process and contribution. They used a verbal category membership task of words referring to typical

Fig. 12.3 Categorization task. The figure shows examples of living and nonliving pictures. The task can be used to map the cortical localization of categorizing

and atypical exemplars. They could show the involvement of both hemispheres in the categorization process of typical exemplars while the right-sided hemisphere was involved in the categorization of atypical exemplars (Passeri et al. 2015). They confirmed the Jung-Beeman theory, which describes a coarser semantic processing in the right-sided compared to the left-sided hemisphere (Jung-Beeman 2005). Another group investigated the cortical localizations of categorization by nrTMS. They used a task of 80 living and nonliving objects and stimulated 52 predefined cortical sites over both hemispheres (Fig. 12.3). In Maurer's study, the highest ER on the left hemisphere was located within the MFG and the supramarginal gyrus (SMG). The highest ER was also found within the parietal lobe for the right hemisphere. Taken together, they showed the feasibility of successfully interfering with the categorization process by nrTMS and the accordance of the results with current literature (Maurer et al. 2021).

As the overview in Table 12.4 shows, several tasks have been used to map the cortical localization of categorizing in the past (Table 12.4). Since the categorization process is complex, it is difficult to recommend the most effective mapping parameters. However, most researchers use repetitive pulses with a 5–10 Hz frequency and a simultaneous pulse and stimulus onset.

12.7 Future Aspects

The studies mentioned above and approaches for mapping brain functions apart from motor and language functions are equally crucial for neurosurgery and basic neuroscientific research. Mainly nrTMS confirmed the feasibility of mapping further brain functions like arithmetic processing and visuospatial attention in healthy subjects and similarly represents a more accurate technique for basic researchers. Its application in patients for preoperative mapping seems justified and enables neurosurgeons to transfer preoperative nrTMS data to the operating room. This might shorten the intraoperative DES mapping procedure and decrease the patient's burden. Additionally, it allows the evaluation of different tasks preoperatively to choose the most effective task for the intraoperative procedure.

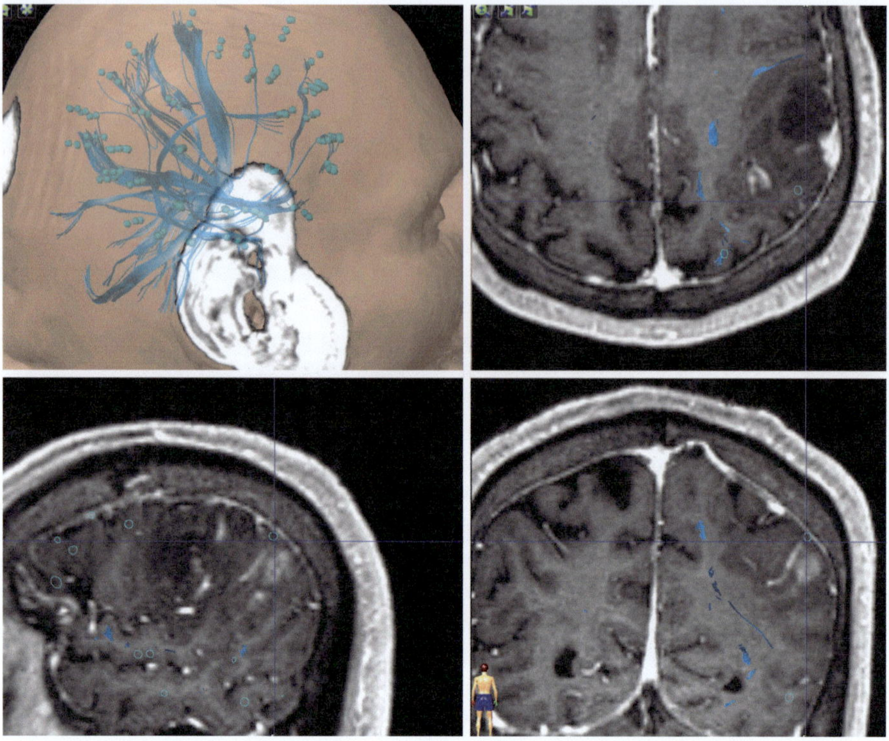

Fig. 12.4 nrTMS-based DTI FT. The figure shows the results of the preoperative mapping of arithmetic processing by nrTMS as transferred to the intraoperative neuronavigation system (iPlanNet Cranial 3.0.1, Brainlab AG, Munich, Germany). Arithmetic processing-positive cortical sites in terms of nrTMS are used as ROI for nrTMS-DTI FT. The figure shows the preoperative neuronavigation of a 76-year-old male with a glioblastoma within the right parietal lobe (Fierro et al. 2000)

However, as resection probability maps have shown when functionality should be preserved, subcortical fibers must be spared during resection (De Witt Hamer et al. 2013; De Benedictis and Duffau 2011; Duffau 2014). Thus, the nrTMS-based DTI FT approach also seems reasonable for the use combined with these data. It has already been shown that nTMS motor mapping data and nrTMS language mapping data can be used as ROI to perform preoperative DTI-FT (Krieg et al. 2012; Frey et al. 2012; Sollmann et al. 2015a). Thereby, nrTMS-based DTI-FT enables the visualization of subcortical networks, which are associated with further cortical functions apart from language. Transferring the DTI-FT data to neuronavigation systems can also be used for preoperative planning and intraoperative guidance (Fig. 12.4).

However, the technique still has to be further refined. Until now, nrTMS studies investigating neuropsychological brain functions apart from language have often been performed using a stimulation frequency of 5–10 Hz (Tables 12.1, 12.2, 12.3, and 12.4) (Giglhuber et al. 2016; Maurer et al. 2016). These parameters have been chosen due to experiences from language mapping. Still, earlier studies have shown that using different stimulation frequencies affects the distribution of language-positive sites such as Hauck et al. (2015a). Likewise, choosing the proper task is crucial for obtaining relevant results when mapping higher cortical functions, such as language (Hauck et al. 2015b). These two issues might also be appreciable for mapping neuropsychological brain functions. Especially the right choice of tasks and the dependency of results on different tasks seem to be crucial (Coello et al. 2013). As a further step, the reliability of results must be confirmed for the application in patients.

As described, mapping further cortical functions by rTMS makes a tremendous preoperative mapping potential available for neurosurgeons. However, it is not sensible to map each possible brain function in each patient. The choice of tasks for the preoperative and intraoperative mapping must be tailored to each patient's individuality and the location of the pathology. In this case, rTMS contributes to a functional neurocognitive assessment (Duffau 2013).

References

Ahrens MM, Veniero D, Freund IM, Harvey M, Thut G. Both dorsal and ventral attention network nodes are implicated in exogenously driven visuospatial anticipation. Cortex. 2019;117:168–81. https://doi.org/10.1016/j.cortex.2019.02.031.

Ambrus GG, Dotzer M, Schweinberger SR, Kovács G. The occipital face area is causally involved in the formation of identity-specific face representations. Brain Struct Funct. 2017a;222(9):4271–82. https://doi.org/10.1007/s00429-017-1467-2.

Ambrus GG, Windel F, Burton AM, Kovács G. Causal evidence of the involvement of the right occipital face area in face-identity acquisition. NeuroImage. 2017b;148:212–8. https://doi.org/10.1016/j.neuroimage.2017.01.043.

Ambrus GG, Amado C, Krohn L, Kovács G. TMS of the occipital face area modulates cross-domain identity priming. Brain Struct Funct. 2019;224(1):149–57. https://doi.org/10.1007/s00429-018-1768-0.

Andres M, Pelgrims B, Michaux N, Olivier E, Pesenti M. Role of distinct parietal areas in arithmetic: an fMRI-guided TMS study. NeuroImage. 2011;54(4):3048–56. https://doi.org/10.1016/j.neuroimage.2010.11.009.

Atkinson AP, Adolphs R. The neuropsychology of face perception: beyond simple dissociations and functional selectivity. Philos Trans R Soc Lond Ser B Biol Sci. 2011;366(1571):1726–38. https://doi.org/10.1098/rstb.2010.0349.

Bagattini C, Mele S, Brignani D, Savazzi S. No causal effect of left hemisphere hyperactivity in the genesis of neglect-like behavior. Neuropsychologia. 2015;72:12–21. https://doi.org/10.1016/j.neuropsychologia.2015.04.010.

Bahun I, Ivezic E, Đogas VV, Filipcic IS, Filipcic I. A pilot study of accelerated deep transcranial magnetic stimulation effects on cognitive functioning in patients diagnosed with treatment resistant major depressive disorder. Psychiatry Res. 2022;316:114750. https://doi.org/10.1016/j.psychres.2022.114750.

Bartolomeo P, Thiebaut de Schotten M, Duffau H. Mapping of visuospatial functions during brain surgery: a new tool to prevent unilateral spatial neglect. Neurosurgery. 2007;61:E1340.

Barton JJ. Higher cortical visual deficits. Continuum (Minneap Minn). 2014;20(4 neuro-ophthalmology):922–41. https://doi.org/10.1212/01.con.0000453311.29519.67.

Bhat P, Goyal V, Kumaran SS, Srivastava AK, Behari M, Dwivedi SN. Mechanisms of 1 Hz inhibitory and 5 Hz excitatory repetitive transcranial magnetic stimulations in Parkinson's disease: a

functional magnetic resonance imaging study. Brain Connect. 2023a;13(4):247–63. https://doi.org/10.1089/brain.2022.0043.
Bhat P, Kumaran SS, Goyal V, Srivastava AK, Behari M. Effect of rTMS at SMA on task-based connectivity in PD. Behav Brain Res. 2023b;452:114602. https://doi.org/10.1016/j.bbr.2023.114602.
Bjoertomt O, Cowey A, Walsh V. Spatial neglect in near and far space investigated by repetitive transcranial magnetic stimulation. Brain. 2002;125(Pt 9):2012–22.
Blankenburg F, Ruff CC, Bestmann S, Bjoertomt O, Josephs O, Deichmann R, et al. Studying the role of human parietal cortex in visuospatial attention with concurrent TMS-fMRI. Cereb Cortex. 2010;20(11):2702–11. https://doi.org/10.1093/cercor/bhq015.
Bona S, Cattaneo Z, Silvanto J. The causal role of the occipital face area (OFA) and lateral occipital (LO) cortex in symmetry perception. J Neurosci. 2015;35(2):731–8. https://doi.org/10.1523/jneurosci.3733-14.2015.
Bona S, Silvanto J, Cattaneo Z. TMS over right OFA affects individuation of faces but not of exemplars of objects. Neuropsychologia. 2018;117:364–70. https://doi.org/10.1016/j.neuropsychologia.2018.06.024.
Bonato M. Neglect and extinction depend greatly on task demands: a review. Front Hum Neurosci. 2012;6:195. https://doi.org/10.3389/fnhum.2012.00195.
Brandling-Bennett EM, Bookheimer SY, Horsfall JL, Moftakhar P, Sedrak M, Barkulis CT, et al. A paradigm for awake intraoperative memory mapping during forniceal stimulation. Neurocase. 2012;18(1):26–38. https://doi.org/10.1080/13554794.2010.547509.
Burbaud P, Camus O, Guehl D, Bioulac B, Caille JM, Allard M. A functional magnetic resonance imaging study of mental subtraction in human subjects. Neurosci Lett. 1999;273(3):195–9.
Busigny T, Van Belle G, Jemel B, Hosein A, Joubert S, Rossion B. Face-specific impairment in holistic perception following focal lesion of the right anterior temporal lobe. Neuropsychologia. 2014;56:312–33. https://doi.org/10.1016/j.neuropsychologia.2014.01.018.
Campanella S, Joassin F, Rossion B, De Volder A, Bruyer R, Crommelinck M. Association of the distinct visual representations of faces and names: a PET activation study. NeuroImage. 2001;14(4):873–82. https://doi.org/10.1006/nimg.2001.0877.
Capotosto P, Sulpizio V, Galati G, Baldassarre A. Visuo-spatial attention and semantic memory competition in the parietal cortex. Sci Rep. 2023;13(1):6218. https://doi.org/10.1038/s41598-023-33533-0.
Caramazza A, Mahon BZ. The organization of conceptual knowledge: the evidence from category-specific semantic deficits. Trends Cogn Sci. 2003;7(8):354–61.
Caramazza A, Shelton JR. Domain-specific knowledge systems in the brain the animate-inanimate distinction. J Cogn Neurosci. 1998;10(1):1–34.
Cattaneo Z, Silvanto J, Pascual-Leone A, Battelli L. The role of the angular gyrus in the modulation of visuospatial attention by the mental number line. NeuroImage. 2009;44(2):563–8. https://doi.org/10.1016/j.neuroimage.2008.09.003.
Cattaneo Z, Devlin JT, Salvini F, Vecchi T, Silvanto J. The causal role of category-specific neuronal representations in the left ventral premotor cortex (PMv) in semantic processing. NeuroImage. 2010;49(3):2728–34. https://doi.org/10.1016/j.neuroimage.2009.10.048.
Cattaneo Z, Bona S, Silvanto J. Different neural representations for detection of symmetry in dot-patterns and in faces: a state-dependent TMS study. Neuropsychologia. 2020;138:107333. https://doi.org/10.1016/j.neuropsychologia.2020.107333.
Chambers CD, Payne JM, Stokes MG, Mattingley JB. Fast and slow parietal pathways mediate spatial attention. Nat Neurosci. 2004;7(3):217–8. https://doi.org/10.1038/nn1203.
Chao LL, Haxby JV, Martin A. Attribute-based neural substrates in temporal cortex for perceiving and knowing about objects. Nat Neurosci. 1999;2(10):913–9. https://doi.org/10.1038/13217.
Ciricugno A, Ferrari C, Rusconi ML, Cattaneo Z. The left posterior cerebellum is involved in orienting attention along the mental number line: an online-TMS study. Neuropsychologia. 2020;143:107497. https://doi.org/10.1016/j.neuropsychologia.2020.107497.

Coello AF, Moritz-Gasser S, Martino J, Martinoni M, Matsuda R, Duffau H. Selection of intraoperative tasks for awake mapping based on relationships between tumor location and functional networks. J Neurosurg. 2013;119:1380. https://doi.org/10.3171/2013.6.JNS122470.

Cohen Kadosh R, Cohen Kadosh K, Schuhmann T, Kaas A, Goebel R, Henik A, et al. Virtual dyscalculia induced by parietal-lobe TMS impairs automatic magnitude processing. Curr Biol. 2007;17(8):689–93. https://doi.org/10.1016/j.cub.2007.02.056.

Cohen L, Dehaene S, Chochon F, Lehericy S, Naccache L. Language and calculation within the parietal lobe: a combined cognitive, anatomical and fMRI study. Neuropsychologia. 2000;38(10):1426–40.

Corbetta M, Kincade MJ, Lewis C, Snyder AZ, Sapir A. Neural basis and recovery of spatial attention deficits in spatial neglect. Nat Neurosci. 2005;8(11):1603–10. https://doi.org/10.1038/nn1574.

Corrivetti F, Herbet G, Moritz-Gasser S, Duffau H. Prosopagnosia induced by a left anterior temporal lobectomy following a right Temporo-occipital resection in a multicentric diffuse low-grade glioma. World Neurosurg. 2016;97:756.e1. https://doi.org/10.1016/j.wneu.2016.10.025.

Cowell SF, Egan GF, Code C, Harasty J, Watson JD. The functional neuroanatomy of simple calculation and number repetition: a parametric PET activation study. NeuroImage. 2000;12(5):565–73. https://doi.org/10.1006/nimg.2000.0640.

Damasio H, Grabowski TJ, Tranel D, Hichwa RD, Damasio AR. A neural basis for lexical retrieval. Nature. 1996;380(6574):499–505. https://doi.org/10.1038/380499a0.

Dambeck N, Sparing R, Meister IG, Wienemann M, Weidemann J, Topper R, et al. Interhemispheric imbalance during visuospatial attention investigated by unilateral and bilateral TMS over human parietal cortices. Brain Res. 2006;1072(1):194–9. https://doi.org/10.1016/j.brainres.2005.05.075.

De Benedictis A, Duffau H. Brain hodotopy: from esoteric concept to practical surgical applications. Neurosurgery. 2011;68(6):1709–23; discussion 1723. https://doi.org/10.1227/NEU.0b013e3182124690.

De Benedictis A, Duffau H, Paradiso B, Grandi E, Balbi S, Granieri E, et al. Anatomo-functional study of the temporo-parieto-occipital region: dissection, tractographic and brain mapping evidence from a neurosurgical perspective. J Anat. 2014;225(2):132–51. https://doi.org/10.1111/joa.12204.

de Graaf TA, Jacobs C, Roebroeck A, Sack AT. FMRI effective connectivity and TMS chronometry: complementary accounts of causality in the visuospatial judgment network. PLoS One. 2009;4(12):e8307. https://doi.org/10.1371/journal.pone.0008307.

De Witt Hamer PC, Robles SG, Zwinderman AH, Duffau H, Berger MS. Impact of intraoperative stimulation brain mapping on glioma surgery outcome: a meta-analysis. J Clin Oncol. 2012;30(20):2559–65. https://doi.org/10.1200/jco.2011.38.4818.

De Witt Hamer PC, Hendriks EJ, Mandonnet E, Barkhof F, Zwinderman AH, Duffau H. Resection probability maps for quality assessment of glioma surgery without brain location bias. PLoS One. 2013;8(9):e73353. https://doi.org/10.1371/journal.pone.0073353.

Dehaene S. Electrophysiological evidence for category-specific word processing in the normal human brain. Neuroreport. 1995;6(16):2153–7.

Dehaene S, Tzourio N, Frak V, Raynaud L, Cohen L, Mehler J, et al. Cerebral activations during number multiplication and comparison: a PET study. Neuropsychologia. 1996;34(11):1097–106.

Della Puppa A, De Pellegrin S, d'Avella E, Gioffre G, Munari M, Saladini M, et al. Right parietal cortex and calculation processing: intraoperative functional mapping of multiplication and addition in patients affected by a brain tumor. J Neurosurg. 2013;119(5):1107–11. https://doi.org/10.3171/2013.6.jns122445.

Della Puppa A, De Pellegrin S, Lazzarini A, Gioffre G, Rustemi O, Cagnin A, et al. Subcortical mapping of calculation processing in the right parietal lobe. J Neurosurg. 2015a;122(5):1038–41. https://doi.org/10.3171/2014.10.jns14261.

Della Puppa A, De Pellegrin S, Rossetto M, Rustemi O, Saladini M, Munari M, et al. Intraoperative functional mapping of calculation in parietal surgery. New insights and clinical impli-

cations. Acta Neurochir. 2015b;157(6):971–7; discussion 977,. https://doi.org/10.1007/s00701-015-2426-4.
Devlin JT, Gonnerman LM, Andersen ES, Seidenberg MS. Category-specific semantic deficits in focal and widespread brain damage: a computational account. J Cogn Neurosci. 1998;10(1):77–94.
Di Gregorio F, La Porta F, Casanova E, Magni E, Bonora R, Ercolino MG, et al. Efficacy of repetitive transcranial magnetic stimulation combined with visual scanning treatment on cognitive and behavioral symptoms of left hemispatial neglect in right hemispheric stroke patients: study protocol for a randomized controlled trial. Trials. 2021;22(1):24. https://doi.org/10.1186/s13063-020-04943-6.
Druzgal TJ, D'Esposito M. Dissecting contributions of prefrontal cortex and fusiform face area to face working memory. J Cogn Neurosci. 2003;15(6):771–84. https://doi.org/10.1162/089892903322370708.
Duecker F, Sack AT. The hybrid model of attentional control: new insights into hemispheric asymmetries inferred from TMS research. Neuropsychologia. 2014;74:21. https://doi.org/10.1016/j.neuropsychologia.2014.11.023.
Duffau H. Cognitive assessment in glioma patients. J Neurosurg. 2013;119(5):1348–9. https://doi.org/10.3171/2012.2.jns112372.
Duffau H. The huge plastic potential of adult brain and the role of connectomics: new insights provided by serial mappings in glioma surgery. Cortex. 2014;58:325–37. https://doi.org/10.1016/j.cortex.2013.08.005.
Duffau H, Denvil D, Lopes M, Gasparini F, Cohen L, Capelle L, et al. Intraoperative mapping of the cortical areas involved in multiplication and subtraction: an electrostimulation study in a patient with a left parietal glioma. J Neurol Neurosurg Psychiatry. 2002;73(6):733–8.
Dzhelyova MP, Ellison A, Atkinson AP. Event-related repetitive TMS reveals distinct, critical roles for right OFA and bilateral posterior STS in judging the sex and trustworthiness of faces. J Cogn Neurosci. 2011;23(10):2782–96. https://doi.org/10.1162/jocn.2011.21604.
Eick CM, Kovács G, Rostalski SM, Röhrig L, Ambrus GG. The occipital face area is causally involved in identity-related visual-semantic associations. Brain Struct Funct. 2020;225(5):1483–93. https://doi.org/10.1007/s00429-020-02068-9.
Ellison A, Schindler I, Pattison LL, Milner AD. An exploration of the role of the superior temporal gyrus in visual search and spatial perception using TMS. Brain. 2004;127(Pt 10):2307–15. https://doi.org/10.1093/brain/awh244.
Faye A, Jacquin-Courtois S, Reynaud E, Lesourd M, Besnard J, Osiurak F. Numerical cognition: a meta-analysis of neuroimaging, transcranial magnetic stimulation and brain-damaged patients studies. Neuroimage Clin. 2019;24:102053. https://doi.org/10.1016/j.nicl.2019.102053.
Ferrari C, Lega C, Vernice M, Tamietto M, Mende-Siedlecki P, Vecchi T, et al. The dorsomedial prefrontal cortex plays a causal role in integrating social impressions from faces and verbal descriptions. Cereb Cortex. 2016;26(1):156–65. https://doi.org/10.1093/cercor/bhu186.
Ferrari C, Gamond L, Gallucci M, Vecchi T, Cattaneo Z. An exploratory TMS study on prefrontal lateralization in valence categorization of facial expressions. Exp Psychol. 2017;64(4):282–9. https://doi.org/10.1027/1618-3169/a000363.
Ferrari C, Oldrati V, Gallucci M, Vecchi T, Cattaneo Z. The role of the cerebellum in explicit and incidental processing of facial emotional expressions: a study with transcranial magnetic stimulation. NeuroImage. 2018;169:256–64. https://doi.org/10.1016/j.neuroimage.2017.12.026.
Fierro B, Brighina F, Oliveri M, Piazza A, La Bua V, Buffa D, et al. Contralateral neglect induced by right posterior parietal rTMS in healthy subjects. Neuroreport. 2000;11(7):1519–21.
Fierro B, Brighina F, Piazza A, Oliveri M, Bisiach E. Timing of right parietal and frontal cortex activity in visuo-spatial perception: a TMS study in normal individuals. Neuroreport. 2001;12(11):2605–7.
Fresnoza S, Christova M, Purgstaller S, Jehna M, Zaar K, Hoffermann M, et al. Dissociating arithmetic operations in the parietal cortex using 1 Hz repetitive transcranial magnetic stimulation: the importance of strategy use. Front Hum Neurosci. 2020;14:271. https://doi.org/10.3389/fnhum.2020.00271.

Frey D, Strack V, Wiener E, Jussen D, Vajkoczy P, Picht T. A new approach for corticospinal tract reconstruction based on navigated transcranial stimulation and standardized fractional anisotropy values. NeuroImage. 2012;62(3):1600–9. https://doi.org/10.1016/j.neuroimage.2012.05.059.

Fruhholz S, Godde B, Lewicki P, Herzmann C, Herrmann M. Face recognition under ambiguous visual stimulation: fMRI correlates of "encoding styles". Hum Brain Mapp. 2011;32(10):1750–61. https://doi.org/10.1002/hbm.21144.

Fuggetta G, Rizzo S, Pobric G, Lavidor M, Walsh V. Functional representation of living and non-living domains across the cerebral hemispheres: a combined event-related potential/transcranial magnetic stimulation study. J Cogn Neurosci. 2009;21(2):403–14. https://doi.org/10.1162/jocn.2008.21030.

Fusar-Poli P, Placentino A, Carletti F, Landi P, Allen P, Surguladze S, et al. Functional atlas of emotional faces processing: a voxel-based meta-analysis of 105 functional magnetic resonance imaging studies. J Psychiatry Neurosci. 2009;34(6):418–32.

Gainotti G, Marra C. Differential contribution of right and left temporo-occipital and anterior temporal lesions to face recognition disorders. Front Hum Neurosci. 2011;5:55. https://doi.org/10.3389/fnhum.2011.00055.

Gamond L, Cattaneo Z. The dorsomedial prefrontal cortex plays a causal role in mediating in-group advantage in emotion recognition: a TMS study. Neuropsychologia. 2016;93(Pt A):312–7. https://doi.org/10.1016/j.neuropsychologia.2016.11.011.

Garcia-Sanz S, Ghotme KA, Hedmont D, Arévalo-Jaimes MY, Cohen Kadosh R, Serra-Grabulosa JM, et al. Use of transcranial magnetic stimulation for studying the neural basis of numerical cognition: a systematic review. J Neurosci Methods. 2022;369:109485. https://doi.org/10.1016/j.jneumeth.2022.109485.

Gauthier I, Tarr MJ, Moylan J, Skudlarski P, Gore JC, Anderson AW. The fusiform "face area" is part of a network that processes faces at the individual level. J Cogn Neurosci. 2000;12(3):495–504.

Gießing C, Alavash M, Herrmann CS, Hilgetag CC, Thiel CM. Individual differences in local functional brain connectivity affect TMS effects on behavior. Sci Rep. 2020;10(1):10422. https://doi.org/10.1038/s41598-020-67162-8.

Giglhuber K, Maurer S, Zimmer C, Meyer B, Krieg SM. Evoking visual neglect-like deficits in healthy volunteers – an investigation by repetitive navigated transcranial magnetic stimulation. Brain Imaging Behav. 2016;11:17. https://doi.org/10.1007/s11682-016-9506-9.

Giglhuber K, Maurer S, Zimmer C, Meyer B, Krieg SM. Mapping visuospatial attention: the greyscales task in combination with repetitive navigated transcranial magnetic stimulation. BMC Neurosci. 2018;19(1):40. https://doi.org/10.1186/s12868-018-0440-1.

Giussani C, Pirillo D, Roux FE. Mirror of the soul: a cortical stimulation study on recognition of facial emotions. J Neurosurg. 2010a;112(3):520–7. https://doi.org/10.3171/2009.5.jns081522.

Giussani C, Roux FE, Ojemann J, Sganzerla EP, Pirillo D, Papagno C. Is preoperative functional magnetic resonance imaging reliable for language areas mapping in brain tumor surgery? Review of language functional magnetic resonance imaging and direct cortical stimulation correlation studies. Neurosurgery. 2010b;66(1):113–20. https://doi.org/10.1227/01.NEU.0000360392.15450.C9.

Gobel SM, Walsh V, Rushworth MF. The mental number line and the human angular gyrus. NeuroImage. 2001;14(6):1278–89. https://doi.org/10.1006/nimg.2001.0927.

Gobel SM, Calabria M, Farne A, Rossetti Y. Parietal rTMS distorts the mental number line: simulating 'spatial' neglect in healthy subjects. Neuropsychologia. 2006a;44(6):860–8. https://doi.org/10.1016/j.neuropsychologia.2005.09.007.

Gobel SM, Rushworth MF, Walsh V. Inferior parietal rtms affects performance in an addition task. Cortex. 2006b;42(5):774–81.

Gomez J, Pestilli F, Witthoft N, Golarai G, Liberman A, Poltoratski S, et al. Functionally defined white matter reveals segregated pathways in human ventral temporal cortex associated with category-specific processing. Neuron. 2015;85(1):216–27. https://doi.org/10.1016/j.neuron.2014.12.027.

Gothard KM, Battaglia FP, Erickson CA, Spitler KM, Amaral DG. Neural responses to facial expression and face identity in the monkey amygdala. J Neurophysiol. 2007;97(2):1671–83. https://doi.org/10.1152/jn.00714.2006.

Grafton ST, Fadiga L, Arbib MA, Rizzolatti G. Premotor cortex activation during observation and naming of familiar tools. NeuroImage. 1997;6(4):231–6. https://doi.org/10.1006/nimg.1997.0293.

Guidali G, Bagattini C, De Matola M, Brignani D. Influence of frontal-to-parietal connectivity in pseudoneglect: a cortico-cortical paired associative stimulation study. Cortex. 2023;169:50–64. https://doi.org/10.1016/j.cortex.2023.08.012.

Guo K, Calver L, Soornack Y, Bourke P. Valence-dependent disruption in processing of facial expressions of emotion in early visual cortex-A transcranial magnetic stimulation study. J Cogn Neurosci. 2020;32(5):906–16. https://doi.org/10.1162/jocn_a_01520.

Hadjikhani N, de Gelder B. Neural basis of prosopagnosia: an fMRI study. Hum Brain Mapp. 2002;16(3):176–82. https://doi.org/10.1002/hbm.10043.

Haglund MM, Berger MS, Shamseldin M, Lettich E, Ojemann GA. Cortical localization of temporal lobe language sites in patients with gliomas. Neurosurgery. 1994;34(4):567-576; discussion 576.

Handwerker DA, Ianni G, Gutierrez B, Roopchansingh V, Gonzalez-Castillo J, Chen G, et al. Theta-burst TMS to the posterior superior temporal sulcus decreases resting-state fMRI connectivity across the face processing network. Netw Neurosci. 2020;4(3):746–60. https://doi.org/10.1162/netn_a_00145.

Harris JA, Clifford CW, Miniussi C. The functional effect of transcranial magnetic stimulation: signal suppression or neural noise generation? J Cogn Neurosci. 2008;20(4):734–40. https://doi.org/10.1162/jocn.2008.20048.

Hauck T, Tanigawa N, Probst M, Wohlschlaeger A, Ille S, Sollmann N, et al. Stimulation frequency determines the distribution of language positive cortical regions during navigated transcranial magnetic brain stimulation. BMC Neurosci. 2015a;16:5. https://doi.org/10.1186/s12868-015-0143-9.

Hauck T, Tanigawa N, Probst M, Wohlschlaeger A, Ille S, Sollmann N, et al. Task type affects location of language-positive cortical regions by repetitive navigated transcranial magnetic stimulation mapping. PLoS One. 2015b;10(4):e0125298. https://doi.org/10.1371/journal.pone.0125298.

Hayashi N, Ishii K, Kitagaki H, Kazui H. Regional differences in cerebral blood flow during recitation of the multiplication table and actual calculation: a positron emission tomography study. J Neurol Sci. 2000;176(2):102–8.

Heilman KM. Right hemisphere dominance for attention: the mechanism underlying hemispheric asymmetries of inattention (neglect). Neurology. 1980;30(3):327. https://doi.org/10.1212/wnl.30.3.327.

Heinen K, Ruff CC, Bjoertomt O, Schenkluhn B, Bestmann S, Blankenburg F, et al. Concurrent TMS-fMRI reveals dynamic interhemispheric influences of the right parietal cortex during exogenously cued visuospatial attention. Eur J Neurosci. 2011;33(5):991–1000. https://doi.org/10.1111/j.1460-9568.2010.07580.x.

Heinen K, Feredoes E, Ruff CC, Driver J. Functional connectivity between prefrontal and parietal cortex drives visuo-spatial attention shifts. Neuropsychologia. 2017;99:81–91. https://doi.org/10.1016/j.neuropsychologia.2017.02.024.

Hier DB, Mondlock J, Caplan LR. Recovery of behavioral abnormalities after right hemisphere stroke. Neurology. 1983;33(3):345–50.

Hilgetag CC, Theoret H, Pascual-Leone A. Enhanced visual spatial attention ipsilateral to rTMS-induced 'virtual lesions' of human parietal cortex. Nat Neurosci. 2001;4(9):953–7. https://doi.org/10.1038/nn0901-953.

Hommet C, Bardet F, de Toffol B, Perrier D, Biraben A, Vignal JP, et al. Unilateral spatial neglect following right inferior parietal cortectomy. Epilepsy Behav. 2004;5(3):416–9. https://doi.org/10.1016/j.yebeh.2004.02.008.

Humphreys GW, Forde EM. Hierarchies, similarity, and interactivity in object recognition: "category-specific" neuropsychological deficits. Behav Brain Sci. 2001;24(3):453–76. discussion 476-509

Hung J, Driver J, Walsh V. Visual selection and posterior parietal cortex: effects of repetitive transcranial magnetic stimulation on partial report analyzed by Bundesen's theory of visual attention. J Neurosci. 2005;25(42):9602–12. https://doi.org/10.1523/jneurosci.0879-05.2005.

Ille S, Sollmann N, Hauck T, Maurer S, Tanigawa N, Obermueller T, et al. Impairment of preoperative language mapping by lesion location: a functional magnetic resonance imaging, navigated transcranial magnetic stimulation, and direct cortical stimulation study. J Neurosurg. 2015;123:314. https://doi.org/10.3171/2014.10.JNS141582.

Ille S, Drummer K, Giglhuber K, Conway N, Maurer S, Meyer B, et al. Mapping of arithmetic processing by navigated repetitive transcranial magnetic stimulation in patients with parietal brain tumors and correlation with postoperative outcome. World Neurosurg. 2018;114:e1016–30. https://doi.org/10.1016/j.wneu.2018.03.136.

Jacquet PO, Avenanti A. Perturbing the action observation network during perception and categorization of actions' goals and grips: state-dependency and virtual lesion TMS effects. Cereb Cortex. 2015;25(3):598–608. https://doi.org/10.1093/cercor/bht242.

Janssens SEW, Sack AT, Jessen S, de Graaf TA. Can processing of face trustworthiness bypass early visual cortex? A transcranial magnetic stimulation masking study. Neuropsychologia. 2020;137:107304. https://doi.org/10.1016/j.neuropsychologia.2019.107304.

Jehkonen M, Ahonen JP, Dastidar P, Koivisto AM, Laippala P, Vilkki J, et al. Visual neglect as a predictor of functional outcome one year after stroke. Acta Neurol Scand. 2000;101(3):195–201.

Jehkonen M, Laihosalo M, Kettunen JE. Impact of neglect on functional outcome after stroke—a review of methodological issues and recent research findings. Restor Neurol Neurosci. 2006;24:209.

Jeschke L, Mathias B, von Kriegstein K. Inhibitory TMS over visual area V5/MT disrupts visual speech recognition. J Neurosci. 2023;43(45):7690–9. https://doi.org/10.1523/jneurosci.0975-23.2023.

Jin Z, Xie K, Ni X, Jin DG, Zhang J, Li L. Transcranial magnetic stimulation over the right dorsolateral prefrontal cortex modulates visuospatial distractor suppression. Eur J Neurosci. 2021;53(10):3394–403. https://doi.org/10.1111/ejn.15164.

Jung-Beeman M. Bilateral brain processes for comprehending natural language. Trends Cogn Sci. 2005;9(11):512–8. https://doi.org/10.1016/j.tics.2005.09.009.

Kadosh KC, Walsh V, Kadosh RC. Investigating face-property specific processing in the right OFA. Soc Cogn Affect Neurosci. 2011;6(1):58–65. https://doi.org/10.1093/scan/nsq015.

Katz N, Hartman-Maeir A, Ring H, Soroker N. Functional disability and rehabilitation outcome in right hemisphere damaged patients with and without unilateral spatial neglect. Arch Phys Med Rehabil. 1999;80(4):379–84.

Keenan JP, Wheeler MA, Gallup GG Jr, Pascual-Leone A. Self-recognition and the right prefrontal cortex. Trends Cogn Sci. 2000;4(9):338–44. https://doi.org/10.1016/S1364-6613(00)01521-7.

Kiefer M. Perceptual and semantic sources of category-specific effects: event-related potentials during picture and word categorization. Mem Cogn. 2001;29(1):100–16.

Kiefer M. Repetition-priming modulates category-related effects on event-related potentials: further evidence for multiple cortical semantic systems. J Cogn Neurosci. 2005;17(2):199–211. https://doi.org/10.1162/0898929053124938.

Kim YH, Min SJ, Ko MH, Park JW, Jang SH, Lee PK. Facilitating visuospatial attention for the contralateral hemifield by repetitive TMS on the posterior parietal cortex. Neurosci Lett. 2005;382(3):280–5. https://doi.org/10.1016/j.neulet.2005.03.043.

Kinsbourne M. Hemi-neglect and hemisphere rivalry. Adv Neurol. 1977;18:41–9.

Kitada R, Johnsrude IS, Kochiyama T, Lederman SJ. Functional specialization and convergence in the occipito-temporal cortex supporting haptic and visual identification of human faces and body parts: an fMRI study. J Cogn Neurosci. 2009;21(10):2027–45. https://doi.org/10.1162/jocn.2009.21115.

Koch G, Oliveri M, Torriero S, Caltagirone C. Modulation of excitatory and inhibitory circuits for visual awareness in the human right parietal cortex. Exp Brain Res. 2005;160(4):510–6. https://doi.org/10.1007/s00221-004-2039-2.

Koch G, Bonni S, Giacobbe V, Bucchi G, Basile B, Lupo F, et al. Theta-burst stimulation of the left hemisphere accelerates recovery of hemispatial neglect. Neurology. 2012;78(1):24–30. https://doi.org/10.1212/WNL.0b013e31823ed08f.

Krieg SM, Buchmann NH, Gempt J, Shiban E, Meyer B, Ringel F. Diffusion tensor imaging fiber tracking using navigated brain stimulation–a feasibility study. Acta Neurochir. 2012;154(3):555–63. https://doi.org/10.1007/s00701-011-1255-3.

Kurimoto M, Asahi T, Shibata T, Takahashi C, Nagai S, Hayashi N, et al. Safe removal of glioblastoma near the angular gyrus by awake surgery preserving calculation ability–case report. Neurol Med Chir (Tokyo). 2006;46(1):46–50.

Landis T, Cummings JL, Christen L, Bogen JE, Imhof HG. Are unilateral right posterior cerebral lesions sufficient to cause prosopagnosia? Clinical and radiological findings in six additional patients. Cortex. 1986;22(2):243–52.

Lee KM. Cortical areas differentially involved in multiplication and subtraction: a functional magnetic resonance imaging study and correlation with a case of selective acalculia. Ann Neurol. 2000;48(4):657–61.

Leonard CM, Rolls ET, Wilson FA, Baylis GC. Neurons in the amygdala of the monkey with responses selective for faces. Behav Brain Res. 1985;15(2):159–76.

Lunven M, Thiebaut de Schotten M, Bourlon C, Duret C, Migliaccio R, Rode G, et al. White matter lesional predictors of chronic visual neglect: a longitudinal study. Brain. 2015;138(Pt 3):746–60. https://doi.org/10.1093/brain/awu389.

Mahayana IT, Liu CL, Chang CF, Hung DL, Tzeng OJ, Juan CH, et al. Far-space neglect in conjunction but not feature search following transcranial magnetic stimulation over right posterior parietal cortex. J Neurophysiol. 2014a;111(4):705–14. https://doi.org/10.1152/jn.00492.2013.

Mahayana IT, Tcheang L, Chen CY, Juan CH, Muggleton NG. The precuneus and visuospatial attention in near and far space: a transcranial magnetic stimulation study. Brain Stimul. 2014b;7(5):673–9. https://doi.org/10.1016/j.brs.2014.06.012.

Maldonado IL, Moritz-Gasser S, de Champfleur NM, Bertram L, Moulinie G, Duffau H. Surgery for gliomas involving the left inferior parietal lobule: new insights into the functional anatomy provided by stimulation mapping in awake patients. J Neurosurg. 2011;115(4):770–9. https://doi.org/10.3171/2011.5.jns112.

Martin A, Wiggs CL, Ungerleider LG, Haxby JV. Neural correlates of category-specific knowledge. Nature. 1996;379(6566):649–52. https://doi.org/10.1038/379649a0.

Martins IP, Ferreira J, Borges L. Acquired procedural dyscalculia associated to a left parietal lesion in a child. Child Neuropsychol. 1999;5(4):265–73. https://doi.org/10.1076/0929-7049(199912)05:04;1-r;ft265.

Matejko AA, Price GR, Mazzocco MM, Ansari D. Individual differences in left parietal white matter predict math scores on the preliminary scholastic aptitude test. NeuroImage. 2013;66:604–10. https://doi.org/10.1016/j.neuroimage.2012.10.045.

Mattavelli G, Cattaneo Z, Papagno C. Transcranial magnetic stimulation of medial prefrontal cortex modulates face expressions processing in a priming task. Neuropsychologia. 2011;49(5):992–8. https://doi.org/10.1016/j.neuropsychologia.2011.01.038.

Mattavelli G, Rosanova M, Casali AG, Papagno C, Romero Lauro LJ. Top-down interference and cortical responsiveness in face processing: a TMS-EEG study. NeuroImage. 2013;76:24–32. https://doi.org/10.1016/j.neuroimage.2013.03.020.

Mattavelli G, Pisoni A, Romero Lauro LJ, Marino BF, Bonomi M, Rosanova M, et al. TMS-EEG approach unveils brain mechanisms underlying conscious and unconscious face perception. Brain Stimul. 2019;12(4):1010–9. https://doi.org/10.1016/j.brs.2019.02.022.

Maurer S, Tanigawa N, Sollmann N, Hauck T, Ille S, Boeckh-Behrens T, et al. Non-invasive mapping of calculation function by repetitive navigated transcranial magnetic stimulation. Brain Struct Funct. 2016;221(8):3927–47. https://doi.org/10.1007/s00429-015-1136-2.

Maurer S, Giglhuber K, Sollmann N, Kelm A, Ille S, Hauck T, et al. Non-invasive mapping of face processing by navigated transcranial magnetic stimulation. [Original Research]. Front Hum Neurosci. 2017;11:*4*.

Maurer S, Butenschoen VM, Meyer B, Krieg SM. Non-invasive mapping of cortical categorization function by repetitive navigated transcranial magnetic stimulation. Sci Rep. 2021;11(1):24480. https://doi.org/10.1038/s41598-021-04071-4.

Mayer E, Martory MD, Pegna AJ, Landis T, Delavelle J, Annoni JM. A pure case of Gerstmann syndrome with a subangular lesion. Brain. 1999;122(Pt 6):1107–20.

McGraw P, Mathews VP, Wang Y, Phillips MD. Approach to functional magnetic resonance imaging of language based on models of language organization. Neuroimaging Clin N Am. 2001;11(2):343–53. x

Meister IG, Wienemann M, Buelte D, Grunewald C, Sparing R, Dambeck N, et al. Hemiextinction induced by transcranial magnetic stimulation over the right temporo-parietal junction. Neuroscience. 2006;142(1):119–23. https://doi.org/10.1016/j.neuroscience.2006.06.023.

Mevorach C, Humphreys GW, Shalev L. Attending to local form while ignoring global aspects depends on handedness: evidence from TMS. Nat Neurosci. 2005;8(3):276–7. https://doi.org/10.1038/nn1400.

Middag-van Spanje M, Duecker F, Gallotto S, de Graaf TA, van Heugten C, Sack AT, et al. Transcranial magnetic stimulation over posterior parietal cortex modulates alerting and executive control processes in attention. Eur J Neurosci. 2022;56(10):5853–68. https://doi.org/10.1111/ejn.15830.

Miniussi C, Ruzzoli M, Walsh V. The mechanism of transcranial magnetic stimulation in cognition. Cortex. 2010;46(1):128–30. https://doi.org/10.1016/j.cortex.2009.03.004.

Moore CJ, Price CJ. A functional neuroimaging study of the variables that generate category-specific object processing differences. Brain. 1999;122(Pt 5):943–62.

Muggleton NG, Postma P, Moutsopoulou K, Nimmo-Smith I, Marcel A, Walsh V. TMS over right posterior parietal cortex induces neglect in a scene-based frame of reference. Neuropsychologia. 2006;44(7):1222–9. https://doi.org/10.1016/j.neuropsychologia.2005.10.004.

Muri RM, Buhler R, Heinemann D, Mosimann UP, Felblinger J, Schlaepfer TE, et al. Hemispheric asymmetry in visuospatial attention assessed with transcranial magnetic stimulation. Exp Brain Res. 2002;143(4):426–30. https://doi.org/10.1007/s00221-002-1009-9.

Neggers SF, Huijbers W, Vrijlandt CM, Vlaskamp BN, Schutter DJ, Kenemans JL. TMS pulses on the frontal eye fields break coupling between visuospatial attention and eye movements. J Neurophysiol. 2007;98(5):2765–78. https://doi.org/10.1152/jn.00357.2007.

Nyffeler T, Cazzoli D, Wurtz P, Luthi M, von Wartburg R, Chaves S, et al. Neglect-like visual exploration behaviour after theta burst transcranial magnetic stimulation of the right posterior parietal cortex. Eur J Neurosci. 2008;27(7):1809–13. https://doi.org/10.1111/j.1460-9568.2008.06154.x.

Ojemann GA, Whitaker HA. Language localization and variability. Brain Lang. 1978;6(2):239–60.

Ojemann G, Ojemann J, Lettich E, Berger M. Cortical language localization in left, dominant hemisphere. An electrical stimulation mapping investigation in 117 patients. J Neurosurg. 1989;71(3):316–26. https://doi.org/10.3171/jns.1989.71.3.0316.

Oliver R, Bjoertomt O, Driver J, Greenwood R, Rothwell J. Novel 'hunting' method using transcranial magnetic stimulation over parietal cortex disrupts visuospatial sensitivity in relation to motor thresholds. Neuropsychologia. 2009;47(14):3152–61. https://doi.org/10.1016/j.neuropsychologia.2009.07.017.

Paracampo R, Tidoni E, Borgomaneri S, di Pellegrino G, Avenanti A. Sensorimotor network crucial for inferring amusement from smiles. Cereb Cortex. 2017;27(11):5116–29. https://doi.org/10.1093/cercor/bhw294.

Pascual-Leone A, Gates JR, Dhuna A. Induction of speech arrest and counting errors with rapid-rate transcranial magnetic stimulation. Neurology. 1991;41(5):697–702.

Pascual-Leone A, Gomez-Tortosa E, Grafman J, Alway D, Nichelli P, Hallett M. Induction of visual extinction by rapid-rate transcranial magnetic stimulation of parietal lobe. Neurology. 1994;44(3 Pt 1):494–8.

Passeri A, Capotosto P, Di Matteo R. The right hemisphere contribution to semantic categorization: a TMS study. Cortex. 2015;64:318–26. https://doi.org/10.1016/j.cortex.2014.11.014.

Paz-Caballero D, Cuetos F, Dobarro A. Electrophysiological evidence for a natural/artifactual dissociation. Brain Res. 2006;1067(1):189–200. https://doi.org/10.1016/j.brainres.2005.10.046.

Perani D, Cappa SF, Bettinardi V, Bressi S, Gorno-Tempini M, Matarrese M, et al. Different neural systems for the recognition of animals and man-made tools. Neuroreport. 1995;6(12):1637–41.

Peters L, De Smedt B. Arithmetic in the developing brain: a review of brain imaging studies. Dev Cogn Neurosci. 2018;30:265–79. https://doi.org/10.1016/j.dcn.2017.05.002.

Picht T, Krieg SM, Sollmann N, Rosler J, Niraula B, Neuvonen T, et al. A comparison of language mapping by preoperative navigated transcranial magnetic stimulation and direct cortical stimulation during awake surgery. Neurosurgery. 2013;72(5):808–19. https://doi.org/10.1227/NEU.0b013e3182889e01.

Pitcher D. Facial expression recognition takes longer in the posterior superior temporal sulcus than in the occipital face area. J Neurosci. 2014;34(27):9173–7. https://doi.org/10.1523/jneurosci.5038-13.2014.

Pitcher D, Walsh V, Yovel G, Duchaine B. TMS evidence for the involvement of the right occipital face area in early face processing. Curr Biol. 2007;17(18):1568–73. https://doi.org/10.1016/j.cub.2007.07.063.

Pitcher D, Garrido L, Walsh V, Duchaine BC. Transcranial magnetic stimulation disrupts the perception and embodiment of facial expressions. J Neurosci. 2008;28(36):8929–33. https://doi.org/10.1523/jneurosci.1450-08.2008.

Pitcher D, Charles L, Devlin JT, Walsh V, Duchaine B. Triple dissociation of faces, bodies, and objects in extrastriate cortex. Curr Biol. 2009;19(4):319–24. https://doi.org/10.1016/j.cub.2009.01.007.

Pitcher D, Duchaine B, Walsh V, Yovel G, Kanwisher N. The role of lateral occipital face and object areas in the face inversion effect. Neuropsychologia. 2011;49(12):3448–53. https://doi.org/10.1016/j.neuropsychologia.2011.08.020.

Pitcher D, Pilkington A, Rauth L, Baker C, Kravitz DJ, Ungerleider LG. The human posterior superior temporal sulcus samples visual space differently from other face-selective regions. Cereb Cortex. 2020;30(2):778–85. https://doi.org/10.1093/cercor/bhz125.

Plaza M, Gatignol P, Cohen H, Berger B, Duffau H. A discrete area within the left dorsolateral prefrontal cortex involved in visual-verbal incongruence judgment. Cereb Cortex. 2008;18(6):1253–9. https://doi.org/10.1093/cercor/bhm169.

Plow EB, Cattaneo Z, Carlson TA, Alvarez GA, Pascual-Leone A, Battelli L. The compensatory dynamic of inter-hemispheric interactions in visuospatial attention revealed using rTMS and fMRI. Front Hum Neurosci. 2014;8:226. https://doi.org/10.3389/fnhum.2014.00226.

Pourtois G, Sander D, Andres M, Grandjean D, Reveret L, Olivier E, et al. Dissociable roles of the human somatosensory and superior temporal cortices for processing social face signals. Eur J Neurosci. 2004;20(12):3507–15. https://doi.org/10.1111/j.1460-9568.2004.03794.x.

Pu S, Li YN, Wu CX, Wang YZ, Zhou XL, Jiang T. Cortical areas involved in numerical processing: an intraoperative electrostimulation study. Stereotact Funct Neurosurg. 2011;89(1):42–7. https://doi.org/10.1159/000321186.

Rapcsak SZ. Face memory and its disorders. Curr Neurol Neurosci Rep. 2003;3(6):494–501.

Rapcsak SZ, Nielsen L, Littrell LD, Glisky EL, Kaszniak AW, Laguna JF. Face memory impairments in patients with frontal lobe damage. Neurology. 2001;57(7):1168–75.

Rawji V, Latorre A, Sharma N, Rothwell JC, Rocchi L. On the use of TMS to investigate the pathophysiology of neurodegenerative diseases. Front Neurol. 2020;11:584664. https://doi.org/10.3389/fneur.2020.584664.

Renzi C, Schiavi S, Carbon CC, Vecchi T, Silvanto J, Cattaneo Z. Processing of featural and configural aspects of faces is lateralized in dorsolateral prefrontal cortex: a TMS study. NeuroImage. 2013;74:45–51. https://doi.org/10.1016/j.neuroimage.2013.02.015.

Ricci R, Salatino A, Li X, Funk AP, Logan SL, Mu Q, et al. Imaging the neural mechanisms of TMS neglect-like bias in healthy volunteers with the interleaved TMS/fMRI technique: preliminary evidence. Front Hum Neurosci. 2012;6:326. https://doi.org/10.3389/fnhum.2012.00326.

Rochas V, Gelmini L, Krolak-Salmon P, Poulet E, Saoud M, Brunelin J, et al. Disrupting pre-SMA activity impairs facial happiness recognition: an event-related TMS study. Cereb Cortex. 2013;23(7):1517–25. https://doi.org/10.1093/cercor/bhs133.
Rossion B. Twenty years of investigation with the case of prosopagnosia PS to understand human face identity recognition. Part II: Neural basis. Neuropsychologia. 2022;173:108279. https://doi.org/10.1016/j.neuropsychologia.2022.108279.
Roux FE, Boetto S, Sacko O, Chollet F, Tremoulet M. Writing, calculating, and finger recognition in the region of the angular gyrus: a cortical stimulation study of Gerstmann syndrome. J Neurosurg. 2003;99(4):716–27. https://doi.org/10.3171/jns.2003.99.4.0716.
Roux FE, Borsa S, Demonet JF. "The mute who can sing": a cortical stimulation study on singing. J Neurosurg. 2009a;110(2):282–8. https://doi.org/10.3171/2007.9.17565.
Roux FE, Boukhatem L, Draper L, Sacko O, Demonet JF. Cortical calculation localization using electrostimulation. J Neurosurg. 2009b;110(6):1291–9. https://doi.org/10.3171/2008.8.jns17649.
Roux FE, Dufor O, Lauwers-Cances V, Boukhatem L, Brauge D, Draper L, et al. Electrostimulation mapping of spatial neglect. Neurosurgery. 2011;69(6):1218–31. https://doi.org/10.1227/NEU.0b013e31822aefd2.
Rusconi E, Walsh V, Butterworth B. Dexterity with numbers: rTMS over left angular gyrus disrupts finger gnosis and number processing. Neuropsychologia. 2005;43(11):1609–24. https://doi.org/10.1016/j.neuropsychologia.2005.01.009.
Russell SM, Elliott R, Forshaw D, Kelly PJ, Golfinos JG. Resection of parietal lobe gliomas: incidence and evolution of neurological deficits in 28 consecutive patients correlated to the location and morphological characteristics of the tumor. J Neurosurg. 2005;103(6):1010–7. https://doi.org/10.3171/jns.2005.103.6.1010.
Ruwaili A, Fatima R, Hussain A, Uzair M, Abualait T, Imdad K, et al. The effectiveness of transcranial magnetic stimulation in treating apraxia. CNS Neurol Disord Drug Targets. 2023;23:1030. https://doi.org/10.2174/0118715273249412231010171926.
Ruzzoli M, Marzi CA, Miniussi C. The neural mechanisms of the effects of transcranial magnetic stimulation on perception. J Neurophysiol. 2010;103(6):2982–9. https://doi.org/10.1152/jn.01096.2009.
Sack AT. Using non-invasive brain interference as a tool for mimicking spatial neglect in healthy volunteers. Restor Neurol Neurosci. 2010;28(4):485–97. https://doi.org/10.3233/rnn-2010-0568.
Sack AT, Kohler A, Bestmann S, Linden DE, Dechent P, Goebel R, et al. Imaging the brain activity changes underlying impaired visuospatial judgments: simultaneous FMRI, TMS, and behavioral studies. Cereb Cortex. 2007;17(12):2841–52. https://doi.org/10.1093/cercor/bhm013.
Salatino A, Poncini M, George MS, Ricci R. Hunting for right and left parietal hot spots using single-pulse TMS: modulation of visuospatial perception during line bisection judgment in the healthy brain. Front Psychol. 2014;5:1238. https://doi.org/10.3389/fpsyg.2014.01238.
Salatino A, Chillemi G, Gontero F, Poncini M, Pyasik M, Berti A, et al. Transcranial magnetic stimulation of posterior parietal cortex modulates line length estimation but not illusory depth perception. Front Psychol. 2019;10:1169. https://doi.org/10.3389/fpsyg.2019.01169.
Sanai N, Mirzadeh Z, Berger MS. Functional outcome after language mapping for glioma resection. N Engl J Med. 2008;358(1):18–27. https://doi.org/10.1056/NEJMoa067819.
Sanai N, Martino J, Berger MS. Morbidity profile following aggressive resection of parietal lobe gliomas. J Neurosurg. 2012;116(6):1182–6. https://doi.org/10.3171/2012.2.jns111228.
Sandrini M, Rossini PM, Miniussi C. The differential involvement of inferior parietal lobule in number comparison: a rTMS study. Neuropsychologia. 2004;42(14):1902–9. https://doi.org/10.1016/j.neuropsychologia.2004.05.005.
Sauseng P, Feldheim JF, Freunberger R, Hummel FC. Right prefrontal TMS disrupts interregional anticipatory EEG alpha activity during shifting of visuospatial attention. Front Psychol. 2011;2:241. https://doi.org/10.3389/fpsyg.2011.00241.

Schintu S, Cunningham CA, Freedberg M, Taylor P, Gotts SJ, Shomstein S, et al. Callosal anisotropy predicts attentional network changes after parietal inhibitory stimulation. NeuroImage. 2021;226:117559. https://doi.org/10.1016/j.neuroimage.2020.117559.

Schwab PJ, Miller A, Raphail AM, Levine A, Haslam C, Coslett HB, et al. Virtual reality tools for assessing unilateral spatial neglect: a novel opportunity for data collection. J Vis Exp. 2021;169 https://doi.org/10.3791/61951.

Siebner HR, Hartwigsen G, Kassuba T, Rothwell JC. How does transcranial magnetic stimulation modify neuronal activity in the brain? Implications for studies of cognition. Cortex. 2009;45(9):1035–42. https://doi.org/10.1016/j.cortex.2009.02.007.

Sim EJ, Kiefer M. Category-related brain activity to natural categories is associated with the retrieval of visual features: evidence from repetition effects during visual and functional judgments. Brain Res Cogn Brain Res. 2005;24(2):260–73. https://doi.org/10.1016/j.cogbrainres.2005.02.006.

Sliwinska MW, Pitcher D. TMS demonstrates that both right and left superior temporal sulci are important for facial expression recognition. NeuroImage. 2018;183:394–400. https://doi.org/10.1016/j.neuroimage.2018.08.025.

Sliwinska MW, Elson R, Pitcher D. Dual-site TMS demonstrates causal functional connectivity between the left and right posterior temporal sulci during facial expression recognition. Brain Stimul. 2020;13(4):1008–13. https://doi.org/10.1016/j.brs.2020.04.011.

Sollmann N, Giglhuber K, Tussis L, Meyer B, Ringel F, Krieg SM. nTMS-based DTI fiber tracking for language pathways correlates with language function and aphasia – a case report. Clin Neurol Neurosurg. 2015a;136:25–8. https://doi.org/10.1016/j.clineuro.2015.05.023.

Sollmann N, Ille S, Hauck T, Maurer S, Negwer C, Zimmer C, et al. The impact of preoperative language mapping by repetitive navigated transcranial magnetic stimulation on the clinical course of brain tumor patients. BMC Cancer. 2015b;15:261. https://doi.org/10.1186/s12885-015-1299-5.

Solomon-Harris LM, Rafique SA, Steeves JK. Consecutive TMS-fMRI reveals remote effects of neural noise to the "occipital face area". Brain Res. 2016;1650:134–41. https://doi.org/10.1016/j.brainres.2016.08.043.

Stengel C, Vernet M, Amengual JL, Valero-Cabré A. Causal modulation of right hemisphere fronto-parietal phase synchrony with transcranial magnetic stimulation during a conscious visual detection task. Sci Rep. 2021;11(1):3807. https://doi.org/10.1038/s41598-020-79812-y.

Studer B, Cen D, Walsh V. The angular gyrus and visuospatial attention in decision-making under risk. NeuroImage. 2014;103:75–80. https://doi.org/10.1016/j.neuroimage.2014.09.003.

Suchan J, Umarova R, Schnell S, Himmelbach M, Weiller C, Karnath HO, et al. Fiber pathways connecting cortical areas relevant for spatial orienting and exploration. Hum Brain Mapp. 2014;35(3):1031–43. https://doi.org/10.1002/hbm.22232.

Tang Y, Xu L, Zhu T, Cui H, Qian Z, Kong G, et al. Visuospatial learning selectively enhanced by personalized transcranial magnetic stimulation over Parieto-hippocampal network among patients at clinical high-risk for psychosis. Schizophr Bull. 2023;49(4):923–32. https://doi.org/10.1093/schbul/sbad015.

Tarapore PE, Findlay AM, Honma SM, Mizuiri D, Houde JF, Berger MS, et al. Language mapping with navigated repetitive TMS: proof of technique and validation. NeuroImage. 2013;82:260–72. https://doi.org/10.1016/j.neuroimage.2013.05.018.

Thiebaut de Schotten M, Urbanski M, Duffau H, Volle E, Levy R, Dubois B, et al. Direct evidence for a parietal-frontal pathway subserving spatial awareness in humans. Science. 2005;309(5744):2226–8. https://doi.org/10.1126/science.1116251.

Thut G, Nietzel A, Pascual-Leone A. Dorsal posterior parietal rTMS affects voluntary orienting of visuospatial attention. Cereb Cortex. 2005;15(5):628–38. https://doi.org/10.1093/cercor/bhh164.

Umarova RM, Reisert M, Beier TU, Kiselev VG, Kloppel S, Kaller CP, et al. Attention-network specific alterations of structural connectivity in the undamaged white matter in acute neglect. Hum Brain Mapp. 2014;35(9):4678–92. https://doi.org/10.1002/hbm.22503.

Valero-Cabre A, Rushmore RJ, Payne BR. Low frequency transcranial magnetic stimulation on the posterior parietal cortex induces visuotopically specific neglect-like syndrome. Exp Brain Res. 2006;172(1):14–21. https://doi.org/10.1007/s00221-005-0307-4.

van Eimeren L, Niogi SN, McCandliss BD, Holloway ID, Ansari D. White matter microstructures underlying mathematical abilities in children. Neuroreport. 2008;19(11):1117–21. https://doi.org/10.1097/WNR.0b013e328307f5c1.

van Eimeren L, Grabner RH, Koschutnig K, Reishofer G, Ebner F, Ansari D. Structure-function relationships underlying calculation: a combined diffusion tensor imaging and fMRI study. NeuroImage. 2010;52(1):358–63. https://doi.org/10.1016/j.neuroimage.2010.04.001.

Van Ettinger-Veenstra HM, Huijbers W, Gutteling TP, Vink M, Kenemans JL, Neggers SF. fMRI-guided TMS on cortical eye fields: the frontal but not intraparietal eye fields regulate the coupling between visuospatial attention and eye movements. J Neurophysiol. 2009;102(6):3469–80. https://doi.org/10.1152/jn.00350.2009.

Walsh V, Cowey A. Transcranial magnetic stimulation and cognitive neuroscience. Nat Rev Neurosci. 2000;1(1):73–9. https://doi.org/10.1038/35036239.

Wang J, Tian Y, Wang M, Cao L, Wu H, Zhang Y, et al. A lateralized top-down network for visuospatial attention and neglect. Brain Imaging Behav. 2016;10(4):1029–37. https://doi.org/10.1007/s11682-015-9460-y.

Warrington EK, Shallice T. Category specific semantic impairments. Brain. 1984;107(Pt 3):829–54.

Whalen J, McCloskey M, Lesser RP, Gordon B. Localizing arithmetic processes in the brain: evidence from a transient deficit during cortical stimulation. J Cogn Neurosci. 1997;9(3):409–17. https://doi.org/10.1162/jocn.1997.9.3.409.

Wu Y, Wang J, Zhang Y, Zheng D, Zhang J, Rong M, et al. The neuroanatomical basis for posterior superior parietal lobule control lateralization of visuospatial attention. Front Neuroanat. 2016;10:32. https://doi.org/10.3389/fnana.2016.00032.

Yang H, Susilo T, Duchaine B. The anterior temporal face area contains invariant representations of face identity that can persist despite the loss of right FFA and OFA. Cereb Cortex. 2014;26:1096. https://doi.org/10.1093/cercor/bhu289.

Young AW, Newcombe F, de Haan EH, Small M, Hay DC. Face perception after brain injury. Selective impairments affecting identity and expression. Brain. 1993;116(Pt 4):941–59.

Yu X, Chen C, Pu S, Wu C, Li Y, Jiang T, et al. Dissociation of subtraction and multiplication in the right parietal cortex: evidence from intraoperative cortical electrostimulation. Neuropsychologia. 2011;49(10):2889–95. https://doi.org/10.1016/j.neuropsychologia.2011.06.015.

Zachariou V, Nikas CV, Safiullah ZN, Gotts SJ, Ungerleider LG. Spatial mechanisms within the dorsal visual pathway contribute to the Configural processing of faces. Cereb Cortex. 2017;27(8):4124–38. https://doi.org/10.1093/cercor/bhw224.

Zago L, Pesenti M, Mellet E, Crivello F, Mazoyer B, Tzourio-Mazoyer N. Neural correlates of simple and complex mental calculation. NeuroImage. 2001;13(2):314–27. https://doi.org/10.1006/nimg.2000.0697.

Ziegler JC, Montant M, Briesemeister BB, Brink TT, Wicker B, Ponz A, et al. Do words stink? Neural reuse as a principle for understanding emotions in Reading. J Cogn Neurosci. 2018;30(7):1023–32. https://doi.org/10.1162/jocn_a_01268.

Zuanazzi A, Cattaneo L. The right hemisphere is independent from the left hemisphere in allocating visuospatial attention. Neuropsychologia. 2017;102:197–205. https://doi.org/10.1016/j.neuropsychologia.2017.06.005.

nTMS in Pediatrics: Special Issues and Solutions

13

Adrianna Giuffre and Alexander Rotenberg

13.1 Introduction

Neuronavigated transcranial magnetic stimulation (nTMS) is a safe and well-tolerated noninvasive technique for accurate functional mapping of the eloquent cortex. Despite its clinical utility as a tool for presurgical planning of brain surgery, nTMS applications have been largely restricted to adult populations. There is a clinical unmet need to establish accurate and reliable nTMS methods in children with neurological diseases as current invasive (cortical stimulation) and noninvasive (the Wada test and fMRI) presurgical planning methods are challenging and have limited success in younger patients. In the last decade, nTMS applications in pediatric neurosurgery have been well-received, particularly those focused on presurgical planning for children with brain tumors and/or intractable epilepsy. Below, we address special considerations for nTMS in pediatric populations, focusing on children with intractable focal epilepsy.

A. Giuffre
Neuromodulation Program, Department of Neurology, Division of Epilepsy and Clinical Neurophysiology, Boston Children's Hospital, Harvard Medical School, Boston, MA, USA

Boston Children's Hospital F M Kirby Neurobiology Center, Boston, MA, USA
e-mail: Adrianna.Giuffre@childrens.harvard.edu

A. Rotenberg (✉)
Department of Neurology, Boston Children's Hospital, Harvard Medical School, Boston, MA, USA
e-mail: Alexander.Rotenberg@childrens.harvard.edu

S. M. Krieg, T. Picht (eds.), *Navigated Transcranial Magnetic Stimulation in Neurosurgery*, https://doi.org/10.1007/978-3-031-97155-6_13

13.2 Epidemiology

An appreciable fraction of epilepsy surgeries is destined to be in children and drives the need to establish noninvasive functional presurgical mapping techniques in this specialized population. Epilepsy is common, affecting 0.5 to 1% of children (Aaberg et al. 2017). Pharmacotherapy cannot control seizures in approximately one-third of children, despite numerous antiepilepsy drugs available (Bartolini et al. 2022; Löscher 2017; Sultana et al. 2021; Wirrell 2013). The prevalence of focal pharmacoresistant epilepsy, characterized by seizures arising from extratemporal foci, is notably higher in children (Berg et al. 2012; Liava et al. 2012). In part, this discrepancy can be attributed to the predominance of syndromes associated with cortical malformations, tuberous sclerosis, and developmental encephalopathies in pediatrics (Bartolini et al. 2022; Sultana et al. 2021), syndromes which are more likely to manifest symptomatically and require surgical resection of a seizure focus (Dorfmüller et al. 2014; Jayalakshmi et al. 2014; Weiner et al. 2004). Additional neuroclastic syndromes, such as sequelae of perinatal stroke, Sturge-Weber syndrome, and Rasmussen's encephalitis, are more commonly identified in early life due to their propensity to trigger seizures (Duchowny 1989; Jayalakshmi et al. 2014).

For select patients with intractable focal epilepsy, neocortical resective surgery offers a potential curative option, and successful surgical outcomes are influenced by complete excision of the epileptogenic zone, including but not limited to the seizure onset zone (Widjaja et al. 2020). Accurate localization of the eloquent cortex is critical to prevent possible functional deficits, especially in pediatric populations where potential deficits may be lifelong. Supratentorial brain tumors in children, while less common than in adults, also account for an appreciable fraction of pediatric neocortical resective surgeries and cases of intractable seizures (Clusmann et al. 2004; Jayalakshmi et al. 2014; Nickels et al. 2012; Starnes et al. 2021).

13.3 Conventional Preoperative Methods for Localization of Function: The Wada Test and fMRI

The challenge of localizing motor and language function in children cannot be overstated. Many conventional preoperative functional mapping methods that are widely used in adults often encounter substantial hurdles when applied to the pediatric demographic. These challenges are rooted in the unique requirements and limitations inherent to children, necessitating the need for alternative approaches, such as nTMS, for functional preoperative mapping.

One well-established method, Dr. Wada's intracarotid amobarbital procedure (IAP), is routinely used in presurgical evaluation of patients with intractable focal epilepsy to localize language and memory dominance (Masters et al. 2000; Perrine et al. 1995; Sperling et al. 1994). However, the IAP is employed less frequently in pediatric neurosurgical evaluations given concerns for radiation exposure and potential risk for moderate (0.3%, allergic reaction to contrast; 0.1%, bleeding from the catheter insertion site; 0.1%, infection) or severe (7.2%, encephalopathy; 1.2%, seizure; 0.6%, stroke; 0.6%, transient ischemic attack; 0.6%, hemorrhage at catheter insertion site; 0.4%, carotid artery dissections) complications (Loddenkemper et al. 2008). Moreover, behavioral management and adaptation for age-appropriate test items such that simplified testing batters for children, often coupled with reduced amobarbital dose, can compromise the IAP's effectiveness (Jayalakshmi et al. 2014), such that the IAP establishes language dominance in fewer than two-thirds of preadolescent children (Bahn et al. 1997).

Clinical fMRI is commonly employed to localize motor and language function prior to resective surgery in children with epilepsy (Krieg et al. 2022; Starnes et al. 2021). fMRI in children offers a promising alternative yielding lateralization results comparable to the gold standards, including the Wada test and cortical direct electrical stimulation (DES) (Beers and Federico 2012; Garrett et al. 2012; Jayalakshmi et al. 2014; Kwan et al. 2010; Starnes et al. 2021). Expressive language mapping tasks, such as antonym generation, and receptive/whole language tasks, such as auditory response naming, exhibit robust activation patterns in the frontal and temporal lobes in adults and older children (Szaflarski et al. 2017). However, fMRI in young children introduces its own set of challenges. Foremost, cooperation inside the MRI scanner poses a significant hurdle, as the likelihood of poor task compliance or confounding due to excessive patient motion increases in children. Such strongly implicated regions of activation require full patient cooperation, which may not always be feasible (Price et al. 2006). Motor mapping by fMRI in children is similarly compromised by issues related to patient cooperation. Performing active movement tasks without physical constraints cannot account for factors such as the pace at which movements are initiated, the force applied, and the length or duration of the movements, all of which can vary significantly among pediatric patients (Price et al. 2006).

In some cases, sedation is employed to mitigate the challenges posed by young age and behavioral dysregulation during fMRI. While sedated, the operator manually manipulates specific joints, and the resulting activation patterns during passive movement are used to map motor function. Similarly, passive language tasks are employed with sedated children, as they do not require overt patient participation and can be completed in a short period of time (approximately 7 min). However, the interpretation of the resultant passive/sedated motor and language maps, both of which possess appreciable volitional behavioral components, remains problematic in the young patient age group.

In light of these challenges, alternative approaches, such as nTMS, have gained prominence in the realm of presurgical functional mapping, particularly in pediatric populations with epilepsy. Its reliability and safety have been well-established, making it a valuable tool for improving the precision of presurgical functional mapping in pediatric patients where conventional methods may fall short.

13.4 Special Considerations for nTMS in Children

Children are not small adults. Developmental factors and cognitive limitations play a crucial role in the effectiveness of nTMS. Aspects such as head circumference, skull thickness, degree of myelination, brain development, and other factors should be carefully considered during nTMS in pediatrics. With some creativity, more customary factors, such as subject cooperation, technical equipment, and electrode placement, are surmountable determinants of nTMS success in pediatric populations (Hameed et al. 2017).

Cortical functional testing in children presents distinctive challenges. A common problem is the need to adapt adult-based nTMS accessories to children. For instance, the mechanics of chairs available are adequate for the intended use in larger patients but are suboptimal when working with a small child. Supportive cushions, tightly fitted to the chair to remain in place, may be needed to prevent the child from sliding or moving during stimulation. Booster seats are often necessary to allow for children to reach a necessary height to rest their heads in the headrests. For children who are not able to sit on their own, or for very young patients (<3 years old), a family member may sit in the chair to hold the child during the session. However, as nTMS functional mapping is performed with traditional manual handheld of the TMS coil, maintaining precise coil positioning may be difficult in such instances and requires careful attention from the TMS operator.

With some nTMS equipment, a head reflector is utilized for stereo camera tracking of head motion and is affixed around the child's head using an elastic headband. Once the nTMS registration is completed, maintaining the stability of this head tracker on the forehead throughout the stimulation is essential to ensure precise neuronavigation. However, due to the tension exerted by the elastic band and the pressure from the plastic on the forehead, children frequently report headaches during the nTMS session. To alleviate the discomfort caused by the band and the resulting headaches, a wax pen can be employed to mark the exact position of the reflectors. This allows for their temporary removal if the patient expresses discomfort and subsequent reattachment without the need for reregistering the relative positions of scalp landmarks. This marking technique serves to ensure that any reflector movement during the session remains visible to the operator and can be corrected as necessary. To avoid headache induced by the elastic headband, more modern head reflectors come affixed to the forehead using adhesive tape, and its positioning can similarly be marked with a wax pen. In some instances, the forehead of some children may be too small for adhesive subject tackers, and in the authors' experience, may be affixed to the cheek on the face. In any event, it is best practice to reperform the registration step in the event the head tracker slips or is temporarily removed.

A comparable technical challenge in pediatric nTMS is the effective attachment of surface EMG electrodes in a manner that prevents young patients from dislodging them during the session. Medical tape is often applied to minimize this likelihood, although its subsequent removal can prove uncomfortable for children, occasionally leading to transient skin erythema. Smaller EMG surface electrodes may also be necessary to adequately capture reliable MEPs in extremity muscles.

Developmental considerations and cognitive limitations of pediatric patients are factors pertinent to the success of nTMS mapping. When behavioral dysregulation poses challenges during nTMS, the team must exercise creativity to address the issue effectively. Frequently, a member of the stimulation team or a patient's family member is called upon to hold a tablet or mobile phone in front of the child, to play a favorite video. This approach helps divert the child's attention from the stimulation or encourages them to maintain proper positioning for the stereo camera during the nTMS session. In instances where patients struggle to complete a session, or when such distractions are required, it can be beneficial to divide a single mapping session into multiple visits. While this necessitates additional planning on the part of the family and additional scheduling coordination from the hospital administrators, the advantages of obtaining comprehensive motor and/or language maps for presurgical patients far outweigh any inconveniences.

13.5 nTMS Stimulation Parameters

13.5.1 Age and Motor Threshold

Single-pulse TMS is routinely used to examine corticospinal pathways and correlations with neuromotor development prior to preoperative functional mapping. TMS can quantify electrophysiological myelination maturation and synaptic efficacy, reflecting changes in cortical excitability patterns are observable throughout the course of development (Garvey et al. 2003). During maturation, ongoing myelination is a contributing factor to facilitated cortical excitability and is reflected by rMT decreasing with age (Garvey et al. 2003). As such, the success rate for activating the motor cortex is likely lowest in neonates (Eyre et al. 2001; Koh and Eyre 1988; Lin and Pascual-Leone 2002) due to high motor thresholds in this population. For instance, age is reliably negatively correlated with rMT, operationally defined as the minimum machine stimulator output (%MO) or quantified electric field strength (V/m) required to elicit MEP responses of 50 μV, on ≥50% of trials from intrinsic hand muscles, contralateral to the stimulated hemisphere (Fig. 13.1).

Although motor thresholds are relatively stable within an individual, multiple factors influence variations in motor threshold across individuals. Within-subject variations include medications, genetics, alertness, muscle or neurological disorders, and age. Motor thresholds are most often quantified while a muscle is in a resting state. Due to the high motor thresholds in young children, rMT may be difficult to obtain where a stimulus intensity of 100%MO fails to produce reliable MEPs from target muscles. Additional factors such as medications, genetics, alertness, movement disorders, and neurological illnesses may also contribute to altered motor cortex excitability in children. In the authors' experience of obtaining rMT in children aged <1 year up to young adults, patients younger than 4 years of age will likely require stimulation to be performed at 100%MO to reliably elicit MEPs from target muscles, particularly when mapping proximal upper (e.g., deltoid) and lower extremity (e.g., tibialis anterior) muscles.

Some children may have difficulty remaining staying still during nTMS sessions and may be unable to maintain a relaxed, rested state of a target muscle. In such instances, or in the event 100% MO does not elicit reliable MEPS at rest, acquiring active motor thresholds (AMT) may be suitable. In such instances, the authors instruct participants to maintain a muscle contraction in their target muscle by holding a pinch grip, squeezing the hand of their parent, or having younger children grasp a toy, providing verbal or visual feedback of the EMG. While the muscle is contracting at 10–20% of its maximum voluntary contraction (MVC), AMT is defined as the minimum stimulus intensity required to elicit MEP responses of 200 μV peak-to-peak amplitude in five of ten stimulations. During muscle contractions, the excitability of motor cortex is increased, and therefore, a lower stimulus intensity can elicit an MEP response.

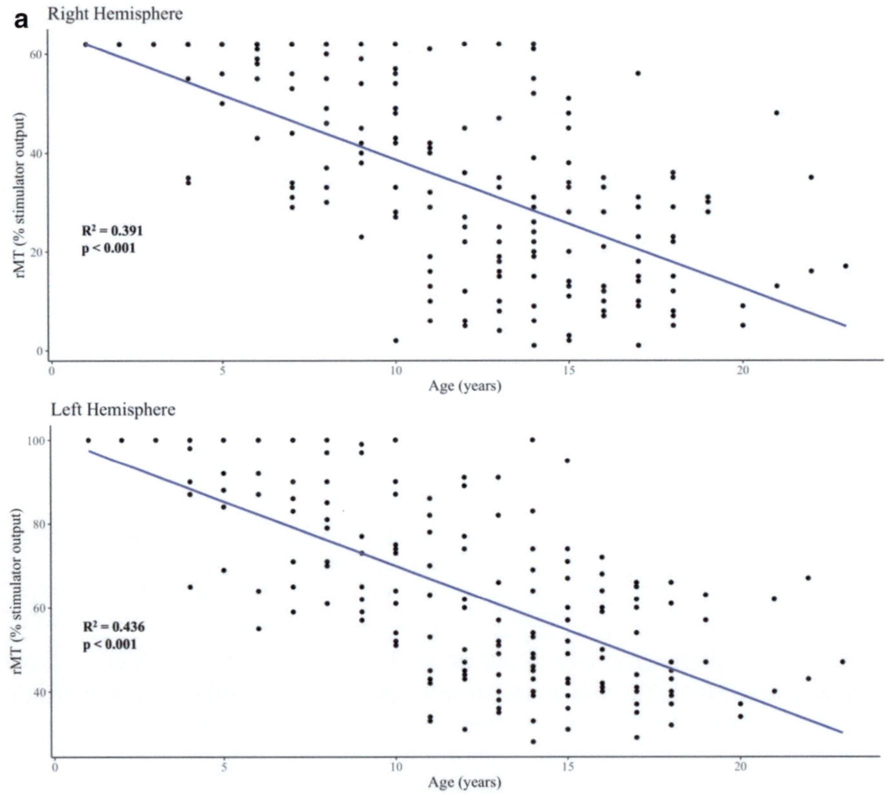

Fig. 13.1 Correlations between rMT and age. Patient data (Giuffre et al., unpublished) from children with epilepsy where age (x-axis; in years) is negatively correlated with rMT. (**a**) % stimulator output in the right hemisphere and left hemisphere and (**b**) electric field strength (V/m) in the right hemisphere and left hemisphere. *rMT = resting motor threshold*

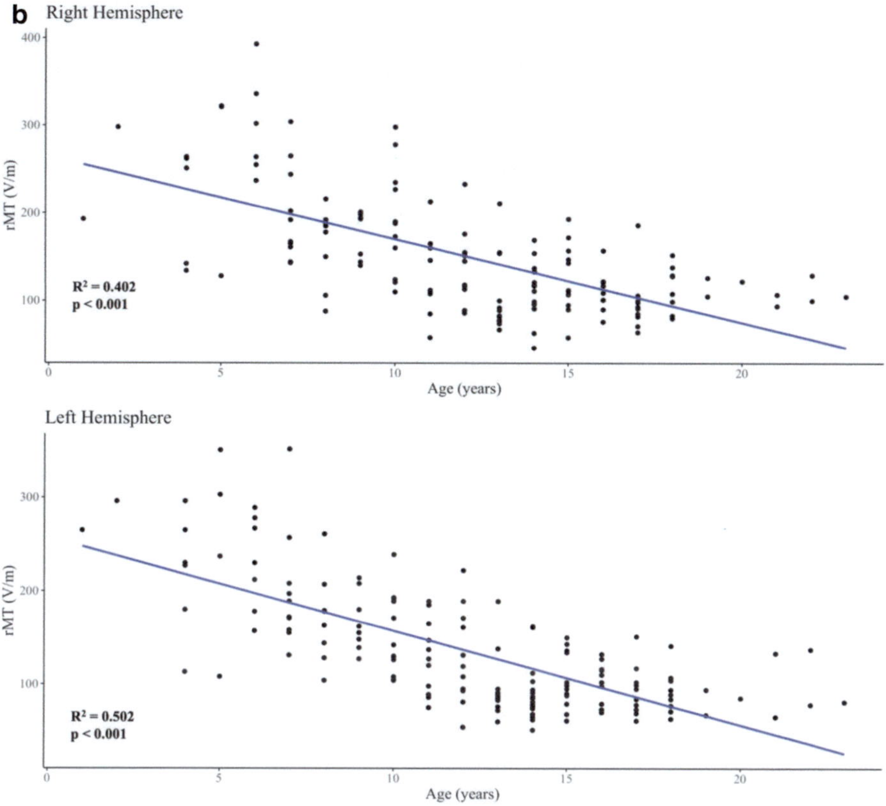

Fig. 13.1 (continued)

AMT and RMT can also be extrapolated from a single-pulse TMS paradigm to generate a stimulus response curve (SRC). At 100% MSO, the TMS intensity is decreased in 5–10% increments until no MEP are detectable in the target muscle (~30% MSO). Peak-to-peak MEP amplitude and percentage of MSO generate a SRC of which RMT can be extrapolated (Ridding and Rothwell 1997). Single-pulse SRC paradigms are useful for obtaining RMT/AMT in a relatively short time period and are indicative of neurophysiological changes (Ridding and Rothwell 2007). In healthy individuals, SRCs are generally sigmoidal, with MEP amplitudes plateauing at higher stimulation intensities. SRC fluctuations reflect changes in cortical excitability and connectivity patters. For example, higher maximum MEP, earlier plateau, or steeper rate of rise reflect increased cortical excitability.

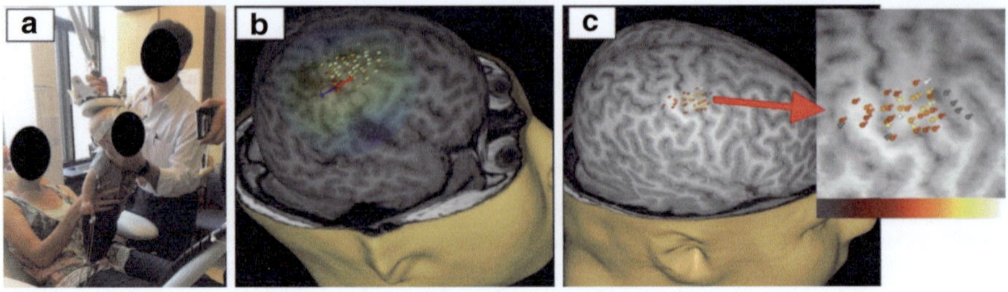

Fig. 13.2 nTMS motor mapping in children. Electrodes are placed bilaterally on the abductor pollicis brevis (APB) muscles; single-pulse TMS is applied to M1 (110%rMT) to elicit MEPs quantifiable with surface EMG (**a**). An approximation of stimulating electric field induced by a single TMS pulse is displayed on a 3D reconstruction of an individual's anatomic magnetic MRI, where field center is indicated by the junction between the red and blue arrows, indicating the direction of induced current, corresponding to electric field strength (V/m) at the stimulation site (**b**). Examples of the composite map of right hemispheric stimulation sites evoking MEPs (> 50 μV) of the left APB muscle, where intensity of response is color-coded from lowest (gray) to highest (white), are displayed on cortical surface rendering motor map reflective of responsive cortical sites (**c**)

13.5.2 nTMS for Motor Mapping

nTMS (and robotic nTMS) can safely produce motor maps (Giuffre et al. 2019, 2021; Grab et al. 2018) in children with neurological disorders such as perinatal stroke (Kirton et al. 2021; Zewdie et al. 2020). More recent applications in children demonstrate the feasibility and clinical utility for nTMS motor mapping prior to resective epilepsy surgery (Krieg et al. 2022; Narayana et al. 2021; Rejnö-Habte Selassie et al. 2018; Rosenstock et al. 2020; Schramm et al. 2021; Starnes et al. 2021; Vitikainen et al. 2013). Preoperative nTMS motor mapping correlates well with intraoperative DCS with protocols established in adults and healthy subjects (Krieg et al. 2017) and offers the possibility to adapt and optimize the surgical approach (Jeltema et al. 2021; Lefaucheur and Picht 2016; Picht et al. 2013)—though refinement of nTMS protocols is necessary for young children. An example of nTMS motor mapping is shown in Figs. 13.2 and 13.3. Overall, nTMS is well-received, with the most common side effect reported as headache.

13.5.3 Navigated Repetitive TMS (nrTMS) for Language Mapping

nrTMS offers an alternative to fMRI as a more flexible noninvasive functional mapping method and requires specialized considerations in children. In the authors' clinical program, the NexSpeech® software (Nexstim Plc., Helsinki, Finland) is used for functional mapping of expressive language as part of routine presurgical evaluation for children with refractory focal epilepsy (Pasichnik et al. 2022) that many other pediatric centers are implementing (Lehtinen et al. 2018; Narayana et al. 2021). Children perform an object-naming task, where a series of black and white images are presented visually at baseline for time-limited blocks. The images correctly identified are then represented coupled

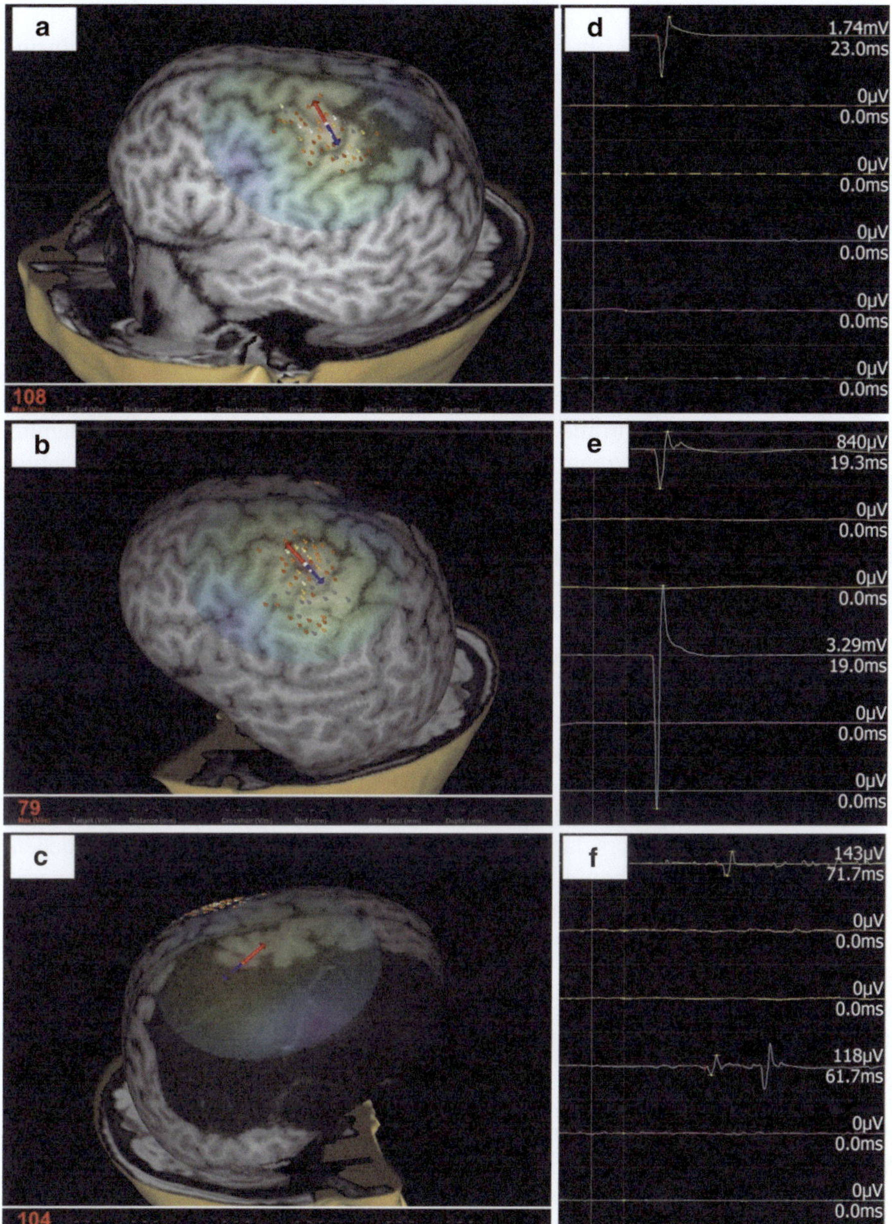

Fig. 13.3 Clinical nTMS motor cortex activation in children. A computed stimulating electric field induced by a single TMS pulse is displayed on a 3D patient MRI reconstruction refractory focal epilepsy with (**a**) stimulus delivered near the seizure onset zone. In a second patient with perinatal stroke, the stimulus is delivered to the unaffected hemisphere (**b**) and to the affected hemisphere (**c**). Representative right abductor pollicis brevis (APB) muscle sample (green deflection) showing right APB resultant from left hemispheric stimulation (**d** and **e**) and ipsilateral right hemispheric stimulation (**f**), where the vertical line (white) corresponds to stimulus time. MEP amplitude and latency are indicated on the right. Note bilateral MEP activation with stimulus to the intact hemisphere. *Tracing: top (green) = right APB; fourth from to (blue) = left APB*

with repetitive stimulation typically delivered in 1-sec 5 Hz trains. rTMS is applied to the vertex to obtain a secondary baseline of object-naming during stimulation, prior to targeting the language centers, such as the pars triangularis, approximating Broca's region and its homolog in both hemispheres. Language mapping is performed at a starting point of 120% rMT to the maximal stimulator output and may be lowered depending on tolerability. The outcome measures during the nrTMS mapping protocol are manually reviewed and scored as correct (no error), complete speech arrest (no response), or paraphasic error (semantic error). Lateralized expressive language is characterized by exclusive presence of either error type in one hemisphere only.

Establishing an accurate baseline of subject's expressive language is crucial and requires an appreciable flexibility in the object-naming settings to accommodate the range of development across pediatric populations. Most often, a long latency in verbal response relative to the timing of the visual stimulus in the object-naming task requires either an increase of the interval between successive images or a prolongation of the stimulation train. The relatively slow verbal response of pediatric patients to presented visual stimuli also modifies the post hoc analysis such that delays in response are difficult to interpret, and in the authors' practice, only speech arrest or paraphasic errors are scored to generate the pediatric language map in young children.

Language mapping in children may be limited by the relatively high rMT in children which obligates stimulation intensities beyond a threshold patients can tolerate and can lead to false-positive responses. Compared to temporal lobe stimulation, stimulation of inferior frontal lobes is often less tolerable due to peripheral activation of the facial muscles. The stimulus intensity is often lowered from 120% rMT according to the subject's tolerability; however, this may lower the success rate for observing speech errors during TMS. TMS intensities of 70–100 V/m independent of individual rMT should be carefully considered to successfully elicit speech and language errors while minimizing the discomfort associated with high intensities.

13.6 Special Population: Children with Epilepsy

As previously mentioned, a significant portion of children who stand to benefit from resective surgery and antecedent nTMS functional mapping are patients with refractory epilepsy. The majority of patients with intractable seizures are often on one or more antiepileptic drugs, which tend to alter rMT. In the authors' patient population, voltage-gated sodium channel blockers such as oxcarbazepine, lamotrigine, and lacosamide contribute to increased rMT in our patients (Paulus et al. 2008; Tsuboyama et al. 2020; Tsuboyama et al. 2019). While the effects of many prevalent central nervous system medications on TMS have been well-documented (Paulus et al. 2008), the precise interaction of these agents with developmentally regulated physiological processes that influence TMS outcome measures remains to be elucidated.

13.7 nTMS Safety in Children

TMS (navigated or otherwise) has well-documented safety in pediatric populations. Further investigations in the developing brain are required to increase translation toward clinical applications. Particularly, repetitive stimulation (i.e., nrTMS) in language mapping raises concerns for seizure induction, which may be of greater significance in children due to their potentially lower seizure thresholds compared to adults (Pohlmann-Eden et al. 2006). Fortunately, the safety profile of TMS in children is encouraging, extending to healthy children and children with other neurological disorders (Allen et al. 2017; Zewdie et al. 2020). In the authors' population of children with epilepsy, a population that is by definition seizure-prone, only a small percentage (2–3% of over 200 patients) experienced seizures during nrTMS language mapping. However, all observed seizures were the patients' habitual seizures, and it remains unclear whether the seizures were causally induced by nTMS/nrTMS or occurred coincidentally during the mapping sessions. Encouragingly, no incidents of atypical seizures or status epilepticus have been reported in any child undergoing nTMS, underscoring the overall safety of nTMS in this vulnerable population. Nevertheless, ongoing vigilance and close monitoring during TMS in children, particularly those with a history of epilepsy, remain essential to ensure their safety and well-being.

13.8 Conclusion

Neocortical resective surgery is as a curative option for children with intractable focal epilepsy or brain tumors, and accurate localization of the eloquent cortex in these young patients is critical to prevent functional deficits. nTMS appears strikingly safe in children and overcomes behavioral and physiologic limitations that are inherent in other mapping techniques. Particularly in pediatric focal epilepsy surgery evaluations, preoperative nTMS testing offers functional mapping that is otherwise unavailable in this fragile population.

References

Aaberg KM, Gunnes N, Bakken IJ, Lund Søraas C, Berntsen A, Magnus P, et al. Incidence and prevalence of childhood epilepsy: a Nationwide cohort study. Pediatrics. 2017;139(5):e20163908.

Allen CH, Kluger BM, Buard I. Safety of transcranial magnetic stimulation in children: a systematic review of the literature. Pediatr Neurol. 2017 Mar;68:3–17.

Bahn MM, Lin W, Silbergeld DL, Miller JW, Kuppusamy K, Cook RJ, et al. Localization of language cortices by functional MR imaging compared with intracarotid amobarbital hemispheric sedation. AJR Am J Roentgenol. 1997 Aug;169(2):575–9.

Bartolini E, Ferrari AR, Lattanzi S, Pradella S, Zaccara G. Drug-resistant epilepsy at the age extremes: disentangling the underlying etiology. Epilepsy Behav. 2022 Jul;132:108739.

Beers CA, Federico P. Functional MRI applications in epilepsy surgery. Can J Neurol Sci. 2012 May;39(3):271–85.

Berg AT, Zelko FA, Levy SR, Testa FM. Age at onset of epilepsy, pharmacoresistance, and cognitive outcomes: a prospective cohort study. Neurology. 2012 Sep 25;79(13):1384–91.
Clusmann H, Kral T, Gleissner U, Sassen R, Urbach H, Blümcke I, et al. Analysis of different types of resection for pediatric patients with temporal lobe epilepsy. Neurosurgery. 2004;54(4):847–59. discussion 859-860
Dorfmüller G, Ferrand-Sorbets S, Fohlen M, Bulteau C, Archambaud F, Delalande O, et al. Outcome of surgery in children with focal cortical dysplasia younger than 5 years explored by stereo-electroencephalography. Childs Nerv Syst. 2014 Nov;30(11):1875–83.
Duchowny MS. Surgery for intractable epilepsy: issues and outcome. Pediatrics. 1989 Nov;84(5):886–94.
Eyre JA, Taylor JP, Villagra F, Smith M, Miller S. Evidence of activity-dependent withdrawal of corticospinal projections during human development. Neurology. 2001 Nov 13;57(9):1543–54.
Garrett MC, Pouratian N, Liau LM. Use of language mapping to aid in resection of gliomas in eloquent brain regions. Neurosurg Clin N Am. 2012 Jul;23(3):497–506.
Garvey MA, Ziemann U, Bartko JJ, Denckla MB, Barker CA, Wassermann EM. Cortical correlates of neuromotor development in healthy children. ClinNeurophysiol. 2003 Sep;114:1662–70.
Giuffre A, Cole L, Kuo H-C, Carlson HL, Grab J, Kirton A, et al. Non-invasive modulation and robotic mapping of motor cortex in the developing brain. JoVE (J Vis Exp). 2019 Jul 1;149:e59594.
Giuffre A, Zewdie E, Carlson HL, Wrightson JG, Kuo H-C, Cole L, et al. Robotic transcranial magnetic stimulation motor maps and hand function in adolescents. Physiol Rep. 2021;9(7):e14801.
Grab JG, Zewdie E, Carlson HL, Kuo H-C, Ciechanski P, Hodge J, et al. Robotic TMS mapping of motor cortex in the developing brain. J Neurosci Methods. 2018 Nov;1(309):41–54.
Hameed MQ, Dhamne SC, Gersner R, Kaye HL, Oberman LM, Pascual-Leone A, et al. Transcranial magnetic and direct current stimulation in children. Curr Neurol Neurosci Rep. 2017 Feb;17(2):11.
Jayalakshmi S, Panigrahi M, Nanda SK, Vadapalli R. Surgery for childhood epilepsy. Ann Indian Acad Neurol. 2014 Mar;17(Suppl 1):S69–79.
Jeltema H-R, Ohlerth A-K, de Wit A, Wagemakers M, Rofes A, Bastiaanse R, et al. Comparing navigated transcranial magnetic stimulation mapping and "gold standard" direct cortical stimulation mapping in neurosurgery: a systematic review. Neurosurg Rev. 2021 Aug;44(4):1903–20.
Kirton A, Metzler MJ, Craig BT, Hilderley A, Dunbar M, Giuffre A, et al. Perinatal stroke: mapping and modulating developmental plasticity. Nat Rev Neurol. 2021 Jul;17(7):415–32.
Koh TH, Eyre JA. Maturation of corticospinal tracts assessed by electromagnetic stimulation of the motor cortex. Arch Dis Child. 1988 Nov;63(11):1347–52.
Krieg SM, Lioumis P, Mäkelä JP, Wilenius J, Karhu J, Hannula H, et al. Protocol for motor and language mapping by navigated TMS in patients and healthy volunteers; workshop report. Acta Neurochir. 2017 Jul;159(7):1187–95.
Krieg SM, Bernhard D, Ille S, Meyer B, Combs S, Rotenberg A, et al. Neurosurgery for eloquent lesions in children: state-of-the-art rationale and technical implications of perioperative neurophysiology. Neurosurg Focus. 2022 Dec;53(6):E4.
Kwan A, Ng WH, Otsubo H, Ochi A, Snead OC, Tamber MS, et al. Hemispherectomy for the control of intractable epilepsy in childhood: comparison of 2 surgical techniques in a single institution. Neurosurgery. 2010 Dec;67(2 Suppl Operative):429–36.
Lefaucheur J-P, Picht T. The value of preoperative functional cortical mapping using navigated TMS. Neurophysiol Clin. 2016 Apr;46(2):125–33.
Lehtinen H, Mäkelä JP, Mäkelä T, Lioumis P, Metsähonkala L, Hokkanen L, et al. Language mapping with navigated transcranial magnetic stimulation in pediatric and adult patients undergoing epilepsy surgery: comparison with extraoperative direct cortical stimulation. Epilepsia Open. 2018 Jun;3(2):224–35.
Liava A, Francione S, Tassi L, Lo Russo G, Cossu M, Mai R, et al. Individually tailored extratemporal epilepsy surgery in children: anatomo-electro-clinical features and outcome predictors in a population of 53 cases. Epilepsy Behav. 2012 Sep 1;25(1):68–80.

Lin K-L, Pascual-Leone A. Transcranial magnetic stimulation and its applications in children. Chang Gung Med J. 2002 Jul;25(7):424–36.

Loddenkemper T, Morris HH, Möddel G. Complications during the Wada test. Epilepsy Behav. 2008 Oct;13(3):551–3.

Löscher W. Animal models of seizures and epilepsy: past, present, and future role for the discovery of Antiseizure drugs. Neurochem Res. 2017 Jul;42(7):1873–88.

Masters LT, Perrine K, Devinsky O, Nelson PK. Wada testing in pediatric patients by use of Propofol anesthesia. AJNR Am J Neuroradiol. 2000 Aug;21(7):1302–5.

Narayana S, Gibbs SK, Fulton SP, McGregor AL, Mudigoudar B, Weatherspoon SE, et al. Clinical utility of transcranial magnetic stimulation (TMS) in the Presurgical evaluation of motor, speech, and language functions in young children with refractory epilepsy or brain tumor: preliminary evidence. Front Neurol. 2021;12:650830.

Nickels KC, Wong-Kisiel LC, Moseley BD, Wirrell EC. Temporal lobe epilepsy in children. Epilepsy Res Treat. 2012;2012:849540.

Pasichnik A, Tsuboyama M, Jannati A, Vega C, Kaye HL, Damar U, et al. Discrepant expressive language lateralization in children and adolescents with epilepsy. Ann Clin Transl Neurol. 2022 Sep;9(9):1459–64.

Paulus W, Classen J, Cohen LG, Large CH, Di Lazzaro V, Nitsche M, et al. State of the art: pharmacologic effects on cortical excitability measures tested by transcranial magnetic stimulation. Brain Stimul. 2008 Jul;1(3):151–63.

Perrine K, Westerveld M, Sass KJ, Devinsky O, Dogali M, Spencer DD, et al. Wada memory disparities predict seizure laterality and postoperative seizure control. Epilepsia. 1995 Sep;36(9):851–6.

Picht T, Krieg SM, Sollmann N, Rösler J, Niraula B, Neuvonen T, et al. A comparison of language mapping by preoperative navigated transcranial magnetic stimulation and direct cortical stimulation during awake surgery. Neurosurgery. 2013 May;72(5):808–19.

Pohlmann-Eden B, Beghi E, Camfield C, Camfield P. The first seizure and its management in adults and children. BMJ. 2006 Feb 11;332(7537):339–42.

Price CJ, Crinion J, Friston KJ. Design and analysis of fMRI studies with neurologically impaired patients. J Magn Reson Imaging. 2006 Jun;23(6):816–26.

Rejnö-Habte Selassie G, Pegenius G, Viggedal G, Hallböök T, Thordstein M. Navigated transcranial magnetic stimulation for preoperative cortical mapping of expressive language in children: development of a method. Epilepsy Behav. 2018 Oct;87:180–7.

Ridding MC, Rothwell JC. Stimulus/response curves as a method of measuring motor cortical excitability in man. Electroencephalogr Clin Neurophysiol Electromyogr Mot Control. 1997 Oct;105(5):340–4.

Ridding MC, Rothwell JC. Is there a future for therapeutic use of transcranial magnetic stimulation? Nat Rev Neurosci. 2007 Jul;8(7):559–67.

Rosenstock T, Picht T, Schneider H, Vajkoczy P, Thomale U-W. Pediatric navigated transcranial magnetic stimulation motor and language mapping combined with diffusion tensor imaging tractography: clinical experience. J Neurosurg Pediatr. 2020 Jul 24;26(5):583–93.

Schramm S, Mehta A, Auguste KI, Tarapore PE. Navigated transcranial magnetic stimulation mapping of the motor cortex for preoperative diagnostics in pediatric epilepsy. J Neurosurg Pediatr. 2021 Jun 25;28(3):287–94.

Sperling MR, Saykin AJ, Glosser G, Moran M, French JA, Brooks M, et al. Predictors of outcome after anterior temporal lobectomy: the intracarotid amobarbital test. Neurology. 1994 Dec;44(12):2325–30.

Starnes K, Depositario-Cabacar D, Wong-Kisiel L. Presurgical evaluation strategies for intractable epilepsy of childhood. Semin Pediatr Neurol. 2021 Oct;39:100915.

Sultana B, Panzini M-A, Veilleux Carpentier A, Comtois J, Rioux B, Gore G, et al. Incidence and prevalence of drug-resistant epilepsy: a systematic review and meta-analysis. Neurology. 2021 Apr 27;96(17):805–17.

Szaflarski JP, Gloss D, Binder JR, Gaillard WD, Golby AJ, Holland SK, et al. Practice guideline summary: use of fMRI in the presurgical evaluation of patients with epilepsy: report of the guideline development, dissemination, and implementation Subcommittee of the American Academy of neurology. Neurology. 2017 Jan 24;88(4):395–402.

Tsuboyama M, Lee Kaye H, Rotenberg A. Biomarkers obtained by transcranial magnetic stimulation of the motor cortex in epilepsy. Front Integr Neurosci. 2019;13:57.

Tsuboyama M, Kaye HL, Rotenberg A. Review of transcranial magnetic stimulation in epilepsy. Clin Ther. 2020 Jul;42(7):1155–68.

Vitikainen A-M, Salli E, Lioumis P, Mäkelä JP, Metsähonkala L. Applicability of nTMS in locating the motor cortical representation areas in patients with epilepsy. Acta Neurochir. 2013 Mar;155(3):507–18.

Weiner HL, Ferraris N, LaJoie J, Miles D, Devinsky O. Epilepsy surgery for children with tuberous sclerosis complex. J Child Neurol. 2004 Sep;19(9):687–9.

Widjaja E, Jain P, Demoe L, Guttmann A, Tomlinson G, Sander B. Seizure outcome of pediatric epilepsy surgery: systematic review and meta-analyses. Neurology. 2020 Feb 18;94(7):311–21.

Wirrell EC. Predicting pharmacoresistance in pediatric epilepsy. Epilepsia. 2013 May;54(Suppl 2):19–22.

Zewdie E, Ciechanski P, Kuo HC, Giuffre A, Kahl C, King R, et al. Safety and tolerability of transcranial magnetic and direct current stimulation in children: prospective single center evidence from 3.5 million stimulations. Brain Stimul. 2020 May 1;13(3):565–75.

nTMS in Epilepsy: Functional Mapping and Therapeutic Applications

14

Shalini Narayana, Negar Noorizadeh, and James W. Wheless

14.1 Introduction

Epilepsy is one of the most common chronic neurological disorders and is associated with significant morbidity and nearly three times increased risk of mortality. Worldwide, approximately 50 million individuals suffer from epilepsy (www.who.int/news-room/fact-sheets/detail/epilepsy). Nearly 3.4 million living in the United States have been diagnosed with epilepsy, and children make up about 12% of this group (www.cdc.gov). While the majority of individuals with epilepsy become seizure free with treatment with appropriate antiseizure medication (ASM), about 30% of patients continue to have seizures and develop refractory or pharmacoresistant epilepsy. For many of these individuals, and in particular those with focal seizures,

S. Narayana (✉)
Division of Pediatric Neurology, Department of Pediatrics, The University of Tennessee Health Science Center, Memphis, TN, USA

Department Anatomy and Neurobiology, The University of Tennessee Health Science Center, Memphis, TN, USA

Transcranial Magnetic Stimulation Laboratory, Neuroscience Institute, Le Bonheur Children's Hospital, Memphis, TN, USA
e-mail: snaraya2@uthsc.edu

N. Noorizadeh
Division of Pediatric Neurology, Department of Pediatrics, The University of Tennessee Health Science Center, Memphis, TN, USA

Transcranial Magnetic Stimulation Laboratory, Neuroscience Institute, Le Bonheur Children's Hospital, Memphis, TN, USA

J. W. Wheless
Division of Pediatric Neurology, Department of Pediatrics, The University of Tennessee Health Science Center, Memphis, TN, USA

Neuroscience Institute, Le Bonheur Children's Hospital, Memphis, TN, USA

S. M. Krieg, T. Picht (eds.), *Navigated Transcranial Magnetic Stimulation in Neurosurgery*, https://doi.org/10.1007/978-3-031-97155-6_14

epilepsy surgery has been shown to be an effective option providing good seizure control and improved quality of life (Barba et al. 2021, 2022; Dwivedi et al. 2017; Engel et al. 2013; Spencer et al. 2007; Wiebe et al. 2001). The presurgical evaluation therefore includes accurate localization of the ictal onset zone and precise identification of eloquent cortex in its vicinity. An accurate evaluation is critical in assisting surgical planning and minimizing postoperative deficits while maximizing the resection of the seizure focus. The anatomical lesions causing the seizures can be identified by anatomical magnetic resonance imaging (MRI), and several noninvasive methods are available to approximate seizure onset zone including electroencephalogram (EEG), video-EEG, high-density EEG, single-photon emission computed tomography (SPECT), positron emission tomography (PET), and magnetoencephalography (MEG), often used in combination (Sebastiano et al. 2020). The ictal onset zone is often confirmed by invasive stereo-EEG (sEEG) or subdural grid electrodes.

14.1.1 Functional Mapping with Navigated Transcranial Magnetic Stimulation

Both invasive and noninvasive modalities are available to localize eloquent function, also termed functional mapping. Electrical cortical stimulation (ESM) has long been considered the invasive "gold standard" for localizing motor, somatosensory, speech, and language areas and the intracarotid sodium amobarbital (Wada) procedure as the benchmark for lateralizing language-related cortical areas. However, the realization of the limitations of these invasive procedures (Jayakar et al. 2018; Papanicolaou et al. 2018) has led to the development of safer noninvasive alternatives. These include MEG, functional MRI (fMRI), PET, and more recently, navigated transcranial magnetic stimulation (nTMS). Despite other mapping modalities having wider acceptance, the risks of invasive procedures and limitations of other noninvasive methods mean that successful functional mapping remains a significant challenge in critical groups of patients with intractable epilepsy. Children make up one such group (de Ribaupierre et al. 2012; Jayakar 2018; Wang et al. 2018), in whom ESM risks are significant (i.e., after discharges provoked seizures due to higher amperage motor threshold (Chou et al. 2018)) with only a 50% success rate in children under 10 years (Krieg et al. 2022; Schevon et al. 2007) and both MEG and fMRI are noninformative in ~30% (Birg et al. 2013; Yerys et al. 2009). The lack of comprehensive sensorimotor and language mapping for these patients frequently prevents timely surgery that could significantly improve cognitive function (Braun 2020; Skirrow et al. 2019) and quality of life (Engel et al. 2013) or surgery proceeds without a good functional map, resulting in postoperative motor or language deficits (Sherman et al. 2011). Additionally, noninvasive imaging methods are often unsuccessful even in older children and adults who have cranial metal, claustrophobia, cannot remain still, or cannot comply with testing demands. Accurate localization of motor, somatosensory, speech, and language areas in MEG is challenging in patients who have frequent interictal epileptiform discharges in these areas. The merits of MEG, fMRI, and nTMS are discussed in detail in the chapter titled "nTMS, MEG, and fMRI: Comparing and Contrasting Three Functional Mapping Techniques" by Tarapore and Nagarajan. For all of these patient groups, in whom other methods are not successful, nTMS is especially promising as it aligns closely with the neurophysiological mechanism of ESM, is safe, well-tolerated, relatively inexpensive, and readily performed in even young children (Narayana et al. 2015a, b, 2021).

14.1.2 Clinical Population

Although temporal lobe epilepsy (TLE) is overall the most common type of refractory seizure referred for epilepsy surgery, the causes of refractory seizures treatable by surgery differ between children and adults. The most common childhood epilepsies that undergo surgical evaluation result from injury or insult to the brain (perinatal injury/stroke), infection, abnormalities in brain development (cortical dysplasia), inflammation (Rasmussen syndrome), or genetic syndromes (tuberous sclerosis complex (TSC)) (Çataltepe and Jallo 2010). An extratemporal seizure focus is more common in children (Englot et al. 2013). Epilepsy related to brain tumor is an important etiology in both children and adults. In adults, the other causes of intractable epilepsy include traumatic brain injury and stroke. The most common pathology noted in adults who undergo epilepsy surgery is hippocampal sclerosis, seen in approximately 45%, while focal cortical dysplasia is the most common finding observed in children, seen approximately 40% of the time (Baumgartner et al. 2019).

Of the nearly 800 patients that have undergone nTMS mapping at our institution, 80% were 18 years or younger with a mean age of 13 ± 8.6 years. Eighty-five percent of patients had a diagnosis of refractory seizures, and the remaining patients were evaluated for a newly diagnosed brain tumor. The main causes of their seizures included nonlesional epilepsy (39%), cortical dysplasia (19%), previous history of brain tumor (11%), perinatal stroke/hemorrhage (10%), and tuberous sclerosis (7%). Almost all patients underwent motor mapping of upper extremity (95%), while 78% underwent lower extremity mapping and 62% underwent speech and language mapping.

14.2 TMS Motor Mapping in Epilepsy

While nTMS motor mapping has been integrated into the presurgical evaluation of patients with brain tumors (Haddad et al. 2021; Jung et al. 2019; Krieg et al. 2017; Lefaucheur and Picht 2016), its application in epilepsy surgery is still in its early stages. Few centers, including ours, have demonstrated the feasibility of nTMS motor mapping in this group, showing its ability to provide valuable information and aid in surgical planning (Narayana et al. 2015a; b, 2021; Säisänen et al. 2010; Schramm et al. 2021; Vitikainen et al. 2009, 2013).

14.2.1 TMS Motor Mapping Procedure

The motor mapping procedure in patients with epilepsy is similar to the recommended protocol for adult brain tumor patients and healthy volunteers (Krieg et al. 2017). A navigated TMS system is used to register the TMS coil to the patient's MRI to visualize the stimulation parameters and sites in real time and display the results of the mapping study. Routinely, cortical representations of small muscles of the hand (abductor pollicis brevis or first dorsal interosseus), forearm (brachioradialis), and leg (tibialis anterior) are mapped. Other muscles including orbicularis oris, biceps brachii, deltoid, or abductor hallucis can be mapped if the seizure focus is thought to be in these areas as indicated by other imaging methods or seizure semiology. The process recommended by Krieg and colleagues (Krieg et al. 2017) is followed for localizing the hot spot, determining the resting motor threshold (rMT), and mapping extent of the motor cortex in both hemispheres. Figure 14.1 illustrates examples of nTMS motor mapping in patients with the epileptogenic focus in its vicinity.

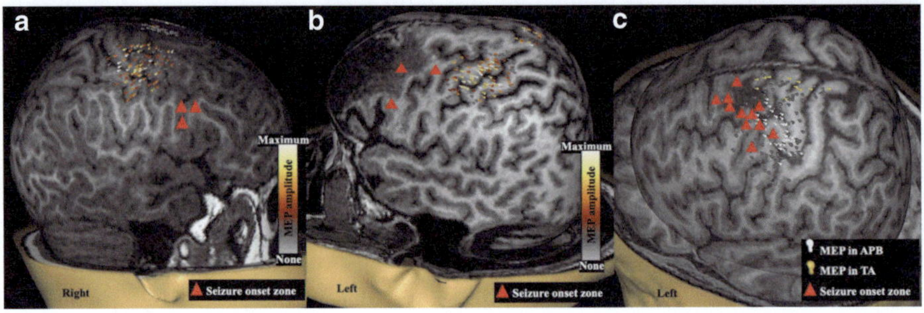

Fig. 14.1 Examples of nTMS motor mapping in patients with refractory epilepsy. The nTMS-derived motor cortex is shown as a heat map of MEP amplitude (patients **a** and **b**) or as as MEPs in individual muscles of upper and lower extremities (patient **c**). In patients **a** and **b**, *White pegs* represent brain areas where nTMS resulted in MEPs with the greatest amplitude, and *gray pegs* indicate no MEP response, while *yellow, orange, and red pegs* represent intermediate MEP amplitudes (patients **a** and **b**). The seizure onset zones are shown as *solid red triangles*. (**a**) A 5-year-old female with history of intractable epilepsy secondary to TSC. The seizure onset zone was localized to the right frontal region and motor mapping by nTMS demonstrated the hand motor cortex to be normally located in the precentral gyrus and posterior to the seizure focus. (**b**) A 36-year-old male with history of seizures since age 9, status post left frontal pole resection, corpus callosotomy, and VNS, presented with ongoing refractory seizures. His seizure foci were localized along the posterior margins of the resection cavity. nTMS localized his motor cortex to be over the precentral gyrus and posterior to the seizure onset zone. The patient underwent left frontal polar lobectomy and post operatively did not experience any right upper extremity motor deficits. (**c**) A 31-year-old female with history of cerebral cavernous malformation over the left medial frontal cortex, status post resection, presented with ongoing seizures. Seizure onset zone was localized around the surgical cavity, while her hand and leg motor cortices were localized along the posterior margin of the lesion. Abbreviations: nTMS: navigated transcranial magnetic stimulation, TSC: tuberous sclerosis complex, MEP: motor evoked potential, APB: abductor pollicis brevis, TA: tibialis anterior, VNS: vagus nerve stimulator

14.2.2 Procedures Unique to Epilepsy Cohort

At most centers, including ours, the epilepsy cohort undergoing nTMS motor mapping include children and adults, many of whom have comorbidities such as developmental delay, attention disorders, anxiety, etc. In these individuals where assessment of rMT may not be feasible or in those whose rMT is at 100% of maximum stimulator output (MSO), active MT measurement (aMT) can be a useful alternative. Furthermore, if aMT cannot be measured, then the extent of the motor cortex can still be mapped using a nTMS intensity that results in motor evoked potentials (MEPs) > 50 μV amplitude. In instances where nTMS fails to elicit a clear MEP in the target muscles, as in very young children and those with aMT >100% MSO, we utilize the cortical silent periods (cSPs), a short period of electromyography suppression observed in active muscles, to approximate the location of the motor cortex (Narayana et al. 2015b, 2021). Examples of nTMS-elicited cSP in young children are shown in Fig. 14.2. The cSP duration reflects the gamma-aminobutyric acid (GABA)–mediated motor-cortex inhibition, and the nTMS intensity required to elicit cSP is lower than that required to elicit an MEP. There is clear evidence that the cSP is generated predominantly within the motor cortex (Wolters et al. 2011), although cSP-only maps were more lateral when compared to maps where cSP was preceded by a MEP (Wassermann et al. 1993). Additional details of considerations of nTMS motor mapping in pediatric epilepsy patients are addressed in the chapter titled: “nTMS in Pediatrics: Special Issues and Solutions” by Giuffre and Rotenberg.

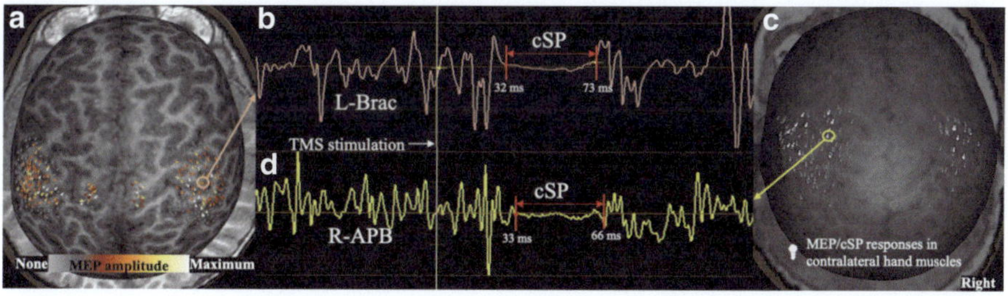

Fig. 14.2 Examples of cSP elicited by nTMS in young children with refractory epilepsy. (**a**, **b**) A 4.6-year-old male with history of right subependymal heterotopia and developmental delay. He could not maintain a quiet EMG baseline and nTMS applied over the central sulcus in the right hemisphere elicited cSPs with and without associated MEPs in the contralateral facilitated Brac muscle. nTMS-derived motor cortex is shown as a *heat map* of MEP amplitude. An example of cSP without an associated MEP is shown in panel **b** with its cortical location indicated in panel **a**. The cSP onset is identified as the time when the EMG activity reduces to baseline, which occurs later than the MEP onset. In this example, the cSP onset was at 32 ms and its duration was 41 ms. (**c**, **d**) An 18-month-old female with history of left hemisphere perinatal stroke involving left temporal lobe and subsequent infantile spasms. The motor areas identified by nTMS are shown as *white pegs*. Panel **d** demonstrated an example of cSP without an associated MEP recorded in the right APB muscle following nTMS to the left pre central gyrus with its cortical location indicated in panel **c**. The cSP onset was at 33 ms with a duration of 33 ms. Abbreviations: nTMS: navigated transcranial magnetic stimulation, cSP: cortical silent period, MEP: motor evoked potential, APB: abductor pollicis brevis, Brac: brachioradialis, EMG: electromyography, L: left, R: right

Not infrequently, the epilepsy cohort requiring mapping includes individuals with epilepsia partialis continua (EPC). At our institution, patients with EPCs secondary to Rassmusen's encephalitis, tumor, and refractory epilepsy have been referred for nTMS motor mapping. In all these patients, motor mapping of upper extremities, lower extremities, and lips and mouth was crucial for determining its spatial relationship to the seizure focus as well as for surgical planning. Because of the ongoing motor seizures on electromyography (EMG), the frequent and high amplitude contractions in the affected muscles may appear to have similar shape, amplitude, and timing as a nTMS-evoked MEP, making it difficult to distinguish them apart. Therefore, the nTMS motor map may not represent the complete extent of the motor cortex. The safety of nTMS mapping in patients with EPC and other serious seizure types is further discussed in Sect. 14.4 under safety of TMS.

14.2.3 Parameters Unique to Epilepsy Cohort

Differing from patients with brain tumor, those with refractory seizures are on often two or more ASMs. Therefore, it is expected that rMT in this cohort would be higher than that observed in brain tumor group. Although rMT in patients with refractory seizures taking ASM is higher than the nonepilepsy control group, the rMT levels among groups with different degrees of seizure control are not that clear cut. Both lower (Badawy et al. 2010, 2013) and higher (Pawley et al. 2017) rMTs have been reported in patients with refractory seizures when compared to those whose seizures are well-controlled. Furthermore, GABA agonist ASMs including lorazepam and vigabatrin have shown to have no influence on rMT (Ziemann 2004). While increases in rMT have been observed in healthy individuals after receiving ASMs that block the voltage-gated Na + channels such as oxcarbazepine, carbamazepine, lamotrigine, and lacosamide (Lang et al. 2013; Paulus et al. 2008) and in children with new-onset epilepsy after treatment with sodium valproate (Andreasson et al. 2020), chronic administration of these drugs in patients with intractable epilepsy has shown variable or no effect on the rMT (Badawy et al. 2013; Tsuboyama et al. 2020). In our center, the majority of patients tested do have intractable epilepsy, and the three most commonly prescribed ASMs are levetiracetam, lacosamide, and oxcarbazepine. We believe that these ASMs do not adversely affect rMT measurements. However, in this patient group, the timing of nTMS mapping in relationship to ASM tapering and a seizure event is important, as tapering ASM has been reported to decrease rMT, as seen in postictal measurements in those who had focal seizures with impaired awareness (Helling et al. 2022). However, rMT has been observed to be stable in patients who had bilateral tonic–clonic seizures prior to nTMS mapping (Helling et al. 2022). It is therefore recommended that each patient be screened for type of seizure, ASMs, and time of last seizure prior to nTMS mapping (Braden et al. 2022).

In addition to localizing motor cortices, nTMS motor mapping in epilepsy patients may shed light on the underlying pathophysiology and thereby assist in lateralizing the seizure focus particularly in those with nonlesional epilepsy.

For instance, prior to treatment initiation, the hemisphere with seizure focus or the epileptic hemisphere was found to have higher rMT compared to contralateral hemisphere in adults with focal epilepsy. However, these differences do not appear to persist in patients who become refractory to ASM as observed in a 3-year follow-up study (Badawy et al. 2010, 2013). This is consistent with our experience in our predominantly pediatric cohort. We find about half the group had higher rMT in the epileptic hemisphere and the other half had higher rMT in the nonepileptic hemisphere, although the rMT in the two hemispheres showed a strong correlation ($r = 0.73$) (Yelemali et al. 2020). Another parameter that can assist in identifying the epileptic hemisphere is the persistence of uncrossed corticospinal tracts in that hemisphere (Kaye et al. 2017). In a small case series in pediatric patients with an unilateral seizure focus, the authors found that preserved uncrossed corticospinal tracts for the foot were significantly more likely in the epileptic hemisphere than in the nonepileptic hemisphere (Kaye et al. 2017). Finally, differences in corticomotor latencies can also aid in identifying the epileptic hemisphere, as recently a shortened MEP latency in the epileptic hemisphere of children with focal epilepsy has been reported (Liu et al. 2023).

14.2.4 Comparison of nTMS-Derived Motor Mapping Against "Gold Standard" Electrical Stimulation

Only one study has examined the accuracy of nTMS motor mapping in patients with epilepsy against electrical stimulation mapping (ESM) using subdural grids (Vitikainen et al. 2013). Thirteen patients aged between 9 years and 35 years were studied, and the mean distance between the hand motor cortex localized by nTMS and that identified by ESM was 11 ± 4 mm (range 7–17 mm). In our patient cohort, we identified six patients (12.2 ± 8 years, range 2.3–25.9 years, 2 females) who underwent motor mapping with nTMS and subdural grid ESM. Examples of nTMS and ESM motor mapping in two patients who underwent phase II evaluation for epilepsy surgery are shown in Fig. 14.3. The average distance between the hand motor cortices localized by nTMS and ESM was 7.3 ± 2 mm (range 3–10 mm). When compared to ESM, nTMS demonstrated 100% sensitivity, 57% specificity, 80% positive predictive value, 100% negative predictive value, and 84% overall accuracy. The Cohen's kappa was calculated to be 0.63, indicating substantial agreement between ESM and nTMS results. All six patients underwent epilepsy surgery, and no nTMS or ESM identified motor areas were resected. Following resection, two patients experienced transient motor deficits which resolved in a matter of weeks, and none experienced lasting motor deficits. The accuracy of nTMS motor mapping in patients with epilepsy is comparable to that reported in tumor patients where nTMS was compared against intraoperative electrical stimulation (Jeltema et al. 2021). Based on extensive evidence for the validity of nTMS motor mapping in brain tumor patients, and the emerging confirmation in patients with epilepsy, we consider nTMS motor mapping to be a reliable and effective substitute to ESM in patients of all ages.

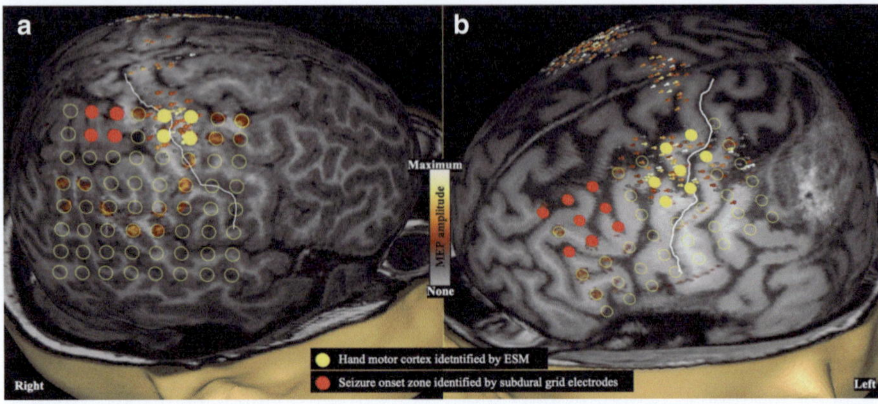

Fig. 14.3 Validation of nTMS motor maps against ESM in patients with refractory epilepsy. The nTMS-derived motor cortex is shown as a heat map of MEP amplitude with *white pegs* representing brain areas where nTMS resulted in MEPs with the greatest amplitude. *Gray pegs* indicate no MEP response, while *yellow, orange, and red pegs* represent intermediate MEP amplitudes. The subdural grid electrodes (4.5 mm diameter, 10 mm interelectrode distance) are shown as *open yellow circles*. The electrodes where ESM resulted in hand movement are represented by *solid yellow circles*. The seizure onset zones are shown as *solid red circles*. The *white line* indicates the central sulcus. (**a**) A 13-year-old male with history of intractable, cryptogenic complex partial seizures of right parietal lobe origin. A subdural grid array covered frontal, parietal, and temporal areas in the right hemisphere. The nTMS and ESM motor locations showed excellent overlap. The seizure onset zone was posterior to the sensorimotor cortex. The patient underwent right parietal topectomy, and his left hand function remained unaffected. (**b**) A 25.9-year-old male with history of intractable cryptogenic complex partial seizure of left frontal lobe origin. A subdural grid array covered frontal, parietal, and temporal areas in the left hemisphere. nTMS and ESM motor locations demonstrated excellent overlap. The seizure onset zone was in the frontal lobe anterior to motor cortex. The patient underwent topectomy of the left superior and middle frontal gyri. Postoperatively, he had no motor deficits. Abbreviations: nTMS: navigated transcranial magnetic stimulation, MEP: motor evoked potential, ESM: electrical stimulation mapping

14.3 nTMS Speech and Language Mapping in Epilepsy

Similar to nTMS motor mapping, nTMS speech and language mapping is not routinely used to localize speech and language areas in patients undergoing epilepsy surgery. Only a handful of studies have shown that nTMS speech and language mapping is feasible in this cohort, providing valuable data, and assisting in planning subdural grid or sEEG placement and the surgical procedure (Narayana et al. 2015a, 2021; Babajani-Feremi et al. 2016; Babajani-Feremi et al. 2018; Lehtinen et al. 2018; Pasichnik et al. 2022; Rezaie et al. 2020; Schiller et al. 2020; Zhang et al. 2020).

14.3.1 nTMS Speech and Language Mapping Procedure

The nTMS speech and language mapping procedure in patients with epilepsy is similar to the recommended protocol for adult brain tumor patients and healthy volunteers (Krieg et al. 2017). Routinely, temporal and frontal lobes in both hemispheres are stimulated using repetitive nTMS (nrTMS) while patients perform a naming task. As part of the mapping protocol adopted at our institution, nrTMS is applied starting from the supramarginal and angular gyri and extending to the superior, middle, and inferior temporal gyri, as anteriorly as the patient can tolerate. Then stimulation of the middle and inferior frontal gyri, including the pars opercularis and pars triangularis, as well as premotor regions is attempted. The interstimulus interval is set between 3000 and 5000 ms based on patient performance, and two runs of baseline performance are obtained. nrTMS is time-locked to stimulus onset without any delay, and 5 Hz is the most frequently used rate. In addition to object naming, which is the most commonly used task, other tasks including color naming (Narayana et al. 2021), spontaneous speech (Zhang et al. 2020), and verb generation (Greiner et al. 2023) have also been used during nrTMS. Examples of nrTMS language mapping in patients having the seizure onset zone in the vicinity of speech and language cortices are shown in Fig. 14.4. In patients a–c, nrTMS localized speech and language cortices to be away from the seizure onset zone which were then resected. However, in patient D, the speech and language cortices were localized within the seizure onset zone, and therefore, the patient underwent responsive neurostimulator (RNS) placement.

14.3.2 Procedures Unique to Epilepsy Cohort

At most centers, including ours, the epilepsy cohort undergoing nrTMS speech and language mapping includes preschool children and individuals with developmental delay. While these patients may not consistently name objects, many of them can accurately name colors. We have previously shown that the color-naming task can be used to successfully map speech and language cortices in this cohort (Narayana et al. 2020, 2021; Rezaie et al. 2020; Schiller et al. 2020). Another important factor is the patient's ability to gauge, tolerate, and report pain or discomfort during nrTMS, which is especially important during speech and language mapping. Consistent with the recommendations for nrTMS mapping procedure in brain tumor patients (Krieg et al. 2017), we evaluate pain/discomfort during nrTMS using the visual analog scale (VAS) for pain. This assessment is recorded at baseline, four times throughout mapping, and whenever a patient spontaneously expressed discomfort. The nrTMS intensity is decreased if participants rate pain ≥ 3 on the VAS. However, a large number of the epilepsy cohort include young children and individuals with decreased cognitive ability, often unable to accurately report pain or discomfort they may experience during nrTMS. Therefore, nrTMS mapping for these individuals must be approached differently than in adults. At our center, we include cartoon faces corresponding to different pain levels along with a numerical

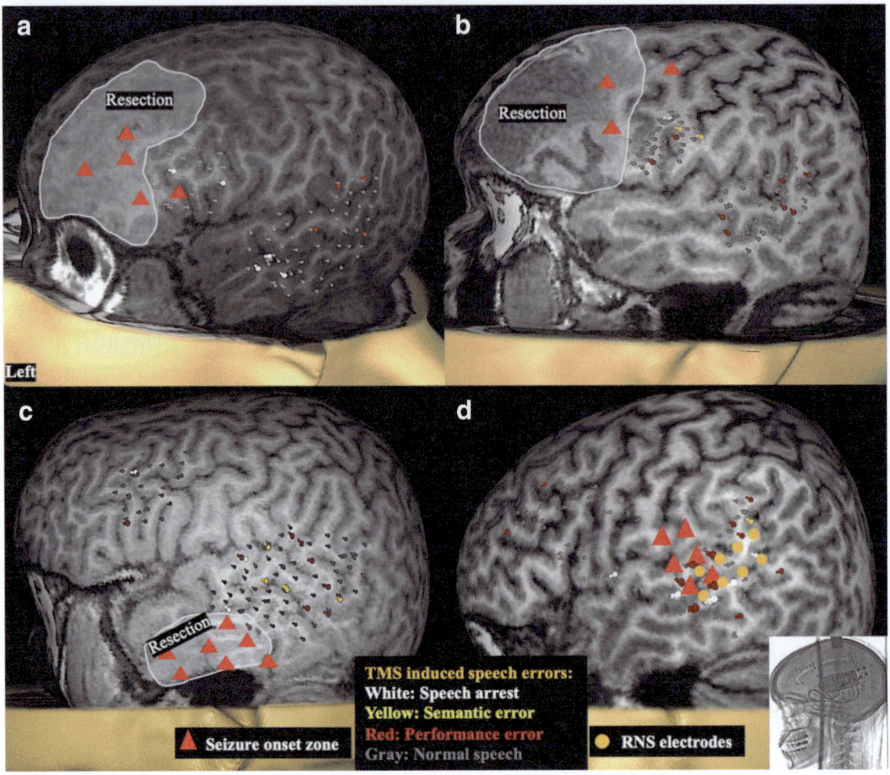
a
Resection
b
Resection
Left
c
Resection
d
TMS induced speech errors:
White: Speech arrest
Yellow: Semantic error
Red: Performance error
Gray: Normal speech
Seizure onset zone
RNS electrodes

Fig. 14.4 Examples of nrTMS speech and language mapping in patients with intractable epilepsy. nrTMS mapping was performed using 5 Hz rTMS, surveying temporal and frontal areas in both hemispheres. The locations of nrTMS-induced speech errors are projected onto the patient's MRI: *white pegs* for speech arrest, *yellow pegs* for sematic errors, and *red pegs* for performance errors, including hesitation. Seizure onset zones are shown as *solid red triangles*, and surgical resection areas are *outlined in blue* on the MRI. (**a**) A 5-year-old left-handed female with a history of symptomatic refractory focal epilepsy related to linear morphea en coup de sabre focal cortical dysplasia with gliosis. A color-naming task was used, and although she performed well at baseline (without nrTMS) during the first mapping session, she could not comply with study demands and was unable to name colors in a consistent manner during nrTMS application. Therefore, the study was discontinued. The mapping session was attempted 1 month later, and the left hemisphere was successfully mapped. nrTMS elicited speech arrest in the inferior frontal cortex in the left hemisphere, posterior to the lesion. The seizure onset zone was within the lesion. Following surgery, the patient experienced some decreased movements on the right as well as decreased spontaneous speech initially. However, at 6-month follow-up, her face was symmetric, and she had fluid speech and normal strength in her extremities. (**b**) A 36-year-old right-handed male with a history of seizure onset at age 9 years underwent left frontal pole resection, corpus callosotomy, and VNS placement. He presented with ongoing refractory seizures. His seizure foci were localized along the posterior margins of the resection cavity. An object-naming task was used, and nrTMS applied over the Broca's area resulted in speech arrest, semantic errors, and performance errors, all posterior to the seizure foci. Stimulation of homologous regions in the right hemisphere also resulted in speech disruption in the form of speech arrest and performance errors, suggestive of bilateral representation for expressive language. He underwent frontal lobectomy and postoperatively; he initially showed slight right facial weakness, decreased spontaneous speech, and decreased dorsiflexion of his right foot. However, all these symptoms improved within days, and at the time of discharge, his speech was back to normal, face was symmetrical, and he could elevate his right foot almost all the way back to normal. (**c**) A 17-year-old right-handed male with intractable epilepsy secondary to TSC underwent nrTMS during an object-naming task. Language areas were found in the posterior parts of superior and middle temporal gyri, posterior to the seizure onset zone. Presence of speech and language areas in the anterior temporal lobe (over the seizure onset zone) could not be surveyed due to excessive muscle stimulation during nrTMS. Stimulation of homologous regions in the right hemisphere resulted in fewer speech errors, suggesting left hemisphere dominance for expressive language. The patient underwent anterior temporal lobectomy, and postoperatively, no speech or language deficits were observed. (**d**) A 25-year-old right-handed female with seizure onset at 13 years of age presenting with intractable epilepsy. MRI of brain was normal. Video EEG, MEG, and subdural grid electrocorticography localized her seizure onset zone to the posterior part of superior and middle temporal gyri and temporoparietal junction. Wada suggested bilateral language, left > right. An object-naming task was used, and nrTMS found language areas in the posterior parts of superior and middle temporal gyri, and temporoparietal junction within the seizure onset zone in the left hemisphere. Stimulation of homologous regions in the right hemisphere resulted in fewer speech errors, suggesting a left hemisphere dominance for expressive language. fMRI also indicated left hemisphere dominance with speech and localized language areas in the posterior temporal lobe and temporoparietal junction in the left hemisphere. Due to the overlap between critical language areas in the dominant hemisphere and the seizure onset zone, the resection of the seizure onset zone was considered risky for language deficits. Therefore, the patient wished to pursue RNS placement. The RNS leads were placed over the posterior part of superior and middle temporal gyri (shown by solid orange circles). The inset is the X-ray showing the RNS leads and the stimulator position. At an 8-year follow-up, she has no speech and language deficits and remaining seizure free on lacosamide, zonisamide, and RNS. Abbreviations: nrTMS: navigated repetitive transcranial magnetic stimulation, VNS: vegus nerve stimulator, TSC: tuberous sclerosis complex, EEG: electroencephalography, MEG: magnetoencephalography, fMRI: functional magnetic resonance imaging, RNS: responsive neurostimulator scale to accommodate different abilities. Additionally, it is also essential to monitor body language for indications of pain and discomfort such as grimacing, moving the head away from the TMS coil, or decreased cooperation. To increase compliance in this population, we avail the expertise of Certified Child Life Specialists. Their purpose is to help the patients cope with stressful healthcare experiences and procedures, thereby promoting safety and compliance in the pediatric setting. Furthermore, in patients where compliance or tolerability is of concern, the hemisphere to be stimulated first should be decided by prioritizing clinical need, and therefore, the lesional hemisphere should be mapped first.

Finally, a patient's seizure type and seizure frequency can impact the results of nrTMS speech and language mapping. For example, in a patient whose seizure semiology includes behavioral arrest, their seizure semiology like hesitation or inability to name the object may mimic nrTMS effects. It is therefore important to distinguish the speech errors caused by nrTMS from the patient's seizure semiology. Therefore, during mapping and post hoc analysis, attention should be paid to the overall performance and visible signs of seizure activity. If the performance during nrTMS is suspected to be influenced by seizure activity, a baseline without nrTMS should be repeated or sham nrTMS (nrTMS applied to nonspeech and nonlanguage areas) attempted. During analysis of the study, the patient's behavior during each stimulus should be compared against several preceding stimuli and baseline performance. nrTMS results in such cases should be interpreted with caution or confirmed by other modalities. Rarely, nrTMS speech and language mapping may be warranted in patients with ongoing seizures, i.e., EPC or electrographic seizures. While this is uniquely challenging, modification of nrTMS procedure and parameters can result in safe and successful mapping in such individuals (see (Narayana et al. 2020) for an example). Safety of nrTMS speech and language mapping in patients with epilepsy is further discussed in Sect. 14.4 under safety of TMS. Additional details regarding considerations for speech and language mapping in pediatric epilepsy patients are in addressed in the chapter titled "nTMS in Pediatrics: Special Issues and Solutions" by Giuffre and Rotenberg.

14.3.3 Parameters Unique to Epilepsy Cohort

The intensity of nrTMS during speech and language mapping is a critical parameter. Higher intensities are not well tolerated (especially during frontal lobe mapping) and may increase the chances of inducing a seizure. On the other hand, low intensities may not elicit any speech or language errors, resulting in a false-negative finding. The recommendation to set the starting stimulation intensity at ipsilateral rMT (Krieg et al. 2017) is feasible for most adults as their thresholds are around 30–40% of the MSO. However, as discussed in the chapter titled "nTMS in Pediatrics: Special Issues and Solutions" by Giuffre and Rotenberg, and based our own experience, the epilepsy cohort tend to have higher rMT due to their younger age and being treated with ASM. In our cohort, the average rMT in children aged 4–5 years who underwent speech and language mapping was 83% of MSO. The rMT decreased with increasing age and was on average 66% of MSO in 8- to 9-year-olds, 53% of MSO in 12- to 13-year-olds, and 48% of MSO in 16- to 17-year-olds. The average rMT in adults was 46% of MSO. We therefore suggest that the stimulation intensity for mapping should instead be adjusted to deliver an electric field (E-field) of 70–100 V/m at the cortical surface (~20 mm peeling depth). We routinely set the stimulation intensity to be 35–40% of MSO which delivers an E-field of 80–100 V/m. In our experience with nearly 400 patients, the average stimulation intensity was 36 ± 5% of MSO, which corresponded to an E-field of 78 ± 17 V/m. In approximately 25% of patients, the mapping intensity was ≥90% rMT, and not surprisingly, these patients were mostly adolescents and adults (average age 19.3 years, average rMT 36% MSO). The mapping intensity was <50% of rMT in 18% of patients who were younger (average age 11 years, average rMT 83% MSO). However, there were no significant differences between the two groups with respect to E-field delivered, number and types of errors elicited, error rate, or the laterality indices. Moreover, we have found that the stimulation intensity does not differ between hemispheres in each individual or types of speech errors (Rezaie et al. 2020; Schiller et al. 2020). Our data suggest that using nrTMS intensities of 70–100 V/m, independent of individual rMT, can maximize clinically relevant results while minimizing unsuccessful results from patient discomfort due to too high intensity or failure to elicit errors due to too low intensity. Other nrTMS parameters including stimulus duration, interstimulus interval, and the rate of stimulation may need to be modified on an individual basis based on patients' baseline performance and their cognitive ability.

14.3.4 Determining Hemispheric Dominance with TMS

In addition to localizing speech and language areas, nrTMS mapping can also be used to estimate hemispheric dominance (HD) for language in a manner similar to MEG and fMRI. The number and types of errors in the two hemispheres can be used to calculate a laterality index (LI). The LI can be calculated as $100 \times [E_{left} - E_{right}/E_{left} + E_{right}]$, where E_{left} and E_{right} are the sum of speech errors in the left and right hemispheres, respectively. However, we propose a weighted LI metric where speech arrest is weighted more than semantic and performance errors. Since performance error judgments were the most subjective, they should be weighted the least. The weighted laterality index (wLI) can be calculated as $100 \times [wE_{left} - wE_{right}/wE_{left} + wE_{right}]$, where wE_{left} and wE_{right} are the weighted sum of speech errors in the left and right hemispheres, respectively. Similar to fMRI and MEG, values of 100 to ≤10 indicate a left HD, −100 to ≤ −10 indicate a right HD, and values between −10 and + 10 imply a balanced bilateral representation of language. Alternatively, the error rates can also be used to determine the HD ratio, where values >1 indicate left hemisphere dominance and values <1 imply right hemisphere dominance (Ille et al. 2016). The nrTMS-derived laterality index allows for comparison with other mapping modalities including fMRI, MEG, and Wada. Examples of patients with epilepsy exhibiting a left HD and a balanced bilateral representation for speech and language are shown in Fig. 14.5.

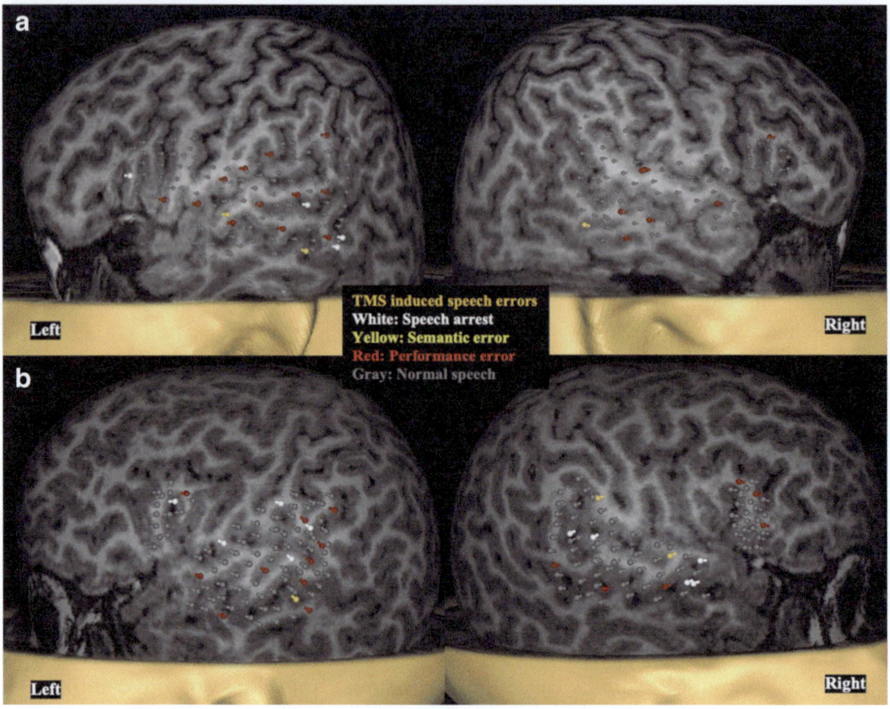

Fig. 14.5 Hemispheric dominance determined by nrTMS in patients with intractable epilepsy. The locations of nrTMS-induced speech errors are projected onto the patient's MRI as *white pegs* for speech arrest, *yellow pegs* for sematic errors, and *red pegs* for performance errors including hesitation. (**a**) A 17-year-old left-handed female with a history of nonlesional epilepsy since the age of 13. She was found to have independent seizure foci in both hemispheres (right > left). Her rMT in the left hemisphere was 35% of MSO, and in the right hemisphere, it was 38% of MSO. Speech and language mapping was performed using an intensity of 31% of MSO. nrTMS induced more speech arrest, semantic errors, and performance errors in the left temporal lobe when compared to the right hemisphere. The wLI was 51 indicating left hemisphere dominance for speech and language. (**b**) A 7-year-old right-handed female with a history of seizures since age of 5. Her MRI showed left hippocampal sclerosis, and the seizure foci were localized in the left temporal lobe. Her rMT in the left hemisphere was 70% of MSO, and in the right hemisphere, it was 54% of MSO. Speech and language mapping was performed using an intensity of 33% of MSO. nrTMS resulted in similar number of speech errors in both hemispheres. The wLI was 3, indicating a balanced bilateral representation of language function. She underwent left temporal lobectomy with no postoperative speech or language deficits. Abbreviations: nrTMS: navigated repetitive transcranial magnetic stimulation, MRI: magnetic resonance imaging, rMT: resting motor threshold, MSO: maximum stimulator output, wLI: weighted laterality index

14.3.5 Validation of nrTMS-Derived Speech and Language Mapping

Despite the many advantages of nrTMS mapping, it is not yet fully integrated into the standard-of-care practice by epilepsy surgery programs. One of the limiting factors is the lack of comparative data against other validated and accepted mapping methods, both invasive and noninvasive. Toward this objective, it is important to evaluate the validity of HD estimates and the localization of speech and language areas using nrTMS against other validated methods specifically in patients with epilepsy. Hence, we compared nrTMS-derived HD estimates with those obtained from fMRI (Schiller et al. 2020), MEG (Rezaie et al. 2020), and "gold standard" Wada in our epilepsy cohort (Noorizadeh et al. 2022). The accuracy of nrTMS in localizing critical speech and language areas in the epilepsy cohort has also been compared against fMRI (Schiller et al. 2020), "gold standard" ESM (Babajani-Feremi et al. 2016; Lehtinen et al. 2018; Zhang et al. 2020), and postoperative outcomes (Babajani-Feremi et al. 2018).

14.3.6 Hemispheric Dominance Estimates: nrTMS vs. fMRI, MEG, and Wada

We compared HD estimated by nrTMS against that derived from fMRI in 40 patients with epilepsy (65% children; 55% females) (Schiller et al. 2020). nrTMS demonstrated moderate agreement with fMRI with an overall sensitivity of 76% and a specificity of 52% (see Table 14.1 for detailed statistical performance metrics). nrTMS performance improved when the tasks used in nrTMS and fMRI were concordant (i.e., object naming) and in the two modalities were more concordant in children. However, when compared to fMRI, the HD determined by nrTMS underestimated the incidence of left dominance and overestimated bilateral representation (Schiller et al. 2020). These results are very similar to that reported in adult brain tumor cohort where nrTMS and fMRI were concordant in 67% of patients (Durner et al. 2019). Some of the discrepancy observed could be a result of fundamental differences in the methods wherein fMRI measures a hemodynamic response, and nrTMS is a stimulation method. Therefore, we examined if nrTMS would be more concordant with a modality such as MEG that measures neuronal activity. We compared nrTMS-derived HD with MEG in 67 patients with epilepsy (81% children, 54% female) (Rezaie et al. 2020). Language was mapped in MEG during an auditory word recognition paradigm, and LI was calculated using the number of dipoles and their spatial extent in the two hemispheres. nrTMS demonstrated moderate agreement with a sensitivity of 74% and a specificity of 72% (see Table 14.1 for other statistical performance metrics). Finally, we compared nrTMS derived language dominance with the "gold standard" Wada test in 12 epilepsy patients (Noorizadeh et al. 2022). Similar to the fMRI and MEG comparisons, a moderate concordance was observed between the nrTMS-derived HD compared to that estimated by the Wada (see Table 14.1 for details).

The performance of nrTMS against fMRI, MEG, and Wada highlights the methodological differences between these techniques. Unlike fMRI and MEG, which are whole brain imaging methods, and Wada, which evaluates an entire hemisphere, nrTMS involves an iterative survey of individual brain areas, potentially leading to a sampling bias. While every attempt should be made to thoroughly survey potential speech and language cortices in both hemispheres, excessive muscle stimulation and pain during nrTMS frequently preclude complete mapping. A common finding

Table 14.1 Statistical performance metrics of nrTMS compared to other noninvasive and invasive mapping methods

TMS vs.	fMRI-HD[a] (Object naming)	fMRI-HD[a] (Verb generation)	MEG-HD[b]	WADA-HD[c]	fMRI Localization[a] (Left hemisphere)	fMRI Localization[a] (Right hemisphere)	ESM-Localization[d,e,f] (Left hemisphere)
Patients	12	34	67	12	40	40	45
Children	58%	62%	81%	50%	65%	65%	42%
Adults	42%	38%	19%	50%	35%	35%	58%
Sensitivity	75%	72%	74%	75%	27%	10.5%	74%
Specificity	67%	50%	72%	75%	82%	85%	81%
PPV	69%	62%	78%	86%	31%	11%	33%
NPV	73%	62%	68%	60%	79%	84%	96%
Accuracy	71%	62%	73%	75%	69%	74%	80%
Cohen's kappa	0.42	0.22	0.46	0.47	0.09	−0.05	0.35
Level of agreement	Moderate	Fair	Moderate	Moderate	Slight/none	Slight/none	Moderate

Sources: [a]Schiller et al. (2020)
[b]Rezaie et al. (2020)
[c]Noorizadeh et al. (2022)
[d]Babajani-Feremi et al. (2016)
[e]Lehtinen et al. (2018)
[f]Zhang et al. (Zhang et al. 2020)
Abbreviations: *nrTMS* navigated repetitive transcranial magnetic stimulation, *MEG* magnetoencephalography, *fMRI* functional magnetic resonance imaging, *ESM* electrical stimulation mapping, *HD* hemispheric dominance, *PPV* positive predictive value, *NPV* negative predictive value

across these comparisons in the epilepsy cohort is that nrTMS identifies a number of speech and language areas in the right (nondominant) hemisphere. This aligns with similar finding observed in adult brain tumor patients as well (Rösler et al. 2014). Being mechanistically different than other mapping methods (causative rather than correlational), nrTMS may uniquely be suited to identify language reorganization or a "reserve" in the nondominant hemisphere. For instance, patients in whom nrTMS had identified more speech and language areas in the nondominant hemisphere were less likely to develop postoperative deficits following brain tumor surgery in the dominant hemisphere (Ille et al. 2016). Similarly, in our epilepsy population, we have observed that even when crucial eloquent cortex in the dominant hemisphere was resected, only a small percentage of patients had speech deficits, as in majority of them, nrTMS had identified speech and language areas bilaterally (Narayana et al. 2021; Boardman et al. 2021). However, whether the brain areas in the nondominant hemisphere identified by nrTMS are truly eloquent cortices or mere false-positive localization is yet to be confirmed. Another difference between nrTMS and other modalities is that assessment of nrTMS-induced speech errors is rather subjective. In some patients, such as children and those already experiencing speech and language deficits, distinguishing nrTMS-induced errors from their baseline performance is challenging. Although patient factors including handedness, lesion type, lesion location, cognitive abilities, and disease-induced language reorganization can influence nrTMS results, it does not explain why nrTMS results were divergent from other modalities, which were equally affected by the same patient characteristics. These comparative studies suggest that the choice of nrTMS parameters may be influencing nrTMS results. To further facilitate the use of nrTMS to characterize language function, especially in pediatrics, future studies should prioritize improving the accuracy of nrTMS by optimizing nrTMS parameters.

14.3.7 Localization of Speech and Language Cortices: nrTMS vs. fMRI and Electrical Stimulation Mapping

We examined the accuracy of nrTMS in localizing critical speech and language areas in the epilepsy cohort by comparing it against fMRI. Although speech and language areas in superior and middle temporal, supramarginal gyri, and pars opercularis are identified in most patients, at an individual level, there was only a slight agreement between nrTMS- and fMRI-identified areas (Schiller et al. 2020). The details of the statistical performance metrics are listed in Table 14.1. Overall, nrTMS showed low sensitivity (20%) but high specificity (83%); however, sensitivity was higher for left hemisphere regions. In particular, nrTMS had nearly 70% sensitivity in identifying eloquent cortex in the supramarginal gyrus in the left hemisphere (Schiller et al. 2020). The high specificity observed in this study confirmed that nrTMS can identify cortical areas that do not contain essential speech or language function with high degree of confidence.

Three studies have compared the accuracy of language areas identified by nrTMS in the frontal and temporal lobes of the left hemisphere against subdural grid ESM in epilepsy patients (Babajani-Feremi et al. 2016; Lehtinen et al. 2018; Zhang et al. 2020) (see Table 14.1 for statistical performance metrics). Altogether, fewer than 50 epilepsy patients were included in these validation studies, and less than half were children. Overall, the pooled sensitivity (74%) and specificity (81%) of nrTMS in identifying speech and language areas in epilepsy patients were similar to its HD performance (against fMRI, MEG, and Wada) but worse when compared to its performance in brain tumor patients (reviewed in Ref. (Jeltema et al. 2021)). These differences may be stemming from methodological differences (extraoperative grid ESM vs. intraoperative ESM), analytical differences (different criteria for defining overlap between nrTMS and ESM), and/or disease process (epilepsy acquired in childhood during language development vs. tumor onset in adulthood). nrTMS suffers from lower sensitivity and positive predictive values (PPVs) because of excessive false-positive findings; however, its performance in detecting nonlanguage areas (i.e., negative predictive value (NPV)) in the epilepsy cohort is high, much like in the brain tumor population. Examples of nrTMS maps of speech and language in two adolescents with intractable epilepsy, showing a range of sensitivity, specificity, PPV, and NPV, are shown in Fig. 14.6.

As presented above, the accuracy and reliability of nrTMS cannot be precisely defined, primarily due to the inherent methodological differences between the various mapping modalities (both invasive and noninvasive) and their unique weaknesses. Another way to evaluate the accuracy of the mapping modalities including nrTMS is to examine postoperative speech and language function. With respect to the accuracy of localization of speech and language areas, we examined whether nrTMS along with other modalities (fMRI, MEG and high gamma electrocorticography) could predict postoperative speech and language outcomes in 11 patients who underwent epilepsy surgery (Babajani-Feremi et al. 2018). Although no nrTMS-positive areas were resected in six out of seven patients, all seven exhibited postoperative decline in naming. However, in these patients, the resected cortex consisted of the anterior third of the temporal lobe, frontal lobe anterior to pars orbicularis, and superior and middle frontal gyri, and none of these areas were surveyed by nrTMS. However, only one out of four patients with no postoperative language deficit had nrTMS-positive area resected, corresponding to a good specificity of 75%. In this study, the combined use of fMRI and MEG was optimal in predicting postoperative deficits, and adding nrTMS and high gamma electrocorticography provided higher accuracy (Babajani-Feremi et al. 2018). In a larger epilepsy population (n = 43), we have observed that although resections in the dominant hemisphere resulted in more speech deficits than those in the nondominant hemisphere (24% vs. 12%), the number of participants overall with deficits was low (Boardman et al. 2021). Although nrTMS-identified language areas were resected, many patients did not suffer from any postoperative deficits, and in these patients, nrTMS had localized additional language areas away from the lesion in the same hemisphere or in the contralesional hemisphere. Thus, we believe that nrTMS mapping demonstrated considerable functional reserve in these patients (Narayana et al. 2021; Boardman et al. 2021). Future large-scale multicenter studies should be designed to optimize the nrTMS protocol to decrease false-positive results and clearly delineate whether the brain areas identified by nrTMS are truly speech and language cortices.

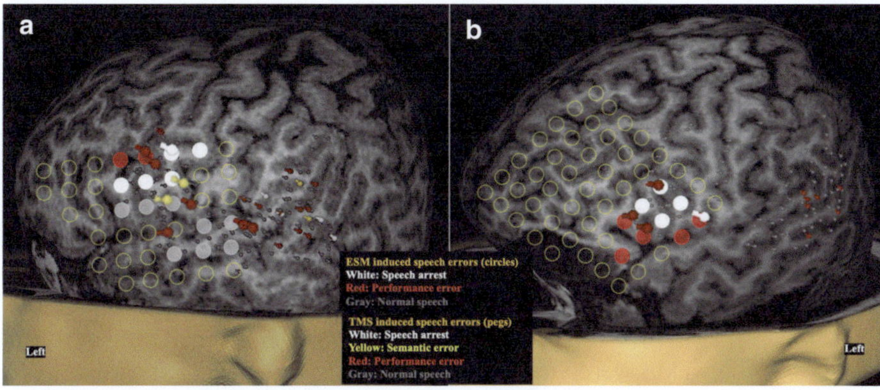

Fig. 14.6 Validation of nrTMS-derived speech and language maps against ESM-confirmed locations in patients with refractory epilepsy. Speech and language areas identified by nrTMS and ESM are projected on the patient's MRI. The nrTMS and ESM sites were localized in the same space by registering the grid CT to the anatomical MRI used for nrTMS mapping. The subdural grid electrodes (4.5 mm diameter, 10 mm interelectrode distance) are shown as *open yellow* circles (**a**) A 14-year-old male with a history of nonlesional epilepsy since the age of 7. He underwent phase II epilepsy surgery evaluation. A 6 × 8 grid (8 contacts were removed) was placed over left frontal and temporal lobes. Speech arrest (*white peg and solid white circles*) and performance errors (*red pegs and solid red circles*) were elicited in the same areas by both nrTMS and ESM. ESM of the area where nrTMS elicited a semantic error (*yellow pegs*) resulted in speech arrest (*solid white circle*). Other performance errors elicited by nrTMS were not confirmed by ESM (*gray circles*). Compared to ESM, nrTMS had a sensitivity of 88%, specificity of 57%, PPV of 41%, and NPV of 93%. nrTMS also identified several language areas in the right hemisphere, and the patient was deemed to have bilateral representation for speech and language, which was confirmed by Wada. Patient underwent resection of anterior temporal lobectomy and did not have any speech or language deficits following surgery. (**b**) A 14-year-old female with a history of dysembryoplastic neuroepithelial tumor (WHO grade 1) in the left frontal lobe with recurrence with associated intractable symptomatic partial onset seizures. She underwent phase II epilepsy surgery evaluation. A 7 × 8 grid (3 contacts were removed) was placed over left frontal lobe. Performance errors (*red peg and solid red circle*) were elicited in the same areas by both nrTMS and ESM. One nrTMS-elicited performance error was noted under the electrode where ESM caused speech arrest (*solid white circle*). One nrTMS-elicited speech arrest (*white peg*) was noted under the electrode where ESM caused performance error. ESM resulted in speech arrest over an area where nrTMS did not elicit any errors. Compared to ESM, nrTMS had a sensitivity of 31%, specificity of 100%, PPV of 100%, and NPV of 10%. TMS identified few language areas in the right hemisphere, and she was deemed to be left hemisphere dominant for speech and language. The patient underwent microsurgical resection of the tumor and did not have any speech or language deficits following surgery. nrTMS: navigated repetitive transcranial magnetic stimulation, ESM: electrical stimulation mapping, MRI: magnetic resonance imaging, CT: computed tomography, PPV: positive predictive value, NPV: negative predictive value, WHO: World Health Organization

14.4 Safety of TMS

Safety of TMS in patients with epilepsy is discussed in great depth in the chapter titled "Safety Considerations of the use of TMS" by Di Iorio and Rossini, and safety of TMS in children is addressed in the chapter titled "nTMS in Pediatrics: Special Issues and Solutions" by Giuffre and Rotenberg. Consistent with these reports, our experience also contributes to the safety of TMS in patients with epilepsy, particularly in children. Out of 500 attempted TMS sessions at our institution, involving 429 individual patients (51% males, 82% $\leq$ 18 y), seizures occurred in 29 sessions (5.8%) during or after TMS, with semiology consistent with their typical presentation (Braden et al. 2022). The patients who experienced a seizure during or immediately following TMS were younger, mapped using higher TMS intensity, had a history of daily seizures, and were on more ASMs compared to those who did not have a seizure during mapping. In all patients, mapping was either completed in the same session or at a later date. Mapping was also completed safely in 276 patients with extracranial (dental metal, facial metal, and VNS), epicranial (titanium mesh, screws, plates, and ventriculoperitoneal shunt), and intracranial (encephalocele clips and aneurysm coils) metal. Mapping had to be discontinued in 11% of patients who continued to experience pain despite reducing the TMS intensity. The tolerance for motor mapping was much higher (mapping could not be completed due to pain in only 0.4% of total motor mapping sessions) compared to speech and language mapping and (14% of speech and language mapping sessions discontinued due to pain). Another unusual population included in our study were individuals with EPC, including one patient with Rassmusen's encephalitis involving right upper extremity, one with a tumor in the leg motor cortex, one with epilepsy with EPC involving right upper extremity, and one with epilepsy with EPC of left lower half of the face. No EPC patients had any adverse events during their TMS study and were able to successfully complete all TMS studies. Our data suggest that TMS parameters used during mapping are safe and do not worsen EPC in such patients. To further improve the safety of TMS studies, standardized screening and adverse event documentation (see the Appendix in (Braden et al. 2022) for a sample format) should be universally adopted.

14.5 Emerging Applications: Low Frequency rTMS for Treating Intractable Epilepsy

14.5.1 Low-Frequency rTMS for Treating Focal Epilepsy

The increased cortical excitability in patients with epilepsy can also be a target of therapeutic TMS. Most commonly, low-frequency rTMS is used since it is thought to decrease cortical excitability by mediating long-term depression and has an excellent safety profile. More importantly, the inhibitory effects of transcranial magnetic stimulation are long lasting. Therefore, TMS is attractive as a treatment alternative in patients with refractory focal epilepsy. Over the last two decades, several of studies ranging from case reports to randomized, double-blind, sham-controlled series have investigated the efficacy of low-frequency TMS in treating refractory epilepsy (see meta-analyses by Cooper and colleagues (Cooper et al. 2018) and a Cochrane review by Walton and colleagues (Walton et al. 2021)). These studies evaluating the efficacy of rTMS in epilepsy treatment have high interstudy heterogeneity with respect to inclusion criteria, and TMS parameters including rate, intensity, target site, number of stimulations, and duration of treatment (Cooper et al. 2018; Walton et al. 2021). They were also found to have moderate to high levels of study bias regarding patient sample representation, recruitment, characterizing, and reporting predictive outcomes, as well as significant publication bias (Cooper et al. 2018; Walton et al. 2021). However, the meta-analyses indicate an overall 30% reduction in seizure frequency following low-frequency rTMS treatment (Cooper et al. 2018), although TMS was shown to be highly effective in patients with cortical dysplasia or neocortical epilepsy (effect size of 0.7) (Hsu et al. 2011). Other salient findings from the meta-analyses were that rTMS treatment was more effective in individuals younger than 21 years (vs. older individuals), with a figure-eight coil (vs. other coil types), when TMS was applied to the seizure focus (vs. vertex or along the midline), and when the seizure focus was in the temporal lobe (vs. extratemporal) (Cooper et al. 2018). Across all studies, rTMS was found to be safe, with few and minor adverse events reported. Because TMS is easy and safe to administer and does not result in the systemic adverse effects common to pharmacotherapy, every effort should be made to improve the efficacy of low-frequency rTMS in treating epilepsy. Future studies should consider improving the accuracy of localization of epileptogenic focus, systematically investigating TMS parameters, and optimizing TMS targeting using image guidance.

14.5.2 rTMS for Treating Status Epilepticus

Beyond treating refractory epilepsy, rTMS has also been used to treat more challenging epilepsy syndromes including EPC (also called focal status epilepticus) and status epilepticus (SE). In such cases, neuromodulation with TMS has been attempted as an adjuvant to pharmacological treatments. Overall, 21 patients (13 adults and 8 children) with various etiologies of focal and generalized SE have been reported to have undergone treatment with TMS (Zeiler et al. 2015). The stimulation parameters, including the number of trains (range 1–15), train duration (2–1800 s), frequency of stimulation (0.5–20 Hz), and treatment duration (single treatment to 2 weeks), varied between the studies. Seizure reduction or control with rTMS occurred in 15 of the 21 patients (71%) with 10 (48%) showing a complete response and 5 (24%) having a partial response (Zeiler et al. 2015). Specifically, all patients with generalized refractory SE, 80% of patients with focal SE, and 50% of patients with focal refractory SE responded to rTMS. However, the therapeutic effects were not long lasting as seizures recurred in 73% of the patients. These preliminary findings should be confirmed in future large-scale studies before rTMS can be suggested as a viable option for treating various SE syndromes.

14.6 Conclusions

Presurgical functional mapping in epilepsy surgery is critical, and nTMS is a well-suited noninvasive tool to evaluate motor, speech, and language functions in the individuals with refractory epilepsy of various etiologies. nTMS is especially useful in young children in whom most other modalities are frequently unsuccessful. The functional mapping procedures in refractory epilepsy are similar to the recommended protocol for adult brain tumor patients and healthy volunteers. However, modifications are warranted to accommodate younger age, ASM, seizure type and frequency, and range of cognitive/compliance abilities unique to this cohort. nTMS-derived localization of motor cortex has excellent agreement with ESM, leading us to consider nTMS motor mapping a reliable and effective substitute to ESM in patients of all ages and etiologies. Along with localization of critical eloquent cortices, speech and language mapping with nrTMS allows for determining HD for language. nrTMS-derived HD measures exhibit moderate agreement with fMRI, MEG, and Wada. Speech and language areas identified by nrTMS show only slight agreement with fMRI, a method based on hemodynamic response, but exhibited moderate agreement with ESM, a stimulation method similar to nrTMS. Consistent with the reports in brain tumor patients, speech and language mapping with nrTMS in epilepsy cohort suffers from excessive false-positive but excellent true-negative results. Future large-scale multicenter studies should be designed to optimize the nrTMS protocol in order to decrease false-positive results and truly identify speech and language cortices. TMS in epilepsy cohort has excellent safety profile and has the potential for being an effective treatment for refractory epilepsy.

References

Andreasson A, Sigurdsson GV, Pegenius G, Thordstein M, Hallböök T. Cortical excitability measured with transcranial magnetic stimulation in children with epilepsy before and after antiepileptic drugs. Dev Med Child Neurol. 2020;62(7):793–8.

Babajani-Feremi A, Narayana S, Rezaie R, Choudhri AF, Fulton SP, Boop FA, et al. Language mapping using high gamma electrocorticography, fMRI, and TMS versus electrocortical stimulation. Clin Neurophysiol. 2016;127(3):1822–36.

Babajani-Feremi A, Holder CM, Narayana S, Fulton SP, Choudhri AF, Boop FA, et al. Predicting postoperative language outcome using presurgical fMRI, MEG, TMS, and high gamma ECoG. Clin Neurophysiol. 2018;129(3):560–71.

Badawy RAB, Macdonell RAL, Berkovic SF, Newton MR, Jackson GD. Predicting seizure control: cortical excitability and antiepileptic medication. Ann Neurol. 2010;67(1):64–73.

Badawy RAB, Jackson GD, Berkovic SF, Macdonell RAL. Cortical excitability and refractory epilepsy: a three-year longitudinal transcranial magnetic stimulation study. Int J Neur Syst. 2013;23(01):1250030.

Barba C, Cossu M, Guerrini R, Di Gennaro G, Villani F, De Palma L, et al. Temporal lobe epilepsy surgery in children and adults: a multicenter study. Epilepsia. 2021;62(1):128–42.

Barba C, Giometto S, Lucenteforte E, Pellacani S, Matta G, Bettiol A, et al. Seizure outcome of temporal lobe epilepsy surgery in adults and children: a systematic review and meta-analysis. Neurosurgery. 2022;91(5):676–83.

Baumgartner C, Koren JP, Britto-Arias M, Zoche L, Pirker S. Presurgical epilepsy evaluation and epilepsy surgery. F1000Res. 2019;8:1818.
Birg L, Narayana S, Rezaie R, Papanicolaou A. Technical tips: MEG and EEG with sedation. Neurodiagnostic J. 2013;53(3):229–40.
Boardman T, Shepard N, Gibbs S, Boop FA, Wheless JW, Narayana S. Using transcranial magnetic stimulation (TMS) to map language reorganization prior to brain surgery. Epilepsia. 2021;62(S3):265.
Braden AA, Weatherspoon SE, Boardman T, Williard T, Adkins A, Gibbs SK, et al. Image-guided TMS is safe in a predominately pediatric clinical population. Clin Neurophysiol. 2022;137:193–206.
Braun KPJ. Influence of epilepsy surgery on developmental outcomes in children. Eur J Paediatr Neurol. 2020;24:40–2.
Çataltepe O, Jallo GI. Pediatric epilepsy surgery: introduction. In: Çataltepe O, Jallo GI, editors. Pediatric epilepsy surgery: preoperative assessment and surgical treatment. New York/Stuttgart: Thieme; 2010. p. 1–6.
Chou N, Serafini S, Muh CR. Cortical language areas and plasticity in pediatric patients with epilepsy: a review. Pediatr Neurol. 2018;78:3–12.
Cooper YA, Pianka ST, Alotaibi NM, Babayan D, Salavati B, Weil AG, et al. Repetitive transcranial magnetic stimulation for the treatment of drug-resistant epilepsy: a systematic review and individual participant data meta-analysis of real-world evidence. Epilepsia Open. 2018;3(1):55–65.
de Ribaupierre S, Wang A, Hayman-Abello S. Language mapping in temporal lobe epilepsy in children: special considerations. Epilepsy Res Treat. 2012;2012:1–11.
Durner G, Pala A, Federle L, Grolik B, Wirtz CR, Coburger J. Comparison of hemispheric dominance and correlation of evoked speech responses between functional magnetic resonance imaging and navigated transcranial magnetic stimulation in language mapping. J Neurosurg Sci. 2019;63(2):106–13.
Dwivedi R, Ramanujam B, Chandra PS, Sapra S, Gulati S, Kalaivani M, et al. Surgery for drug-resistant epilepsy in children. N Engl J Med. 2017;377(17):1639–47.
Engel J, Wiebe S, Radhakrishnan K, Palmini A. Surgical treatment for epilepsy. Neurologisch (Wien). 2013;(September):12–4.
Englot DJ, Breshears JD, Sun PP, Chang EF, Auguste KI. Seizure outcomes after resective surgery for extra–temporal lobe epilepsy in pediatric patients: a systematic review. J Neurosurg Pediatr. 2013;12(2):126–33.
Greiner HM, Maue E, Horn PS, Vannest J, Vedala K, Leach JL, et al. Tolerability of transcranial magnetic stimulation language mapping in children. Epilepsy Res. 2023;194:107183.
Haddad AF, Young JS, Berger MS, Tarapore PE. Preoperative applications of navigated transcranial magnetic stimulation. Front Neurol. 2021;11:628903.
Helling RM, Shmuely S, Bauer PR, Tolner EA, Visser GH, Thijs RD. Tracking cortical excitability dynamics with transcranial magnetic stimulation in focal epilepsy. Ann Clin Transl Neurol. 2022;9(4):540–51.
Hsu WY, Cheng CH, Lin MW, Shih YH, Liao KK, Lin YY. Antiepileptic effects of low frequency repetitive transcranial magnetic stimulation: a meta-analysis. Epilepsy Res. 2011;96(3):231–40.
Ille S, Kulchytska N, Sollmann N, Wittig R, Beurskens E, Butenschoen VM, et al. Hemispheric language dominance measured by repetitive navigated transcranial magnetic stimulation and postoperative course of language function in brain tumor patients. Neuropsychologia. 2016;91:50–60.
Jayakar P. Cortical electrical stimulation mapping: special considerations in children. J Clin Neurophysiol. 2018;35(2):106–9.
Jayakar P, Jayakar A, Libenson M, Arzimanoglou A, Rydenhag B, Cross JH, et al. Epilepsy surgery near or in eloquent cortex in children-practice patterns and recommendations for minimizing and reporting deficits. Epilepsia. 2018;59(8):1484–91.

Jeltema HR, Ohlerth AK, De Wit A, Wagemakers M, Rofes A, Bastiaanse R, et al. Comparing navigated transcranial magnetic stimulation mapping and "gold standard" direct cortical stimulation mapping in neurosurgery: a systematic review. Neurosurg Rev. 2021;44(4):1903–20.

Jung J, Lavrador JP, Patel S, Giamouriadis A, Lam J, Bhangoo R, et al. First United Kingdom experience of navigated transcranial magnetic stimulation in preoperative mapping of brain tumors. World Neurosurg. 2019;122:e1578–87.

Kaye HL, Gersner R, Boes AD, Pascual-Leone A, Rotenberg A. Persistent uncrossed corticospinal connections in patients with intractable focal epilepsy. Epilepsy Behav. 2017;75:66–71.

Krieg SM, Lioumis P, Mäkelä JP, Wilenius J, Karhu J, Hannula H, et al. Protocol for motor and language mapping by navigated TMS in patients and healthy volunteers; workshop report. Acta Neurochir. 2017;159(7):1187–95.

Krieg SM, Bernhard D, Ille S, Meyer B, Combs S, Rotenberg A, et al. Neurosurgery for eloquent lesions in children: state-of-the-art rationale and technical implications of perioperative neurophysiology. Neurosurg Focus. 2022;53(6):E4.

Lang N, Rothkegel H, Peckolt H, Deuschl G. Effects of lacosamide and carbamazepine on human motor cortex excitability: a double-blind, placebo-controlled transcranial magnetic stimulation study. Seizure. 2013;22(9):726–30.

Lefaucheur JP, Picht T. The value of preoperative functional cortical mapping using navigated TMS. Neurophysiol Clin. 2016;46(2):125–33.

Lehtinen H, Mäkelä JP, Mäkelä T, Lioumis P, Metsähonkala L, Hokkanen L, et al. Language mapping with navigated transcranial magnetic stimulation in pediatric and adult patients undergoing epilepsy surgery: comparison with extraoperative direct cortical stimulation. Epilepsia Open. 2018;3(2):224–35.

Liu J, Tsuboyama M, Jannati A, Kaye HL, Hipp JF, Rotenberg A. Shortened motor evoked potential latency in the epileptic hemisphere of children with focal epilepsy. J Clin Neurophysiol [Internet] 2023 [cited 2023 Oct 22]; Available from: https://journals.lww.com/10.1097/WNP.0000000000001022

Narayana S, Papanicolaou AC, McGregor A, Boop FA, Wheless JW. Clinical applications of transcranial magnetic stimulation in pediatric neurology. J Child Neurol. 2015a;30(9):1111–24.

Narayana S, Rezaie R, McAfee SS, Choudhri AF, Babajani-Feremi A, Fulton S, et al. Assessing motor function in Young children with transcranial magnetic stimulation. Pediatr Neurol. 2015b;52(1):94–103.

Narayana S, Embury LM, Shah N, Weatherspoon S, Choudhri AF, Boop FA. Noninvasive localization of language cortex in an awake 4-year-old child with Rasmussen encephalitis: a case report. Operat Neurosurg. 2020;18(5):E175–80.

Narayana S, Gibbs SK, Fulton SP, McGregor AL, Mudigoudar B, Weatherspoon SE, et al. Clinical utility of transcranial magnetic stimulation (TMS) in the Presurgical evaluation of motor, speech, and language functions in Young children with refractory epilepsy or brain tumor: preliminary evidence. Front Neurol. 2021;12:650830.

Noorizadeh N, Wheless JW, Fulton S, Mudigoudar B, Nevill L, Holder CM, et al. Concordance between TMS and Wada in determining hemispheric dominance for language. In Annual Meeting of the American Epilepsy Society in Nashville TN, USA; 2022.

Papanicolaou AC, Rezaie R, Narayana S, Choudhri AF, Abbas-Babajani-Feremi, Boop FA, et al. On the relative merits of invasive and non-invasive pre-surgical brain mapping: new tools in ablative epilepsy surgery. Epilepsy Res. 2018;142:153–5.

Pasichnik A, Tsuboyama M, Jannati A, Vega C, Kaye HL, Damar U, et al. Discrepant expressive language lateralization in children and adolescents with epilepsy. Ann Clin Transl Neurol. 2022;9(9):1459–64.

Paulus W, Classen J, Cohen LG, Large C, Di Lazzaro V, Nitsche MA, et al. State of the art: pharmacologic effects on cortical excitability measures tested by transcranial magnetic stimulation. Brain Stimul. 2008;1:151–63.

Pawley AD, Chowdhury FA, Tangwiriyasakul C, Ceronie B, Elwes RDC, Nashef L, et al. Cortical excitability correlates with seizure control and epilepsy duration in chronic epilepsy. Ann Clin Transl Neurol. 2017;4(2):87–97.

Rezaie R, Schiller KK, Embury L, Boop FA, Wheless JW, Narayana S. The clinical utility of transcranial magnetic stimulation in determining hemispheric dominance for language: a magnetoencephalography comparison study. J Clin Neurophysiol. 2020;37(2):90–103.

Rösler J, Niraula B, Strack V, Zdunczyk A, Schilt S, Savolainen P, et al. Language mapping in healthy volunteers and brain tumor patients with a novel navigated TMS system: evidence of tumor-induced plasticity. Clin Neurophysiol. 2014;125(3):526–36.

Säisänen L, Könönen M, Julkunen P, Määttä S, Vanninen R, Immonen A, et al. Non-invasive preoperative localization of primary motor cortex in epilepsy surgery by navigated transcranial magnetic stimulation. Epilepsy Res. 2010;92(2–3):134–44.

Schevon CA, Carlson C, Zaroff CM, Weiner HJ, Doyle WK, Miles D, et al. Pediatric language mapping: sensitivity of Neurostimulation and Wada testing in epilepsy surgery. Epilepsia. 2007;48(3):539–45.

Schiller K, Choudhri AF, Jones T, Holder C, Wheless JW, Narayana S. Concordance between transcranial magnetic stimulation and functional magnetic resonance imaging (MRI) derived localization of language in a clinical cohort. J Child Neurol. 2020;35(6):363–79.

Schramm S, Mehta A, Auguste KI, Tarapore PE. Navigated transcranial magnetic stimulation mapping of the motor cortex for preoperative diagnostics in pediatric epilepsy. J Neurosurg Pediatr. 2021;28:287–94.

Sebastiano DR, Tassi L, Duran D, Visani E, Gozzo F, Cardinale F, et al. Identifying the epileptogenic zone by four non-invasive imaging techniques versus stereo-EEG in MRI-negative presurgery epilepsy patients. Clin Neurophysiol. 2020;131(8):1815–23.

Sherman EMS, Wiebe S, Fay-McClymont TB, Tellez-Zenteno J, Metcalfe A, Hernandez-Ronquillo L, et al. Neuropsychological outcomes after epilepsy surgery: systematic review and pooled estimates: cognitive change after epilepsy surgery. Epilepsia. 2011;52(5):857–69.

Skirrow C, Cross JH, Owens R, Weiss-Croft L, Martin-Sanfilippo P, Banks T, et al. Determinants of IQ outcome after focal epilepsy surgery in childhood: a longitudinal case-control neuroimaging study. Epilepsia. 2019;60(5):872–84.

Spencer SS, Berg AT, Vickrey BG, Sperling MR, Bazil CW, Haut S, et al. Health-related quality of life over time since resective epilepsy surgery. Ann Neurol. 2007;62(4):327–34.

Tsuboyama M, Kaye HL, Rotenberg A. Review of transcranial magnetic stimulation in epilepsy. Clin Ther. 2020;42(7):1155–68.

Vitikainen AM, Lioumis P, Paetau R, Salli E, Komssi S, Metsähonkala L, et al. Combined use of non-invasive techniques for improved functional localization for a selected group of epilepsy surgery candidates. NeuroImage. 2009;45(2):342–8.

Vitikainen AM, Salli E, Lioumis P, Mäkelä JP, Metsähonkala L. Applicability of nTMS in locating the motor cortical representation areas in patients with epilepsy. Acta Neurochir. 2013;155(3):507–18.

Walton D, Spencer DC, Nevitt SJ, Michael BD. Transcranial magnetic stimulation for the treatment of epilepsy. Cochrane Epilepsy Group, editor. Cochrane Database Syst Rev. 2021;2021(4 Art. No.: CD011025.):1–33.

Wang TR, Bailey RC, Goodkin HP, Mahaney KB. Trends in intracranial monitoring for pediatric medically intractable epilepsy: 2000–2012. Neurology. 2018;90(9):e771–8.

Wassermann EM, Pascual-Leone A, Valls-Solé J, Toro C, Cohen LG, Hallett M. Topography of the inhibitory and excitatory responses to transcranial magnetic stimulation in a hand muscle. Electroencephalogr Clin Neurophysiol. 1993;89(6):424–33.

Wiebe S, Blume W, Girvin J, Wliasziw M. A randomized, controlled trial of surgery for temporal-lobe epilepsy. N Engl J Med. 2001;345(5):311–8.

Wolters A, Ziemann U, Benecke R. The cortical silent period. In: Wassermann EM, Epstein CM, Ziemann U, Walsh V, Paus T, Lisanby SH, editors. The Oxford handbook of transcranial stimulation. New York: Oxford University Press; 2011. p. 91–102.

Yelemali P, Schiller KK, Gibbs S, Wheless JW, Narayana S. Clinical, demographic, and behavioral correlates of TMS cortical excitability measure in focal motor epilepsy and tumor patients. Rhodes J Biol Sci. 2020;34(Spring):54–9.

Yerys BE, Jankowski KF, Shook D, Rosenberger LR, Barnes KA, Berl MM, et al. The fMRI success rate of children and adolescents: typical development, epilepsy, attention deficit/hyperactivity disorder, and autism spectrum disorders. Hum Brain Mapp. 2009;30(10):3426–35.

Zeiler FA, Matuszczak M, Teitelbaum J, Gillman LM, Kazina CJ. Transcranial magnetic stimulation for status epilepticus. Epilepsy Res Treat. 2015;2015:1–10.

Zhang X, Zhang G, Yu T, Xu C, Yan X, Ma K, et al. Multitask preoperative language mapping in epilepsy surgery: a combination of navigated transcranial magnetic stimulation and extraoperative electrical cortical stimulation. J Clin Neurosci. 2020;79:259–65.

Ziemann U. TMS and drugs. Clin Neurophysiol. 2004;115(8):1717–29.

Assessment of Chronic Cervical Myelopathy

15

Zdunczyk

Degenerative cervical myelopathy (DCM) is a chronic and progressive spinal cord injury affecting the cervical spine. The pathophysiology is characterized by a persistent and overlying compression of the cervical spinal cord, resulting from degenerative disc disease, spondyloarthropathy, and/or other degenerative pathologies. This compression initiates a cascade of pathological changes. The compression of the spinal cord reduces blood flow and impairs the supply of nutrients, which in turn causes mechanical and metabolic injury to the neural tissues. Over time, these factors result in neuronal dysfunction, demyelination, and neuronal cell death, which contribute to the progressive neurological deficits characteristic of DCM.

15.1 Epidemiology

DCM is primarily caused by age-related changes in the cervical spine, although genetic predisposition and environmental factors may also play a role. The clinical presentation is marked by a broad variety of neurological symptoms, including hypesthesia and dysesthesia, loss of manual dexterity, gait deficits, motor paresis, and bladder and bowel dysfunction. Overall, DCM is a major cause of disability in the elderly.

Zdunczyk (✉)
Department of Neurosurgery, Charité – Universitätsmedizin Berlin, Freie Universität Berlin, Humboldt-Universität zu Berlin, and Berlin Institute of Health, Berlin, Germany
e-mail: anna.zdunczyk@charite.de

S. M. Krieg, T. Picht (eds.), *Navigated Transcranial Magnetic Stimulation in Neurosurgery*, https://doi.org/10.1007/978-3-031-97155-6_15

15.2 Conventional Preoperative Methods

Diagnosing DCM is generally a clinical diagnosis combined with the proof of spinal cord compression by neuroimaging studies. Magnetic resonance imaging (MRI) is the gold standard for visualizing spinal cord compression and identifying structural abnormalities such as disc herniation, osteophyte formation, and ligamentum flavum hypertrophy. Additionally, electrophysiological studies addressing the central conduction SSEP and MEP as well as EMG/ENG may be performed to assess nerve function and rule out other neurological conditions. However, it is still poorly understood why the disease course sometimes presents with severely and progressive, and in other cases slowly and remittent, symptoms, even though MRI and electrophysiology reveal comparable results. Here we face a diagnostic gap to early detect patients who carry a high risk for clinical deterioration and who will benefit from early decompressive surgery.

15.3 Corticospinal Reserve Capacity Assessment

NTMS has been shown to be a useful tool for noninvasively characterizing the pattern of motor network dysfunction in patients with DCM. This assessment is defined by changes in corticocortical and corticospinal excitability as reflected by resting motor threshold (RMT), MEP latency, cortical silent period (CSP), and motor area.

Due to specific, adaptive changes in the motor network observed in the three clinical severity stages of DCM (mild, moderate, severe), the concept of corticospinal reserve capacity was introduced (Zdunczyk et al. 2018). In this three-stage model, adaptive mechanisms, including recruitment of secondary motor areas and corticospinal disinhibition, allow patients to remain adequately functional. In an early stage of the disease, the state of "compensation" (mild, JOA >15) is defined by an initial decline in RC, but unchanged CSP and preserved motor area. As the disease progresses, the first symptoms of corticospinal dysfunction appear (moderate, JOA 12–15). This stage is characterized by a maximum expansion of the neuronal reserve to maintain motor function, reflected by a progressive decrease in RC with a simultaneous reactive decrease in CSP (disinhibition) and increase in motor area (unmasking of preexisting latent lateral connections and reintegration of adjacent neuronal populations). As the disease continues and a critical limit of reserve expansion is passed, severe neurological symptoms accumulate (severe, JOA <12). The state of "exhaustion" is characterized by reduced RC, increased CSP (high cortical inhibition), and severely reduced motor area.

15.3.1 Resting Motor Threshold

RMT reflects the cortical excitability of pyramidal axons and their corticospinal connections. An increase in RMT was observed in pathologies characterized by a reduced density of corticospinal neurons. A higher stimulation intensity was required to discharge resting spinal motoneurons (Di Lazzaro et al. 2004). In patients with DCM suffering from mild to moderate symptoms (Zdunczyk et al. 2018), no relevant difference was found when RMT was compared with healthy subjects. Net corticocortical excitability remains unchanged until a critical level of neurodegeneration is reached. However, in the early stages of the disease, most axotomized corticospinal neurons survive and regain functional access to sensorimotor circuits through axonal sprouting into unaffected projections (Schwab et al. 2006).

15.3.2 Recruitment Curve

The MEP RC serves indirectly as a measure of axonal integrity excitability and corticospinal conductivity. The slope of the curve depends not only on the distribution of excitability within the corticospinal pathway but also on the spatial distribution of excitable elements in the cortex under the stimulation coil (Nicotra et al. 2013; Siebner et al. 2003). In previous investigations in stroke patients, the severity of axonal damage has been attributed to a reduced and flattened RC.

To obtain the RC of each hemisphere, single TMS pulses are delivered over the hotspot at varying stimulus intensities between 80% and 140% RMT (5–7 pulses, determined automatically by the system software for each subject). Each MEP amplitude is then plotted as a function of stimulus intensity (in % stimulator power) with the parameters RC slope, V50, and R according to the variable slope analysis.

15.3.3 MEP Latency

MEP latency, in particular, serves as a valuable parameter in evaluating the functional integrity of the motor pathways and is also used to describe impairment in signal conduction due to demyelination and axonal degeneration. Motor evoked potentials (MEPs) are electrical signals generated in response to transcranial magnetic or electrical stimulation of the motor cortex. These signals travel through the corticospinal tract and are recorded from various muscles. The time taken for these signals to travel from the cortex to the peripheral muscles is referred to as MEP latency. This latency reflects the conduction speed of neural impulses along the motor pathways (Barker et al. 1987; Day et al. 1989).

In cervical myelopathy, the compression of the spinal cord leads to demyelination and axonal damage, which in turn slows down neural conduction. This pathological process is reflected in the prolonged MEP latency observed in affected patients. Several studies have demonstrated that patients with cervical myelopathy exhibit significantly increased MEP latency compared to healthy individuals (Nakamae et al. 2009; Nicotra et al. 2012; Lanza et al. 2020).

The measurement of MEP latency holds considerable clinical significance in the context of cervical myelopathy. Increased MEP latency can indicate the presence of spinal cord compression even in cases where conventional imaging may not reveal significant abnormalities. Further, MEP latency is useful in monitoring disease progression and response to treatment. Finally a reduction in MEP latency following surgical decompression or other therapeutic interventions suggests improvement in neural conduction and, consequently, clinical recovery (Nakanishi et al. 2014; Li et al. 2022; Barthélemy et al. 2015).

In our protocol we express it as the median value in ms of five stimuli applied over the hotspot using the individual RMT intensity for each hemisphere.

15.3.4 Cortical Silent Period

Intracortical and corticospinal inhibition are estimated by the duration of the CSP. The shift in CSP is thought to be mediated by gamma-aminobutyric acid (GABA)B receptors on pyramidal cells (McCormick et al. 1989). The early phase of CSP (first 50–75 ms) is mainly mediated by spinal inhibition, whereas the late phase is characterized by long-lasting cortical inhibition mediated by GABAB (Inghilleri et al. 1993; Werhahn et al. 1999). A reduced afferent excitatory cortico-cortical and thalamocortical drive could be due to degeneration of corticospinal axons and soma shrinkage of pyramidal cells (Inghilleri et al. 1993; Werhahn et al. 1999). In our previous investigations, a longer CSP was observed in patients with severe symptoms (JOA ≤12) and was mainly associated with a reduced motor area (Zdunczyk et al. 2018), and high cortical inhibition could therefore be associated with functional impairment and reduced motor area activation.

To measure cortical and corticospinal inhibition in DCM, the absolute cortical silent period (CSP) over the hotspot is measured for both hemispheres. The subject is instructed to clench both fists with maximum force while 10 stimuli are delivered at 140% RMT. To estimate the absolute duration of the CSP, the latency (ms) from MEP offset to the end of resumption of sustained electromyographic (EMG) activity was determined post hoc.

15.3.5 Motor Area

Cortical reorganization plays a critical role in functional recovery (Ghosh et al. 2009; Kaas et al. 2008; Kao et al. 2009). During recovery, enlargement of the original motor area has been observed through unmasking of preexisting latent lateral connections, facilitation of ineffective synapses, and/or formation of new synaptic connections within the precentral gyrus (Jurkiewicz et al. 2007; Lotze et al. 2006). New corticospinal projections have also been detected. Axonal sprouting may be the underlying mechanism for the reintegration of injured axons into sensorimotor circuits (Jurkiewicz et al. 2010; Darian-Smith et al. 2014). In patients following traumatic SCI, fMRI studies have shown an increased volume of M1 activation as well as increased activation of nonprimary motor areas. (Jurkiewicz et al. 2007; Kao et al. 2009) Motor area enlargement was associated with favorable motor recovery (Duggal et al. 2010) and was also observed in an fMRI study of patients with cervical spinal stenosis. This finding suggests that adaptive mechanisms such as unmasking of latent connections, disinhibition, and formation of new synapses may lead to functional reorganization of the motor area to maintain sufficient corticospinal output.

To assess the magnitude of motor cortical activation, motor mapping is performed at 105% RMT of the FDI for both hands. The TMS coil stimulates the area of cortex closest to the coil, and therefore, the muscle representation area can be simplified to a surface following the shape of the skull. Using the normal vector approach, we project the original stimulus locations for the FDI from three-dimensional space onto the two-dimensional surface. Using the convex hull method (Julkunen et al. 2014), the transformed coordinates are used to calculate the cortical representation area.

15.3.6 Implications for Clinical Practice

Based on the results of our previous investigations, we propose a three-step model of corticospinal reserve capacity. Our findings indicate that in patients with cervical myelopathy and mild symptoms, adaptive mechanisms, including recruitment of secondary motor areas and disinhibition, enable patients to remain adequately functional This state of "compensation" is marked by preserved motor area, a beginning decrease in RC, and unchanged CSP. Due to the preserved functional "corticospinal reserve," the JOA >12 group may have a greater likelihood of benefiting from surgery by recovering from neurological symptoms or at least maintaining their current status.

When compensatory mechanisms reach their limit, there is an imminent risk of neurological deterioration. This stage of "exhaustion" is marked by a highly increased motor area, decreased cortical inhibition, and further reduced recruitment curve. Thus, this patient group should undergo short-term surgery to prevent further clinical impairment. At an advanced stage of the disease, progressing neurological damage directly translates into further clinical impairment. This stage of "deterioration" is characterized by a reduced motor area, high cortical inhibition with increased CSP, and low recruitment curve.

The evaluation of corticocortical and corticospinal excitability by nTMS revealed fundamental adaptive mechanisms in the motor network. The new concept of func tional compensation for DCM on the cortical and spinal level is an individual capacity to adapt to the state of chronic spinal cord injury during DCM in order to maintain motor functionality.

It is hypothesized that by detecting the degree of compensatory reorganization, it is possible to differentiate between a preserved and an exhausted functional "corticospinal reserve," which could potentially stratify future patients in terms of risk for further neurological deterioration on the one hand and potential for recovery after surgical decompression on the other hand. The novel findings will assist physicians in providing more objective counsel to their patients and informing them about the natural history of cervical myelopathy and the risks associated with various treatment options.

15.4 Future Directions

Future research should aim to conduct multicenter studies to enhance the generalizability of the corticospinal reserve concept. Long-term follow-up assessment is critical to understanding the progression of compensatory mechanisms and their sustained impact on clinical outcomes in a chronic disease like DCM. Studies to explore the potential of neurorehabilitative neuromodulatory strategies to enhance the corticospinal reserve capacity and improve functional recovery are another promising direction.

References

Zdunczyk A, Schwarzer V, Mikhailov M, Bagley B, Rosenstock T, Picht T, Vajkoczy P. The Corticospinal Reserve Capacity: Reorganization of Motor Area and Excitability As a Novel Pathophysiological Concept in Cervical Myelopathy. Neurosurgery. 2018;83(4):810–18. https://doi.org/10.1093/neuros/nyx437.

Part V

Neuromodulation and Therapeutic Applications in Neurosurgery

Navigated rTMS for the Treatment of Pain

16

Jean-Pascal Lefaucheur and Benjamin Bardel

16.1 Introduction

Chronic pain syndromes can have diverse origins but present similar sensory, cognitive, or affective aspects, more or less intertwined in the foreground of clinical symptomatology (Garcia-Larrea and Peyron 2013). Classically, three pathophysiological mechanisms of chronic pain can be distinguished: nociceptive pain (due to tissue injury and excessive activation of nociceptors), neuropathic pain (due to injury or impaired function of the peripheral or central sensory nervous system), or nociplastic pain (due to primary central sensitization) (Cohen et al. 2021). Pharmacotherapy may prove insufficient to control chronic pain syndromes, leaving room for non-drug solutions for intractable pain, including neuromodulation techniques (Knotkova et al. 2021). Neuromodulation for analgesic purposes can be invasive or non-invasive (using or not using surgically implanted electrodes and stimulators) applied to various peripheral or central nervous structures (Nizard et al. 2012a, b). The cerebral cortex is one of the potential targets of analgesic neuromodulation, since the first clinical results were reported in the early nineties (Tsubokawa et al. 1991a, b). Invasive epidural stimulation of the primary motor cortex (M1) was shown to produce long-term analgesia in about half of patients with chronic neuropathic pain resistant to medication (Fontaine et al. 2009; Nguyen et al. 2009; Cruccu et al. 2016). In the mid-nineties, therapeutic strategies of non-invasive cortical stimulation were developed using repetitive transcranial magnetic stimulation (rTMS), first for treating depression (George et al. 1996). At the end of the nineties (Lefaucheur et al. 1998, 2001a, b), rTMS therapy delivered to M1 has

J.-P. Lefaucheur (✉) · B. Bardel
Clinical Neurophysiology Unit, UR 4391, Nervous Excitability and Therapeutics, Henri Mondor Hospital, Public Assistance – Paris Hospitals, Paris-Est-Créteil University, Créteil, France
e-mail: jean-pascal.lefaucheur@hmn.aphp.fr; benjamin.bardel@aphp.fr

S. M. Krieg, T. Picht (eds.), *Navigated Transcranial Magnetic Stimulation in Neurosurgery*, https://doi.org/10.1007/978-3-031-97155-6_16

also been successfully applied in patients suffering from chronic neuropathic pain, as well as in patients with non-neuropathic pain syndromes (Lefaucheur 2006, 2016; Lefaucheur et al. 2008a; Leung et al. 2009). However, in this chapter we will deal almost exclusively with the treatment of refractory neuropathic pain, as it is mainly in this context that neuronavigated rTMS procedures have been used.

16.2 Clinical Evidence

A group of international experts found a level A of evidence for the efficacy of rTMS to relieve focal neuropathic pain when applied with a focal figure-of-eight coil as trains of magnetic pulses delivered at high frequency (HF, 10–20 Hz) over M1 (precentral gyrus), on the contralateral hemisphere to the painful side (Lefaucheur et al. 2020). Numerous sham-controlled studies have reported a beneficial analgesic effect of repeated sessions of HF rTMS over M1 (3–15 daily sessions) in patients with neuropathic pain (including phantom limb pain and orofacial neuropathic pain) in the short term (1 week to less than 1 month beyond the time of stimulation) (Khedr et al. 2005, 2015; Fricová et al. 2013; Hasan et al. 2014; Attal et al. 2016; Malavera et al. 2016; Nurmikko et al. 2016; Mori et al. 2021b; Zhao et al. 2021) or in the medium term (lasting effect for more than 1 month up to 3 months) (Ahmed et al. 2011; Yılmaz et al. 2014; Ma et al. 2015; Pei et al. 2019). Naturalistic studies (Hodaj et al. 2015; Kobayashi et al. 2015; Pommier et al. 2016; Quesada et al. 2018; Hodaj et al. 2020) have shown that the analgesic effect of rTMS could even be prolonged for at least 6 months in patients with neuropathic pain responding to treatment by means of maintenance sessions, which has also been confirmed by sham-controlled studies (Quesada et al. 2020; Attal et al. 2021).

Thus, repetition of daily rTMS sessions is capable of producing cumulative effects in terms of intensity and duration of pain relief, reducing pain scores by 20–45% on average with an overall responder rate that can be estimated between 35% and 60% (Lefaucheur et al. 2014). A series of 4–10 rTMS sessions (6–7 on average) is normally sufficient to determine whether or not a patient responds to the procedure on a given stimulation target (Quesada et al. 2018; Lefaucheur and Nguyen 2019). Analgesic efficacy must be evaluated on clinical grounds, with regard to the reduction of pain intensity and also its interference with activities of daily living and quality of life. To prolong and amplify the analgesic effects of M1 HF-rTMS, it is necessary to perform repeated maintenance sessions. The frequency of these maintenance sessions can be weekly (Kobayashi et al. 2015), monthly (Lefaucheur et al. 2004a; Pommier et al. 2016; Quesada et al. 2018, 2020), or spaced according to the duration of the clinical post-effect (Lefaucheur and Nguyen 2019). A model of increasing spacing of intersession intervals (Hodaj et al. 2015, 2020; Attal et al. 2021) can be proposed based on Ebbinghaus' spaced repetition theory, which has been shown to improve the consolidation of memorization or learning (and therefore brain plasticity) compared to a repetition with intervals of equal duration (Ebbinghaus 2013).

16.3 Optimal Parameters of Stimulation

Stimulation frequency and number of pulses per session were found to be crucial rTMS parameters to ensure analgesic efficacy. In patients with chronic neuropathic pain, whether of peripheral or central origin, analgesia can be produced by the administration of HF-rTMS (10–20 Hz) over M1, but not of low-frequency (LF) rTMS (0.5–1 Hz) (Lefaucheur et al. 2001a; André-Obadia et al. 2006; Saitoh et al. 2007). Furthermore, two studies demonstrated that rTMS at 10 Hz was more effective than rTMS at 5 Hz (Saitoh et al. 2007; Mori et al. 2022), which supports the preferential use of a frequency of 10 Hz or more. Analgesic efficacy could be even better at a frequency of 20 Hz (Fricová et al. 2013), as also suggested by a recent meta-analysis (Naik et al. 2024), but this may be due to a bias of publication and is not supported by a comparative study (Säisänen et al. 2022). Regarding the number of pulses per session, it should not be less than 1000 pulses (Mori et al. 2022), but there is no evidence of the benefit of increasing this value to more than 2000 pulses (Säisänen et al. 2022).

Among the stimulation parameters that still require consensus, the question arises of how to optimally position the stimulation coil over the motor cortex, especially since almost all reported analgesic effects of rTMS of the motor cortex to date have been obtained with a figure-of-eight coil, which delivers a relatively focal electric field in the brain. The use of circular and double-cone coils delivering a larger electric field has been shown to be ineffective (Rollnik et al. 2002), with the exception of studies using a double-cone coil to target the posterior insula (Dongyang et al. 2021; da Cunha et al. 2022). Other types of coils, producing even wider and deeper cortical stimulation, such as H coils, have shown certain analgesic effects in chronic pain patients (Onesti et al. 2013; Shimizu et al. 2017; Bouhassira et al. 2024), but these results remain to be confirmed.

In fact, the use of a figure-of-eight coil raises various specific targeting questions and supports the need for a neuronavigated procedure for rTMS treatment of pain patients (Lefaucheur 2010).

16.4 Navigated rTMS Procedure for Pain Therapy

A navigation system integrates high resolution structural T1-weighted magnetic resonance imaging (MRI) data and can generate a three-dimensional reconstruction of the brain of an individual patient. Using a camera that detects markers placed on the patient's head and on the rTMS coil, these two elements are co-registered in the patient's individual MRI space. This allows real-time visualization of the location and orientation of the coil on the patient's head with an accuracy of approximately 5 mm in the motor cortical region (Ruohonen and Karhu 2010; Schmidt et al. 2015).

When using a navigation system to guide the targeting of rTMS for pain treatment, there are two different issues depending on whether one considers the rostrocaudal axis or the mediolateral axis from the perspective of cortical anatomy.

First of all, regarding the rostrocaudal axis, the question concerns distinguishing the central sulcus and the precentral gyrus in the functional anatomy of M1 in order to define the optimal stimulation site to produce analgesic effects. It is known that M1 occupies the anterior wall of the central sulcus and only a limited part of the exposed surface of the precentral gyrus (Geyer et al. 2000; Rademacher et al. 2001). This fact would justify the stimulation of the anterior bank of the central sulcus and the use of a navigation system for this purpose. Indeed, the "standard," non-navigated procedure of targeting M1 is based on the determination of the motor hotspot location. The motor hotspot is, by definition, the scalp position where a single-pulse TMS generates the largest motor evoked potentials (MEPs) in a given muscle. However, there is an important interobserver variability (up to 1 cm) of the motor hotspot (Sollmann et al. 2013; Nazarova et al. 2021), and it has been demonstrated that, in at least half of the subjects, the motor hotspot was located near to the precentral sulcus and thus very far anterior to the central sulcus (Ahdab et al. 2016). This anterior shift was reported by several groups (Denslow et al. 2005; Teitti et al. 2008; Ahdab et al. 2010; Diekhoff et al. 2011; Julkunen et al. 2011) and can be explained by two factors: (i) the presence of pyramidal neurons at the origin of corticospinal tracts within the premotor cortical areas (He et al. 1993, 1995) and (ii) the indirect activation of corticospinal tracts in M1 via the dense projections that connect the premotor cortex and M1 (Ghosh and Porter 1988; Shimazu et al. 2004). Furthermore, under pathological conditions, this anatomical relationship between the motor hotspot and central sulcus location becomes even more ambiguous. This is due to disease-induced cortical plasticity processes, as can be shown, for example, by preoperative motor cortical mapping (Lefaucheur and Picht 2016).

On the other hand, there is strong evidence to conclude that the analgesic effects of rTMS delivered to the precentral gyrus are not directly related to pyramidal neuron activation or motor cortical output drive. Modeling studies have shown that the analgesic effects are related to the activation of axons, which run in the superficial cortical layers at the top of the crown of the precentral gyrus, tangential to the cortical surface and close to the central sulcus (Holsheimer et al. 2007a, b; Manola et al. 2007; Nguyen et al. 2011; Saitoh et al. 2013). To produce analgesic effects, these superficial tangential fibers must be preferentially stimulated, and if a figure-of-eight coil is used, the coil handle must be oriented in an anteroposterior direction, parallel to the interhemispheric fissure (André-Obadia et al. 2008; Lefaucheur et al. 2010). Indeed, the rotation of the coil from an anteroposterior axis to a mediolateral axis is capable of modifying the type of activated axons from a tangential to radial orientation in the gyral crown of the cortex (Siebner et al. 2022).

Thus, for various reasons, it appears that the location of the motor hotspot could not be considered a reliable marker to precisely locate the central sulcus. Therefore, it is essential to use a navigation system if the theoretical objective is to precisely target motor cortex rTMS on the superficial grey matter at the level of the anterior bank of the central sulcus. However, the superiority of an anatomically defined target using navigation to produce pain relief remains to be formally demonstrated on clinical grounds.

Indeed, few studies compared the analgesic effect of using navigated rTMS compared to non-navigated rTMS. One study reported similar analgesic effects of navigated versus non-navigated rTMS targeted to the face motor cortical area to treat facial pain (Hodaj et al. 2015). Conversely, another study reported more prolonged analgesic effects of navigated rTMS targeting the somatotopic region of pain compared to non-navigated rTMS targeting the hand motor hotspot in patients with upper or lower limb neuropathic pain (Ayache et al. 2016). Finally, one study showed that the addition of rTMS sessions targeted to an alternative motor cortical region (defined by motor task-related functional MRI, fMRI) increased the rate of responders compared to stimulation of the standard motor hotspot site in a series of patients suffering from chronic pain of various origins (Nurmikko et al. 2016).

The second anatomical issue of rTMS navigation concerns the mediolateral axis along which the different parts of the body (lower limb, upper limb, face) are represented in an ordered manner (motor homunculus). The question is to know whether the analgesic effects of stimulation of the motor cortex are somatotopic (i.e., related to a precise anatomical cortical targeting within the motor homunculus) or not. Furthermore, even if the analgesic effects of stimulation are somatotopic, it remains to be determined whether they are homotopic (i.e., rTMS should be delivered to the specific region of the motor cortical representation of the pain location) or not. This poses a problem regarding the necessity and interest of specifically stimulating the cortical region that corresponds to the painful area.

In most rTMS studies based on targeting the motor hotspot, analgesic effects were obtained by stimulating the motor cortical representation of the hand contralateral to the painful side, regardless of the pain location (Lefaucheur et al. 2004b, 2008b, 2012; Khedr et al. 2005; André-Obadia et al. 2006, 2011; Matsumura et al. 2013; Attal et al. 2016; Andre-Obadia et al. 2018). Similarly, navigated HF-rTMS of M1 has often been targeted to the motor cortical representation of the hand (Hosomi et al. 2013a; Pommier et al. 2016; Quesada et al. 2018, 2020; Attal et al. 2021; Zhao et al. 2021; André-Obadia et al. 2023; Bouhassira et al. 2024). However, other studies have reported the analgesic effects of homotopically delivered rTMS sessions on the motor cortical representation of the pain site (Hirayama et al. 2006; Saitoh et al. 2007; Hosomi et al. 2013b, 2020; Hasan et al. 2014; Hodaj et al. 2015, 2020, 2023; Ayache et al. 2016; Nurmikko et al. 2016).

In the few studies that have compared homotopic versus heterotopic stimulation, similar analgesic effects have generally been reported between the two types of targeting procedures. For example, one study showed that stimulation of the hand motor cortical region improved patients with either facial or hand pain, while stimulation of the face motor cortical region was ineffective in both cases (Andre-Obadia et al. 2018). Another study showed that facial pain was even better improved by stimulation of the hand than the face motor cortical representation (and vice versa for hand pain, which benefited more from stimulation of the face motor cortical representation) (Lefaucheur et al. 2006). For lower limb pain, although debatable (Mori et al. 2021a), the analgesic effect of HF-rTMS did not appear to differ between stimulation of the hand or foot motor cortical region (Jetté et al. 2013; Mori et al. 2021b). Thus, rTMS therapy could be initiated in all cases by stimulating the hand motor region (rather defined on anatomical data using a navigation system), then moved to other cortical motor regions in the absence of analgesic effects (Lefaucheur and Nguyen 2019).

If the outcome of motor cortex rTMS is influenced by somatotopy, the anatomy of the motor cortical representations should be taken into consideration for the application of navigated HF-rTMS of M1. The segment of the central sulcus facing the superior frontal gyrus (F1) corresponds to lower limb motor representation, the middle frontal gyrus (F2) to the upper limbs, and the inferior frontal gyrus (F3) to the face (Fig. 16.1) (Penfield and Boldrey 1937; Nguyen et al. 1999; Ahdab et al. 2014). These anatomical considerations require the use of a navigation system to accurately target these specific regions (Lefaucheur et al. 2024).

However, there is clear evidence that the analgesic effects produced by HF-rTMS of M1 are not always somatotopic. For example, it has clearly been shown that focal stimulation of the motor cortex (hand representation) can be beneficial for diffuse pain syndromes, such as fibromyalgia (Passard et al. 2007; Mhalla et al. 2011; Boyer et al. 2014).

In conclusion, it remains to be demonstrated that navigated HF-rTMS of M1 leads to a better outcome when a specific region of the anterior bank of the central

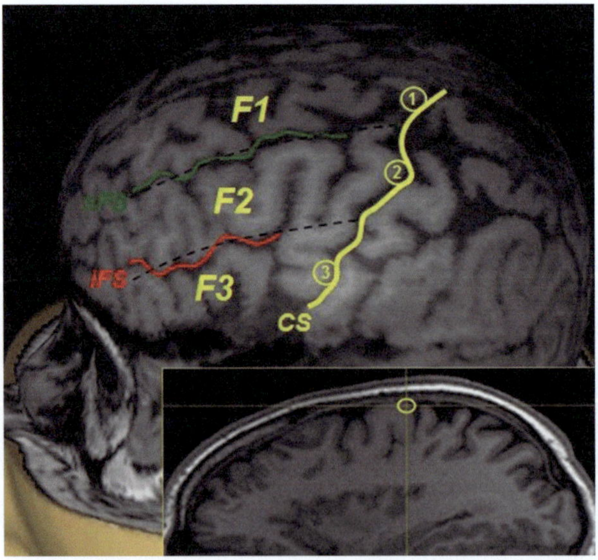

Fig. 16.1 Motor cortex targets for navigated rTMS pain therapy, anatomically located on the anterior bank of the central sulcus (CS). On the mediolateral axis: (1) lower limb target facing the superior frontal gyrus (F1), above the level of the superior frontal sulcus (sFS); (2) upper limb target facing the middle frontal gyrus (F2), above the level of the inferior frontal sulcus (iFS); (3) face target facing the inferior frontal gyrus (F3), below the level of the iFS. On the rostrocaudal axis (sagittal view, inferior part of the figure): the target is located at the top of the anterior wall of the CS, i.e. at the posterior border of the precentral gyrus. (Adapted from Ayache et al. 2016)

sulcus of the homunculus is targeted rather than the hand motor cortical representation (hand knob), regardless of the location of pain.

In all cases, navigation is mandatory to ensure that the stimulation is delivered to the same scalp site and with the same coil orientation during repeated sessions in long-term rTMS therapy practice. Indeed, the use of a navigation system storing the individual coordinates of the optimal stimulation target for a given patient facilitates the performance of repeated rTMS sessions over long periods (Lefaucheur 2010). Compared to conventional targeting based on cranial landmarks drawn on a cap, the advantage of using a navigation system to improve the reproducibility (repeatability) of precise placement of the figure-of-eight coil during a session and between different rTMS sessions has been clearly demonstrated (Caulfield et al. 2022). Navigated HF-rTMS of M1 has clearly been shown to produce sustained analgesic effects in the long term (Lawson McLean et al. 2018).

16.5 Alternative Targets Outside Motor Cortex

The first study evaluating the analgesic effects of rTMS using image-guided navigation showed that only stimulation of M1 was able to relieve pain, but not stimulation of the postcentral gyrus, premotor area, or supplementary motor area (Hirayama et al. 2006). However, navigation could be useful for cortical targets located outside of M1 in the context of pain therapy, e.g., the dorsolateral prefrontal cortex (DLPFC), the right operculum region (S2), or the posterior insula. Beneficial results of prefrontal stimulation (either LF-rTMS or HF-rTMS) have been reported in (i) in neuropathic pain (Borckardt et al. 2009; Sampson et al. 2011; Nardone et al. 2017), although it remains controversial (de Oliveira et al. 2014; Attal et al. 2021; Wang et al. 2023) and (ii) non-neuropathic pain syndromes (Umezaki et al. 2016), including diffuse pain syndromes, such as fibromyalgia (Sampson et al. 2006; Short et al. 2011; Lee et al. 2012). However, the definition of the DLPFC target appears quite imprecise, in particular due to the significant extent of this cortical area (Nauczyciel et al. 2011; Mylius et al. 2013; Pommier et al. 2017).

Other studies have reported the beneficial effect of HF-rTMS targeting the S2 region to relieve neuropathic pain (Lindholm et al. 2015; Ojala et al. 2022). Finally, there are contradictory results on the effect of HF-rTMS targeting the posterior insula (Galhardoni et al. 2019; Dongyang et al. 2021; da Cunha et al. 2022). Thus, it remains to be determined whether cortical regions outside of M1 could serve as alternative rTMS targets for treating patients with neuropathic pain.

16.6 Navigation to Target Brain Networks

The mechanisms underlying the analgesic effects of rTMS are not well known, but they probably involve a modulation of various pain control networks. Indeed, since rTMS-induced currents activate axons more easily than cell bodies (Nowak and Bullier 1998a, b), the mechanisms of action of cortical stimulation must be modeled in terms of neural circuits rather than local brain activity changes. Therefore, in the future, integration of fiber tracking data provided by diffusion tensor imaging could be of particular interest for the treatment of pain by navigated rTMS. In this regard, it should be mentioned that two studies have shown, using tractography data, that the integrity of the thalamocortical tract predicted the beneficial effects of HF-rTMS of M1 in patients with central post-stroke pain (Goto et al. 2008; Ohn et al. 2012). These observations support the hypothesis of an involvement of antidromic modulation of thalamocortical pathways in the analgesic effect of motor cortex stimulation and therefore the potential interest of targeting these networks by the use of a navigated rTMS technique.

Another imaging approach is to define specific networks of the brain connectome using resting-state fMRI (rs-fMRI). Network-based navigated rTMS procedures have been proposed in two studies for the treatment of pain. In a first case report of a patient with phantom limb pain (Scibilia et al. 2018), HF-rTMS targeted over M1 was found to produce significant pain relief and to increase connectivity in the sensory-motor network as assessed by rs-fMRI. In another case series of 7 patients with central post-stroke pain, navigated rTMS was targeted according to a lesion-based network defined in patients in comparison to a normative connectome (Kim et al. 2022). Such novel methods of targeting rTMS for pain treatment are certainly promising and will obviously require an image-guided navigated approach.

16.7 nTMS for Preoperative Testing

A last remark concerns the practice of navigated TMS as a preoperative test for invasive chronic epidural motor cortex stimulation by means of surgically implanted electrodes. It has been demonstrated that the clinical response to HF-rTMS of M1 was correlated with a favorable outcome of the surgical procedure (André-Obadia et al. 2006; Hosomi et al. 2008; Lefaucheur et al. 2004a, 2011). The absence of response to preoperative rTMS tests was even correlated with a poor result of chronic motor cortex stimulation in the long term (André-Obadia et al. 2014). Moreover, rTMS and epidural motor cortex stimulation likely share common mechanisms of action (Lefaucheur et al. 2010). Also, it could be envisaged to use the coordinates of the target validated by the preoperative rTMS tests to guide the placement of the epidural electrodes during the surgical procedure. Unfortunately, data are lacking in this respect, as preoperative studies published to date were performed using a non-navigated rTMS technique.

16.8 Conclusion

It remains to be confirmed that the analgesic efficacy of motor cortex rTMS depends on precise anatomical targeting, which could benefit from a navigation system using morphological or functional brain imaging of each patient. To date, the clinical relevance of a navigated approach to rTMS pain therapy is not certain (Klein et al.

2015), even if navigation clearly facilitates daily practice, particularly with regard to the reproducibility of the procedure. Regarding the optimization of motor cortex rTMS targeting for pain therapy, the main questions to answer are: (i) Is the anatomical targeting of the anterior bank of the central sulcus better than functional targeting of the motor hotspot, whose location varies between subjects? (ii) Is the anatomical or functional targeting of the painful region better than that of the hand regardless of the location of pain? In the absence of a formal response to these two questions, it seems preferable to evaluate in each patient the respective analgesic efficacy of the different motor cortical targets, e.g., the anatomical representation of the hand (hand knob), the motor hotspot of the hand, and the anatomical representation or motor hotspot of the painful region if the latter is not located at the hand. An image-guided procedure is surely a relevant strategy to obtain a more personalized approach adapted to each chronic pain patient treated with rTMS in clinical practice.

References

Ahdab R, Ayache SS, Brugières P, Goujon C, Lefaucheur J-P. Comparison of "standard" and "navigated" procedures of TMS coil positioning over motor, premotor and prefrontal targets in patients with chronic pain and depression. Neurophysiol Clin. 2010;40:27–36.

Ahdab R, Ayache SS, Farhat WH, Mylius V, Schmidt S, Brugières P, Lefaucheur J-P. Reappraisal of the anatomical landmarks of motor and premotor cortical regions for image-guided brain navigation in TMS practice. Hum Brain Mapp. 2014;35:2435–47.

Ahdab R, Ayache SS, Brugières P, Farhat WH, Lefaucheur J-P. The hand motor hotspot is not always located in the hand knob: a neuronavigated transcranial magnetic stimulation study. Brain Topogr. 2016;29:590–7.

Ahmed MA, Mohamed SA, Sayed D. Long-term antalgic effects of repetitive transcranial magnetic stimulation of motor cortex and serum beta-endorphin in patients with phantom pain. Neurol Res. 2011;33:953–8.

André-Obadia N, Peyron R, Mertens P, Mauguière F, Laurent B, Garcia-Larrea L. Transcranial magnetic stimulation for pain control. Double-blind study of different frequencies against placebo, and correlation with motor cortex stimulation efficacy. Clin Neurophysiol. 2006;117:1536–44.

André-Obadia N, Mertens P, Gueguen A, Peyron R, Garcia-Larrea L. Pain relief by rTMS: differential effect of current flow but no specific action on pain subtypes. Neurology. 2008;71:833–40.

André-Obadia N, Magnin M, Garcia-Larrea L. On the importance of placebo timing in rTMS studies for pain relief. Pain. 2011;152:1233–7.

André-Obadia N, Mertens P, Lelekov-Boissard T, Afif A, Magnin M, Garcia-Larrea L. Is life better after motor cortex stimulation for pain control? Results at long-term and their prediction by preoperative rTMS. Pain Physician. 2014;17:53–62.

Andre-Obadia N, Magnin M, Simon E, Garcia-Larrea L. Somatotopic effects of rTMS in neuropathic pain? A comparison between stimulation over hand and face motor areas. Eur J Pain. 2018;22:707–15.

André-Obadia N, Hodaj H, Hodaj E, Simon E, Delon-Martin C, Garcia-Larrea L. Better fields or currents? A head-to-head comparison of transcranial magnetic (rTMS) versus direct current stimulation (tDCS) for neuropathic pain. Neurotherapeutics. 2023;20:207–19.

Attal N, Ayache SS, Ciampi De Andrade D, Mhalla A, Baudic S, Jazat F, Ahdab R, Neves DO, Sorel M, Lefaucheur JP, Bouhassira D. Repetitive transcranial magnetic stimulation and transcranial direct-current stimulation in neuropathic pain due to radiculopathy: a randomized sham-controlled comparative study. Pain. 2016;157:1224–31.

Attal N, Poindessous-Jazat F, De Chauvigny E, Quesada C, Mhalla A, Ayache SS, Fermanian C, Nizard J, Peyron R, Lefaucheur JP, Bouhassira D. Repetitive transcranial magnetic stimulation for neuropathic pain: a randomized multicentre sham-controlled trial. Brain. 2021;144:3328–39.
Ayache SS, Ahdab R, Chalah MA, Farhat WH, Mylius V, Goujon C, Sorel M, Lefaucheur J-P. Analgesic effects of navigated motor cortex rTMS in patients with chronic neuropathic pain. Eur J Pain. 2016;20:1413–22.
Borckardt JJ, Smith AR, Reeves ST, Madan A, Shelley N, Branham R, Nahas Z, George MS. A pilot study investigating the effects of fast left prefrontal rTMS on chronic neuropathic pain. Pain Med. 2009;10:840–9.
Bouhassira D, Jazat-Poindessous F, Farnes N, Franchisseur C, Stubhaug A, Bismuth J, Lefaucheur JP, Hansson P, Attal N. Comparison of the analgesic effects of "superficial" and "deep" repetitive transcranial magnetic stimulation in patients with central neuropathic pain: a randomized sham-controlled multicenter international crossover study. Pain. 2024;165:884.
Boyer L, Dousset A, Roussel P, Dossetto N, Cammilleri S, Piano V, Khalfa S, Mundler O, Donnet A, Guedj E. rTMS in fibromyalgia: a randomized trial evaluating QoL and its brain metabolic substrate. Neurology. 2014;82:1231–8.
Caulfield KA, Fleischmann HH, Cox CE, Wolf JP, George MS, McTeague LM. Neuronavigation maximizes accuracy and precision in TMS positioning: evidence from 11,230 distance, angle, and electric field modeling measurements. Brain Stimul. 2022;15:1192–205.
Cohen SP, Vase L, Hooten WM. Chronic pain: an update on burden, best practices, and new advances. Lancet. 2021;397:2082–97.
Cruccu G, Garcia-Larrea L, Hansson P, Keindl M, Lefaucheur J-P, Paulus W, Taylor R, Tronnier V, Truini A, Attal N. EAN guidelines on central neurostimulation therapy in chronic pain conditions. Eur J Neurol. 2016;23:1489–99.
da Cunha PHM, Dongyang L, Fernandes AM, Thibes RB, Sato J, Tanaka H, Dale C, Lapa JDDS, de Morais ADS, Soares FHC, da Silva VA, Graven-Nielsen T, Teixeira MJ, de Andrade DC. Non-invasive insular stimulation for peripheral neuropathic pain: influence of target or symptom? Neurophysiol Clin. 2022;52:109–16.
de Oliveira RA, de Andrade DC, Mendonça M, Barros R, Luvisoto T, Myczkowski ML, Marcolin MA, Teixeira MJ. Repetitive transcranial magnetic stimulation of the left premotor/dorsolateral prefrontal cortex does not have analgesic effect on central poststroke pain. J Pain. 2014;15:1271–81.
Denslow S, Bohning DE, Bohning PA, Lomarev MP, George MS. An increased precision comparison of TMS-induced motor cortex BOLD fMRI response for image-guided versus function-guided coil placement. Cogn Behav Neurol. 2005;18:119–26.
Diekhoff S, Uludağ K, Sparing R, Tittgemeyer M, Cavuşoğlu M, von Cramon DY, Grefkes C. Functional localization in the human brain: gradient-echo, spin-echo, and arterial spin-labeling fMRI compared with neuronavigated TMS. Hum Brain Mapp. 2011;32:341–57.
Dongyang L, Fernandes AM, da Cunha PHM, Tibes R, Sato J, Listik C, Dale C, Kubota GT, Galhardoni R, Teixeira MJ, Aparecida da Silva V, Rosi J, Ciampi de Andrade D. Posterior-superior insular deep transcranial magnetic stimulation alleviates peripheral neuropathic pain – a pilot double-blind, randomized cross-over study. Neurophysiol Clin. 2021;51:291–302.
Ebbinghaus H. Memory: a contribution to experimental psychology. Ann Neurosci. 2013;20:155–6.
Fontaine D, Hamani C, Lozano A. Efficacy and safety of motor cortex stimulation for chronic neuropathic pain: critical review of the literature. J Neurosurg. 2009;110:251–6.
Fricová J, Klírová M, Masopust V, Novák T, Vérebová K, Rokyta R. Repetitive transcranial magnetic stimulation in the treatment of chronic orofacial pain. Physiol Res. 2013;62(Suppl 1):S125–34.
Galhardoni R, Aparecida da Silva V, García-Larrea L, Dale C, Baptista AF, Barbosa LM, Menezes LMB, de Siqueira SRDT, Valério F, Rosi J Jr, de Lima Rodrigues AL, Reis Mendes Fernandes DT, Lorencini Selingardi PM, Marcolin MA, Duran FLS, Ono CR, Lucato LT, Fernandes AMBL, da Silva FEF, Yeng LT, Brunoni AR, Buchpiguel CA, Teixeira MJ, Ciampi de Andrade D. Insular and anterior cingulate cortex deep stimulation for central neuropathic pain: disassembling the percept of pain. Neurology. 2019;92:e2165–75.

Garcia-Larrea L, Peyron R. Pain matrices and neuropathic pain matrices: a review. Pain. 2013;154(Suppl 1):S29–43.
George MS, Wassermann EM, Post RM. Transcranial magnetic stimulation: a neuropsychiatric tool for the 21st century. J Neuropsychiatry Clin Neurosci. 1996;8:373–82.
Geyer S, Matelli M, Luppino G, Zilles K. Functional neuroanatomy of the primate isocortical motor system. Anat Embryol (Berl). 2000;202:443–74.
Ghosh S, Porter R. Corticocortical synaptic influences on morphologically identified pyramidal neurones in the motor cortex of the monkey. J Physiol (Lond). 1988;400:617–29.
Goto T, Saitoh Y, Hashimoto N, Hirata M, Kishima H, Oshino S, Tani N, Hosomi K, Kakigi R, Yoshimine T. Diffusion tensor fiber tracking in patients with central post-stroke pain; correlation with efficacy of repetitive transcranial magnetic stimulation. Pain. 2008;140:509–18.
Hasan M, Whiteley J, Bresnahan R, MacIver K, Sacco P, Das K, Nurmikko T. Somatosensory change and pain relief induced by repetitive transcranial magnetic stimulation in patients with central poststroke pain. Neuromodulation. 2014;17:731–6.
He SQ, Dum RP, Strick PL. Topographic organization of corticospinal projections from the frontal lobe: motor areas on the lateral surface of the hemisphere. J Neurosci. 1993;13:952–80.
He SQ, Dum RP, Strick PL. Topographic organization of corticospinal projections from the frontal lobe: motor areas on the medial surface of the hemisphere. J Neurosci. 1995;15:3284–306.
Hirayama A, Saitoh Y, Kishima H, Shimokawa T, Oshino S, Hirata M, Kato A, Yoshimine T. Reduction of intractable deafferentation pain by navigation-guided repetitive transcranial magnetic stimulation of the primary motor cortex. Pain. 2006;122:22–7.
Hodaj H, Alibeu J-P, Payen J-F, Lefaucheur J-P. Treatment of chronic facial pain including cluster headache by repetitive transcranial magnetic stimulation of the motor cortex with maintenance sessions: a naturalistic study. Brain Stimul. 2015;8:801–7.
Hodaj H, Payen J-F, Hodaj E, Dumolard A, Maindet C, Cracowski J-L, Delon-Martin C, Lefaucheur J-P. Long-term treatment of chronic orofacial, pudendal, and central neuropathic limb pain with repetitive transcranial magnetic stimulation of the motor cortex. Clin Neurophysiol. 2020;131:1423–32.
Hodaj H, Payen JF, Hodaj E, Sorel M, Dumolard A, Vercueil L, Delon-Martin C, Lefaucheur JP. Long-term analgesic effect of trans-spinal direct current stimulation compared to non-invasive motor cortex stimulation in complex regional pain syndrome. Brain Commun. 2023;5:fcad191.
Holsheimer J, Lefaucheur J-P, Buitenweg JR, Goujon C, Nineb A, Nguyen J-P. The role of intra-operative motor evoked potentials in the optimization of chronic cortical stimulation for the treatment of neuropathic pain. Clin Neurophysiol. 2007a;118:2287–96.
Holsheimer J, Nguyen J-P, Lefaucheur J-P, Manola L. Cathodal, anodal or bifocal stimulation of the motor cortex in the management of chronic pain. Acta Neurochir Suppl. 2007b;97:57–66.
Hosomi K, Saitoh Y, Kishima H, Oshino S, Hirata M, Tani N, Shimokawa T, Yoshimine T. Electrical stimulation of primary motor cortex within the central sulcus for intractable neuropathic pain. Clin Neurophysiol. 2008;119:993–1001.
Hosomi K, Kishima H, Oshino S, Hirata M, Tani N, Maruo T, Yorifuji S, Yoshimine T, Saitoh Y. Cortical excitability changes after high-frequency repetitive transcranial magnetic stimulation for central poststroke pain. Pain. 2013a;154:1352–7.
Hosomi K, Shimokawa T, Ikoma K, Nakamura Y, Sugiyama K, Ugawa Y, Uozumi T, Yamamoto T, Saitoh Y. Daily repetitive transcranial magnetic stimulation of primary motor cortex for neuropathic pain: a randomized, multicenter, double-blind, crossover, sham-controlled trial. Pain. 2013b;154:1065–72.
Hosomi K, Sugiyama K, Nakamura Y, Shimokawa T, Oshino S, Goto Y, Mano T, Shimizu T, Yanagisawa T, Saitoh Y, TEN-P11-01 investigators. A randomized controlled trial of 5 daily sessions and continuous trial of 4 weekly sessions of repetitive transcranial magnetic stimulation for neuropathic pain. Pain. 2020;161:351–60.
Jetté F, Côté I, Meziane HB, Mercier C. Effect of single-session repetitive transcranial magnetic stimulation applied over the hand versus leg motor area on pain after spinal cord injury. Neurorehabil Neural Repair. 2013;27:636–43.

Julkunen P, Ruohonen J, Sääskilahti S, Säisänen L, Karhu J. Threshold curves for transcranial magnetic stimulation to improve reliability of motor pathway status assessment. Clin Neurophysiol. 2011;122:975–83.

Khedr EM, Kotb H, Kamel NF, Ahmed MA, Sadek R, Rothwell JC. Longlasting antalgic effects of daily sessions of repetitive transcranial magnetic stimulation in central and peripheral neuropathic pain. J Neurol Neurosurg Psychiatry. 2005;76:833–8.

Khedr EM, Kotb HI, Mostafa MG, Mohamad MF, Amr SA, Ahmed MA, Karim AA, Kamal SM. Repetitive transcranial magnetic stimulation in neuropathic pain secondary to malignancy: a randomized clinical trial. Eur J Pain. 2015;19:519–27.

Kim NY, Taylor JJ, Kim YW, Borsook D, Joutsa J, Li J, Quesada C, Peyron R, Fox MD. Network effects of brain lesions causing central poststroke pain. Ann Neurol. 2022;92:834–45.

Klein MM, Treister R, Raij T, Pascual-Leone A, Park L, Nurmikko T, Lenz F, Lefaucheur J-P, Lang M, Hallett M, Fox M, Cudkowicz M, Costello A, Carr DB, Ayache SS, Oaklander AL. Transcranial magnetic stimulation of the brain: guidelines for pain treatment research. Pain. 2015;156:1601–14.

Knotkova H, Hamani C, Sivanesan E, Le Beuffe MFE, Moon JY, Cohen SP, Huntoon MA. Neuromodulation for chronic pain. Lancet. 2021;397:2111–24.

Kobayashi M, Fujimaki T, Mihara B, Ohira T. Repetitive transcranial magnetic stimulation once a week induces sustainable long-term relief of central poststroke pain. Neuromodulation. 2015;18:249–54.

Lawson McLean A, Frank S, Zafar N, Waschke A, Kalff R, Reichart R. Time course of the response to navigated repetitive transcranial magnetic stimulation at 10 Hz in chronic neuropathic pain. Neurol Res. 2018;40(7):564–72.

Lee SJ, Kim DY, Chun MH, Kim YG. The effect of repetitive transcranial magnetic stimulation on fibromyalgia: a randomized sham-controlled trial with 1-mo follow-up. Am J Phys Med Rehabil. 2012;91:1077–85.

Lefaucheur J-P. The use of repetitive transcranial magnetic stimulation (rTMS) in chronic neuropathic pain. Neurophysiol Clin. 2006;36:117–24.

Lefaucheur J-P. Why image-guided navigation becomes essential in the practice of transcranial magnetic stimulation. Neurophysiol Clin. 2010;40:1–5.

Lefaucheur J-P. Cortical neurostimulation for neuropathic pain: state of the art and perspectives. Pain. 2016;157(Suppl 1):S81–9.

Lefaucheur JP, Nguyen JP. A practical algorithm for using rTMS to treat patients with chronic pain. Neurophysiol Clin. 2019;49:301–7.

Lefaucheur J-P, Picht T. The value of preoperative functional cortical mapping using navigated TMS. Neurophysiol Clin. 2016;46:125–33.

Lefaucheur J-P, Nguyen J-P, Drouot X, Pollin B, Keravel Y, Harf A. Chronic pain treated by rTMS of motor cortex. Electroencephalogr Clin Neurophysiol. 1998;107:92.

Lefaucheur J-P, Drouot X, Keravel Y, Nguyen J-P. Pain relief induced by repetitive transcranial magnetic stimulation of precentral cortex. Neuroreport. 2001a;12:2963–5.

Lefaucheur J-P, Drouot X, Nguyen J-P. Interventional neurophysiology for pain control: duration of pain relief following repetitive transcranial magnetic stimulation of the motor cortex. Neurophysiol Clin. 2001b;31:247–52.

Lefaucheur J-P, Drouot X, Ménard-Lefaucheur I, Nguyen J-P. Neuropathic pain controlled for more than a year by monthly sessions of repetitive transcranial magnetic cortical stimulation. Neurophysiol Clin. 2004a;34:91–5.

Lefaucheur J-P, Drouot X, Menard-Lefaucheur I, Zerah F, Bendib B, Cesaro P, Keravel Y, Nguyen J-P. Neurogenic pain relief by repetitive transcranial magnetic cortical stimulation depends on the origin and the site of pain. J Neurol Neurosurg Psychiatry. 2004b;75:612–6.

Lefaucheur J-P, Hatem S, Nineb A, Ménard-Lefaucheur I, Wendling S, Keravel Y, Nguyen J-P. Somatotopic organization of the analgesic effects of motor cortex rTMS in neuropathic pain. Neurology. 2006;67:1998–2004.

Lefaucheur J-P, Antal A, Ahdab R, Ciampi de Andrade D, Fregni F, Khedr EM, Nitsche M, Paulus W. The use of repetitive transcranial magnetic stimulation (rTMS) and transcranial direct current stimulation (tDCS) to relieve pain. Brain Stimul. 2008a;1:337–44.

Lefaucheur J-P, Drouot X, Ménard-Lefaucheur I, Keravel Y, Nguyen J-P. Motor cortex rTMS in chronic neuropathic pain: pain relief is associated with thermal sensory perception improvement. J Neurol Neurosurg Psychiatry. 2008b;79:1044–9.

Lefaucheur J-P, Holsheimer J, Goujon C, Keravel Y, Nguyen J-P. Descending volleys generated by efficacious epidural motor cortex stimulation in patients with chronic neuropathic pain. Exp Neurol. 2010;223:609–14.

Lefaucheur J-P, Ménard-Lefaucheur I, Goujon C, Keravel Y, Nguyen J-P. Predictive value of rTMS in the identification of responders to epidural motor cortex stimulation therapy for pain. J Pain. 2011;12:1102–11.

Lefaucheur J-P, Ayache SS, Sorel M, Farhat WH, Zouari HG, Ciampi de Andrade D, Ahdab R, Ménard-Lefaucheur I, Brugières P, Goujon C. Analgesic effects of repetitive transcranial magnetic stimulation of the motor cortex in neuropathic pain: influence of theta burst stimulation priming. Eur J Pain. 2012;16:1403–13.

Lefaucheur JP, André-Obadia N, Antal A, Ayache SS, Baeken C, Benninger DH, Cantello RM, Cincotta M, de Carvalho M, De Ridder D, Devanne H, Di Lazzaro V, Filipović SR, Hummel FC, Jääskeläinen SK, Kimiskidis VK, Koch G, Langguth B, Nyffeler T, Oliviero A, Padberg F, Poulet E, Rossi S, Rossini PM, Rothwell JC, Schönfeldt-Lecuona C, Siebner HR, Slotema CW, Stagg CJ, Valls-Sole J, Ziemann U, Paulus W, Garcia-Larrea L. Evidence-based guidelines on the therapeutic use of repetitive transcranial magnetic stimulation (rTMS). Clin Neurophysiol. 2014;125:2150–206.

Lefaucheur J-P, Aleman A, Baeken C, Benninger DH, Brunelin J, Di Lazzaro V, Filipović SR, Grefkes C, Hasan A, Hummel FC, Jääskeläinen SK, Langguth B, Leocani L, Londero A, Nardone R, Nguyen JP, Nyffeler T, Oliveira-Maia AJ, Oliviero A, Padberg F, Palm U, Paulus W, Poulet E, Quartarone A, Rachid F, Rektorová I, Rossi S, Sahlsten H, Schecklmann M, Szekely D, Ziemann U. Evidence-based guidelines on the therapeutic use of repetitive transcranial magnetic stimulation (rTMS): an update (2014–2018). Clin Neurophysiol. 2020;131:474–528.

Lefaucheur JP, Nguyen JP, Delmas A, Croci S, Bredoux L, Hodaj H. Targeting lower limb, upper limb, and face representation in the primary motor cortex for the practice of neuronavigated transcranial magnetic stimulation. Neuromodulation. 2024;27:572.

Leung A, Donohue M, Xu R, Lee R, Lefaucheur J-P, Khedr EM, Saitoh Y, André-Obadia N, Rollnik J, Wallace M, Chen R. rTMS for suppressing neuropathic pain: a meta-analysis. J Pain. 2009;10:1205–16.

Lindholm P, Lamusuo S, Taiminen T, Pesonen U, Lahti A, Virtanen A, Forssell H, Hietala J, Hagelberg N, Pertovaara A, Parkkola R, Jääskeläinen S. Right secondary somatosensory cortex-a promising novel target for the treatment of drug-resistant neuropathic orofacial pain with repetitive transcranial magnetic stimulation. Pain. 2015;156:1276–83.

Ma S-M, Ni J-X, Li X-Y, Yang L-Q, Guo Y-N, Tang Y-Z. High-frequency repetitive transcranial magnetic stimulation reduces pain in postherpetic neuralgia. Pain Med. 2015;16:2162–70.

Malavera A, Silva FA, Fregni F, Carrillo S, Garcia RG. Repetitive transcranial magnetic stimulation for phantom limb pain in land mine victims: a double-blinded, randomized, sham-controlled trial. J Pain. 2016;17:911–8.

Manola L, Holsheimer J, Veltink P, Buitenweg JR. Anodal vs cathodal stimulation of motor cortex: a modeling study. Clin Neurophysiol. 2007;118:464–74.

Matsumura Y, Hirayama T, Yamamoto T. Comparison between pharmacologic evaluation and repetitive transcranial magnetic stimulation-induced analgesia in poststroke pain patients. Neuromodulation. 2013;16:349–54.

Mhalla A, Baudic S, de Andrade DC, Gautron M, Perrot S, Teixeira MJ, Attal N, Bouhassira D. Long-term maintenance of the analgesic effects of transcranial magnetic stimulation in fibromyalgia. Pain. 2011;152:1478–85.

Mori N, Hosomi K, Nishi A, Dong D, Yanagisawa T, Khoo HM, Tani N, Oshino S, Saitoh Y, Kishima H. Difference in analgesic effects of repetitive transcranial magnetic stimulation according to the site of pain. Front Hum Neurosci. 2021a;15:786225.

Mori N, Hosomi K, Nishi A, Matsugi A, Dong D, Oshino S, Kishima H, Saitoh Y. Exploratory study of optimal parameters of repetitive transcranial magnetic stimulation for neuropathic pain in the lower extremities. Pain Rep. 2021b;6:e964.

Mori N, Hosomi K, Nishi A, Oshino S, Kishima H, Saitoh Y. Analgesic effects of repetitive transcranial magnetic stimulation at different stimulus parameters for neuropathic pain: a randomized study. Neuromodulation. 2022;25:520–7.

Mylius V, Ayache SS, Ahdab R, Farhat WH, Zouari HG, Belke M, Brugières P, Wehrmann E, Krakow K, Timmesfeld N, Schmidt S, Oertel WH, Knake S, Lefaucheur J-P. Definition of DLPFC and M1 according to anatomical landmarks for navigated brain stimulation: inter-rater reliability, accuracy, and influence of gender and age. NeuroImage. 2013;78:224–32.

Naik A, Bah M, Govande M, Palsgaard P, Dharnipragada R, Shaffer A, Air EL, Cramer SW, Croarkin PE, Arnold PM. Optimal frequency in rTMS for the management of chronic pain: a network meta-analysis of randomized controlled trials. World Neurosurg. 2024;184:e53.

Nardone R, Höller Y, Langthaler PB, Lochner P, Golaszewski S, Schwenker K, Brigo F, Trinka E. rTMS of the prefrontal cortex has analgesic effects on neuropathic pain in subjects with spinal cord injury. Spinal Cord. 2017;55:20–5.

Nauczyciel C, Hellier P, Morandi X, Blestel S, Drapier D, Ferre J-C, Barillot C, Millet B. Assessment of standard coil positioning in transcranial magnetic stimulation in depression. Psychiatry Res. 2011;186:232–8.

Nazarova M, Novikov P, Ivanina E, Kozlova K, Dobrynina L, Nikulin VV. Mapping of multiple muscles with transcranial magnetic stimulation: absolute and relative test-retest reliability. Hum Brain Mapp. 2021;42:2508–28.

Nguyen J-P, Lefaucheur J-P, Decq P, Uchiyama T, Carpentier A, Fontaine D, Brugières P, Pollin B, Fève A, Rostaing S, Cesaro P, Keravel Y. Chronic motor cortex stimulation in the treatment of central and neuropathic pain. Correlations between clinical, electrophysiological and anatomical data. Pain. 1999;82:245–51.

Nguyen J-P, Lefaucheur J-P, Raoul S, Roualdès V, Péréon Y, Keravel Y. Motor cortex stimulation for the treatment of neuropathic pain. In: Krames ES, Hunter Peckham P, Rezai AR, editors. Neuromodulation. Amsterdam: Elsevier; 2009. p. 515–26.

Nguyen J-P, Nizard J, Keravel Y, Lefaucheur J-P. Invasive brain stimulation for the treatment of neuropathic pain. Nat Rev Neurol. 2011;7:699–709.

Nizard J, Lefaucheur JP, Helbert M, de Chauvigny E, Nguyen JP. Non-invasive stimulation therapies for the treatment of refractory pain. Discov Med. 2012a;14:21–31.

Nizard J, Raoul S, Nguyen JP, Lefaucheur JP. Invasive stimulation therapies for the treatment of refractory pain. Discov Med. 2012b;14:237–46.

Nowak LG, Bullier J. Axons, but not cell bodies, are activated by electrical stimulation in cortical gray matter. I. Evidence from chronaxie measurements. Exp Brain Res. 1998a;118:477–88.

Nowak LG, Bullier J. Axons, but not cell bodies, are activated by electrical stimulation in cortical gray matter. II. Evidence from selective inactivation of cell bodies and axon initial segments. Exp Brain Res. 1998b;118:489–500.

Nurmikko T, MacIver K, Bresnahan R, Hird E, Nelson A, Sacco P. Motor cortex reorganization and repetitive transcranial magnetic stimulation for pain-a methodological study. Neuromodulation. 2016;19:669–78.

Ohn SH, Chang WH, Park CH, Kim ST, Lee JI, Pascual-Leone A, Kim YH. Neural correlates of the antinociceptive effects of repetitive transcranial magnetic stimulation on central pain after stroke. Neurorehabil Neural Repair. 2012;26:344–52.

Ojala J, Vanhanen J, Harno H, Lioumis P, Vaalto S, Kaunisto MA, Putaala J, Kangasniemi M, Kirveskari E, Mäkelä JP, Kalso E. A randomized, sham-controlled trial of repetitive transcranial magnetic stimulation targeting M1 and S2 in central poststroke pain: a pilot trial. Neuromodulation. 2022;25:538–48.

Onesti E, Gabriele M, Cambieri C, Ceccanti M, Raccah R, Di Stefano G, Biasiotta A, Truini A, Zangen A, Inghilleri M. H-coil repetitive transcranial magnetic stimulation for pain relief in patients with diabetic neuropathy. Eur J Pain. 2013;17:1347–56.

Passard A, Attal N, Benadhira R, Brasseur L, Saba G, Sichere P, Perrot S, Januel D, Bouhassira D. Effects of unilateral repetitive transcranial magnetic stimulation of the motor cortex on chronic widespread pain in fibromyalgia. Brain. 2007;130:2661–70.

Pei Q, Wu B, Tang Y, Yang X, Song L, Wang N, Li Y, Sun C, Ma S, Ni J. Repetitive transcranial magnetic stimulation at different frequencies for postherpetic neuralgia: a double-blind, sham-controlled, randomized trial. Pain Physician. 2019;22:E303–13.

Penfield W, Boldrey E. Somatic motor and sensory representation in the cerebral cortex of man as studied by electrical stimulation. Brain. 1937;60:389–443.

Pommier B, Créac'h C, Beauvieux V, Nuti C, Vassal F, Peyron R. Robot-guided neuronavigated rTMS as an alternative therapy for central (neuropathic) pain: clinical experience and long-term follow-up. Eur J Pain. 2016;20:907–16.

Pommier B, Vassal F, Boutet C, Jeannin S, Peyron R, Faillenot I. Easy methods to make the neuronavigated targeting of DLPFC accurate and routinely accessible for rTMS. Neurophysiol Clin. 2017;47:35–46.

Quesada C, Pommier B, Fauchon C, Bradley C, Créac'h C, Vassal F, Peyron R. Robot-guided neuronavigated repetitive transcranial magnetic stimulation (rTMS) in central neuropathic pain. Arch Phys Med Rehabil. 2018;99:2203–2215.e1.

Quesada C, Pommier B, Fauchon C, Bradley C, Créac'h C, Murat M, Vassal F, Peyron R. New procedure of high-frequency repetitive transcranial magnetic stimulation for central neuropathic pain: a placebo-controlled randomized crossover study. Pain. 2020;161:718–28.

Rademacher J, Burgel U, Geyer S, Schormann T, Schleicher A, Freund HJ, Zilles K. Variability and asymmetry in the human precentral motor system. A cytoarchitectonic and myeloarchitectonic brain mapping study. Brain. 2001;124:2232–58.

Rollnik JD, Wüstefeld S, Däuper J, Karst M, Fink M, Kossev A, Dengler R. Repetitive transcranial magnetic stimulation for the treatment of chronic pain – a pilot study. Eur Neurol. 2002;48:6–10.

Ruohonen J, Karhu J. Navigated transcranial magnetic stimulation. Neurophysiol Clin. 2010;40:7–17.

Säisänen L, Huttunen J, Hyppönen J, Nissen M, Kotiranta U, Mervaala E, Fraunberg MVUZ. Efficacy and tolerability in patients with chronic facial pain of two consecutive treatment periods of rTMS applied over the facial motor cortex, using protocols differing in stimulation frequency, duration, and train pattern. Neurophysiol Clin. 2022;52:95–108.

Saitoh Y, Hirayama A, Kishima H, Shimokawa T, Oshino S, Hirata M, Tani N, Kato A, Yoshimine T. Reduction of intractable deafferentation pain due to spinal cord or peripheral lesion by high-frequency repetitive transcranial magnetic stimulation of the primary motor cortex. J Neurosurg. 2007;107:555–9.

Saitoh Y, Maruo T, Yokoe M, Matsuzaki T, Sekino M. Electrical or repetitive transcranial magnetic stimulation of primary motor cortex for intractable neuropathic pain. Annu Int Conf IEEE Eng Med Biol Soc. 2013;2013:6163–6.

Sampson SM, Rome JD, Rummans TA. Slow-frequency rTMS reduces fibromyalgia pain. Pain Med. 2006;7:115–8.

Sampson SM, Kung S, McAlpine DE, Sandroni P. The use of slow-frequency prefrontal repetitive transcranial magnetic stimulation in refractory neuropathic pain. J ECT. 2011;27:33–7.

Schmidt S, Bathe-Peters R, Fleischmann R, Rönnefarth M, Scholz M, Brandt SA. Nonphysiological factors in navigated TMS studies; confounding covariates and valid intracortical estimates. Hum Brain Mapp. 2015;36:40–9.

Scibilia A, Conti A, Raffa G, Granata F, Abbritti RV, Priola SM, Sindorio C, Cardali S, Germanò A. Resting-state fMR evidence of network reorganization induced by navigated transcranial magnetic repetitive stimulation in phantom limb pain. Neurol Res. 2018;40:241–8.

Shimazu H, Maier MA, Cerri G, Kirkwood PA, Lemon RN. Macaque ventral premotor cortex exerts powerful facilitation of motor cortex outputs to upper limb motoneurons. J Neurosci. 2004;24:1200–11.

Shimizu T, Hosomi K, Maruo T, Goto Y, Yokoe M, Kageyama Y, Shimokawa T, Yoshimine T, Saitoh Y. Efficacy of deep rTMS for neuropathic pain in the lower limb: a randomized, double-blind crossover trial of an H-coil and figure-8 coil. J Neurosurg. 2017;127:1172–80.

Short EB, Borckardt JJ, Anderson BS, Frohman H, Beam W, Reeves ST, George MS. Ten sessions of adjunctive left prefrontal rTMS significantly reduces fibromyalgia pain: a randomized, controlled pilot study. Pain. 2011;152:2477–84.

Siebner HR, Funke K, Aberra AS, Antal A, Bestmann S, Chen R, Classen J, Davare M, Di Lazzaro V, Fox PT, Hallett M, Karabanov AN, Kesselheim J, Beck MM, Koch G, Liebetanz D, Meunier S, Miniussi C, Paulus W, Peterchev AV, Ugawa Y. Transcranial magnetic stimulation of the brain: what is stimulated? – a consensus and critical position paper. Clin Neurophysiol. 2022;140:59–97.

Sollmann N, Hauck T, Obermüller T, Hapfelmeier A, Meyer B, Ringel F, Krieg SM. Inter- and intraobserver variability in motor mapping of the hotspot for the abductor policis brevis muscle. BMC Neurosci. 2013;14:94.

Teitti S, Määttä S, Säisänen L, Könönen M, Vanninen R, Hannula H, Mervaala E, Karhu J. Non-primary motor areas in the human frontal lobe are connected directly to hand muscles. NeuroImage. 2008;40:1243–50.

Tsubokawa T, Katayama Y, Yamamoto T, Hirayama T, Koyama S. Chronic motor cortex stimulation for the treatment of central pain. Acta Neurochir Suppl. 1991a;52:137–9.

Tsubokawa T, Katayama Y, Yamamoto T, Hirayama T, Koyama S. Treatment of thalamic pain by chronic motor cortex stimulation. Pacing Clin Electrophysiol. 1991b;14:131–4.

Umezaki Y, Badran BW, DeVries WH, Moss J, Gonzales T, George MS. The efficacy of daily prefrontal repetitive transcranial magnetic stimulation (rTMS) for burning mouth syndrome (BMS): a randomized controlled single-blind study. Brain Stimul. 2016;9:234–42.

Wang H, Hu Y, Deng J, Ye Y, Huang M, Che X, Yu L. A randomised sham-controlled study evaluating rTMS analgesic efficacy for postherpetic neuralgia. Front Neurosci. 2023;17:1158737.

Yılmaz B, Kesikburun S, Yaşar E, Tan AK. Effect of repetitive transcranial magnetic stimulation on refractory neuropathic pain in spinal cord injury. J Spinal Cord Med. 2014;37:397–400.

Zhao CG, Sun W, Ju F, Jiang S, Wang H, Sun XL, Mou X, Yuan H. Analgesic effects of navigated repetitive transcranial magnetic stimulation in patients with acute central poststroke pain. Pain Ther. 2021;10:1085–100.

High-PAS: Paired Associative Stimulation with High-Frequency Peripheral and High-Intensity Central Component in Spinal Cord Injury Rehabilitation

17

Anastasia Shulga, Erika Kirveskari, Jyrki P. Mäkelä, and Pantelis Lioumis

Spinal cord injury (SCI) dramatically changes the life of an individual, and it is also a substantial economic burden for society. Many therapies have been developed for this condition, but few have progressed to routine clinical practice. Novel non-invasive, cost-effective, and feasible approaches for SCI rehabilitation are needed. Paired associative stimulation (PAS) is a combination of transcranial magnetic stimulation (TMS) and peripheral electrical stimulation (PNS). A novel variant of paired associative stimulation, high-PAS, consists of non-invasive high-intensity TMS and non-invasive high-frequency PNS. A therapeutic effect of high-PAS in patients with incomplete SCI with a wide range of injury severity, time since injury, and age was observed. Tetraplegic and paraplegic, traumatic, as well as neurological SCI patients benefited from upper- or lower-limb high-PAS. High-PAS settings were optimized in studies in healthy subjects. This chapter is focused on the summary of results,

A. Shulga (✉)
Helsinki University Hospital, Helsinki, Finland
e-mail: anastasia.shulga@helsinki.fi

E. Kirveskari
Department of Clinical Neurophysiology, HUS Diagnostic Center, University of Helsinki and Helsinki University Hospital and BioMag Laboratory, HUS Diagnostic Center, University of Helsinki, Aalto University, and Helsinki University Hospital, Helsinki, Finland
e-mail: erika.haaksiluoto@hus.fi

J. P. Mäkelä
BioMag Laboratory, HUS Diagnostic Center, University of Helsinki, and Helsinki University Hospital, Helsinki, Finland
e-mail: jyrki.makela@hus.fi

P. Lioumis
NBE, Aalto University School of Science, Aalto, Finland
e-mail: pantelis.lioumis@aalto.fi

S. M. Krieg, T. Picht (eds.), *Navigated Transcranial Magnetic Stimulation in Neurosurgery*, https://doi.org/10.1007/978-3-031-97155-6_17

observations, and methodological considerations concerning the use of high-PAS in SCI rehabilitation.

17.1 Introduction

Spinal cord injury (SCI) has a devastating effect on the quality of life of an individual and constitutes a substantial economic burden for society (Chen and DeVivo 2019). Although several therapeutic approaches have been recently developed, only a few have progressed from laboratories to actual clinical practice (Jack et al. 2020; Ramer et al. 2014; Tohda and Kuboyama 2011). Safe, non-invasive, and feasible treatments are needed for both acute and chronic SCI. This chapter focuses on the use of electric field (E-field) navigated TMS (nTMS) combined with high-frequency peripheral stimulation in the rehabilitation of SCI patients.

The majority of SCIs are incomplete (Chen and DeVivo 2019; Ackery et al. 2004; Fawcett et al. 2007). Moreover, non-functional connectivity can also remain in clinically complete injuries (Kirshblum and Solinsky 2019; Squair et al. 2016). Strengthening these residual pathways by non-invasive methods has gained considerable attention in human SCI research (Field-Fote 2015, 2016). Long-term potentiation (LTP; Nicoll 2017), the result of joint pre- and postsynaptic neuronal activation, is one of the central targets in rehabilitating the damaged connections after neuronal trauma or disease. Evidence from animal studies indicates that stimulation protocols inducing spike-time-dependent-like plasticity (STDP) between upper and lower motor neurons hold promise for promoting motor recovery (Ahmed 2013; McPherson et al. 2015; Nishimura et al. 2013).

Transcranial magnetic stimulation (TMS) enables a safe, non-invasive activation of a patch of the cortex (Rossi et al. 2020). Paired associative stimulation (PAS) includes TMS and electrical peripheral nerve stimulation (PNS). PAS for motor rehabilitation combines TMS of the primary motor cortex (M1) with PNS of the contralateral limbs (Chen and Udupa 2009). PAS aims for synchronous activation of pre- and postsynaptic neurons to induce plastic changes by LTP-linked mechanisms (Stefan et al. 2000). PAS-induced changes in neuronal connectivity are suggested to represent a form of STDP (Stefan et al. 2000). STDP can result in either synaptic LTP or synaptic long-term depression (LTD), depending on the timing of the pre- and postsynaptic stimuli (Dan and Poo 2004). In healthy individuals, PAS can induce either facilitation or inhibition of motor-evoked potentials (MEPs; responses to TMS recorded from muscles), reflecting LTP-like or LTD-like plasticity in the corticospinal tract (CST). The interstimulus interval (ISI) between TMS and PNS is essential in determining the polarity of the response (Suppa et al. 2017). The first PAS protocols targeted the cortical level (Stefan et al. 2000; Suppa et al. 2017). However, PAS can also be timed to converge in the spinal cord (Taylor and Martin 2009). PAS can be applied to upper and lower motor neurons; it thus is an attractive possibility to strengthen weakened connectivity between these neuronal populations (Jack et al. 2020; Suppa et al. 2017). A single session of conventional PAS with single TMS and single PNS pulses resulted in a transient increase of MEP amplitudes (Carson and Kennedy 2013) and transiently improved motor function in SCI patients (Bunday and Perez 2012). Most PAS experiments have studied the efficacy of a single PAS session. Ten sessions of PAS, with parameters timed to occur at the spinal cord and combined with exercise, modified motor performance in chronic SCI patients (Jo and Perez 2020). Maximum voluntary contraction in targeted muscles increased after PAS with or without exercise but not after sham-PAS with exercise; the effect was preserved at 6 months in the group receiving PAS with exercise. Outcomes in Graded and Redefined Assessment of Strength, Sensibility, and Prehension (GRASSP) and the 10-m walk test did not differ significantly between the groups (Jo and Perez 2020).

In addition to ISI, numerous other variables affect the presence, the polarity, and the magnitude of MEP amplitude changes after conventional PAS (Carson and Kennedy 2013). The outcome can depend, for example, on genetic and neuroanatomical differences, age, medication, and time of day, as well as alertness and focus of attention during PAS (Suppa et al. 2017; Carson and Kennedy 2013). Conventional PAS is excellent for studying and inducing cortical plasticity in humans. However, the demand for high precision in the timing of the stimulation and dependence on the aforementioned factors might hamper its actual clinical applicability. Measurements of SCI patients can be difficult due to small or absent MEPs, muscle spasticity, and unavoidable medications affecting the central nervous system, and difficulties in scheduling the PAS for a precise point of day may occur. Moreover, as stimulation timing is critical to strengthen plasticity, erroneous timing might lead to weaker connectivity (Jack et al. 2020). When PAS is applied in successive sessions as a tool for long-term rehabilitation, the initially calculated ISI might need to be adjusted, as neuronal conductivity may change over time.

To take these factors into account, we developed a modified version of PAS that combines high-intensity TMS and high-frequency PNS ("high-PAS") and depends less on the precision of the stimulation parameters. The STDP model, where synaptic input to dendrites is active just before a somatic input, is now regarded as highly simplified (Suppa et al. 2017). Spike timing is not the only important parameter in inducing plasticity; it depends also on the firing rate, postsynaptic voltage, and synaptic cooperativity (Feldman 2012). For example, in experiments using brain tissue slices, connections exhibited Hebbian STDP only when presynaptic and postsynaptic spikes occurred at moderate firing rates (10–20 Hz); higher firing rates (>30 Hz) induced LTP independent of spike timing (Feldman 2012). High-PAS consists of TMS pulses given at 100% of stimulator output (SO) and high-frequency PNS trains. The aim is to generate multiple orthodromic volleys in the corticospinal tract by high-intensity TMS pulses and multiple antidromic activations by the peripheral stimulus train. In cellular-level studies, multiple interactions induce LTP-like effects that overcome their long-term depression (LTD)-like counterparts (Sjostrom et al. 2001). Therefore, it is plausible that multiple interactions at the spinal cord level would induce an LTP-like effect. Indeed, we have shown that this protocol effectively potentiates MEPs, indicating LTP-like plasticity, even when errors of ±10 ms are introduced into the calculation of the ISI (Shulga et al. 2016a), and that in the lower limbs, high-PAS with 100-Hz PNS upregulates MEPs in healthy subjects by 100–150% with a responder rate of 100% (Mezes et al. 2020).

17.2 Clinical Evidence

Most recovery after all SCIs occurs within 6 months post-injury and the rate of change plateaus at 9 months post-injury (Oleson and Flanders 2019). Although some motor recovery may continue for 2 years or more, it is generally slight and unlikely to improve function significantly (Oleson and Flanders 2019). Patients at the chronic phase (1–15 years post injury) benefited from 4 to 12 weeks of PAS (Rodionov et al. 2019, 2020; Shulga et al. 2016b, 2020; Tolmacheva et al. 2017, 2019a; Vaalto et al. 2021; Vanhanen et al. 2022). PAS can be applied to upper (Rodionov et al. 2019; Tolmacheva et al. 2017, 2019a) and lower limbs (Rodionov et al. 2020; Shulga et al. 2016b, 2020); it improves manual muscle test (MMT) scores in both traumatic (Tolmacheva et al. 2017) and neurological SCI (Tolmacheva et al. 2019a) and in both paraplegic (Shulga et al. 2020) and tetraplegic (Tolmacheva et al. 2017, 2019a) patients. Apart from MMT, functional hand tests improved; e.g., in non-traumatic tetraplegic patients, Box and Block and palm pinch tests, as well as grip dynamometry, all improved after upper limb PAS at 6-month evaluation (Tolmacheva et al. 2019a). Walking speed improved in ambulatory participants after lower limb PAS (Rodionov et al. 2020). Many patients who underwent 4–12 weeks of PAS also described subjective functional improvements and more versatile hand use (Tolmacheva et al. 2017, 2019a). In a case report of one patient, 12 weeks of PAS 1 year after injury enabled the participation in walking rehabilitation deemed futile before PAS. Another PAS for 12 weeks 2 years after the injury improved walking distance by more than 100%. The patient was originally non-ambulatory; after the whole intervention, he ambulated 50% of the time at home (Shulga et al. 2020). In this patient, PAS was the major factor that restored the ability to walk. A clear increase in spinal cord independence measure (SCIM) was achieved in another case study (Rodionov et al. 2019) where PAS was applied to the hands for as long as they improved (for over a year). Before PAS, the subject needed total or partial assistance in eating, bathing, dressing, and grooming (Table 17.1). During the follow-up, he could perform these tasks independently and without adaptive devices.

17.3 General Methodology

Although the implementation of PAS is individualized for each patient, the same procedures are applied for everyone (Fig. 17.1). nTMS is always delivered at 100% SO, whereas the PNS intensity and the ISI between PNS and TMS are individually adjusted. The nerves to be stimulated are selectively chosen based on the distribution of the muscle weakness and the desired clinical benefit. One or both limbs can be stimulated. Preparation for PAS includes measurements of F-responses (late orthodromic responses produced by a pool of motoneurons that is antidromically

Table 17.1 An example of changes in spinal cord independence measure (SCIM) score in a patient who received high-PAS for as long as improvement was observed, before and after the treatment. See (Rodionov et al. 2019) for details

BEFORE, score and meaning	AFTER, score and meaning
1. *Needs partial assistance for eating and/or drinking, or for wearing adaptive devices*	**3.** *Eats and drinks independently; does not require assistance or adaptive devices*
1. *Requires partial assistance for bathing upper body (soaping, washing, drying body and head, manipulating water tap)*	**3.** *Washes independently; does not require adaptive devices or specific setting*
0. *Requires total assistance with dressing upper body (clothes, shoes, permanent orthoses: dressing, wearing, undressing)*	**3.** *Independent with clothes without buttons, zippers or laces; does not require adaptive devices or specific settings*
0. *Requires total assistance with transfer from bed to wheelchair*	**2.** *Independent with transfer from bed to wheelchair*

Adapted from Rodionov et al. (2019)

activated by peripheral motor or mixed-nerve stimulation) and mapping the motor cortex and measurement of MEP latencies (see more detailed protocol description in ref. (Shulga et al. 2021)).

The selection of muscles for motor cortex mapping depends on the peripheral nerves to be stimulated. The summary of nerves that we have stimulated and the corresponding muscles are presented in ref. (Shulga et al. 2021). Mapping starts at the presumed anatomical location of the corresponding muscle representation, at an intensity slightly above the resting motor threshold (RMT) of this area. The location and direction of the coil are thereafter varied to define the sites where TMS elicits the largest and most consistent MEPs recorded with the surface electrodes from the corresponding muscle belly (hotspots). If the RMT is over 100% of the maximum stimulator output (MSO) of the TMS device, mapping requires motor pre-activation. Motor imagery (MI) can also be used. Mapping requires some experience, as in some patients the need to use pre-activation, the abnormal appearance of MEPs, and contamination of the electromyogram (EMG) by spasticity makes mapping more challenging than in healthy subjects or in some other patient groups. During high-PAS, each TMS pulse is synchronized with the first pulse of the PNS train at a pre-calculated ISI. The ISI is calculated by the formula [$F_{\text{latency}} - MEP_{\text{latency}}$] to make the stimuli coincide in the spinal cord. This formula utilizes minimum F-latency and a mean latency of 15 MEPs, thus emphasizing the fastest conducting peripheral fibers and all central fibers (Shulga et al. 2015). nTMS of each muscle representation area is paired with PNS of the nerve innervating that muscle to activate all major muscle groups of the upper or lower limb. PNS consists of trains of six 1-ms pulses delivered at 100 Hz. In our first protocols, we used 50-Hz PNS with clinically beneficial outcomes (Shulga et al. 2015, 2016a, b; Tolmacheva et al. 2017). We subsequently demonstrated in healthy subjects that 100-Hz PNS potentiates MEPs more than 50-Hz PNS (Tolmacheva et al. 2019b). Since then, we have successfully used 100-Hz PNS in our clinical studies (Shulga et al. 2021). 200-Hz or 400-Hz PNS did not enhance MEPs more than 100-Hz PNS (Mezes et al. 2020). Paired stimulations are delivered at 0.2 Hz; a higher frequency of 0.4 Hz with an identical total number of

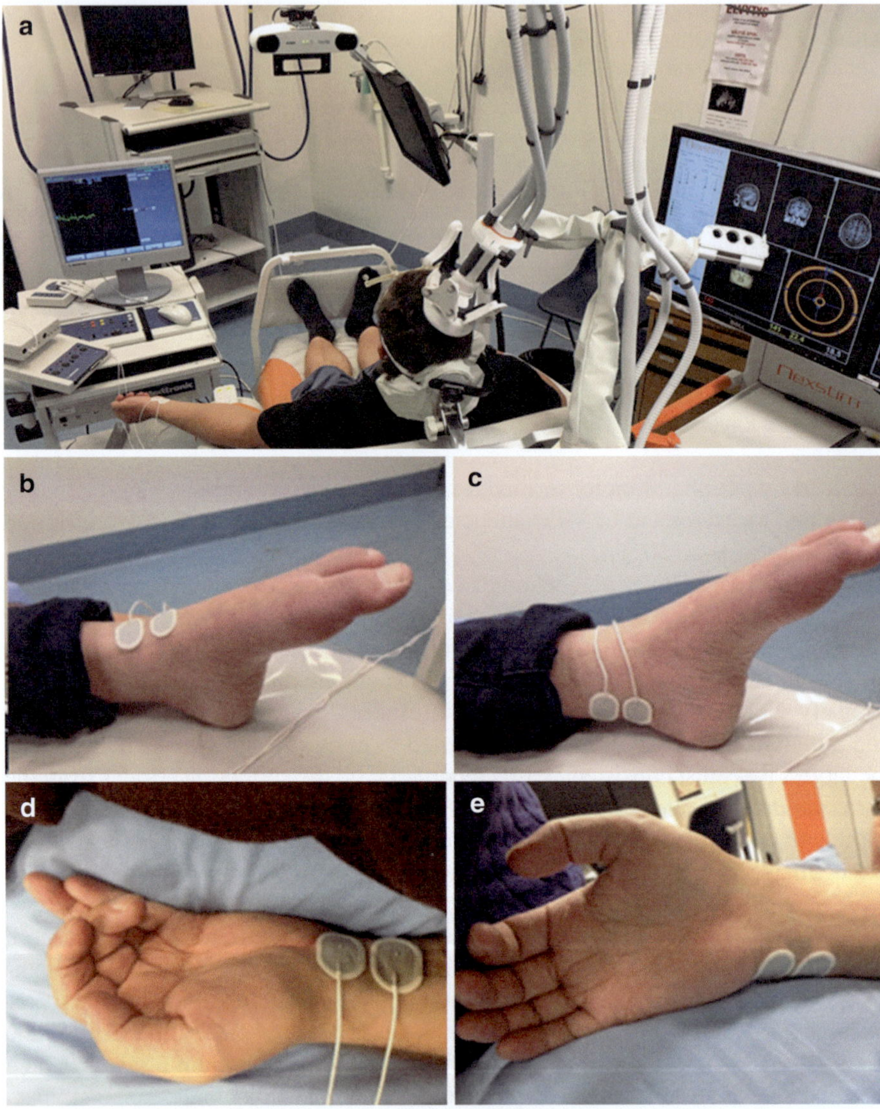

Fig. 17.1 High-PAS setup. (**a**) Preparation for median nerve paired associative stimulation in healthy subject. The setup includes navigated TMS device (right), PNS device (left), and a computer for synchronisation of TMS and PNS (background). (**b–e**) Examples of electrode placements for peripheral nerve stimulations: peroneal (**b**), tibial (**c**), median (**d**), and ulnar (**e**) nerves

pulses is not as effective in potentiating MEPs in healthy subjects (Mezes et al. 2020). We have not investigated the effect of lower frequencies, as prolonging the PAS session would not be beneficial in the clinical setting. The duration of stimulation for each peripheral nerve-M1 hotspot pair is 20 min (240 PAS pulses). The time required for preparations is usually approximately 30 min. We have stimulated patients 5 days per week during the first 2 weeks and 3 days per week thereafter.

Attention is crucial for the success of conventional PAS protocols; stimulation failed to induce plasticity when the subject's attention was directed away from the hand targeted by PAS (Stefan et al. 2004). Therefore, in our first clinical protocols, we combined PAS with motor imagery (MI) (Shulga et al. 2016b; Tolmacheva et al. 2017, 2019a). In addition, decreasing the motor thresholds by MI may enhance generation of required multiple anti- and orthodromic volleys (see "Potential mechanisms"). Furthermore, MI also plausibly activates secondary motor cortical areas, which enhances the stimulation efficacy (Carrillo-de-la-Pena et al. 2008). Although we have shown that MI is not required in healthy subjects, we conduct all healthy subject experiments with MI to closely mimic the situation with SCI patients. Slight motor pre-activation can presumably serve the same purpose as MI. We first used pre-activation in the muscles when RMT exceeded 100% of MSO (to ensure that TMS reaches the spinal cord level) and MI in the remaining muscles (Rodionov et al. 2019). A slight motor pre-activation, when possible, appeared to be easier for patients to accomplish than MI and was also easier to instruct, control, and correct. Consequently, we have used slight pre-activation regardless of RMT in later studies (Rodionov et al. 2020; Shulga et al. 2020). Moreover, pre-activation can potentiate the effects of conventional PAS in SCI patients and healthy subjects (Bunday et al. 2018).

Our patients continued their normal medication during high-PAS and follow-up periods. Medication for pain, spasticity, or other indications used by our patients represented the usual repertoire for this patient group and included, e.g., baclofen, tizanidine, pregabalin, clonazepam, tramadol, zopiclone, mirtazapine, and amitriptyline. These drugs did not prevent the therapeutic effect of high-PAS. It is not possible to estimate in each single case if the drugs diminished, enhanced, or did not influence the outcome.

No serious adverse high-PAS effects were observed. The protocol is generally considered tolerable, including TMS delivered at 100% SO. Peripheral stimulation can be perceived as unpleasant at first, but almost everyone gets used to it. For more detailed discussion on safety issues and tolerability in SCI patients, please see (Shulga et al. 2021).

17.4 Potential Mechanisms

Sensorimotor reorganization that can occur in both supraspinal and spinal neural networks is more extensive than previously assumed. A high level of motor control seems possible with relatively few descending axons extending below the lesion (Edgerton et al. 2019). As outlined above, multiple antidromic activations triggered by high-frequency PNS trains and multiple orthodromic volleys triggered by 100% SO TMS pulses lead to multiple interactions. Since interactions leading to LTP-like effects overcome their long-term depression (LTD)-like counterparts (Sjostrom et al. 2001), the net result of multiple collisions in the spinal cord is an LTP-like effect. Due to the use of the [F_{latency} – MEP $_{\text{latency}}$] formula, the arrival timing between the pre- and postsynaptic volleys at the spinal cord is considerably closer than possible interactions occurring at the cortex; it is thus plausible that the observed MEP potentiation originates mainly from induction of plasticity at corticomotoneuronal synapses of the spinal cord. Corticospinal fibers terminate both on interneurons and directly on alpha and gamma motoneurons of the spinal cord (Goshgarian 2019). Thus, high-PAS might both reroute interneurons (Ling et al. 2020) and directly strengthen synaptic connections between upper and lower motoneurons, which leads to the net effect of improved corticospinal conduction.

High-intensity TMS can induce one D-wave and four I-waves recurring at approximately 1.5-ms intervals within a time window of about 10 ms (Rothwell et al. 1991). A TMS pulse administered at 100% SO activates several neural populations of variable conductivity (Edgley et al. 1997). In a healthy population, 100% SO resulted in significantly stronger MEPs than 120% of MT (Sathyan et al. 2021). Activation also occurs in the wider vicinity of the stimulated cortical site, inducing neuronal firing of the neighboring cortex in a slightly different time frame. In neurological patients, this temporal dispersion of responses of different neural populations can be even broader. Increasing the PNS frequency increases the probability for coincidence of ascending and descending volleys within the effective time window for inducing an LTP-like effect. Moreover, it is plausible that the remaining neural pathways in patients with neurological diseases have a wide range of conductivities due to partial injuries and partial recovery, leading to dispersion of activations.

We have demonstrated in healthy subjects that high-PAS induces a robust potentiation of MEPs at a wide range of intervals between TMS and PNS and tolerates at least ±10 ms errors in ISI (Shulga et al. 2016a), whereas conventional PAS requires a very narrow, precisely defined interval. Moreover, the high-PAS protocol is resilient to small errors in mapping, which may occur in clinical practice (Tolmacheva et al. 2019b). This insensitivity to errors may be one of the reasons underlying our favorable clinical outcomes. Although some functional and anatomical reorganization after SCI occurs spontaneously, the efficacy of the neural pathways to rewire themselves is use-dependent (Edgerton et al. 2019). Temporally and spatially dispersed anti- and orthodromic volleys produced by high-PAS plausibly affect a wide net of connections that are strengthened upon multiple repetitions. Therefore, although our protocol is resilient to initial errors in the exact determination of ISI, exact repetition of the same stimulation setup (ISI, electrode placement, TMS navigation, stimulation intensities) is probably important for the therapeutic effect.

Although 100-Hz PNS is more efficient than 50- and 25-Hz PNS, further increases in PNS frequency up to 200 and 400 Hz do not provide additional efficacy (see above). Thus, bringing the PNS frequency closer to the frequency of I-waves (Rothwell et al. 1991) does not produce stronger potentiation; the exact coincidence of each PNS pulse with each TMS-induced volley does not appear to be the strongest determinant for MEP potentiation. Rather, the specific PNS frequency appears to be important, although PNS by itself does not produce MEP potentiation (Tolmacheva et al. 2019b). LTP is thought to have early and late phases; plasticity-promoting events of the early phase lead to the stable late phase LTP (L-LTP) if sufficiently repeated (Panja and Bramham 2014). L-LTP depends on protein synthesis and is associated with remodeling and enlargement of postsynaptic density

(PSD) and enlargement of pre-existing dendritic spines and de novo synapse formation (Panja and Bramham 2014). Brain-derived neurotrophic factor (BDNF) is thought to play a critical role in stimulating the formation of L-LTP at glutamatergic synapses (Panja and Bramham 2014). The 50-100 Hz stimulation induces activity-dependent release of BDNF in vitro (Balkowiec and Katz 2000; Gartner and Staiger 2002; Lever et al. 2001).

We emphasize that the therapeutic effect originates from dual stimulation and not from TMS or PNS alone. When TMS is used alone, lasting inhibitory aftereffects can be achieved with 1-Hz repetitive TMS and facilitatory aftereffects with high-frequency (>1 Hz) repetitive TMS (Rossi et al. 2009). We have shown that the 0.2-Hz TMS does not lead to MEP increase (Shulga et al. 2016a). Thus, it is highly improbable that TMS alone would have accounted for the obtained results. We have also shown that 50-Hz PNS (Shulga et al. 2016a) or 100-Hz PNS (Tolmacheva et al. 2019b) alone have no effect on MEPs.

17.5 Summary

The specificity of neuromodulation techniques applied to the whole spinal cord is generally limited (Jack et al. 2020). High-PAS is an option that allows a higher degree of selectivity. Navigated TMS has been applied in studies of plasticity induced by cervical myelopathy (Zdunczyk et al. 2018) and in rehabilitation of acute paresis after brain tumor surgery (Ille et al. 2021), among others. We have observed that high-PAS can benefit a wide range of patients with incomplete SCI. Patients with milder and more recent injuries benefit more, improve more quickly, and require shorter interventions than those with more severe and chronic lesions. However, sufficiently long treatment in patients with more severe injuries can also lead to significant increases in independence (Shulga et al. 2021). We have included patients of a wide age range since the incidence of SCI increases in older populations (Chen and DeVivo 2019). As residual connectivity is a prerequisite for high-PAS, this technique would not be beneficial for patients with complete injuries. It remains to be determined if high-PAS can be an adjunct treatment potentiating other approaches that would restore connectivity after complete SCI (such as cell transplantation).

Considering the anatomy and size of the small finger and large lower limb muscles and the fact that MMT and AIS scores measure performance against gravity, it is clear that the therapeutic effects can be detected faster in the upper than lower limbs. Whereas even slight enhancement of the upper limb activity almost immediately benefits everyday life functions (e.g., by enabling more effective grasping), larger changes in the lower limbs are required for improved walking. Therefore, the upper limbs of patients with milder tetraplegia are the easiest target to study and further develop the high-PAS; stronger innervation directly translates into more independent activities of daily living. However, the promising results on lower limb functions are important as well. Even walking part-time or being in an upright position is highly beneficial for the overall health of SCI patients, as blood pressure regulation, bone density, weight control, bowel function, and psychological well-being are improved; pressure sores and heterotopic ossification are prevented; and smoother transfers to and from the wheelchair are feasible (Bauman and Nash 2019).

The equipment for high-PAS is usually available in hospitals and has a well-characterized safety profile. Experience in clinical neurophysiology is required for defining the stimulation parameters at the beginning of a stimulation. A team of a physiatrist or a neurologist together with a physiotherapist is useful for selection of the optimal PAS targets for the stimulation and for evaluation of safety aspects. Thereafter, daily stimulation sessions can be completed by personnel with a diverse educational background, for example, by nurses or physiotherapists. The reduced need for external assistance and increase in quality of life in SCI patients justify the therapy costs. These considerations, together with the results and observations described above, encourage further research of this promising new approach.

References

Ackery A, Tator C, Krassioukov A. A global perspective on spinal cord injury epidemiology. J Neurotrauma. 2004;21(10):1355–70.

Ahmed Z. Electrophysiological characterization of spino-sciatic and cortico-sciatic associative plasticity: modulation by trans-spinal direct current and effects on recovery after spinal cord injury in mice. J Neurosci. 2013;33(11):4935–46.

Balkowiec A, Katz DM. Activity-dependent release of endogenous brain-derived neurotrophic factor from primary sensory neurons detected by ELISA in situ. J Neurosci. 2000;20(19):7417–23.

Bauman W, Nash M. In: Kirshblum S, Lin V, Toimittajat, editors. Endocrinology and metabolism of persons with spinal cord injury. New York: Springer Publishing Company; 2019. 278–279–317. (Spinal Cord Medicine).

Bunday KL, Perez MA. Motor recovery after spinal cord injury enhanced by strengthening corticospinal synaptic transmission. Curr Biol. 2012;22(24):2355–61.

Bunday KL, Urbin MA, Perez MA. Potentiating paired corticospinal-motoneuronal plasticity after spinal cord injury. Brain Stimul. 2018;11(5):1083–92.

Carrillo-de-la-Pena MT, Galdo-Alvarez S, Lastra-Barreira C. Equivalent is not equal: primary motor cortex (MI) activation during motor imagery and execution of sequential movements. Brain Res. 2008;1226:134–43.

Carson RG, Kennedy NC. Modulation of human corticospinal excitability by paired associative stimulation. Front Hum Neurosci. 2013;7:823.

Chen Y, DeVivo M. In: Kirshblum S, Lin V, editors. Epidemiology of spinal cord injury. New York: Springer Publishing Company; 2019. 44–45–62.

Chen R, Udupa K. Measurement and modulation of plasticity of the motor system in humans using transcranial magnetic stimulation. Mot Control. 2009;13(4):442–53.

Dan Y, Poo MM. Spike timing-dependent plasticity of neural circuits. Neuron. 2004;44(1):23–30.

Edgerton R, Gerasimenko Y, Gad P, Sayenko D. In: Kirshblum S, Lin V, editors. Basic concepts underlying activity-dependent mechanisms in the rehabilitation of sensory-motor function after spinal cord injury. New York: Springer Publishing Company; 2019. 897–898–911. (Spinal Cord Medicine).

Edgley SA, Eyre JA, Lemon RN, Miller S. Comparison of activation of corticospinal neurons and spinal motor neurons by magnetic and electrical transcranial stimulation in the lumbosacral cord of the anaesthetized monkey. Brain. 1997;120(Pt 5):839–53.

Fawcett JW, Curt A, Steeves JD, Coleman WP, Tuszynski MH, Lammertse D, Kalichman M, Rask C, Curt A, Ditunno JF, Fehlings MG, Guest JD, Ellaway PH. Guidelines for the conduct of clinical trials for spinal cord injury as developed by the ICCP panel: spontaneous recovery

after spinal cord injury and statistical power needed for therapeutic clinical trials. Spinal Cord. 2007;45(3):190–205.
Feldman DE. The spike-timing dependence of plasticity. Neuron. 2012;75(4):556–71.
Field-Fote EC. Exciting recovery: augmenting practice with stimulation to optimize outcomes after spinal cord injury. Prog Brain Res. 2015;218:103–26.
Field-Fote EC, Yang JF, Basso DM, Gorassini MA. Supraspinal control predicts locomotor function and forecasts responsiveness to training after spinal cord injury. J Neurotrauma. 2016;34:1813.
Gartner A, Staiger V. Neurotrophin secretion from hippocampal neurons evoked by long-term-potentiation-inducing electrical stimulation patterns. Proc Natl Acad Sci USA. 2002;99(9):6386–91.
Goshgarian H. In: Kirshblum S, Lin V, editors. Developement, anatomy and function of the spinal cord. New York: Springer Publishing Company; 2019. 12–13–30. (Spinal Cord Medicine).
Ille S, Kelm A, Schroeder A, Albers LE, Negwer C, Butenschoen VM, ym. Navigated repetitive transcranial magnetic stimulation improves the outcome of postsurgical paresis in glioma patients – a randomized, double-blinded trial. Brain Stimul. 2021;14(4):780–7.
Jack AS, Hurd C, Martin J, Fouad K. Electrical stimulation as a tool to promote plasticity of the injured spinal cord. J Neurotrauma. 2020;37(18):1933–53.
Jo HJ, Perez MA. Corticospinal-motor neuronal plasticity promotes exercise-mediated recovery in humans with spinal cord injury. Brain. 2020;143(5):1368–82.
Kirshblum S, Solinsky R. In: Kirshblum S, Lin V, Toimittajat, editors. Neurological assessment and classification of spinal cord injury. New York: Springer Publishing Company; 2019. 63–64–76. (Spinal Cord Medicine).
Lever IJ, Bradbury EJ, Cunningham JR, Adelson DW, Jones MG, McMahon SB, ym. Brain-derived neurotrophic factor is released in the dorsal horn by distinctive patterns of afferent fiber stimulation. J Neurosci. 2001;21(12):4469–77.
Ling YT, Alam M, Zheng YP. Spinal cord injury: lessons about neuroplasticity from paired associative stimulation. Neuroscientist. 2020;26(3):266–77.
McPherson JG, Miller RR, Perlmutter SI. Targeted, activity-dependent spinal stimulation produces long-lasting motor recovery in chronic cervical spinal cord injury. Proc Natl Acad Sci USA. 2015;112(39):12193–8.
Mezes M, Havu R, Tolmacheva A, Lioumis P, Makela JP, Shulga A. The impact of TMS and PNS frequencies on MEP potentiation in PAS with high-frequency peripheral component. PLoS One. 2020;15(5):e0233999.
Nicoll RA. A brief history of long-term potentiation. Neuron. 2017;93(2):281–90.
Nishimura Y, Perlmutter SI, Eaton RW, Fetz EE. Spike-timing-dependent plasticity in primate corticospinal connections induced during free behavior. Neuron. 2013;80(5):1301–9.
Oleson C, Flanders A. In: Kirshblum S, Lin V, editors. Predicting outcomes following spinal cord injury. New York: Springer Publishing Company; 2019. 149–150–63. (Spinal Cord Medicine).
Panja D, Bramham CR. BDNF mechanisms in late LTP formation: a synthesis and breakdown. Neuropharmacology. 2014;76(Pt C):664–76.
Ramer LM, Ramer MS, Bradbury EJ. Restoring function after spinal cord injury: towards clinical translation of experimental strategies. Lancet Neurol. 2014;13(12):1241–56.
Rodionov A, Savolainen S, Kirveskari E, Makela JP, Shulga A. Restoration of hand function with long-term paired associative stimulation after chronic incomplete tetraplegia: a case study. Spinal Cord Ser Case. 2019;5:81. 019-0225-5. eCollection 2019.
Rodionov A, Savolainen S, Kirveskari E, Makela JP, Shulga A. Effects of long-term paired associative stimulation on strength of leg muscles and walking in chronic tetraplegia: a proof-of-concept pilot study. Front Neurol. 2020;11:397.
Rossi S, Hallett M, Rossini PM, Pascual-Leone A, Group S of TC. Safety, ethical considerations, and application guidelines for the use of transcranial magnetic stimulation in clinical practice and research. Clin Neurophysiol. 2009;120(12):2008–39.
Rossi S, Antal A, Bestmann S, Bikson M, Brewer C, Brockmoller J. Safety and recommendations for TMS use in healthy subjects and patient populations, with updates on training, ethical and regulatory issues: expert guidelines. Clinical Neurophysiol. 2020;132:269–306.

Rothwell JC, Thompson PD, Day BL, Boyd S, Marsden CD. Stimulation of the human motor cortex through the scalp. Exp Physiol. 1991;76(2):159–200.

Sathyan S, Tolmacheva A, Tugin S, Mäkelä JP, Shulga A, Lioumis P. A new paired associative stimulation protocol with high-frequency peripheral component and high-intensity 20 hz repetitive transcranial magnetic stimulation—a pilot study. Int J Environ Res Public Health. 2021;18(21):11224. https://doi.org/10.3390/ijerph182111224.

Shulga A, Lioumis P, Kirveskari E, Savolainen S, Makela JP, Ylinen A. The use of F-response in defining interstimulus intervals appropriate for LTP-like plasticity induction in lower limb spinal paired associative stimulation. J Neurosci Methods. 2015;242C:112–7.

Shulga A, Zubareva A, Lioumis P, Makela JP. Paired associative stimulation with high-frequency peripheral component leads to enhancement of corticospinal transmission at wide range of interstimulus intervals. Front Hum Neurosci. 2016a;10:470.

Shulga A, Lioumis P, Zubareva A, Brandstack N, Kuusela L, Kirveskari E, ym. Long-term paired associative stimulation can restore voluntary control over paralyzed muscles in incomplete chronic spinal cord injury patients. Spinal Cord Ser Cases. 2016b;2:16016.

Shulga A, Savolainen S, Kirveskari E, Makela JP. Enabling and promoting walking rehabilitation by paired associative stimulation after incomplete paraplegia: a case report. Spinal Cord Ser Cases. 2020;6(1):72. 020-0320–7.

Shulga A, Lioumis P, Kirveskari E, Savolainen S, Makela JP. A novel paired associative stimulation protocol with a high-frequency peripheral component: a review on results in spinal cord injury rehabilitation. Eur J Neurosci. 2021;53:3242.

Sjostrom PJ, Turrigiano GG, Nelson SB. Rate, timing, and cooperativity jointly determine cortical synaptic plasticity. Neuron. 2001;32(6):1149–64.

Squair JW, Bjerkefors A, Inglis JT, Lam T, Carpenter MG. Cortical and vestibular stimulation reveal preserved descending motor pathways in individuals with motor-complete spinal cord injury. J Rehabil Med. 2016;48(7):589–96.

Stefan K, Kunesch E, Cohen LG, Benecke R, Classen J. Induction of plasticity in the human motor cortex by paired associative stimulation. Brain. 2000;123(Pt 3):572–84.

Stefan K, Wycislo M, Classen J. Modulation of associative human motor cortical plasticity by attention. J Neurophysiol. 2004;92(1):66–72.

Suppa A, Quartarone A, Siebner H, Chen R, Lazzaro VD, Giudice PD. ym. The associative brain at work: evidence from paired associative stimulation studies in humans. Clinical Neurophysiol. 2017;128(11):2140–64.

Taylor JL, Martin PG. Voluntary motor output is altered by spike-timing-dependent changes in the human corticospinal pathway. J Neurosci. 2009;29(37):11708–16.

Tohda C, Kuboyama T. Current and future therapeutic strategies for functional repair of spinal cord injury. Pharmacol Ther. 2011;132(1):57–71.

Tolmacheva A, Savolainen S, Kirveskari E, Lioumis P, Kuusela L, Brandstack NM. Long-term paired associative stimulation enhances motor output of the tetraplegic hand. J Neurotrauma. 2017;34:2668.

Tolmacheva A, Savolainen S, Kirveskari E, Brandstack N, Makela JP, Shulga A. Paired associative stimulation improves hand function after non-traumatic spinal cord injury: a case series. Clin Neurophysiol Pract. 2019a;4:178–83.

Tolmacheva A, Makela JP, Shulga A. Increasing the frequency of peripheral component in paired associative stimulation strengthens its efficacy. Sci Rep. 2019b;9(1):3849. 019–40474–0.

Vaalto S, Nyman AL, Shulga A. Analgesic effect of paired associative stimulation in a tetraplegic patient with severe drug-resistant neuropathic pain: a case report. Scand J Pain. 2021;21(4):831–8.

Vanhanen J, Parkkonen L, Mäkelä JP, Tolmacheva A, Shulga A, Rodionov A. Effect of long-term paired associative stimulation on the modulation of cortical sensorimotor oscillations after spinal cord injury. Spinal Cord Ser Cases. 2022;8(1):38.

Zdunczyk A, Schwarzer V, Mikhailov M, Bagley B, Rosenstock T, Picht T, ym. The corticospinal reserve capacity: reorganization of motor area and excitability as a novel pathophysiological concept in cervical myelopathy. Neurosurgery. 2018;83(4):810–8.

Treatment of Paresis 18

Melina Engelhardt, Jari Karhu, Petro Julkunen, and Sebastian Ille

18.1 Introduction

Most of the applications of repetitive TMS (rTMS) for the treatment of paresis were initially developed and tested in patients after stroke (Lefaucheur et al. 2020). In recent years, these therapy concepts have also transitioned to neurosurgical applications following resection of a supratentorial brain tumor (Einstein et al. 2022).

Postoperative motor impairments occur in about 25% of brain tumor patients with motor eloquent tumors (Rosenstock et al. 2017). The presence of impairments has a negative impact on patients' quality of life and regaining the ability to work, as well as time to adjuvant treatment (Jakola et al. 2011; Rydén et al. 2020). As the standard of care, physical therapy is applied during the inpatient stay and is recommended to continue after release from the hospital. Yet, many deficits do not fully

M. Engelhardt (✉)
Department of Neurosurgery, Charité – Universitätsmedizin Berlin, Freie Universität Berlin, Humboldt-Universität zu Berlin, and Berlin Institute of Health, Berlin, Germany

Einstein Center for Neurosciences, Charité – Universitätsmedizin Berlin, Freie Universität Berlin, Humboldt-Universität zu Berlin, and Berlin Institute of Health, Berlin, Germany
e-mail: melina.engelhardt@charite.de

J. Karhu
Satucon Inc, Kuopio, Finland
e-mail: jari.karhu@satucon.fi

P. Julkunen
Department of Clinical Neurophysiology, Kuopio University Hospital, Kuopio, Finland

Department of Technical Physics, University of Eastern Finland, Kuopio, Finland
e-mail: petro.julkunen@uef.fi

S. Ille
Department of Neurosurgery, Heidelberg University Hospital, Heidelberg, Germany
e-mail: Sebastian.Ille@med.uni-heidelberg.de

S. M. Krieg, T. Picht (eds.), *Navigated Transcranial Magnetic Stimulation in Neurosurgery*, https://doi.org/10.1007/978-3-031-97155-6_18

recover in the following postoperative months, thus posing a continuing burden for the patient (Rosenstock et al. 2017; Piil et al. 2016). The origins of paresis in this patient group can be manifold, including surgical lesions of functional tissue during resection, intra- or postoperative ischemic events, or edema. Depending on the origin, optimal treatment strategies vary. In recent years, rTMS has been investigated as an adjuvant treatment in these patients due to its ability to induce plastic changes.

In general, rTMS modulates cortical activity by either upregulating or downregulating cortical excitability, depending on the rTMS parameters used. rTMS given at a frequency of ≥3 Hz is likely to lead to cortical excitation, as evidenced by the increased amplitude of the TMS-induced MEPs (Pascual-Leone et al. 1994). In contrast, low-frequency rTMS, that is, a pulse frequency of about 1 Hz, is likely to lead to cortical inhibition, as evidenced by decreased amplitude of the TMS-induced MEP (Chen et al. 1997). In stroke patients, inhibitory 1-Hz-rTMS can be utilized to downregulate the contralesional hemisphere to improve response to motor training by reducing interhemispheric inhibition and potentially facilitating activity in the lesioned hemisphere. Another less investigated option is the excitation of the lesioned side using high-frequency rTMS (Lefaucheur et al. 2020). This chapter summarizes the current state of research using rTMS for the treatment of paresis in brain tumor patients while also giving an outlook on future concepts for personalization of treatment.

18.2 Clinical Evidence

To date, five studies report the use of rTMS in a total of 75 brain tumor patients with postoperative paresis (Einstein et al. 2022). These include two case studies (Kakuda et al. 2010; O'Neal et al. 2020), one non-randomized one-arm trial (Poologaindran et al. 2022), and two randomized controlled trials (Ille et al. 2021; Engelhardt et al. 2024).

Three of these studies (Kakuda et al. 2010; Ille et al. 2021; Engelhardt et al. 2024) applied low-frequency rTMS with a frequency of 1 Hz to the contralesional primary motor cortex (M1), followed by occupational or physical therapy. All of these studies found positive effects on motor rehabilitation following the intervention. A case study (Kakuda et al. 2010) included a patient who underwent resection of a left subfrontal glioma and presented with persisting motor deficits in the right hand after resection. The patient was included in the study 5 years after surgery, as motor recovery was stated to have plateaued with conventional occupational therapy. The patient received 22 sessions of rTMS at 1 Hz for 20 min each to the contralesional M1. RTMS was applied twice daily at 90% of the resting motor threshold (RMT) of the first dorsal interosseous muscle (FDI). Directly after each session, the patient received 60 min of occupational therapy followed by 60 min of self-training. Investigators observed a reduction of the paresis directly after the intervention, which continued to improve until 4 weeks later. Specifically, there was a reduction in spasticity of the affected fingers and increased use of the right hand in activities of daily living.

A randomized controlled trial (Ille et al. 2021) recruited 22 patients with a worsening of motor function directly after resection of a supratentorial brain tumor and ischemia within the corticospinal tract (CST). Patients presented with varying degrees of upper extremity paresis but preserved postoperative motor evoked potentials (MEPs) during an nTMS examination. They received rTMS for 7 days at 1 Hz for 15 min each to the contralesional M1. Stimulation was applied at 110% RMT daily, followed by physical therapy. The study was controlled by a group receiving sham stimulation and physiotherapy. They found a significant improvement in Fugl-Meyer Scores (Fugl-Meyer et al. 1975) 3 months postoperatively compared to sham stimulation with a number needed to treat of only 2.19. This effect was also visible in more comprehensive neurological function as measured with the NIHSS.

Finally, another randomized controlled trial (Engelhardt et al. 2024) tested the same postoperative intervention in 30 patients without the previous inclusion criteria of ischemia in the CST or presence of MEPs. This study found no significant effect on Fugl-Meyer Scores 3 months postoperatively but improved hand motor function measured with the BMRC 1 month postoperatively. When looking at a subgroup of ischemic patients, similar effects to Ille et al. (2021) were observed for the NIHSS, thus suggesting this patient group may be more responsive to rTMS treatment.

One case study (O'Neal et al. 2020) applied continuous theta burst stimulation (cTBS) in a patient with a left insular glioma. The patient presented preoperatively with a slight right hemiparesis and pronator drift, preexisting from a prior tumor resection 6 years before. Additionally, a new deficit in the left lower extremity and right side of the face was observed after the current surgery. Consequently, one session of 5 Hz cTBS (200 pulses total) was applied to the right contralesional M1 with an intensity of 80% of the active motor threshold for the FDI. A slight improvement in the left-sided deficit was observed immediately after cTBS, which continued to resolve until 4 weeks later. There was no effect on the right-sided deficits.

Lastly, one proof-of-concept study (Poologaindran et al. 2022) evaluated the effects of TBS applied to personalized connectome-based target regions in 34 glioma patients with postoperative motor ($n = 21$) or language deficits ($n = 13$). Patients were recruited within 2 weeks after surgery, and stimulation was applied five times daily for 5 days in conjunction with rehabilitative therapy. While stimulation details for individual patients are not given, the authors state that TBS in patients with motor impairments was commonly targeted to modulate sensorimotor network activity. Further, as part of the personalized approach, exact stimulation targets and parameters may also vary between daily stimulation sessions within the same subject. Overall, the authors report an improvement in motor function in 19/21 patients 2 week after the intervention.

18.3 Potential Mechanisms

Contralesional inhibitory 1 Hz stimulation, adapted from studies in stroke patients, relies on interhemispheric connectivity. Application of low-frequency rTMS to the contralesional M1 may restore balance between the two cerebral hemispheres. In healthy individuals at rest, the motor cortices inhibit one another through transcallosal inhibition (TCI) (interhemispheric inhibition; Ferbert et al. 1992). If one side is lesioned, modulatory inhibition to the other side is reduced, leading to increased activation in the contralesional side. The contralesional side still provides inhibitory signals to the lesioned side via the undamaged cortical system, even more than in a healthy, balanced situation. Behaviorally, when healthy subjects perform unilateral limb movements, TCI from the ipsilateral hemisphere is released just prior to the activation of the limb (Murase et al. 2004), a mechanism that may promote more accurate unilateral movements. In the case of stroke, however, TCI from the contralesional M1 is not released with attempted limb movement, further impairing motor recovery (Grefkes et al. 2008).

Inhibitory modulation in the contralesional M1 with LF rTMS may lead to disinhibition, thereby restoring the balance between the two hemispheres and improving the opportunity for neuroplastic change in the perilesional M1. In general, the "release" of the motor system from an overall inhibitory state caused by a lesion or lesion-induced impaired or maladapted feedback, be it sensory or motor, seems beneficial to behavioral recovery (Nowak et al. 2009).

18.4 Alternative Stimulation Strategies

Even though current studies in brain tumor patients focus on low-frequency rTMS approaches, as initially stated, there is also a plausible rationale for treating paretic patients with high-frequency rTMS to the lesioned M1. The pivotal findings of Nudo et al. (1996) revealed that rehabilitative training of skilled hand use after focal cortical infarcts prevented the otherwise frequent loss of hand territory adjacent to the infarct. In some instances, the hand representations expanded into regions formerly occupied by representations of the elbow and shoulder. Functional reorganization in the undamaged motor cortex was accompanied by behavioral recovery of skilled hand function. The authors suggested that rehabilitative training can shape subsequent reorganization in the adjacent intact cortex after local damage to the motor cortex. The undamaged motor cortex may play an important role in motor recovery. Functional neuroimaging in stroke has revealed that neurons in the area of the lesioned M1 surrounding the injury often take over function in patients with good recovery (Zemke et al. 2003), a process sometimes called vicariation. In the acute period after a stroke, the activity in these perilesional neurons is reduced. Thus, excitatory high-frequency rTMS may render perilesional neurons more responsive to therapy, speeding the process of transferring damaged function to nearby neuronal structures. The same effect is gained naturally by unmasking M1 networks, which have the capacity to overtake impaired behavioral function.

To unify the TCI and vicariation model and give individualized therapy recommendations, the bimodal-balance model has been introduced (Di Pino et al. 2014; Ovadia-Caro et al. 2019) (Fig. 18.1). The model differentiates patients based on the integrity of their white matter motor pathways, the so-called structural reserve. In the case of a high structural reserve, an overactivation of the contralesional hemisphere is seen as maladaptive. However, the increased contralesional activity may be compensatory in patients with a low structural reserve. This differentiation is based on studies showing reduced effectiveness of contralesional rTMS in severely affected stroke patients (Bradnam et al. 2012; Ackerley et al. 2010).

It is worth highlighting, however, that postoperative plasticity following motor impairments in brain tumor patients is not yet fully understood (Cargnelutti et al. 2020). Therefore, the mechanisms at play in this group of patients may differ from

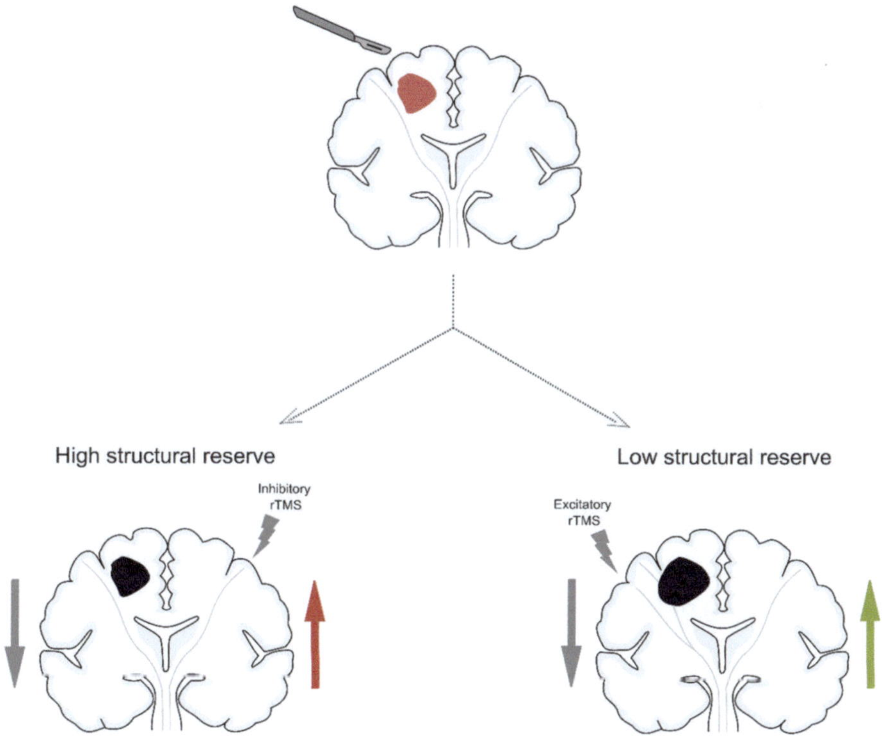

Fig. 18.1 Bimodal-balance model in brain tumor patients. After uni-hemispheric lesions for example, due to tumor resection, an increased activity in the contralesional hemisphere is often observed. The bimodal-balance model postulates that this overactivity is maladaptive in patients with a high structural reserve. In contrast, patients with a low structural reserve due to injuries to the core motor network may benefit from increased contralesional activation. As a consequence, rTMS approaches should be adapted based on this individualized assessment. Given a maladaptive overactivity, low-frequency rTMS can be used to inhibit the contralesional hemisphere. In patients with a low-structural reserve, however, high-frequency rTMS can be used to enhance activity of non-injured motor areas in the lesioned hemisphere

those in other cases. Bryszewski et al. (2013) found increased activation of the ipsilateral premotor cortex and supplementary motor area after resections involving the primary motor cortex. Further, recovery of deficits due to resection of the supplementary motor area seems to depend largely on recruitment of the contralesional SMA (Krainik et al. 2004). Both instances could be seen as examples of damage to the core motor system relating to a low structural reserve. In contrast, patients with preserved postoperative MEPs, where motor deficits seem to be related to ischemia in the CST, may represent a group with high structural reserve due to the small size of the ischemic lesion and, at least in part, intact signaling pathways. This is supported by the observed high responsiveness to contralesional stimulation in this cohort (Ille et al. 2021). However, none of the studies discussed here specifically investigated the structural reserve and TCI in brain tumor patients before the application of rTMS. Thus, conclusions should be drawn with caution.

18.5 Caveats in Current Rehabilitation Approaches

Unarguably, brain tumors occur with a significantly smaller incidence than, for example, strokes. Thus, gathering comparable levels of evidence for treatment of paresis using rTMS will progress proportionally slower. Differences between patients, not only concerning tumor size and location but also tumor histology, molecular characterization, area of resection, and presence of ischemia, among other factors, make this patient group highly heterogeneous (Lauko et al. 2022). Consequently, personalization of stimulation seems necessary to ensure the most effective therapy for each individual patient. Individualizing treatment may also help reduce the variability in responsiveness to rTMS, an issue observed in many therapeutic rTMS trials. Personalization, however, requires a good understanding of underlying mechanisms and levels of efficacy of different intervention types; knowledge must mostly be gained by large, unified clinical trials. This dilemma may also lead researchers to reference back to studies in stroke patients.

18.5.1 Timing of Postoperative rTMS

Most studies in brain tumor patients applied rTMS in the acute postoperative period, starting as early as the first postoperative day (Ille et al. 2021; Engelhardt et al. 2024). It is argued that this reflects the period of highest plasticity and natural recovery and, thus, presents the most significant potential for modulation (Bashir et al. 2010). In support of this idea, the effects of rTMS seem to be more prominent in these studies (Poologaindran et al. 2022; Ille et al. 2021; Engelhardt et al. 2024) compared to case studies that recruited patients in the chronic phase (Kakuda et al. 2010; O'Neal et al. 2020). In stroke patients, low-frequency rTMS of the contralesional M1 is rated with definite efficacy for the acute compared to only possible efficacy for the chronic phase (Lefaucheur et al. 2020). The lower efficacy rating for the chronic stage is mainly driven by a large clinical trial (Harvey et al. 2018) showing no difference between rTMS and sham stimulation in terms of their effect on motor function. These results suggest that early recruitment of patients in the acute phase may be beneficial, thus highlighting the importance of patient screening by neurosurgical departments.

18.5.2 Duration of rTMS Intervention

There is considerable variation between studies in brain tumor patients in terms of the duration of rTMS sessions, the number of daily sessions, and the total number of sessions. The two largest studies referenced here applied stimulation for 15 min daily for a total of 7 days (Ille et al. 2021; Engelhardt et al. 2024). However, accelerated rTMS with up to five daily sessions has also been safely applied (Poologaindran et al. 2022). Overall, studies with a higher number of sessions are regarded as more effective in the rehabilitation of paresis, as they extend the period of active care. Yet, recommendations for an optimal duration are missing. Regarding the duration of individual sessions, it remains uncertain whether extended sessions have a comparable impact on efficacy or if they might even reverse effects, as observed in other non-invasive brain stimulation techniques (Hassanzahraee et al. 2020). Finally, it is proposed that the effectiveness of stimulation remains consistent when administered multiple times per day, compared to over a greater number of days, as long as the total number of sessions remains constant (Modirrousta et al. 2018). From a practical point of view, accelerated rTMS could be a promising way to increase the stimulation dose in neurosurgical applications where patients are only seen for a limited number of days postoperatively. This also helps balance the trade-off between the health economy, patient availability, and resource availability.

18.5.3 Determination of the Stimulation Target

All except for one study in brain tumor patients (Poologaindran et al. 2022) applied rTMS exclusively to the contralesional M1. As discussed previously (see 18.4.), this strategy may be optimal for patients with a high structural reserve as it normalizes a disrupted TCI. In contrast, patients with a low structural reserve may benefit more from ipsilesional stimulation (Di Pino et al. 2014; Ovadia-Caro et al. 2019). Optimal targets for ipsilesional rTMS in brain tumor patients still need to be tested. Generally, premotor areas contain organized neuronal components, which may compensate for lesioned M1 and CST, particularly in the most behaviorally impaired patients. A promising candidate may be a nonprimary cortical area (such as M2), which is activated by nTMS and has an alternate, direct CST providing a functional route to peripheral muscles at the same MEP latencies as M1 proper (Vaalto et al. 2011; Teitti et al. 2008). More remote but densely interconnected areas of the intraparietal sulcus may also be a candidate to suppress—possibly maladaptive—sensory information. Also, a combination of different stimulation targets, as in Poologaindran et al. (2022), may lead to favorable outcomes. Upcoming research will need to gather more data to achieve an adequate stratification of patients, ensuring optimal treatment given the damage in the individual patient.

18.5.4 Importance of Adjuvant Physical Therapy

There is ample evidence that rehabilitation therapy focusing on repetitive and skillful task practice (task-oriented treatment) results in long-term functional recovery. Both animal models (Kleim and Jones 2008) and human clinical trials (Liepert et al. 2000; Wolf 2006) support a use-dependent relationship between task-oriented therapy, neuroplasticity, and functional performance. Interestingly, there is no definite evidence that a specific standardized therapy protocol or a certain threshold of therapy intensity (e.g., (Todhunter-Brown et al. 2014; Coupar et al. 2012; Lang et al. 2009; Winstein et al. 2016)) would be superior to other types of rehabilitation therapy. Notably, a recent multicenter trial demonstrated no differences in long-term functional outcomes between three different types of occupational therapy (Winstein et al. 2016). Only one case study in brain tumor patients did not apply a form of occupational therapy in conjunction with rTMS, and this study failed to report significant effects of the intervention (O'Neal et al. 2020). In contrast, both sham-controlled trials showed the benefits of adding rTMS to physical therapy compared to physical therapy alone (Ille et al. 2021; Engelhardt et al. 2024). Thus, stimulating the brain to drive its adaptive plastic potential seems to also accelerate the rehabilitative potential of task-oriented training in patients with paresis (Lüdemann-Podubecká et al. 2015).

18.5.5 Comparison of Theta Burst Stimulation with rTMS

Theta-burst stimulation (TBS) is a specific form of rTMS, where bursts of 3 pulses at 50 Hz are applied at a frequency of 5 Hz. During continuous TBS (cTBS), these bursts are applied without interruption for the entire stimulation session, often lasting 2–3 min. For intermittent TBS (iTBS), breaks of several seconds are inserted after a defined number of bursts. Like rTMS, TBS can increase the excitability of stimulated brain areas using iTBS (presumably inducing LTP-like phenomena in primary motor networks) or depress the excitability using cTBS (likely causing predominantly LTD-like effects). Further, TBS can improve motor skills and motor learning when applied concurrently with motor practice. It was initially thought that TBS produces more robust and reproducible effects than other rTMS methods in less time, which is yet to be proven to date (Suppa et al. 2016). Studies in brain tumor patients have applied both TBS and rTMS with good effects as long as stimulation was combined with physical therapy (Einstein et al. 2022). Yet, no study to date directly compared both protocols in terms of efficacy. From a practical perspective, it is unarguable that TBS protocols are more time-efficient in delivering stimuli and, thus, might be easier to integrate into clinical practice, especially when multiple sessions per day are applied.

18.6 Future Approaches for Paresis Therapy

18.6.1 Brain-State-Dependent Stimulation

One possible origin of the variability in the response to brain stimulation such as TMS is the variability of the instantaneous brain state at the time of stimulation (Ridding and Ziemann 2010; Zrenner and Ziemann 2024). The 10 Hz μ-rhythm extracted from EEG is a readily and noninvasively available brain state biomarker for characterizing the ongoing state of the sensorimotor system most relevant to the current discussion. NTMS triggered on different phases of the μ-rhythm revealed increased excitability at the early rising phase of the oscillation (Zrenner et al. 2023). Further, high-frequency rTMS applied to the trough induces a strong potentiation of excitability, whereas stimulation at the peak leads showed no effects (Zrenner et al. 2018). Whether these effects also transfer to patients is under investigation in a clinical trial on paretic stroke patients (Lieb et al. 2023).

18.6.2 Multilocus Stimulation

Patients present with a variety of motor deficits, often extending to multiple muscles or limbs. Previous rTMS protocols in brain tumor patients have shown a high specificity of effects to stimulated muscle groups (Ille et al. 2021; Engelhardt et al. 2024), yet the extent of generalization to non-stimulated extremities is still in question. Patients with complex impairments may benefit from stimulation of multiple cortical areas to sufficiently address deficits in different extremities. Moreover, following the approach by Poologaindran et al. (2022), the activity of functional networks may be disrupted at various locations, necessitating the application of stimulation to multiple cortical targets to effectively address these disturbances. In this study, the authors accounted for multiple disruptions by applying TBS to varying targets for different daily sessions. Another piece of evidence for the potential benefit of multilocus stimulation was observed by Long et al. (2018), showing better motor recovery in stroke patients after contralesional 1 Hz rTMS followed by ipsilesional 10 Hz rTMS compared to 1 Hz rTMS alone. One practical concern common to these protocols is the increase in experimental time inevitable with sequential stimulation of multiple cortical targets. Only recently has the development of multilocus TMS (mTMS) enabled the simultaneous stimulation of adjacent cortical targets (Nieminen et al. 2022). MTMS allows for shifting the stimulus focus electronically with millimeter accuracy without physically moving the coil, thus overcoming technological limitations such as overlapping coils to stimulate adjacent brain areas (Ziemann et al. 1996).

18.6.3 Preoperative Stimulation

Research so far has primarily focused on the treatment of existing deficits, as patients are commonly referred to a physician following the occurrence of a lesion. However, neurosurgery offers the unique possibility to access patients before a planned lesioning of the system due to surgery, thus giving the possibility to pre-treat patients before any potential occurrence of impairments (Duffau 2020). This can be achieved by targeting interventions to the preoperative phase, when patients have been identified to be at high risk for postoperative paresis, for example, due to the tumor location (Rosenstock et al. 2017, 2021). Rivera-Rivera et al. (2017) have shown the capability of direct cortical stimulation coupled with intensive motor training to induce a functional reorganization in stimulated areas, leading to a shift of functional areas away from the tumor. In this way, previously inoperable brain areas could be safely resected while motor function was preserved. An upcoming clinical trial (Engelhardt et al. 2023) will investigate the efficacy of preoperative rTMS in replicating these effects noninvasively. The protocol applies low-frequency rTMS for 5–28 days to cortical endings of corticospinal fibers at risk for surgical damage, aiming to increase the amount of tumor resection while preserving motor function postoperatively.

18.7 Conclusion

The ultimate goal of adjuvant therapies in paresis treatment, such as rTMS, is to provide an optimally receptive neuronal environment in which natural recovery can happen and behavioral therapies can be imparted. There is growing evidence for the efficacy of low-frequency rTMS over the contralesional motor cortex to improve rehabilitation of postoperative paresis in the acute phase after tumor resection. Current open questions in the field revolve around patient stratification and treatment personalization to optimize efficacy in the individual patient. To address these points, impairments to patients' structural and functional motor networks have to be quantified and integrated into treatment planning. To reach sufficient patient numbers, multicenter projects and a priori agreements on test protocols are sorely needed. Sufficiently large comparable datasets must be gathered for implementing neuromodulation as an integral part of the growing clinical neurorehabilitation field.

References

Ackerley SJ, Stinear CM, Barber PA, Byblow WD. Combining theta burst stimulation with training after subcortical stroke. Stroke. 2010;41(7):1568–72.

Bashir S, Mizrahi I, Weaver K, Fregni F, Pascual-Leone A. Assessment and modulation of neural plasticity in rehabilitation with transcranial magnetic stimulation. PM&R. 2010;2(12S):S253.

Bradnam LV, Stinear CM, Barber PA, Byblow WD. Contralesional hemisphere control of the proximal paretic upper limb following stroke. Cereb Cortex. 2012;22(11):2662–71.

Bryszewski B, Tybor K, Ormezowska EA, Jaskólski DJ, Majos A. Rearrangement of motor centers and its relationship to the neurological status of low-grade glioma examined on pre- and post-operative fMRI. Clin Neurol Neurosurg. 2013;115(12):2464–70.

Cargnelutti E, Ius T, Skrap M, Tomasino B. What do we know about pre- and postoperative plasticity in patients with glioma? A review of neuroimaging and intraoperative mapping studies. Neuroimage Clin. 2020;28:102435.

Chen R, Classen J, Gerloff C, Celnik P, Wassermann EM, Hallett M, et al. Depression of motor cortex excitability by low-frequency transcranial magnetic stimulation. Neurology. 1997;48(5):1398–403.

Coupar F, Pollock A, Rowe P, Weir C, Langhorne P. Predictors of upper limb recovery after stroke: a systematic review and meta-analysis. Clin Rehabil. 2012;26(4):291–313.

Di Pino G, Pellegrino G, Assenza G, Capone F, Ferreri F, Formica D, et al. Modulation of brain plasticity in stroke: a novel model for neurorehabilitation. Nat Rev Neurol. 2014;10(10):597–608.

Duffau H. Can non-invasive brain stimulation be considered to facilitate reoperation for low-grade glioma relapse by eliciting neuroplasticity? Front Neurol. 2020;11:582489.

Einstein EH, Dadario NB, Khilji H, Silverstein JW, Sughrue ME, D'Amico RS. Transcranial magnetic stimulation for post-operative neurorehabilitation in neuro-oncology: a review of the literature and future directions. J Neuro-Oncol. 2022;157(3):435–43.

Engelhardt M, Schneider H, Reuther J, Grittner U, Vajkoczy P, Picht T, et al. Low-frequency repetitive transcranial magnetic stimulation in patients with motor deficits after brain tumor resection: a randomised, double-blind, sham-controlled trial. Front Oncol. 2024;26;14.

Engelhardt M, Grittner U, Krieg S, Picht T. Preconditioning of the motor network with repetitive navigated transcranial magnetic stimulation (rnTMS) to improve oncological and functional outcome in brain tumor surgery: a study protocol for a randomized, sham-controlled, triple-blind clinical trial. Trials. 2023;24(1):638.

Ferbert A, Priori A, Rothwell JC, Day BL, Colebatch JG, Marsden CD. Interhemispheric inhibition of the human motor cortex. J Physiol. 1992;453(1):525–46.

Fugl-Meyer AR, Jääskö L, Leyman I, Olsson S, Steglind S. The post-stroke hemiplegic patient. 1. A method for evaluation of physical performance. Scand J Rehabil Med. 1975;7(1):13–31.

Grefkes C, Eickhoff SB, Nowak DA, Dafotakis M, Fink GR. Dynamic intra- and interhemispheric interactions during unilateral and bilateral hand movements assessed with fMRI and DCM. NeuroImage. 2008;41(4):1382–94.

Harvey RL, Edwards D, Dunning K, Fregni F, Stein J, Laine J, et al. Randomized sham-controlled trial of navigated repetitive transcranial magnetic stimulation for motor recovery in stroke. Stroke. 2018;49(9):2138–46.

Hassanzahraee M, Nitsche MA, Zoghi M, Jaberzadeh S. Determination of anodal tDCS duration threshold for reversal of corticospinal excitability: an investigation for induction of counter-regulatory mechanisms. Brain Stimul. 2020;13(3):832–9.

Ille S, Kelm A, Schroeder A, Albers LE, Negwer C, Butenschoen VM, et al. Navigated repetitive transcranial magnetic stimulation improves the outcome of postsurgical paresis in glioma patients – a randomized, double-blinded trial. Brain Stimul. 2021;14(4):780–7.

Jakola AS, Gulati S, Weber C, Unsgård G, Solheim O. Postoperative deterioration in health related quality of life as predictor for survival in patients with glioblastoma: a prospective study. PLoS One. 2011;6(12):e28592.

Kakuda W, Abo M, Kobayashi K, Momosaki R, Yokoi A, Ito H, et al. Low-frequency rTMS combined with intensive occupational therapy for upper limb hemiparesis after brain tumour resection. Brain Inj. 2010;24(12):1505–10.

Kleim JA, Jones TA. Principles of experience-dependent neural plasticity: implications for rehabilitation after brain damage. J Speech Lang Hear Res. 2008;51(1):S225.

Krainik A, Duffau H, Capelle L, Cornu P, Boch AL, Mangin JF, et al. Role of the healthy hemisphere in recovery after resection of the supplementary motor area. Neurology. 2004;62(8):1323–32.

Lang CE, MacDonald JR, Reisman DS, Boyd L, Jacobson Kimberley T, Schindler-Ivens SM, et al. Observation of amounts of movement practice provided during stroke rehabilitation. Arch Phys Med Rehabil. 2009;90(10):1692–8.

Lauko A, Lo A, Ahluwalia MS, Lathia JD. Cancer cell heterogeneity & plasticity in glioblastoma and brain tumors. Semin Cancer Biol. 2022;82:162–75.

Lefaucheur JP, Aleman A, Baeken C, Benninger DH, Brunelin J, Di Lazzaro V, et al. Evidence-based guidelines on the therapeutic use of repetitive transcranial magnetic stimulation (rTMS): an update (2014–2018). Clin Neurophysiol. Elsevier Ireland Ltd. 2020;131:474–528.

Lieb A, Zrenner B, Zrenner C, Kozák G, Martus P, Grefkes C, et al. Brain-oscillation-synchronized stimulation to enhance motor recovery in early subacute stroke: a randomized controlled double-blind three- arm parallel-group exploratory trial comparing personalized, non- personalized and sham repetitive transcranial magnetic stimulation (acronym: BOSS-STROKE). BMC Neurol. 2023;23(1):204.

Liepert J, Bauder H, Miltner WHR, Taub E, Weiller C. Treatment-induced cortical reorganization after stroke in humans. Stroke. 2000 Jun;31(6):1210–6.

Long H, Wang H, Zhao C, Duan Q, Feng F, Hui N, et al. Effects of combining high- and low-frequency repetitive transcranial magnetic stimulation on upper limb hemiparesis in the early phase of stroke. Restor Neurol Neurosci. 2018;36(1):21–30.

Lüdemann-Podubecká J, Bösl K, Theilig S, Wiederer R, Nowak DA. The effectiveness of 1Hz rTMS over the primary motor area of the unaffected hemisphere to improve hand function after stroke depends on hemispheric dominance. Brain Stimul. 2015;8(4):823–30.

Modirrousta M, Meek B, Wikstrom S. Efficacy of twice-daily vs once-daily sessions of repetitive transcranial magnetic stimulation in the treatment of major depressive disorder: a retrospective study. Neuropsychiatr Dis Treat. 2018;14:309–16.

Murase N, Duque J, Mazzocchio R, Cohen LG. Influence of interhemispheric interactions on motor function in chronic stroke. Ann Neurol. 2004;55(3):400–9.

Nieminen JO, Sinisalo H, Souza VH, Malmi M, Yuryev M, Tervo AE, et al. Multi-locus transcranial magnetic stimulation system for electronically targeted brain stimulation. Brain Stimul. 2022;15(1):116–24.

Nowak DA, Grefkes C, Ameli M, Fink GR. Interhemispheric competition after stroke: brain stimulation to enhance recovery of function of the affected hand. Neurorehabil Neural Repair. 2009;23(7):641–56.

Nudo RJ, Wise BM, SiFuentes F, Milliken GW. Neural substrates for the effects of rehabilitative training on motor recovery after ischemic infarct. Science. 1996;272(5269):1791–4.

O'Neal CM, Stephens TM, Briggs RG, Sughrue ME, Conner AK. Navigated transcranial magnetic stimulation following awake craniotomy for resection of glioma: description of two cases. Surg Neurol Int. 2020;11:11.

Ovadia-Caro S, Khalil AA, Sehm B, Villringer A, Nikulin VV, Nazarova M. Predicting the response to non-invasive brain stimulation in stroke. Front Neurol. 2019;2:10.

Pascual-Leone A, Valls-Solé J, Wassermann EM, Hallett M. Responses to rapid-rate transcranial magnetic stimulation of the human motor cortex. Brain. 1994;117(4):847–58.

Piil K, Juhler M, Jakobsen J, Jarden M. Controlled rehabilitative and supportive care intervention trials in patients with high-grade gliomas and their caregivers: a systematic review. BMJ Support Palliat Care. 2016;6(1):27–34.

Poologaindran A, Profyris C, Young IM, Dadario NB, Ahsan SA, Chendeb K, et al. Interventional neurorehabilitation for promoting functional recovery post-craniotomy: a proof-of-concept. Sci Rep. 2022;12(1):3039.

Ridding MC, Ziemann U. Determinants of the induction of cortical plasticity by non-invasive brain stimulation in healthy subjects. J Physiol. 2010;588(13):2291–304.

Rivera-Rivera PA, Rios-Lago M, Sanchez-Casarrubios S, Salazar O, Yus M, González-Hidalgo M, et al. Cortical plasticity catalyzed by prehabilitation enables extensive resection of brain tumors in eloquent areas. J Neurosurg. 2017;126(4):1323–33.

Rosenstock T, Grittner U, Acker G, Schwarzer V, Kulchytska N, Vajkoczy P, et al. Risk stratification in motor area–related glioma surgery based on navigated transcranial magnetic stimulation data. J Neurosurg. 2017;126(4):1227–37.

Rosenstock T, Häni L, Grittner U, Schlinkmann N, Ivren M, Schneider H, et al. Bicentric validation of the navigated transcranial magnetic stimulation motor risk stratification model. J Neurosurg. 2021;136:1194.

Rydén I, Carstam L, Gulati S, Smits A, Sunnerhagen KS, Hellström P, et al. Return to work following diagnosis of low-grade glioma. Neurology. 2020;95(7):e856–66.

Suppa A, Huang YZ, Funke K, Ridding MC, Cheeran B, Di Lazzaro V, et al. Ten years of theta burst stimulation in humans: established knowledge, unknowns and prospects. Brain Stimul. 2016;9(3):323–35.

Teitti S, Määttä S, Säisänen L, Könönen M, Vanninen R, Hannula H, et al. Non-primary motor areas in the human frontal lobe are connected directly to hand muscles. NeuroImage. 2008;40(3):1243–50.

Todhunter-Brown A, Baer G, Campbell P, Choo PL, Forster A, Morris J, et al. Physical rehabilitation approaches for the recovery of function and mobility following stroke. Cochrane Database Syst Rev. 2014;2023(8):CD001920.

Vaalto S, Säisänen L, Könönen M, Julkunen P, Hukkanen T, Määttä S, et al. Corticospinal output and cortical excitation-inhibition balance in distal hand muscle representations in nonprimary motor area. Hum Brain Mapp. 2011;32(10):1692–703.

Winstein CJ, Wolf SL, Dromerick AW, Lane CJ, Nelsen MA, Lewthwaite R, et al. Effect of a task-oriented rehabilitation program on upper extremity recovery following motor stroke. JAMA. 2016;315(6):571.

Wolf SL. Believing in brain AND brawn. J Neurol Phys Ther. 2006;30(3):117.

Zemke AC, Heagerty PJ, Lee C, Cramer SC. Motor cortex organization after stroke is related to side of stroke and level of recovery. Stroke. 2003;34(5):e23.

Ziemann U, Rothwell JC, Ridding MC. Interaction between intracortical inhibition and facilitation in human motor cortex. J Physiol. 1996;496(3):873–81.

Zrenner C, Ziemann U. Closed-loop stimulation. Biol Psychiatry. 2024;95:545–52.

Zrenner C, Desideri D, Belardinelli P, Ziemann U. Real-time EEG-defined excitability states determine efficacy of TMS-induced plasticity in human motor cortex. Brain Stimul. 2018;11(2):374–89.

Zrenner C, Kozák G, Schaworonkow N, Metsomaa J, Baur D, Vetter D, et al. Corticospinal excitability is highest at the early rising phase of sensorimotor μ-rhythm. NeuroImage. 2023;266:119805.

Closed-Loop TMS

19

Dania Humaidan, Gábor Kozák, and Ulf Ziemann

19.1 Introduction

The desire to influence brain activity is as old as the realization that the brain controls our body and is the center of consciousness. Being able to selectively activate muscles without direct contact or painful sensation of the subject upon its first demonstration in the mid-1980s (Barker et al. 1985), transcranial magnetic stimulation (TMS) technology carried the promise of fulfilling this dream. However, despite its potential, the TMS has not revolutionized (yet) human neuroscience for many reasons.

Beyond studying the immediate effects of stimulation, one major goal of the TMS research was, from early on, to induce lasting modulation in the human brain in a non-invasive way (Pascual-Leone et al. 1994). This ability is an absolute prerequisite for any therapeutic approach that targets neuropsychiatric network diseases, such as depression, obsessive-compulsive disorder, or stroke.

Based on what we have learned from decades of electrophysiological studies in rodents, it was well known that sufficiently strong, repetitive stimuli can induce long-term changes (plasticity) in the neuronal tissue. Due to its relative experimental simplicity and controllability, TMS was predominantly investigated in the motor system (Cooke and Bliss 2006; Ziemann et al. 2008; Hoogendam et al. 2010), where the different TMS protocols showed high intra- and interindividual variability, with a large proportion of non-responders on a population level (Ridding and Ziemann 2010).

D. Humaidan · G. Kozák · U. Ziemann (✉)
Department of Neurology & Stroke, and Hertie-Institute for Clinical Brain Research, Tübingen, Germany
e-mail: dania.humaidan@uni-tuebingen.de; gabor.kozak@uni-tuebingen.de; ulf.ziemann@uni-tuebingen.de

S. M. Krieg, T. Picht (eds.), *Navigated Transcranial Magnetic Stimulation in Neurosurgery*, https://doi.org/10.1007/978-3-031-97155-6_19

Studies on animal models provided a plausible explanation for this failure. Huerta and Lisman showed on rat hippocampal slices that the phase of ongoing theta-oscillations at the time of stimulation determines the direction of synaptic plasticity (Huerta and Lisman 1993, 1995). The theoretical framework was mostly given since the early 1990s; see, e.g., Jando et al. on artificial neural networks to detect epileptic seizures in rats (Jando et al. 1993). However, due to a lack of computational power for processing electroencephalography (EEG) in real-time, there were very few to no resources invested until very recently to explore these opportunities. As a consequence of these technical difficulties, TMS was boiled down to a very simple one-setup-fits-them-all approach. This meant that apart from slight adjustments of the parameters to target the motor cortex of the person undergoing TMS (such as stimulation intensity and location and orientation of the induced electrical field in the brain), the same set of parameters was being used across subjects. These fixed protocols then encountered a major update with the introduction of brain state-dependent stimulation.

19.2 Brain State-Dependent Stimulation

The work of Zrenner and colleagues challenged the status quo and opened a whole new era for TMS by developing a real-time EEG-triggered TMS paradigm (Zrenner et al. 2018). They demonstrated that instantaneous brain state can be utilized to control the efficacy of plasticity induction by repetitive TMS (rTMS) in the human motor cortex (Zrenner et al. 2018; Baur et al. 2020). Using an autoregressive (AR) model, they forward-predicted the EEG signal, estimating the instantaneous endogenous mu-rhythm phase at time zero using a Hilbert transform, and TMS was triggered when a predefined phase condition was met. This real-time EEG-TMS approach has been used to test the effects of instantaneous endogenous mu-rhythm phase on corticospinal excitability as measured by motor evoked potential (MEP) amplitude and on long-term change of corticospinal excitability induced by phase-specific rTMS. The AR model was also used in the work of Ding and colleagues (Ding et al. 2020) to predict the EEG phase, and they showed the prediction accuracy of the AR model algorithm to be better than that of the conventional algorithm when the model order is lower, such that the prediction accuracy will increase as the signal quality increases. This form of applying TMS while certain features in the EEG signal are satisfied is called brain state-dependent (BSD) stimulation.

While these works used an AR to predict EEG signals, a recent work by Pankka and colleagues (Pankka et al. 2021) used a different method for a more accurate EEG prediction, namely the deep learning model of WaveNet. Compared with an order 30 Yule-Walker AR model, the WaveNet model achieved a significantly improved phase estimation accuracy.

An example of the use of BSD stimulation in a clinical setting was presented in the work of Faller and colleagues (Faller et al. 2022). Here patients with major depressive disorder (MDD) were subjected to rTMS triggered either synchronously or asynchronously timed to their quasi-alpha EEG oscillation in the dorsolateral prefrontal cortex. They noticed an increased entrainment over sessions and a decreased optimal phase difference for the anterior cingulate in patients of the synchronized group. This conclusion can be taken into consideration when performing rTMS in future sessions on these patients. Another study targeted the trough of the alpha-rhythm in the dorsolateral prefrontal cortex in MDD patients and demonstrated a superior decrease of alpha-power, a potential marker of depression severity when the patients were subjected to this BSD stimulation protocol, but not when stimulated by random-phase stimulation or a conventional fixed intermittent theta-burst stimulation protocol (Zrenner et al. 2020). Meanwhile, a multicenter therapeutic trial has been started in Germany (BOSS-STROKE) that tests the superiority of BSD rTMS of the motor cortex in subacute motor stroke patients, targeting the trough of the sensorimotor mu-rhythm vs. random-phase stimulation and vs. brain state-dependent sham stimulation (Lieb et al. 2023).

BSD stimulation was not only investigated based on the prefrontal alpha or sensorimotor mu-rhythm. It has earlier been observed that corticospinal excitability is larger during the peak (up-state) than the trough (down-state) of the slow oscillation (<1 Hz) during non-REM sleep (Bergmann et al. 2012). BDS stimulation was also performed targeting theta oscillations originating from sources in the dorsomedial prefrontal cortex (DMPFC) in the work of Gordon and colleagues (Gordon et al. 2022). They noticed in healthy subjects that rTMS targeting the positive peak decreased TMS-induced theta power, while rTMS targeting the trough increased TMS-induced theta power and gamma-theta phase-amplitude coupling and improved working memory performance.

Although BSD rTMS represents a huge step forward toward individualization of stimulation for achieving stronger effect sizes, the effects are not being updated based on the output in the current session. The BSD parameters (e.g., the target phase of the ongoing EEG oscillation) are predetermined and are kept constant over the sessions. However, this assumes that the conclusions made on previous experiments apply to all upcoming sessions performed on the same or different subjects at different timings and under different conditions. While these observations might be true on the group level, there might be potentially addressable differences via parameter shift among individuals. Hence, a potentially better approach would be to start with an updated model based on previous conclusions and then further adapt the stimulation model parameters based on the current feedback.

19.3 Closed-Loop Stimulation

The concept of a closed-loop system is basically updating the parameters of the system adaptively on later time steps based on the output from earlier ones. A good example of how these systems work is the closed-loop insulin pump. Patients with type 1 diabetes suffer from high blood sugar values because their pancreas produces little or no insulin, and therefore, the treatment is to inject external insulin into the body. Insulin is one of the main factors that governs blood sugar levels, but not the only one. Vice versa, insulin level is not a single variable function of glucose concentration but is also dependent on anticipated food intake, sympathetic activation, and other hormone levels. For simplicity, let's consider a direct relationship between the insulin hormone and blood sugar level and treat all the factors as a black box. How should the required dose of insulin be determined? A very crude approach would be to give all patients the same amount of insulin at fixed times. A better approach would be to give insulin at mealtime since we know based on a priori knowledge that the blood sugar level would increase more after meals. Even better, blood sugar levels can be measured at a certain frequency during the day, and the timing and dose of insulin can be adjusted based on this input. However, the most accurate solution would be to have a measurement tool that monitors blood sugar continuously, similar to how the normal pancreas works, and based on the value of blood sugar at time t_i adjusts the insulin dose at time t_{i+1} in real time. Updating the parameters at time t_{i+1} based on the output at time t_i is the essence of closed-loop systems. It has to be noted that in this advanced case, we are still blind to any other input except the blood glucose level that reaches the pancreas. If the parameter space is expanded (more measurable features of the dynamic system are included), the release of insulin will be even more physiological.

Similarly, in a closed-loop EEG-TMS system, the stimulation parameters at time t_{i+1} would be adjusted based on the output, which is partly the result of performed stimulation at time t_i.

This means that as the BSD stimulation presented a huge step from open-loop TMS that did not take into consideration the internal state of the brain at the time of stimulation, closed-loop TMS would be a major further advancement that would not only consider the internal brain state in general but would target a preferred brain state for each subject at each time step. A very smooth explanation of the difference between these different approaches to performing TMS was presented by Bergmann (2018), in which an excellent illustration of the open-loop brain state-independent, open-loop brain state-dependent, and closed-loop brain state-dependent brain stimulation schematic was provided, which is shown in Fig. 19.1.

Such a concept to update the stimulation parameters has been increasingly used, implementing different methods and targeting different stimulation parameters to be optimized. We will now list some of these methods, from simple pipelines to more sophisticated algorithms.

Basically, a closed-loop system can be as simple as the methods used to determine the motor threshold. As long as the parameters are being updated at the current time step based on the output of the previous time step, this would still be considered a closed-loop system. The matter of determining the motor threshold (MT) has been an important question when it comes to performing TMS, since the intensity of the stimulation, which is one of the few parameters that we have been managing to control, relies on determining the MT in the most optimal way (Groppa et al. 2012; Rossini et al. 2015). Now, if anyone would think about an "adaptive" way of optimizing a value, the answer would most likely be, "I would test a value, then test another one, then a third one, until I find a good value." This is what sequential parameter estimation (SPE) does. A very good comparison of a method that uses sequential testing with those that do not was presented in the work of Silbert et al.

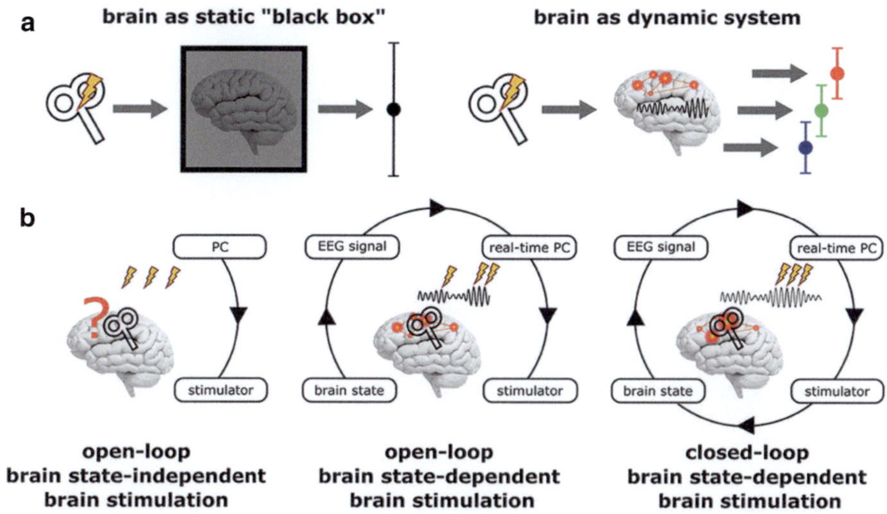

Fig. 19.1 Illustration of the different approaches to performing non-invasive brain stimulation (Bergmann 2018)

(2013). They compared the Rossini–Rothwell (R–R) relative-frequency method (Rossini et al. 2015) with an adaptive threshold-hunting method based on parameter estimation by sequential testing (PEST) (Awiszus 2003) for determining MT, defined as the minimum stimulation intensity to result in an MEP amplitude of at least 50 μV, and the stimulation intensity to target an MEP amplitude of 1 mV ($I_{1\,mV}$). In the PEST method, the planned intensity is displayed before delivering the pulse, and then after it is delivered, the user would observe the output and manually indicate if the achieved MEP was the desired one based on predetermined criteria (for example, if the MEP is larger than 1 mV). An illustration of how a closed-loop PEST algorithm works is shown in Fig. 19.2.

Determining the MT using this adaptive method has been shown to be efficient, such that 12 stimuli would be enough to determine the threshold, compared with 56 stimuli using non-adaptive methods (Dissanayaka et al. 2018). In order to further enhance the efficiency of the method, it was updated to include Bayesian regression to include more prior knowledge on the one hand and to be able to apply a stopping condition on the other hand, thus reducing the number of required stimuli to only seven (Groppa et al. 2012). More recently, these methods have been further updated by adopting a Bayesian estimation method, which iteratively incorporated information from the TMS responses into the probability density function (Wang et al. 2023).

The two keywords in determining the MT, sequential estimation and Bayesian optimization, were subsequently expanded to estimate other TMS measures in an adaptive, closed-loop manner.

On the one hand, the SPE method was used to adaptively optimize the "low-level" parameters related to the activation dynamics of the neurons. Two SPE methods have been developed by Alavi and colleagues (Alavi et al. 2019; Mohammad

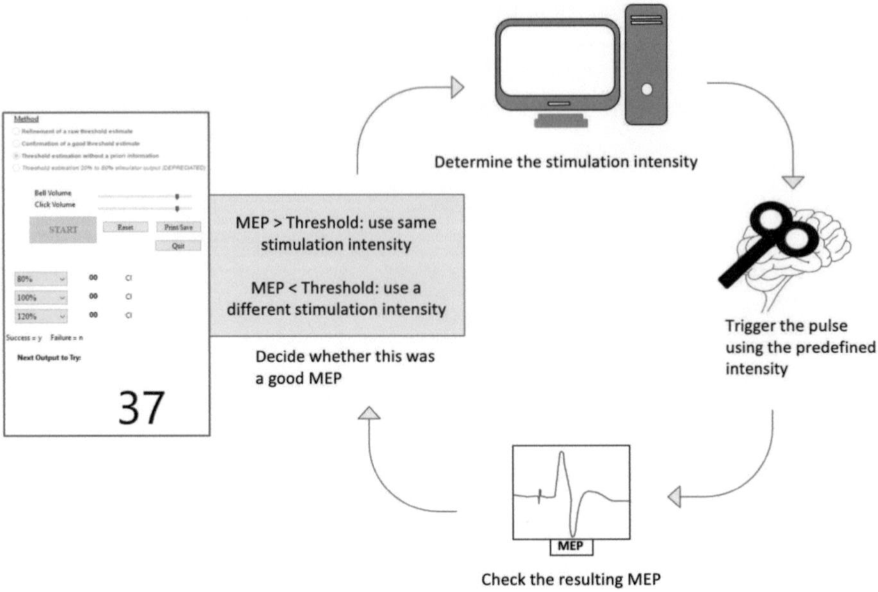

Fig. 19.2 An illustration of the closed-loop PEST algorithm

Mahdi Alavi et al. 2021) for optimal input-output (IO) curve and slope curve estimation in EMG-guided TMS by using the Fisher information matrix maximization. Later in 2022, they presented an SPE method for real-time concurrent estimation of the neural membrane time constant and IO curve parameters with open-loop uniform sampling and closed-loop optimal sampling TMS (Alavi et al. 2022). More recently, this method has been applied for automatic and optimal tuning of pulse amplitude and width for SPE of the neural membrane time constant and IO curve parameters (Alavi et al. 2023), and then to optimize the membrane time constant and coupling gain.

On the other hand, the enhancement provided by Bayesian adaptive methods led to a broader use of this approach to optimize further "high-level" stimulation parameters in a closed-loop manner. In the work of Tervo et al. (2020), an actual closed-loop was constructed to update one TMS parameter. First, the best stimulation target was searched for using the principle of Bayesian optimization. The idea is based on starting at a certain location and testing the resulting MEP amplitude when stimulating at that location, then moving to another nearby location and testing again; as enough responses accumulate, the model can determine the best location to stimulate to achieve the highest MEP amplitude. Updating a different stimulation parameter, the orientation of the coil, was also possible using this method (Tervo et al. 2022). An illustration of this closed-loop TMS pipeline is shown in Fig. 19.3.

Recently, a different approach was proposed to close the loop in TMS by Singh and colleagues (Singh et al. 2023). Instead of letting an algorithm decide when to stimulate, they attempted to close the loop through input from the subject receiving

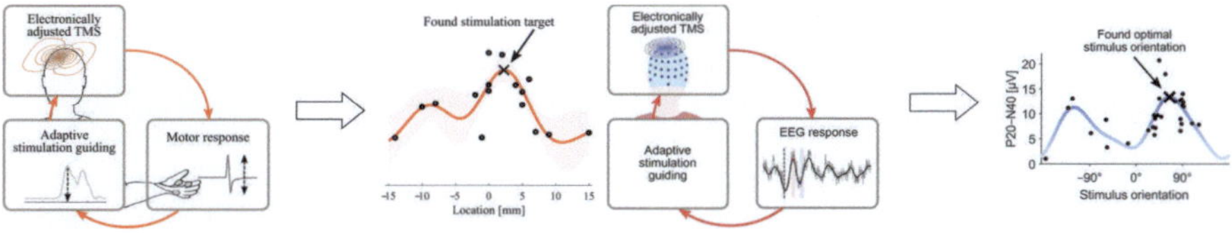

Fig. 19.3 Closed-loop TMS based on Bayesian optimization to optimize the coil location and orientation (Tervo et al. 2020, 2022)

TMS. In this setup, called the TMS Synchronized Exoskeleton Feedback (TSEF) platform, the EMG activity of the subject's affected hand was continuously monitored, and an activation threshold was predetermined. Whenever this threshold was exceeded, it was considered an intention to move, and this would trigger both the TMS pulse to the lesioned motor cortex and the exoskeleton device used to assist the patient in performing the movement by providing proprioceptive feedback to the brain. They tested the model on three patients receiving rTMS for neurorehabilitation and compared the output in terms of resulting ipsilesional cortical excitability (measured by MEP amplitude) and changes in the Fugl-Meyer scale. They found that the patients completed their session in their own time, and compared with the control group receiving physiotherapy, the TSEF group had increased MEPs and an improved Fugl-Meyer score. An illustration of the used setup is shown in Fig. 19.4.

The presented works have stepped closer toward a reliable closed-loop TMS setup, but they were limited in the complexity of the used algorithms, the number of parameters to be optimized, and the lack of application to optimization of plasticity induction. In order to be able to update more parameters, one possible way is to move toward the use of more complex methods, such as deep learning-based methods. Recently, the field of TMS has been hit by two hypes that have revolutionized the medical field: artificial intelligence (AI) in healthcare and personalized medicine, within the framework of a large European project titled "ConnectToBrain." This project aims at developing AI-based personalized multichannel TMS. The idea is to perform TMS targeting many locations simultaneously, i.e., a brain network composed of two or more components, using adaptively learned stimulation parameters based on the features obtained from the electrophysiological signals of the subject and machine learning-based algorithms, with the aim to modify the function of the stimulated network beyond the period of stimulation.

A first-of-its-kind machine learning (ML)-based real-time closed-loop rTMS experiment is being conducted within the ConnectToBrain project. The used algorithm belongs to a subfield in ML, namely reinforcement learning (RL). A first trial in 2021 tested different RL algorithms and reported on the performance metrics using a simulated dataset (Humaidan et al. 2021). The setup was then tested on real data, and preliminary results were presented (Humaidan 2023). Specifically, the algorithm learned to identify the individually optimal phase of the ongoing mu-rhythm to be targeted by paired-pulse TMS targeting the brain network of the supplementary motor area-primary motor cortex (SMA-M1) to maximize the enhancement of facilitatory effective connectivity between SMA and M1.

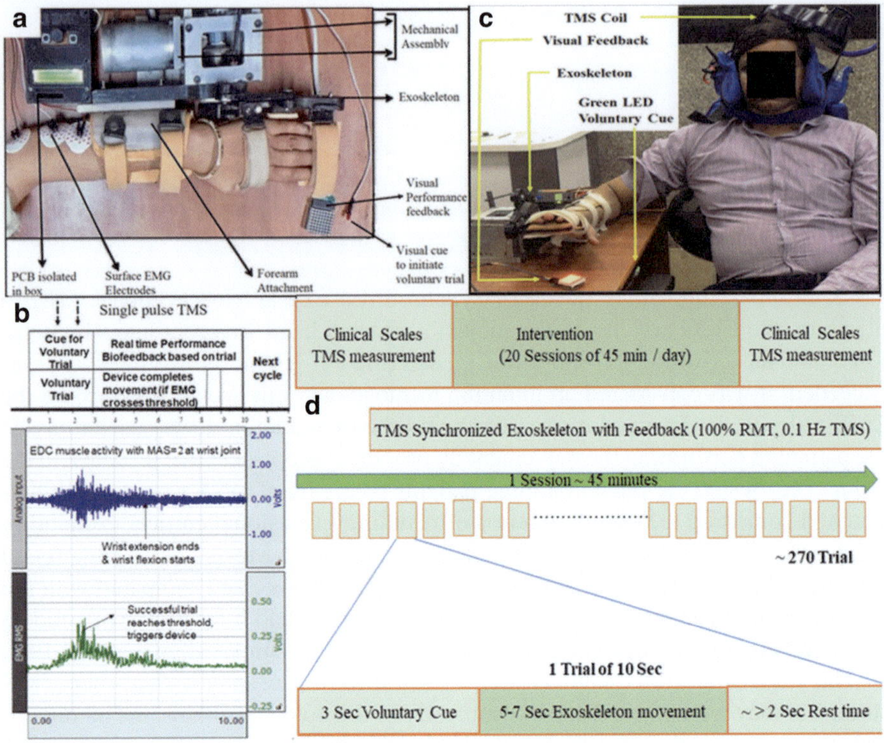

Fig. 19.4 A patient-specific brain stimulation platform with a two-way feedback system (Singh et al. 2023). (**a**) The used exoskeleton, mounted on the hand of a patient. (**b**) Protocol showing EMG activity during a 10-s trial. (**c**) Set-up of the TMS synchronized with the exoskeleton device. (**d**) Study design

This experiment paves the road toward the final goal of the ConnectToBrain project, which can be illustrated in the following setup: The subject walks into the TMS room, a multi-locus TMS helmet is put on their head, which can target any area in the brain, and then their electrophysiological signals are recorded and analyzed so that the suitable parameter set can be used for the stimulation. The suitable set would not only be subject-specific but also day- and time-of-the-day-specific, even for the same subject. The factors affecting and deciding what would be the perfect set of stimulation parameters for this subject in this session are still unclear, and investigating these factors while demonstrating the inter- and intra-subject variability in the parameters and results also belongs to the aims of the ConnectToBrain project. The researchers in this project have already managed to develop a multi-locus TMS coil by integrating a set of 5 large overlapping coils. This allows changing the stimulation site without physically moving the coil, which would prove very useful for targeting multi-node brain networks (Koponen et al. 2018; Nieminen et al. 2022).

In fact, the idea of having a personalized helmet for TMS has been suggested by Badran and colleagues (Badran et al. 2020). They proposed custom, individually tailored TMS helmets, which can be used to administer TMS in diverse settings where trained TMS operators are unavailable or in case the experiment or treatment involves movement. They even went on to suggest that such custom helmets can be used to self-administer stimulation, for example, in an at-home setting. The proposed helmets are shown in Fig. 19.5.

19.4 Future Outlook

So far, there have been very few studies on closed-loop TMS. This can be rooted back to the many limitations facing such a setup. Zrenner and colleagues presented a comprehensive overview of the conceptual and methodological frameworks for closed-loop TMS and addressed many limitations in their review (Zrenner et al. 2016). Recently, Zrenner and Ziemann (2023) published a review of the current state of closed-loop TMS with a special focus on applying closed-loop TMS in psychiatric patients. Here we highlight the main pillars of TMS-EEG research that should be addressed before a realistic closed-loop TMS setup can be established.

First, updating the stimulation parameters based on features extracted from the EEG signal means that an accurate analysis of the signal is required. This represents a huge challenge given that the EEG signal usually contains noise from different sources and comes with a small signal-to-noise ratio (Rashid et al. 2020). However, as seen from the improvement in EEG prediction using deep learning models (Pankka et al. 2021; Shaban and Amara 2022; Ibrahim et al. 2022), we believe the quality of EEG processing will continue to progress significantly to enable real-time analysis of the EEG signal for closed-loop TMS.

Second, whatever algorithm is used, it needs a sufficient amount of data to conclude the best parameter set for this subject in this session. Updating the parameters in real time means that we only have a limited amount of time to collect these data.

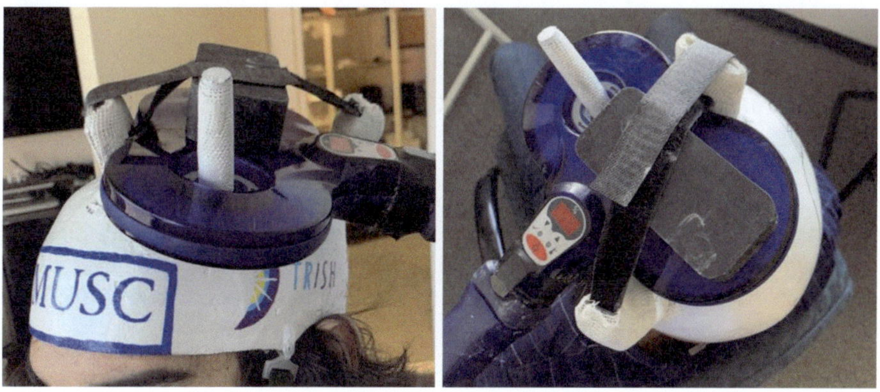

Fig. 19.5 A personalized helmet for TMS (Badran et al. 2020)

Based on the used method, obtaining enough data might take some time, during which the subject needs to be stable and the obtained data need to be usable; the more complex the method, the larger the required amount of data. This can reach a point, especially when deep learning-based algorithms are used, at which it is not feasible to keep stimulating until enough data have been collected.

Third, even if EEG processing became accurate enough and we managed to get enough data, what part of the signal should be fed back to the system to close the loop and update the parameters? This leads to a list of essential questions in closed-loop TMS.

1. What are we trying to modify by adapting the stimulation parameters?

An obvious answer to this question would be the same aim of TMS since we started to apply it 30 years ago, which is inducing a change to an area of the brain to treat a malfunction. However, this does not precisely describe the desired change. Starting from a pathological state A of the brain, we aim to reach a physiological state B. How to best define states A and B? These states might represent a certain degree of excitability of the targeted brain area or a certain degree of connectivity between the targeted area and other areas of the brain, which might be targeted as well via paired-pulse TMS. Measuring these states could be via the potentials evoked by stimulating the area using TMS; a good example is MEP as an index of corticospinal excitability or TMS-evoked EEG potentials (TEPs), but a clear definition is still needed.

2. Which EEG-extracted features play an important role in determining the desired state?

Although we still do not have a clear set of features that can be extracted from EEG to represent a certain brain state, studies conducted so far have managed to identify some parameters, such as the phase and power of the EEG rhythm. However, we still need a clear list of features to be extracted and optimized. The design of the closed-loop setup (used algorithm, number of trials needed, etc.) depends on the number and nature of parameters that need to be optimized.

3. Can further features, apart from the EEG-extracted ones, be used to formulate the desired brain state?

Clearly, we already know some great electrophysiological outputs that can indicate the current brain state, namely the evoked potentials, as well as neuroimaging data, namely functional MRI data. Combined with EEG features, the integrated set of features might represent reasonable feedback data to the system to close the loop and adapt the stimulation parameters.

4. Which stimulation parameters should be adaptively updated?

At least, there is the area being stimulated (where to stimulate) and the orientation of the induced electric field in the brain (in which direction the current flow). Further, there is the intensity, the timing (when to stimulate), and the frequency of the stimulation. If more than one area is targeted, then these parameters need to be determined for each area.

If we can let ourselves peek into the future of personalized, real-time, closed-loop TMS, it might look similar to what we illustrated in Fig. 19.6. Further studies are needed to answer these questions and build the true closed-loop TMS setup.

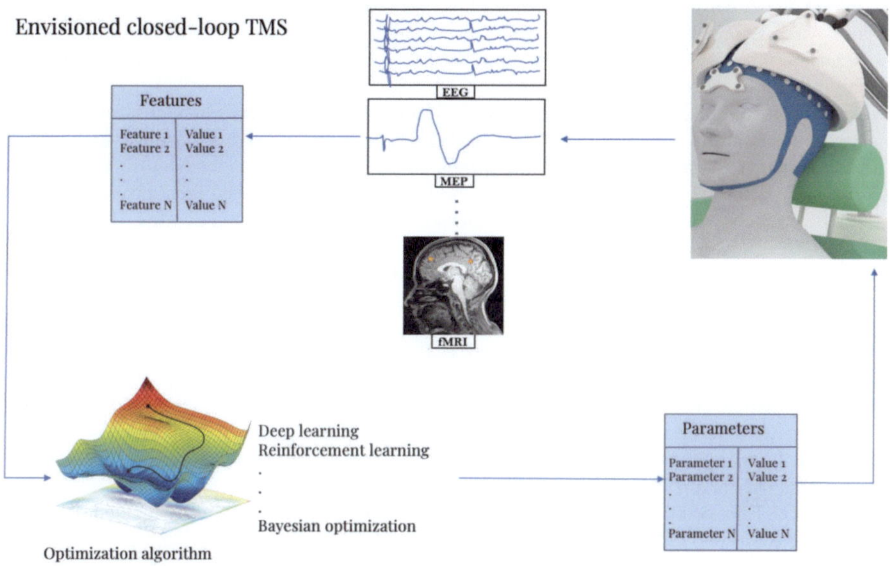

Fig. 19.6 A general illustration of the envisioned closed-loop TMS setup

References

Alavi SMM, Goetz SM, Peterchev AV. Optimal estimation of neural recruitment curves using fisher information: application to transcranial magnetic stimulation. IEEE Trans Neural Syst Rehabil Eng. 2019;27(6):1320–30.

Alavi SMM, Vila-Rodriguez F, Mahdi A, Goetz SM. A formalism for sequential estimation of neural membrane time constant and input-output curve towards selective and closed-loop transcranial magnetic stimulation(). J Neural Eng. 2022;19(5):056017.

Alavi SMM, Vila-Rodriguez F, Mahdi A, Goetz SM. Closed-loop optimal and automatic tuning of pulse amplitude and width in EMG-guided controllable transcranial magnetic stimulation. Biomed Eng Lett. 2023;13(2):119–27.

Awiszus F. TMS and threshold hunting. Suppl Clin Neurophysiol. 2003;56:13–23.

Badran BW, Caulfield KA, Lopez JW, Cox C, Stomberg-Firestein S, DeVries WH, et al. Personalized TMS helmets for quick and reliable TMS administration outside of a laboratory setting. Brain Stimul. 2020;13(3):551–3.

Barker AT, Jalinous R, Freeston IL. Non-invasive magnetic stimulation of human motor cortex. Lancet. 1985;1(8437):1106–7.

Baur D, Galevska D, Hussain S, Cohen LG, Ziemann U, Zrenner C. Induction of LTD-like corticospinal plasticity by low-frequency rTMS depends on pre-stimulus phase of sensorimotor mu-rhythm. Brain Stimul. 2020;13(6):1580–7.

Bergmann TO. Brain state-dependent brain stimulation. Front Psychol. 2018;9:2108.

Bergmann TO, Molle M, Diedrichs J, Born J, Siebner HR. Sleep spindle-related reactivation of category-specific cortical regions after learning face-scene associations. NeuroImage. 2012;59(3):2733–42.

Cooke SF, Bliss TV. Plasticity in the human central nervous system. Brain. 2006;129(Pt 7):1659–73.

Ding Z, Ouyang G, Chen H, Li X. Closed-loop transcranial magnetic stimulation of real-time EEG based on the AR mode method. Biomed Phys Eng Expr. 2020;6(3):035010.

Dissanayaka T, Zoghi M, Farrell M, Egan G, Jaberzadeh S. Comparison of Rossini-Rothwell and adaptive threshold-hunting methods on the stability of TMS induced motor evoked potentials amplitudes. J Neurosci Res. 2018;96(11):1758–65.

Faller J, Doose J, Sun X, McLntosh JR, Saber GT, Lin Y, et al. Daily prefrontal closed-loop repetitive transcranial magnetic stimulation (rTMS) produces progressive EEG quasi-alpha phase entrainment in depressed adults. Brain Stimul. 2022;15(2):458–71.

Gordon PC, Belardinelli P, Stenroos M, Ziemann U, Zrenner C. Prefrontal theta phase-dependent rTMS-induced plasticity of cortical and behavioral responses in human cortex. Brain Stimul. 2022;15(2):391–402.

Groppa S, Oliviero A, Eisen A, Quartarone A, Cohen LG, Mall V, et al. A practical guide to diagnostic transcranial magnetic stimulation: report of an IFCN committee. Clin Neurophysiol. 2012;123(5):858–82.

Hoogendam JM, Ramakers GM, Di Lazzaro V. Physiology of repetitive transcranial magnetic stimulation of the human brain. Brain Stimul. 2010;3(2):95–118.

Huerta PT, Lisman JE. Heightened synaptic plasticity of hippocampal CA1 neurons during a cholinergically induced rhythmic state. Nature. 1993;364(6439):723–5.

Huerta PT, Lisman JE. Bidirectional synaptic plasticity induced by a single burst during cholinergic theta oscillation in CA1 in vitro. Neuron. 1995;15(5):1053–63.

Humaidan D. Enhancing the connectivity in a 2-node motor network using reinforcement learning-based closed-loop rTMS. Brain Stimul. 2023;16(1):120.

Humaidan D, Vetter D-E, Metsomaa J, Ermolova M, Ziemann U. Reinforcement machine learning for closed-loop rTMS stimulation of brain networks. Brain Stimul. 2021;14(6):1696.

Ibrahim FE, Emara HM, El-Shafai W, Elwekeil M, Rihan M, Eldokany IM, et al. Deep-learning-based seizure detection and prediction from electroencephalography signals. Int J Numer Method Biomed Eng. 2022;38(6):e3573.

Jando G, Siegel RM, Horvath Z, Buzsaki G. Pattern recognition of the electroencephalogram by artificial neural networks. Electroencephalogr Clin Neurophysiol. 1993;86(2):100–9.

Koponen LM, Nieminen JO, Ilmoniemi RJ. Multi-locus transcranial magnetic stimulation-theory and implementation. Brain Stimul. 2018;11(4):849–55.

Lieb A, Zrenner B, Zrenner C, Kozak G, Martus P, Grefkes C, et al. Brain-oscillation-synchronized stimulation to enhance motor recovery in early subacute stroke: a randomized controlled double-blind three- arm parallel-group exploratory trial comparing personalized, non- personalized and sham repetitive transcranial magnetic stimulation (acronym: BOSS-STROKE). BMC Neurol. 2023;23(1):204.

Mohammad Mahdi Alavi S, Goetz SM, Saif M. Input-output slope curve estimation in neural stimulation based on optimal sampling principles(). J Neural Eng. 2021;18(4):046071.

Nieminen JO, Sinisalo H, Souza VH, Malmi M, Yuryev M, Tervo AE, et al. Multi-locus transcranial magnetic stimulation system for electronically targeted brain stimulation. Brain Stimul. 2022;15(1):116–24.

Pankka H, Roine T, Lehtinen J, Ilmoniemi RJ. Improving closed-loop TMS timing using the Wavenet model. Brain Stimul. 2021;14(6):1636–7.

Pascual-Leone A, Valls-Sole J, Wassermann EM, Hallett M. Responses to rapid-rate transcranial magnetic stimulation of the human motor cortex. Brain. 1994;117(Pt 4):847–58.

Rashid M, Sulaiman N, Majeed APPA, Musa RM, Ab Nasir AF, Bari BS, et al. Current status, challenges, and possible solutions of EEG-based brain-computer interface: a comprehensive review. Front Neurorobot. 2020;14:25.

Ridding MC, Ziemann U. Determinants of the induction of cortical plasticity by non-invasive brain stimulation in healthy subjects. J Physiol. 2010;588(Pt 13):2291–304.

Rossini PM, Burke D, Chen R, Cohen LG, Daskalakis Z, Di Iorio R, et al. Non-invasive electrical and magnetic stimulation of the brain, spinal cord, roots and peripheral nerves: basic principles and procedures for routine clinical and research application. An updated report from an I.F.C.N. Committee. Clin Neurophysiol. 2015;126(6):1071–107.

Shaban M, Amara AW. Resting-state electroencephalography based deep-learning for the detection of Parkinson's disease. PLoS One. 2022;17(2):e0263159.

Silbert BI, Patterson HI, Pevcic DD, Windnagel KA, Thickbroom GW. A comparison of relative-frequency and threshold-hunting methods to determine stimulus intensity in transcranial magnetic stimulation. Clin Neurophysiol. 2013;124(4):708–12.

Singh N, Saini M, Kumar N, Padma Srivastava MV, Mehndiratta A. Individualized closed-loop TMS synchronized with exoskeleton for modulation of cortical-excitability in patients with stroke: a proof-of-concept study. Front Neurosci. 2023;17:1116273. https://doi.org/10.3389/fnins.2023.1116273.

Tervo AE, Metsomaa J, Nieminen JO, Sarvas J, Ilmoniemi RJ. Automated search of stimulation targets with closed-loop transcranial magnetic stimulation. NeuroImage. 2020;220:117082.

Tervo AE, Nieminen JO, Lioumis P, Metsomaa J, Souza VH, Sinisalo H, et al. Closed-loop optimization of transcranial magnetic stimulation with electroencephalography feedback. Brain Stimul. 2022;15(2):523–31.

Wang B, Peterchev AV, Goetz SM. Three novel methods for determining motor threshold with transcranial magnetic stimulation outperform conventional procedures. J Neural Eng. 2023;20(5):056002.

Ziemann U, Paulus W, Nitsche MA, Pascual-Leone A, Byblow WD, Berardelli A, et al. Consensus: motor cortex plasticity protocols. Brain Stimul. 2008;1(3):164–82.

Zrenner C, Ziemann U. Closed-loop stimulation. Biol Psychiatry. 2023;95:545–52.

Zrenner C, Belardinelli P, Muller-Dahlhaus F, Ziemann U. Closed-loop neuroscience and non-invasive brain stimulation: a tale of two loops. Front Cell Neurosci. 2016;10:92.

Zrenner C, Desideri D, Belardinelli P, Ziemann U. Real-time EEG-defined excitability states determine efficacy of TMS-induced plasticity in human motor cortex. Brain Stimul. 2018;11(2):374–89.

Zrenner B, Zrenner C, Gordon PC, Belardinelli P, McDermott EJ, Soekadar SR, et al. Brain oscillation-synchronized stimulation of the left dorsolateral prefrontal cortex in depression using real-time EEG-triggered TMS. Brain Stimul. 2020;13(1):197–205.

20 nTMS Depression Therapy

Joshua Kuluva and Dabney Warner

20.1 Introduction

Major depressive disorder (MDD) is one of the most prevalent disorders globally, affecting approximately 280 million people worldwide (World Health Organization 2023). In the United States, 21.0 million adults had at least one major depressive episode in 2021, with 14.5 million adults having at least one major depressive episode that caused severe impairment (National Institution of Mental Health 2022). A diagnosis of MDD is characterized by a combination of a persistently low or depressed mood, anhedonia (decreased interest in pleasurable activities), feelings of guilt or worthlessness, lack of energy, poor concentration, appetite changes, psychomotor retardation or agitation, sleep disturbances, or suicidal thoughts. To be diagnosed with MDD, an individual must have at least five of these symptoms, one of which must be a depressed mood or anhedonia that causes social or occupational impairment. Major depressive disorder is associated with an increased risk of all-cause mortality, as a recent study showed that all-cause mortality was more than two times more common in MDD patients (Cuijpers et al. 2014). Suicidal thoughts and behaviors constitute part of the diagnostic criteria for MDD, and suicidal behavior, self-inflicted harm with or without suicidal ideation, was reported in at least half of MDD patients. The longer the suicidal episodes are, the more frequent suicidal behavior occurs (McManus et al. 2019; Nordentoft et al. 2011). Furthermore, suicidal behavior and previous suicide attempts do increase the death rate from suicide, with 700,000 people dying from suicide each year (World Health Organization 2023).

Overall, MDD is the leading cause of disability worldwide as measured by years lived with disability (ADAA 2017). Additionally, MDD is ranked as the third cause of the burden of disease worldwide and is predicted to be the first cause by the year

J. Kuluva (✉) · D. Warner
Piedmont Neuroscience Center, Oakland, CA, USA
e-mail: joshuakuvulamd@piedmontneurosciencecenter.com

S. M. Krieg, T. Picht (eds.), *Navigated Transcranial Magnetic Stimulation in Neurosurgery*, https://doi.org/10.1007/978-3-031-97155-6_20

2030 (National Institution of Mental Health 2022). The economic burden of MDD is substantial as well. In 2018, resultant costs from MDD were estimated to be $236 billion among U.S. adults, which is an increase of more than 35% since 2010 (Psychiatry 2021). MDD expenses include direct costs, such as medical services, including outpatient care, inpatient care, the emergency room, and medications; workplace costs, such as absenteeism (missed days of work) and presenteeism (reduced productivity at work); and suicide-related costs. In addition to being costly, MDD is chronic and highly recurrent. Approximately 50% of MDD patients have a lifetime risk of recurring after their first depressive episode, with an additional 20% chance of recurrence following each subsequent depressive episode (National Institution of Mental Health 2022). Due to the increasing prevalence of the disease and the ramifications of MDD both individually and societally, it is of paramount importance to continue developing innovative treatments to help alleviate global suffering.

20.2 Pathophysiology of Depression

Based on neuroanatomical studies created to explore the physiological discrepancies that may underlie the gross morphological changes and functional deficits present in people with MDD, it is evident that there are various regions within the prefrontal cortex that contribute to the symptomatology of MDD, including the dorsolateral prefrontal cortex, both the rostral and caudal portions of the orbitofrontal cortex and the anterior cingulate cortex (Cohen et al. 1989; Keilp et al. 2001; Koolschijn et al. 2009).

The dorsolateral prefrontal cortex (DLPFC) is the cortical area responsible for motor planning, organization, and regulation. It is involved in both the integration of sensory and mnemonic information and the regulation of intellectual function and action (Bunge et al. 2000). Neuroimaging studies of patients with major depressive disorder have revealed volume reductions in both gray and white matter, sulcal widening, decreases in glucose metabolism, and alterations in blood flow in their prefrontal cortices (Drevets 2000). There were also variations in neuronal densities in layers II, III, and VI of MDD subjects compared with control subjects; the densities of the largest neurons in the aforementioned layers were reduced significantly, and those of the smallest neurons increased, indicating that the pyramidal layers of the DLPFC are likely involved in MDD. There was a reduction in glial cell density observed in layers III and V of the DLPFC in subjects with major depressive disorder as well. As layer VI neurons are an essential source of efferents to subcortical structures such as the basal ganglia and thalamus, dysfunction of them can also potentially lead to mood disorders due to lessened cortical input. In addition to neuronal density discrepancies within the DLPFC, there is also evidence that MDD could be tied to altered levels of gamma-aminobutyric acid (GABA) within the brain. There is a significant increase in the density of GABAergic neurons in the DLPFC in MDD subjects (Bielau et al. 2007). Subpopulations of cortical GABAergic interneurons based on co-expression of

GABAergic markers with calcium-binding proteins found significant reductions in neuron somal size and density of calbindin-IR neurons, involved in memory processes and social interactions, in the DLPFC in patients with MDD (Bjerke et al. 2020). Furthermore, lesion studies of the DLPFC have proven that the region plays a large role in motivation and pursuit of goals (Szczepanski and Knight 2014). Damage to the DLPFC often results in a state of pseudo depression characterized by abulia—loss of initiative, diminished motivation, flattened effect, reduced verbal output, apathy, and behavioral slowness.

The reduction of neuronal density associated with the DLPFC of patients with major depressive disorder was also observed in the orbitofrontal cortex (OFC), which is believed to play a large role in MDD as well (Cohen et al. 1989). There was a significant reduction observed in the neuronal size and density of the largest neurons in layers II-IV, as well as in the neuron somal size and density for calbindin-IR neurons, of the orbitofrontal cortex in subjects with major depressive disorder. Furthermore, postmortem studies have shown there is both cell loss and cell atrophy in the orbitofrontal cortex in patients with MDD (Koolschijn et al. 2009). This area of the brain plays an integral role in decision-making, formulation of expectations, sensory integration, and emotional regulation through the inhibition of limbic regions such as the hippocampus and amygdala; lesions of the orbitofrontal cortex can result in worsened social interaction and ability to empathize, which is a similar symptomology to what can be observed in patients diagnosed with MDD (Kringelbach 2005; Szczepanski and Knight 2014; Koolschijn et al. 2009).

20.3 TMS for Depression

There are several available treatment options for MDD, but the most common methods include a combination of psychotherapeutic, pharmacological, or interventional therapies. Most of the pharmacologic treatments for MDD target reuptake and degradation of specific neurotransmitters, such as selective serotonin reuptake inhibitors (SSRIs), serotonin-norepinephrine reuptake inhibitors (SNRIs), tricyclic antidepressants (TCAs), and atypical antidepressants (Bains and Abdijadid 2022). Despite the multitude of options, however, many patients do not respond to several rounds of antidepressant treatment and are considered to have treatment-resistant depression. In order to investigate the prevalence of treatment-resistant depression among MDD patients, Dr. Gaynes conducted a trial in 2006 aimed toward identifying which of the myriad of available treatments for MDD is the most effective next option for patients that don't achieve remission with initial or subsequent treatment (Gaynes and Rush 2008). In order to evaluate this approach, he created a protocol with four levels of antidepressant treatments consisting of an amalgamation of pharmacological and psychological treatments. Each patient was started at level one, which was a prescription of citalopram, and if they didn't enter remission after 14 weeks, they moved up to the next level; this procedure was followed for each subsequent level after their current treatment was qualified as a failure. It was discovered that after two levels of treatment, approximately 30% of patients in level 1 achieved remission, and of those that progressed to level 2, 30% achieved remission again. From this trial, it was determined that there was approximately a 50% chance that a patient would reach remission after two levels of treatment, a 60% chance of remission after three rounds, and a 70% chance of remission after four rounds. A patient is considered to have treatment-resistant depression after two failed courses of significant antidepressant treatments; thus, the trial proved that almost half of the people diagnosed with MDD have treatment-resistant depression, and approximately 30% won't respond even after four rounds of antidepressant treatment.

Based on our neuroanatomic understanding of depression, researchers began to think about ways to stimulate the prefrontal cortex in order to correct dysfunctional states. The implementation of rTMS as a potentially efficacious treatment for treatment-resistant MDD originated with a clinical trial conducted by Dr. George in 1995, which involved stimulation of the left prefrontal cortex of six medication-resistant subjects diagnosed with depression (George et al. 1995a). Dr. George's trial consisted of one unipolar and five bipolar type II patients who were actively depressed with an average HAMD17 of 23.8 ± 4.2, and he continued treatment if there was a clinical response noted after the first 5 days of the rTMS treatment. Each patient received 20 two-second trains of 20 Hertz rTMS at 80% of their motor threshold every morning for at least 5 days. The trial resulted in a general overall improvement in the group's average HAMD17, which dropped to 17.5 ± 8.4 after treatment, and two patients with very significant clinical improvements despite having been refractory to most antidepressant medications previously. The findings of this initial trial supported the postulation that rTMS treatment of the left prefrontal cortex can yield antidepressant effects.

Treating medication-resistant depression with rTMS was then FDA-approved following the clinical trial led by Dr. O'Reardon in 2007, which was a double-blind, multisite study intended to test the efficacy and safety of rTMS treatment of the left DLPFC for acute treatment of major depression (O'Reardon et al. 2007). The trial was comprised of 301 patients that were diagnosed with medication-resistant MDD and were randomly assigned to active ($n = 155$) or sham ($n = 146$) rTMS conditions; five rTMS sessions were completed weekly for a period of 4–6 weeks. This protocol involved the delivery of 10 pulses per second, totaling 3000 pulses per session at 120% of the motor threshold. Response to the treatment was measured primarily through MADRS scores at week four and secondarily through response rates using MADRS and HAMD values. At week four of the trial, the active rTMS treatment was considered to be significantly more efficacious than the sham treatment, with the baseline-to-endpoint change of the MADRS scores showing a statistical trend favoring the active treatment in the evaluable study population ($p = 0.057$). The response rates of the participants were significantly higher on weeks four and six, and remission rates were significantly higher on both the MADRS and HAMD-24 measures. Not only was the treatment evidently efficacious, but it was well-tolerated; there was only a 4.5% dropout rate for adverse events that were generally mild, such as transient scalp discomfort.

Following the trial in 2007 and the resultant FDA approval, rTMS of the left DLPFC became a widely used treatment for medication-resistant major depression (Johnson et al. 2013). In order to target the same location that the pioneer trials did, it became standard practice to implement the "5-cm method" first described by Dr. George in his 1995 trial. During their initial TMS planning sessions, the rTMS administrator functionally localized the left and right scalp positions that maximally induced a motor-evoked potential of the contralateral abductor pollicis brevis. They were then able to determine the resting motor threshold at these positions utilizing either electromyography or visual monitoring to identify the motor-evoked potentials; at these potentials, the center of the rTMS coil marked both the "left hemisphere motor cortex" and the "right hemisphere motor cortex." From these locations, they identified a sagittal plane five centimeters anterior of each motor site and marked it on the scalp as the prefrontal cortex. Once this process was complete, the rTMS administrators began the treatment of the left prefrontal cortex based on its calculated location. However, while this method of rTMS for depression did demonstrate antidepressant effects, the lack of ability to visualize the exact area that is receiving treatment creates potential ramifications in terms of both treatment efficacy and versatility of available treatments.

20.4 Navigated TMS

As the "5-cm method" only acts as an educated estimation and the brain of every individual differs, there is no way to ensure that the prefrontal cortex is truly being targeted with each patient. For this reason, navigated TMS (nTMS) as a treatment for depression is an enticing option, as it enables the precise localization of the cortical area that is being targeted with the TMS treatment using the patient's individualized MRI. In 2017, the first TMS device that uses MRI-guided navigation was cleared to treat patients with depression in the United States (Nexstim nTMS 2023). Using navigated TMS, clinicians were able to visualize the direction of the coil, distance to the scalp, and resultant electric field of magnetic stimulation superimposed on an individual patient's brain MRI. This opens the opportunity to explore the benefits of more precisely targeting individuals with nrTMS (navigated repetitive TMS).

A 2009 study of 48 people with medication resistant depression investigated the potential benefits of more precisely navigated repetitive TMS (Paillère Martinot et al. 2010). It compared the effect of rTMS administered over the area of the prefrontal cortex with the most observed hypometabolism, nrTMS (n = 16), patients that received a placebo treatment (n = 14), and rTMS using the standard "5-cm" method (n = 18). Each participant received 10 sessions of 1600 pulses delivered at 10 hertz at their 90% motor threshold. Treatment was only considered to be efficacious, measured by the change in their MADRS scores, in patients that received active TMS treatment over the left DLPFC, which was defined as Brodmann's Areas (BA) 9 and 46. It was determined that the nrTMS was able to target BA 9 and 46 with significantly better accuracy and frequency throughout the duration of the treatment. To investigate the potential for a discrepancy in treatment efficacy based on the method of localization, a case study was performed in 2007 on a woman with treatment-resistant MDD during which she was given a two-week treatment of 10 Hz-rTMS (Lefaucheur et al. 2007). The location of the treatment's target was found using the standard "5-cm" procedure during the first week of treatment and a navigated system during the second week. It was observed that at the conclusion of the first week of stimulation, the patient improved mostly in speech production, but there was no significant improvement in mood. Once nrTMS was implemented, however, there were significantly more antidepressant effects noticed at the end of the second week. The navigational system was able to determine that the target for the standard localization corresponded with Broca's area rather than the DLPFC; the prefrontal cortex was 8.3 cm anterior to the hand motor cortex, which indicates that the standard "5-cm" procedure may lead to inaccurate targeting of the prefrontal cortex on some rTMS recipients and potentially be less efficacious.

A study conducted in 2009 sought to determine whether or not nrTMS was truly more effective than the standard treatment; it consisted of 51 patients with treatment-resistant depression that were randomized to receive either the standard "5-cm" treatment ($n = 24$) or the neuro-navigated high-frequency, defined as 10 hertz, treatment ($n = 24$) of the left DLPFC (Fitzgerald et al. 2009). The protocol for each patient was the application of 30 5-second trains each day for 5 days per week at the patient's 100% motor threshold. The trial included 15 daily sessions with a 5-day extension period for subjects with a greater than 25% reduction in their MADRS score. At the conclusion of the study, it was determined that while there was an overall significant reduction in the subjects' MADRS scores over the duration of the trial, there were better outcomes observed in the nrTMS group than the standard group after 4 weeks ($p = 0.02$). Ten patients in the nrTMS group (42%) met the response criteria, while five patients in the standard rTMS group (18%) met the response criteria, and seven patients in the nrTMS group (30%) met the remission criteria, while three patients in the standard rTMS group (11%) met the remission criteria. The secondary outcomes, which were the patients' scores on the Beck Depression Inventory (BDI), Brief Psychiatric Rating Scale, CORE rating of psychomotor disturbance, Global Assessment of Functioning (GAF), and Clinical Global Improvements Scale, were also observed to be better for the group that received nrTMS instead of the standard "5-cm procedure."

In addition to being more precise and accurate in its localization of the DLPFC, nTMS also enables physicians to better target other areas of the brain that are associated with depression that aren't adequately targeted in standard rTMS, or with the use of the 5 cm rule, such as the orbitofrontal cortex (OFC). In 2017, a case series of 42 patients with unipolar MDD who received inhibitory right OFC-rTMS in open-label treatment was composed in order to investigate the safety and efficacy of TMS that targeted the orbitofrontal region (Feffer et al. 2018). Of the 42 participants, 30 patients had previously failed treatment of the DLPFC; it has been found that patients with chronic features of high anhedonia are less likely to respond to treatment of the DLPFC, so the OFC was targeted instead. The procedure involved the delivery of six trains of 60 1-hertz pulses, with 30-second intertrain intervals, for a total of 360 pulses in 8 min and 30 s. The participants were initially offered a course of 20 sessions of treatment, scheduled to be completed once daily 5 days each week, and if they achieved a clinical improvement or response, they were given an extension to 30 sessions. They were assessed primarily through the changes in their BDI-II throughout the course of the treatment; response to the TMS treatment was defined as an improvement of more than 50% of their baseline score, and remission was defined as a final treatment BDI-II score of less than or equal to 12. 35.7% of the patients achieved the response criterion, 23.8% of the patients achieved the remission criterion, and the average overall improvement in depressive symptoms was 36.5 ± 26.5%. Of the 30 patients that had previously not responded to TMS treatment of the DLPFC, nine responded to the treatment of their OFC, and seven patients achieved remission. Thus, given that nrTMS allows the operator to more easily target regions outside of the DLPFC, it stands to reason that nrTMS is an attractive therapeutic intervention as we start to think about TMS protocols outside of the conventional high-frequency DLPFC treatments.

20.5 Registries: nrTMS Versus Conventional rTMS

A California-based clinic found superior results when treating the DLPFC using nrTMS rather than with the standard procedure (Piedmont Neuroscience Center 2023). The response rate of MDD patients was 71.4%, which was defined as having a PHQ score of less than 10 or a BDI score of less than 19 at the conclusion of nrTMS treatment, and the remission rate was 42.9%, which was defined as having a PHQ score of less than 5 or a BDI score of less than 10 at the conclusion of treatment; the average percentage decrease of their nrTMS patients' PHQ scores was 48.2%. Similarly, a multi-site international registry of nrTMS data also showed impressive response and remission data from several clinical sites across the world (Nexstim 2022). Of 403 patients treated with a course of nrTMS, 76.2% of patients met response criteria, and 49.6% of the patients had a complete remission of depressive symptoms. When compared to centers that use non-navigated TMS, these results were also superior. Registry data was collected from 103 practice sites that performed standard rTMS using the "5-cm" method, which included 3814 patients with primary MDD diagnoses that had completed TMS treatment; completion was defined as either responding to the treatment or receiving more than 20 sessions and completing an end-of-acute-treatment PHQ-9 assessment. The average baseline PHQ was 19.8 ± 4.1, which was reduced to an average PHQ of 8.6 ± 6.4. The overall response rate was 65.0%, and the overall remission rate was 31.7%. When comparing these rates to those gathered for nrTMS, it is evident that nrTMS does have the potential to yield better antidepressant results than conventional non-navigated rTMS treatment.

	California-based nrTMS clinic registry (nrTMS)	Multi-center registry (nrTMS)	“5-cm” method (standard rTMS)
Response rate	71.4%	76.2%	65.0%
Remission rate	42.9%	49.6%	31.7%

20.6 Technical Aspects of nrTMS

In addition to whether nrTMS is more efficacious as a treatment of MDD than the standard “5-cm method,” there are additional factors that must be considered when creating a complete comparison of the two methodologies.

From a treatment team’s perspective, nrTMS may be considered a more desirable way of delivering the treatment than rTMS, as it enables technicians to establish daily consistency with each patient. With standard rTMS machines, a cap or mark is used to approximate the position of the coil relative to the cortical area that is being targeted; however, there is no way to determine if the coil slips off the mark during the treatment. When initially localizing the area to target, there is no estimation involved in determining where to target and how to ensure that the patient is receiving the most accurate stimulation in the correct region. These day-to-day discrepancies in daily treatment locations likely have a negative impact on the outcome of the treatment.

It is also important to acknowledge that nTMS may have an extra barrier of entrance for patients compared to rTMS due to the inherent necessity for an MRI to be completed before the treatment. With the widespread adoption of MRI technology in most healthcare systems, this barrier is rapidly becoming less of an obstacle. Furthermore, once the patient has received the MRI, there is an added benefit that the patient is more easily able to visualize and understand the treatment. This can give the patient more confidence that the treatment is targeting the appropriate location. Further, using the MRI guidance makes it easier for the physician and their staff to find the motor threshold and treatment location, as they can directly visualize the relevant cortical structures.

When evaluating the totality of its favorable characteristics, it is evident that neuro-navigated TMS is essential to not only the current administration of TMS but also the advancement of TMS as a treatment for refractory major depressive disorder. As we will see in our next section, some of the most cutting-edge and exciting advancements in the treatment of depression rely heavily on individualized neuro-navigated TMS.

20.7 Individualized Neuronavigation Utilizing fMRI Data

The treatment of previously recognized structures is only one way that nTMS creates the potential for additional benefits and clinical improvements for MDD patients. The most recent nrTMS advancements utilize more sophisticated brain scans, such as fMRIs, of individual patients to target specific regions in their brains that can only be localized using neuro-navigational technology.

A meta-analysis that integrated several brain volumetric magnetic resonance imaging studies compared the differences between 2418 patients with MDD and 1974 healthy individuals from 64 studies and found that the largest difference between the brain of an MDD subject and a healthy subject was the volume reduction in the anterior cingulate cortex ($p = 0.006$) (Koolschijn et al. 2009). This was most pronounced in the left anterior cortex volume ($p = 0.005$) but also evident in the right anterior cortex as well. The reduction of gray matter in the left subgenual anterior cingulate cortex (sgACC) ventral to the corpus callosum is the most prominent volumetric abnormality in the brains of people diagnosed with MDD (Drevets 2000). Many regions of the brain that are associated with structural abnormalities are also found to concurrently have atypical blood flow and glucose metabolism (Drevets and Price 2005). In the sgACC, metabolic activity appears to be increased in patients with MDD, and depression severity correlates positively with metabolism in the sgACC; sgACC metabolism appears elevated in the depressed vs. remitted phases of the same MDD subjects, and reduction to both metabolism and blood flow to this region is found with stimulation of the sgACC (Drevets et al. 1997; George et al. 1995b; Neumeister et al. 2004; Osuch et al. 2000).

The most novel protocol for the treatment of MDD utilizing TMS is the Stanford Accelerated Intelligent Neuromodulation Therapy for Treatment-Resistant Depression, which was piloted by Dr. Williams in 2020 with the goal of determining the safety, tolerability, and efficacy of accelerated high-dose intermittent theta burst stimulation (iTBS) using functional connectivity MRIs (fcMRIs) guided targeting (Cole et al. 2020). Their patients were required to either be experiencing a nonpsychotic major depressive episode as part of either MDD or bipolar disorder as characterized by the DSM-5 criteria and have not responded to at least one antidepressant medication, as well as have a HAMD score greater than 20. Before beginning the treatment course, each patient had a structural and resting-state functional MRI scan with a 3-T GE Discovery MR750 scanner with a 32-channel imaging coil, and simultaneous multi-slice imaging sequences were used to collect a total of 29 sets of three slices of each patient's MRIs. Each patient received five consecutive days of ten iTBS treatments, which is a shortened TMS session from 3 to 37 min with similar results to the other methods of stimulation. The patients received sessions of iTBS with 60 cycles of 10 bursts of three pulses at 50 hertz at the patient's 90% motor threshold; they received 18,000 pulses each day, which totaled 90,000 pulses at the end of the treatment's duration. The Localite Neuronavigation System was utilized to position the nTMS coil over the predetermined individualized stimulation target each session. They focused the treatment on the portion of the left DLPFC with the strongest functional anticorrelation to the sgACC, which indicates the area where the stimulation of the left DLPFC can create the strongest inhibition of the sgACC. Their baseline resting fcMRIs were used to generate specified left DLPFC for each individual participant, as the analyses were conducted in the patient's own brain space. Their resting-state scans were preprocessed according to typical methodologies implemented by the Statistical Parametric Mapping (SPM) program. Normalization parameters that were established by the SPM segmentation based on tissue probability maps were applied to the Montreal Neurological Institute's space regions of interest for the left DLPFC to map these areas onto the patient's individual brain; the patient's space region of interest for the left DLPFC formed the search area for the TMS coil placement.

The effectiveness of the S.A.I.N.T. protocol was quantified primarily through changes in their MADRS scores from baseline to immediately after the protocol and secondarily through reductions in their HAMD6 and BDI-II scores. Response to the treatment was defined as a reduction of the score that is equal to or greater than 50% on the scales; 90.5% of their patients responded to the treatment. Remission was defined as having a score of less than 11 on the MADRS, a score of less than 5 on the HAMD6, and a score of less than 13 on the BDI-II; there was an 86.4% remission rate, and after 1 month, 70% of the patients continued to meet the response criteria. It took an average of 2.3 ± 1.13, or 23 10-minute treatments, for participants to meet the response criterion, and an average of 2.63 ±, or 26 10-minute treatments, for participants to meet the mission criterion. The S.A.I.N.T. procedure was formulated based on the postulation that conventional rTMS doesn't deliver sufficient stimulation to induce both a response and remission from depression for certain patients, and this was tested by including six patients that had a history of not responding to conventional TMS. These patients had similar baseline scores and ended with similar results to the rest of the trial's subjects; it was proven that patients with a history of not responding to conventional TMS take slightly longer to respond to the treatment, but the nrTMS ultimately had a comparable effect on them and the rest of the participants. Thus, it may be extrapolated that the lack of response to conventional TMS may have been related to using non-navigated targeting rather than a failure of TMS. In addition to the antidepressant effects observed subsequent to the treatment, there was also a significant decrease in suicidality as well. All of the participants reported some degree of suicidal ideation in relation to item 10 of MADRS, and after the protocol was completed, there was a significant decrease in the suicidal ideation reported by the participants; 1 month after the protocol, 80–100% of the subjects remained in remission on that measure. At the conclusion of the trial, it was found that the S.A.I.N.T. protocol yielded unprecedented response and remission rates concurrently with decreased suicidal ideation and improved cognitive inhibition, with no adverse side effects or events.

20.8 Conclusion

The S.A.I.N.T. protocol was fast-tracked for FDA approval in September of 2022 due to its revolutionary success with treating refractory MDD, which evidences the fact that neuro-navigational technology is of paramount importance as TMS protocols continue to evolve, as the S.A.I.N.T. procedure relies on the fcMRI of the participants (FDA 2022). In order to provide the most effective treatment, they were able to localize not just the DLPFC, but the specific part of that region that had the strongest anti-correlation with the anterior cingulate cortex, which is believed to play a large role in MDD and isn't able to be targeted by standard rTMS treatment.

With rational, evidence-based medicine, our interventions are tailored to affect and track individual biomarkers. Whether it is treating patients with cancer, diabetes, or neurobehavioral disorders, our ability to target specific biomarkers makes our treatments more effective. With this approach, we can make TMS more effective at treating patients with MDD. As imaging studies develop to more precisely localize pathology in individuals with MDD, our ability to manipulate this pathology can lead to better treatments. The pioneer TMS trials for MDD crafted the tools to begin treating depression in a different way, but the neuro-navigational technology now available has the potential to create more individually based, rational treatments for patients with MDD. Thus, it stands to reason that if we are able to find individual neuroanatomical biomarkers to target depression, nrTMS is essential to creating more efficacious protocols for treating refractory MDD and alleviating some of the global suffering that is caused by major depressive disorder.

References

ADAA. Facts & Statistics | Anxiety and Depression Association of America, ADAA. 2017. Adaa.org. https://adaa.org/understanding-anxiety/depression/facts-statistics.

Bains N, Abdijadid S. Major depressive disorder. StatPearls Publishing; 2022. https://www.ncbi.nlm.nih.gov/books/NBK559078/.

Bielau H, Steiner J, Mawrin C, Trubner K, Brisch R, Meyer-Lotz G, et al. Dysregulation of GABAergic neurotransmission in mood disorders: a postmortem study. Ann N Y Acad Sci. 2007;1096:157–69.

Bjerke IE, Yates SC, Laja A, Witter MP, Puchades MA, Bjaalie JG, Leergaard TB. Densities and numbers of calbindin and parvalbumin positive neurons across the rat and mouse brain. iScience. 2020;24(1):101906. https://doi.org/10.1016/j.isci.2020.101906.

Bunge SA, Klingberg T, Jacobsen RB, Gabrieli JD. A resource model of the neural basis of executive working memory. Proc Natl Acad Sci U S A. 2000;97:3573–8.

Cohen RM, Semple WE, Gross M, Nordahl TE, King AC, Pickar D, et al. Evidence for common alterations in ceebral glucose metabolism in major affective disorders and schizophrenia. Neuropsychopharmacology. 1989;2:241–54.

Cole EJ, Stimpson KH, Bentzley BS, Gulser M, Cherian K, Tischler C, Nejad R, Pankow H, Choi E, Aaron H, Espil FM, Pannu J, Xiao X, Duvio D, Solvason HB, Hawkins J, Guerra A, Jo B, Raj KS, Phillips AL, Barmak F, Bishop JH, Coetzee JP, DeBattista C, Keller J, Schatzberg AF, Sudheimer KD, Williams NR. Stanford accelerated intelligent neuromodulation therapy for treatment-resistant depression. Am J Psychiatry. 2020;177(8):716–26. https://doi.org/10.1176/appi.ajp.2019.19070720. Epub 2020 Apr 7.

Cuijpers P, Vogelzangs N, Twisk J, Kleiboer A, Li J, Penninx BW. Comprehensive meta-analysis of excess mortality in depression in the general community versus patients with specific illnesses. Am J Psychiatry. 2014;171(4):453–62. https://doi.org/10.1176/appi.ajp.2013.13030325.

Drevets WC. Neuroimaging studies of mood disorders. Biol Psychiatry. 2000;48:813–29.

Drevets WC, Price JL. Neuroimaging and neuropathological studies of mood disorders. In: Licinio JWM, editor. Biology of depression: from novel insights to therapeutic strategies. Weinheim: Wiley-VCH Verlag GmbH & Co.; 2005. p. 427–66.

Drevets WC, Price JL, Simpson JR Jr, Todd RD, Reich T, Vannier M, et al. Subgenual prefrontal cortex abnormalities in mood disorders. Nature. 1997;386:824–7.

FDA clears SAINT rapid-acting brain stimulation approach for those suffering from resistant major depression. Brain & Behavior Research Foundation. 2022. https://www.bbrfoundation.org/content/fda-clears-saint-rapid-acting-brain-stimulation-approach-those-suffering-resistant-major.

Feffer K, Fettes P, Giacobbe P, Daskalakis ZJ, Blumberger DM, Downar J. 1Hz rTMS of the right orbitofrontal cortex for major depression: safety, tolerability and clinical outcomes. Eur Neuropsychopharmacol. 2018;28(1):109–17. https://doi.org/10.1016/j.euroneuro.2017.11.011. Epub 2017 Nov 17.

Fitzgerald PB, Hoy K, McQueen S, Maller JJ, Herring S, Segrave R, Bailey M, Been G, Kulkarni J, Daskalakis ZJ. A randomized trial of rTMS targeted with MRI based neuro-navigation in treatment-resistant depression. Neuropsychopharmacology. 2009;34(5):1255–62. https://doi.org/10.1038/npp.2008.233. Epub 2009 Jan 14.

Gaynes B, Rush A. Treatment Alternatives to Relieve Depression (STAR*D) study. The STAR*D study: treating depression in the real world. Cleve Clin J Med. 2008;75(1). https://www.ccjm.org/content/ccjom/75/1/57.full.pdf.

George MS, Ketter TA, Parekh PI, Horwitz B, Herscovitch P, Post RM. Brain activity during transient sadness and happiness in healthy women. Am J Psychiatry. 1995a;152:341–51.

George MS, Wassermann EM, Williams WA, Callahan A, Ketter TA, Basser P, Hallett M, Post RM. Daily repetitive transcranial magnetic stimulation (rTMS) improves mood in depression. Neuroreport. 1995b;6(14):1853–6. https://doi.org/10.1097/00001756-199510020-00008.

Johnson KA, Baig M, Ramsey D, Lisanby SH, Avery D, McDonald WM, Li X, Bernhardt ER, Haynor DR, Holtzheimer PE 3rd, Sackeim HA, George MS, Nahas Z. Prefrontal rTMS for treating depression: location and intensity results from the OPT-TMS multi-site clinical trial. Brain Stimul. 2013;6(2):108–17. https://doi.org/10.1016/j.brs.2012.02.003. Epub 2012 Mar 14. PMID: 22465743; PMCID: PMC3573221.

Keilp JG, Sackeim HA, Brodsky BS, Oquendo MA, Malone KM, Mann JJ. Neuropsychological dysfunction in depressed suicide attempters. Am J Psychiatry. 2001;158:735–41.

Koolschijn PC, van Haren NE, Lensvelt-Mulders GJ, Hulshoff Pol HE, Kahn RS. Brain volume abnormalities in major depressive disorder: a meta-analysis of magnetic resonance imaging studies. Hum Brain Mapp. 2009;30(11):3719–35. https://doi.org/10.1002/hbm.20801.

Kringelbach ML. The human orbitofrontal cortex: linking reward to hedonic experience. Nat Rev Neurosci. 2005;6(9):691–702.

Lefaucheur JP, Brugières P, Ménard-Lefaucheur I, Wendling S, Pommier M, Bellivier F. The value of navigation-guided rTMS for the treatment of depression: an illustrative case. Neurophysiol Clin. 2007;37(4):265–71. https://doi.org/10.1016/j.neucli.2007.07.003. Epub 2007 Aug 2. PMID: 17996815.

McManus S, Gunnell D, Cooper C, et al. Prevalence of non-suicidal self-harm and service contact in England, 2000–14: repeated cross-sectional surveys of the general population. Lancet Psychiatry. 2019;6(7):573–81. https://doi.org/10.1016/S2215-0366(19)30188-9.

National Institute of Mental Health. Major depression. National Institute of Mental Health; 2022. https://www.nimh.nih.gov/health/statistics/major-depression.

Neumeister A, Nugent AC, Waldeck T, Geraci M, Schwarz M, Bonne O, et al. Neural and behavioral responses to tryptophan depletion in unmedicated patients with remitted major depressive disorder and controls. Arch Gen Psychiatry. 2004;61:765–73.

Nexstim. Nexstim reports excellent treatment outcomes of over 400 MDD Patients in Patient Registry. 2022. www.nexstim.com; Nexstim. https://www.nexstim.com/news-and-events/news/press-release/nexstim-reports-excellent-treatment-outcomes-of-over-400-mdd-patients-in-patient-registry.

Nexstim nTMS. Neuroclast. 2023. Retrieved September 22, 2023, from https://www.neuroclast.com/nexstim#:~:text=The%20Nexstim%20NBT%C2%AE%20system%20was%20launched%20in%20the.

Nordentoft M, Mortensen PB, Pedersen CB. Absolute risk of suicide after first hospital contact in mental disorder. Arch Gen Psychiatry. 2011;68(10):1058–64. https://doi.org/10.1001/archgenpsychiatry.2011.113.

O'Reardon JP, Solvason HB, Janicak PG, Sampson S, Isenberg KE, Nahas Z, McDonald WM, Avery D, Fitzgerald PB, Loo C, Demitrack MA, George MS, Sackeim HA. Efficacy and safety of transcranial magnetic stimulation in the acute treatment of major depression: a multisite randomized controlled trial. Biol Psychiatry. 2007;62(11):1208–16. https://doi.org/10.1016/j.biopsych.2007.01.018. Epub 2007 Jun 14. PMID: 17573044.

Osuch EA, Ketter TA, Kimbrell TA, George MS, Benson BE, Willis MW, et al. Regional cerebral metabolism associated with anxiety symptoms in affective disorder patients. Biol Psychiatry. 2000;48:1020–3.

Paillère Martinot ML, Galinowski A, Ringuenet D, Gallarda T, Lefaucheur JP, Bellivier F, Picq C, Bruguière P, Mangin JF, Rivière D, Willer JC, Falissard B, Leboyer M, Olié JP, Artiges E, Martinot JL. Influence of prefrontal target region on the efficacy of repetitive transcranial magnetic stimulation in patients with medication-resistant depression: a [(18)F]-fluorodeoxyglucose PET and MRI study. Int J Neuropsychopharmacol. 2010;13(1):45–59. https://doi.org/10.1017/S146114570900008X. Epub 2009 Mar 9. PMID: 19267956.

Piedmont Neuroscience Center. TMS for depression. Retrieved November 15, 2023., from https://piedmontneurosciencecenter.com/tms.

Psychiatry.org. The economic cost of depression is increasing; direct costs are only a small part. 2021. Psychiatry.org. https://www.psychiatry.org/News-room/APA-Blogs/The-Economic-Cost-of-Depression-is-Increasing.

Rajkowska G. Postmortem studies in mood disorders indicate altered numbers of neurons and glial cells. Biol Psychiatry. 2000;48:766–77.

Szczepanski SM, Knight RT. Insights into human behavior from lesions to the prefrontal cortex. Neuron. 2014;83(5):1002–18. https://doi.org/10.1016/j.neuron.2014.08.011.

World Health Organization. Depressive disorder (depression). World Health Organisation; 2023. https://www.who.int/news-room/fact-sheets/detail/depression.

Part VI

Further Potential of nTMS

Harnessing Advanced Tractography in Neurosurgical Practice

21

Ahmad Beyh, Ann-Katrin Ohlerth, and Stephanie J. Forkel

21.1 Introduction

In the dynamic field of neurosurgery, the need for cutting-edge methods is increasingly critical. As neurosurgical approaches become more multimodal and technologically advanced, the central role of white matter tracts in the brain and the imperative of their precise preoperative mapping are gaining recognition. Advanced tractography is a vital tool for neurosurgeons, offering unparalleled in vivo delineation and visualisation of these complex neural pathways.

This chapter addresses the urgent need for advanced tractography in modern neurosurgery and discusses how its application can profoundly improve surgical planning, execution, and patients' motor, cognitive, and behavioural outcomes. Initially, we will examine the basic concepts of diffusion-weighted imaging (DWI), the foundational bedrock for advanced tractography techniques. Subsequently, we will venture beyond the limitations of traditional tractography, critically examining

A. Beyh
Department of Psychiatry, Brain Health Institute, Rutgers University, Piscataway, NJ, USA

Donders Institute for Brain Cognition Behaviour, Radboud University, Nijmegen, The Netherlands
e-mail: ahmad.beyh@donders.ru.nl

A.-K. Ohlerth
Neurobiology of Language Department, Max Planck Institute for Psycholinguistics, Nijmegen, The Netherlands
e-mail: Ann-Katrin.Ohlert@mpi.nl

S. J. Forkel (✉)
Donders Institute for Brain Cognition Behaviour, Radboud University, Nijmegen, The Netherlands

Max Planck Institute for Psycholinguistics, Nijmegen, The Netherlands

Brain Connectivity and Behaviour Laboratory, Sorbonne Universities, Paris, France

S. M. Krieg, T. Picht (eds.), *Navigated Transcranial Magnetic Stimulation in Neurosurgery*, https://doi.org/10.1007/978-3-031-97155-6_21

the shortcomings of diffusion tensor imaging (DTI) and advocating for the integration of more sophisticated imaging techniques in surgical practice. Books and reviews have been written on this topic (e.g., Jones 2010; Le Bihan and Johansen-Berg 2012), and we do not aim to cover the theory in depth, but we will offer a brief overview of the concepts we believe are central to understanding the methods and their applications in the context of neurosurgery.

21.2 Neuroimaging

21.2.1 From Diffusion-Weighted Imaging to Advanced Tractography

21.2.1.1 Diffusion-Weighted Imaging (DWI)

Diffusion refers not to the large-scale movement of water molecules but to their small-scale, erratic movements akin to a 'random walk' pattern (see Fig. 21.1). In the brain, water molecules within various tissue types engage in this diffusion process, randomly 'exploring' their immediate surroundings at a microscopic level. The speed of this diffusion process is influenced by several factors, including temperature and the characteristics of the surrounding environment, and is quantified as the apparent diffusion coefficient (ADC; Le Bihan 2013). This concept is

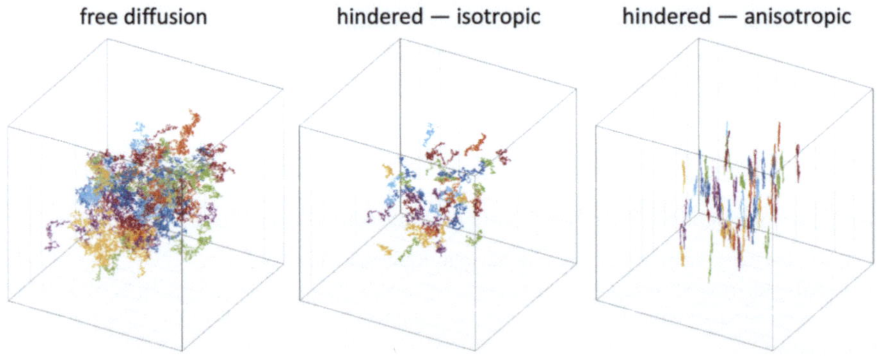

Fig. 21.1 Patterns of water diffusion. In each panel, different colours represent the trajectories of different water molecules diffusing under certain conditions. Over time, a group of water molecules will move and randomly explore their surroundings in a jittery manner. In unrestricted or free diffusion, each molecule is free to move in all directions without any hindrance, resulting in an isotropic (i.e., equal in all directions) diffusion pattern. However, if these molecules are in an environment with other materials that restrict their movement, their diffusion pattern will be influenced by the arrangement of these materials. In a setting with isotropic hindrance, although each molecule can still move equally in all directions, the distance it can travel is limited due to obstructions from evenly dispersed barriers (hindered—isotropic). In an environment with uneven hindrance, the diffusion of water molecules is influenced by a spatial arrangement that creates more obstacles along certain directions, resulting in distinct anisotropic diffusion patterns (hindered—anisotropic)

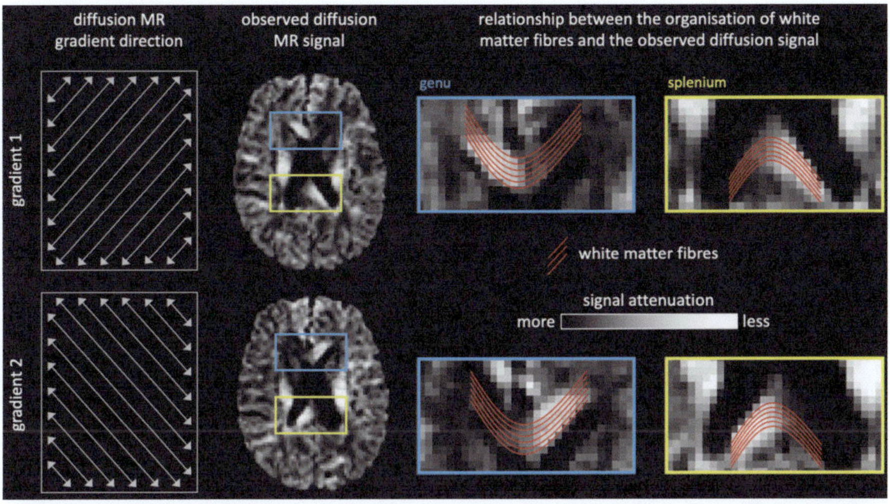

Fig. 21.2 Diffusion-weighted imaging. In DWI, the MRI sequence includes diffusion-weighting gradients that cause a signal attenuation if the imaged tissue contains water molecules that diffuse along the gradient's direction. However, if the primary axis of water diffusion is perpendicular to the gradient's direction, no attenuation is observed. This is what allows us to derive a measure of angular contrast within each voxel. Notice, e.g., that the genu and splenium of the corpus callosum look drastically different under the influence of each gradient

fundamental to diffusion imaging, which utilises the variable nature of diffusion to investigate tissue microstructure and non-invasively chart white matter pathways (i.e., tractography).

DWI is typically acquired using a spin-echo echo planar imaging (SE-EPI) MRI pulse sequence (Basser et al. 1994). This technique quickly applies and removes diffusion-weighting magnetic gradients that affect the imaged protons. When a diffusion gradient is applied, protons diffusing parallel to its primary direction experience a reduction in signal. In contrast, those that are stationary or moving in other directions do not or experience less attenuation. Consequently, DWI extracts information about water diffusion from the amount of signal reduction caused by each diffusion gradient (Fig. 21.2). The characteristics of the applied diffusion gradients are encapsulated in a term called the '*b*-value', measured in s/mm^2. Higher *b*-values suppress more signal from rapid diffusion, e.g., in cerebrospinal fluid (CSF) or in grey matter, and enhance the detection of slow or restricted water diffusion, e.g., in deep white matter. However, as this is contingent upon greater overall signal attenuation, it also lowers the signal-to-noise ratio (SNR). As increasing image resolution also reduces SNR, the use of high *b*-values can be a challenge for some 1.5T scanners.

The observed signal attenuation due to protons diffusing along the diffusion gradient direction can be formalised according to the simplified Stejskal-Tanner equation (Stejskal and Tanner 1965):

$$S = S_0 \cdot e^{-b \cdot \mathrm{ADC}} \tag{21.1}$$

where S is the measured diffusion-weighted signal; S_0 is the signal acquired without diffusion weighting (equivalent to a standard T2-weighted signal) and typically referred to as the b0 image; b is the b-value; and ADC, i.e., the apparent diffusion coefficient, is the variable whose value will change depending on the tissue being imaged. ADC values are highest in CSF, followed by grey matter, then white matter (Le Bihan 2013), and can be calculated directly from the measured DWI data without resorting to complex models:

$$\mathrm{ADC} = \frac{-\ln\left(\frac{S}{S_0}\right)}{b} \tag{21.2}$$

where ln is the natural logarithm. In this context, S and S_0 are the measured reference signal (b0 image) and diffusion-weighted signal (diffusion series), respectively.

21.2.1.2 The Tensor Model

Basser, Mattiello, and Le Bihan (1994) proposed the diffusion tensor model as a simultaneous representation of the ADC and the direction of water diffusion. The rationale is that probing water diffusion along six or more non-collinear directions yields enough information to mathematically model the displacement of the water molecules. The diffusion tensor is thus a three-dimensional ellipsoid with three principal axes, each characterised by a vector of a specific orientation (eigenvector), whose length (eigenvalue) is equivalent to the ADC along that axis. Several metrics are usually derived from the tensor in each voxel, the most common being fractional anisotropy (FA) and mean diffusivity (MD; Forkel and Catani 2017). FA characterises the shape of the tensor, with high values in the white matter and near-zero values in CSF, while MD has the highest values in CSF and the smallest in the deep white matter where diffusion is most hindered.

Though widely successful, the tensor model has inherent limitations that impede the interpretation of derived microstructural measures and the accuracy of resultant tractography reconstructions (Fig. 21.3). DTI tractography assumes that water diffusion in any voxel is accurately represented by a single direction, as indicated by the tensor's principal eigenvector. However, this assumption is often violated—it is estimated that 60–90% of white matter voxels contain multiple fibre populations travelling along different orientations (Dell'Acqua et al. 2013; Tournier et al. 2011). As a result, the diffusion tensor would appear spherical in a voxel where three fibre populations intersect at right angles. This representation obscures the true orientations of the individual fibre populations and yields a low FA value. This not only leads to an inaccurate representation of the water diffusion direction within the voxel but could also cause the premature termination of tractography tracking due to the resultant low FA, as DTI-based algorithms are constrained by an FA threshold usually close to 0.15 during tracking (see Fig. 21.3 for examples of these limitations).

Despite its limitations, DTI continues to be favoured for its simplicity and its relatively low demands on acquisition and computational resources. This is

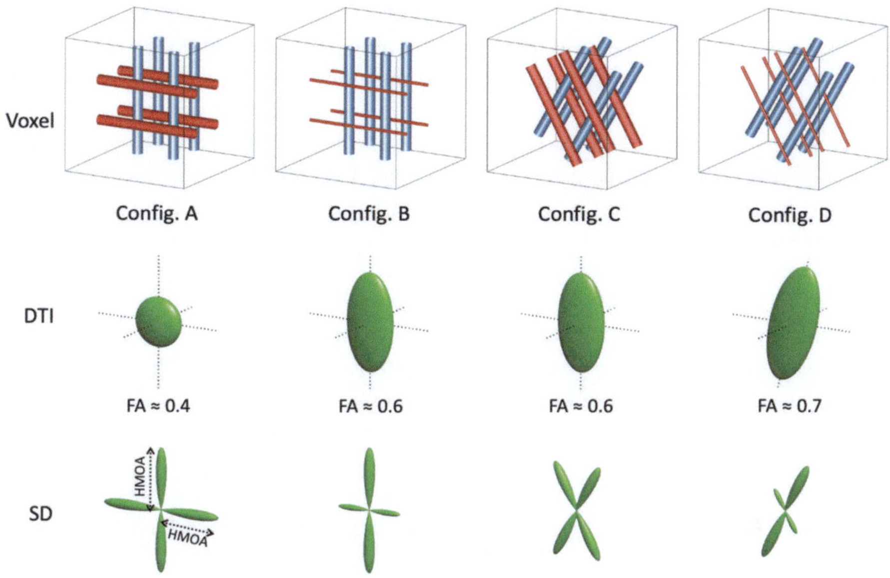

Fig. 21.3 Examples of multifibre anatomy and their reconstruction using DTI and SD. The tensor model has a notable shortcoming in its inability to distinguish crossing fibres. For example, in configuration A, two fibre populations intersecting at right angles are represented by a flat tensor with a low fractional anisotropy (FA) value and with no correct principal axis, leading to erroneous tracking or to stopping the process altogether. Moreover, identical tensors can result from two distinct fibre configurations, as seen in configurations B and C. In configuration D, the deterioration of one set of crossing fibres alters the tensor's shape, FA, and orientation, leading to a different tractography result. While DTI fails to differentiate these scenarios, spherical deconvolution (SD) can. This technique is capable of distinguishing between intersecting fibre populations within voxels and detecting fibre-specific changes, such as the degeneration of one fibre group in configurations B and D. This is achieved using the hindrance-modulated orientational anisotropy (HMOA) metric, which provides an indication of fibre density

particularly the case in clinical settings, where practical requirements often precede the use of more sophisticated acquisition techniques and complex modelling.

21.2.1.3 Spherical Deconvolution (SD)

High angular resolution diffusion imaging (HARDI) refers to a family of advanced imaging tools developed by various research groups to address the main limitations of the tensor model. Although many HARDI methods have been proposed over time, such as Q-ball imaging (Tuch 2004) and diffusion spectrum imaging (DSI; Wedeen et al. 2005), we focus here on spherical deconvolution (SD; Dell'Acqua et al. 2010, 2013, Tournier et al. 2007), which has been optimised with anatomical models and whose technical requirements are close to those of DTI.

SD operates under the premise that the diffusion signal measured in a voxel that contains multiple fibre populations is the sum of the fibre-specific signals within that voxel. SD conceptualises these populations as a continuous distribution of fibre orientations. In this framework, the signal is modelled as a spherical convolution (akin to a 'shape multiplication') between the fibre orientation density function (fODF) and a fibre response function. Therefore, when a diffusion signal is measured along multiple directions using MRI, it can be deconvolved using that fibre response function to estimate fibre orientations. The spherical deconvolution process yields the fibre orientation distribution (FOD). This multi-peak spherical function represents the orientations of the fibre populations traversing a voxel and their respective contributions to the signal (Fig. 21.4).

The FOD yields valuable insights into microstructure due to its capacity to represent multiple fibre orientations. A key measure the FOD provides is the number of fibre orientations (NuFO) within each voxel (Dell'Acqua et al. 2013), which is particularly useful in studies tracking degeneration over time. Another significant measure derived from the FOD is the hindrance-modulated orientational anisotropy (HMOA; Dell'Acqua et al. 2013), which is specific to each lobe of the FOD and relates to its magnitude, representing the apparent density of the represented fibre population. Lastly, the FOD facilitates multi-fibre tractography, which is a significant advancement over the conventional single-fibre tensor model.

SD necessitates the acquisition of diffusion data at elevated *b*-values to achieve high angular resolution and resolve intersecting fibres. While higher *b*-values enhance angular contrast, they also reduce SNR, so values in the range of 2000–3000 s/mm^2 are a good middleground and considered optimal for most tractography studies (Dell'Acqua and Tournier 2019). Additionally, SD requires measuring the diffusion signal across numerous non-collinear diffusion directions. In contrast to DTI, which can be theoretically modelled with as few as six directions (though this is far from recommended), SD typically requires at least 45 directions to yield meaningful results, with 60 being the recommended number for most studies. While most current research studies gather data suitable for SD, this practice has not yet become standard in clinical settings.

21.2.1.4 SD Tractography

SD tractography operates on principles similar to DTI tractography but uses directional information obtained from the FOD instead of the diffusion tensor. When streamline propagation begins, several checks are performed at each step to ensure the accurate reconstruction of fibre paths. A notable difference is that in areas where the FOD shows multiple distinct peaks, an equivalent number of streamlines are initiated, matching the number of peaks. A critical distinction between SD and DTI tractography lies in the criteria that halt streamline propagation. An FA threshold, angular threshold, and step size guide tensor tractography; SD tractography uses HMOA as one of its four stopping criteria. Thus, the four guiding parameters for SD tractography include streamline length, step size, the maximum angle threshold, and the minimum HMOA value.

The technique outlined above is known as deterministic tracking, where a single streamline originates from each lobe of the FOD in every voxel. However, probabilistic tracking methods also offer certain benefits (as noted by Behrens et al. 2007;

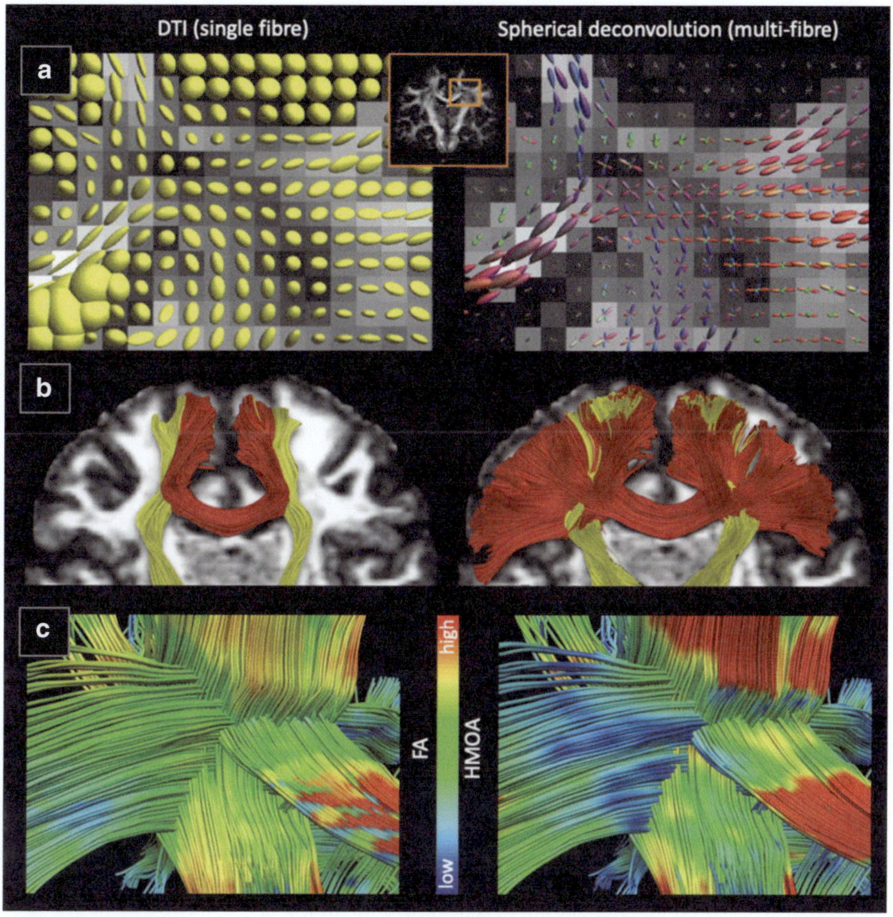

Fig. 21.4 Single fibre vs. multi-fibre tractography. The tensor model (left) can only represent one fibre direction in each voxel, while spherical deconvolution (SD, right) can represent multiple fibre orientations as a fibre orientation distribution function (fODF). These butterfly-shaped 3D functions are then used by tracking algorithms to explore multiple white matter trajectories that can cross the same physical location in the brain. The region highlighted in panel (**a**) is the centrum semiovale, which houses at least three orthogonally crossing white matter pathways: the corticospinal tract, the corpus callosum, and the superior longitudinal fasciculus. All three fibre populations are captured by the fODFs in this region. Panel (**b**) clearly demonstrates DTI's inability to capture such complex information, with the classical, limited reconstruction of the corpus callosum that only connects the medial cortical areas. In contrast, SD captures the full fanning of the callosal fibres. Panel (**c**) depicts the microstructural specificity of the multi-fibre approaches, which can attribute a different microstructural measure to each of the fibre populations crossing the same voxel. (The image in panel (**a**) is adapted from Dell'Acqua and Tournier (2019) with permission (CC-BY); the images in panels (**b**) and (**c**) are inspired by Dell'Acqua and Catani (2012))

Descoteaux et al. 2009). In these methods, multiple streamlines are initiated from each FOD lobe, each following slightly varied paths. One method to establish these paths involves using the shape of the FOD to indicate the dispersion in the underlying fibre populations (Descoteaux et al. 2009). Narrower FOD lobes, resembling sticks, suggest the dominance of a single fibre population in the signal, while broader lobes imply the potential merging of multiple populations at close angles (<35°). These populations may not be distinctly resolvable without using very high *b*-values, but their presence broadens the overall shape of the FOD lobe. In such cases, instead of only following the main FOD peak, which would lie between the two real ones, dispersion SD allows the tractography algorithm to take advantage of the information about anatomical uncertainty, thereby following the real directions. This approach effectively follows the true fibre directions and can be particularly advantageous for tracing smaller fibre bundles or those with highly curved trajectories.

Two of the most striking examples of the advantages of using SD tractography over standard DTI are its ability to visualise the lateral projections of the corpus callosum and the first and second branches of the superior longitudinal fasciculus (SLF; Thiebaut de Schotten et al. 2011)—all of which are impossible to reconstruct with the standard DTI tractography methods (Fig. 21.4).

21.2.1.5 Advantages of Using SD Tractography

An alternative perspective on the distinction between DTI and SD techniques is that SD prioritises anatomical sensitivity, though this may lead to a higher rate of false positives, especially if probabilistic tracking is used. In the context of pre-surgical planning, this enhanced sensitivity can be more beneficial for guiding surgical decisions. For example, when surgeons use tractography to determine the extent of tumour removal, SD tractography provides more detailed information about the underlying white matter anatomy, including how fibre bundles spread as they approach grey matter. This detailed mapping can guide a surgeon to opt for less aggressive resection near crucial white matter areas, especially if the patient prioritises functional preservation. This approach also influences the use of distance metrics between a tumour and relevant white matter bundles to predict surgical outcomes (Fig. 21.5). It is crucial to emphasise that the gain in sensitivity offered by advanced methods also results in fibre bundles with larger volume (they occupy more white matter space), which will likely lead to a reduction of this distance. This nuanced understanding is pivotal for improving the accuracy of predictive models and, consequently, enhancing the reliability of distance-based assessments in neurosurgical planning.

Advanced multi-fibre tractography methods not only offer increased sensitivity but also yield more precise microstructural metrics (Fig. 21.4). This precision arises from the ability of these methods to recognise multiple fibre orientations within a single voxel, each with distinct microstructural characteristics. A significant advantage of this approach is that the metrics are specific to individual fibres rather than to the voxel as a whole, making them highly suitable for tract-based analyses. For instance, in a study of a patient cohort with tumours near the corticospinal tract

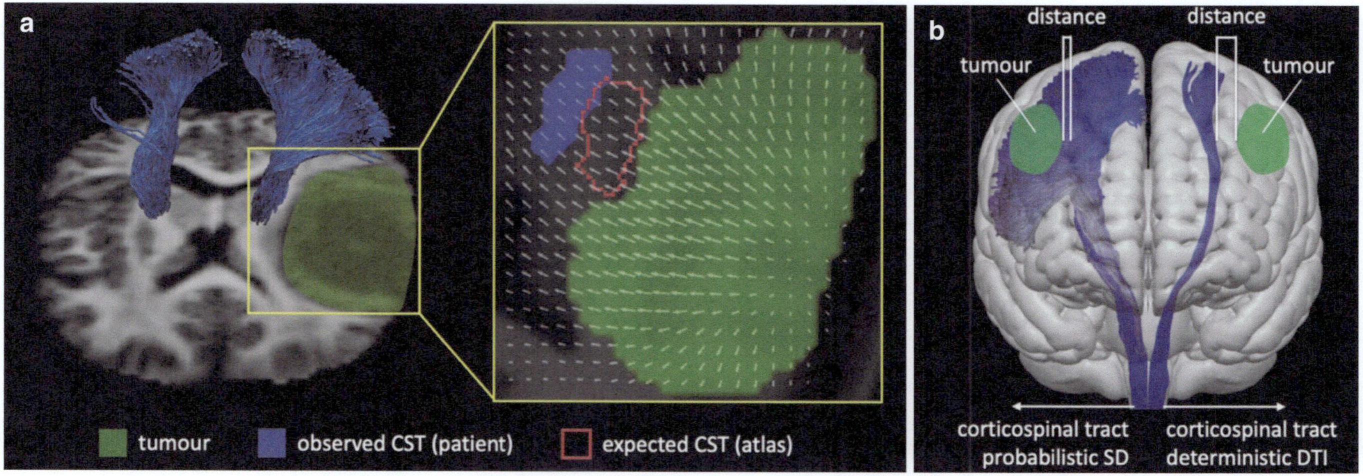

Fig. 21.5 Technical considerations involved in tractography reconstructions for brain tumours. Panel (**a**) emphasises the limitations of atlas-based approaches due to white matter displacement. In this example, the tumour's expansion caused surrounding structures, including the corticospinal tract (CST), to shift medially, moving them away from their expected atlas-based location. Individualised tractography can accurately visualise these displaced connections, aiding in pre-surgical planning. The white arrows in the cutout panel represent the displacement field resulting from the tumour's expansion. Panel (**b**) underscores the impact of algorithm selection on reconstruction reliability. The left-hand side image represents a probabilistic HARDI reconstruction, while the right-hand side panel shows a deterministic reconstruction. The core portion of the CST is visible in both, but its size is more prominent in the probabilistic version, which also shows the entire cortical extent of the CST's projections. (Image modified with permission from Forkel et al. 2025)

(CST), DTI-derived metrics like the voxel-based FA failed to reveal any microstructural differences between the ipsilateral and contralateral CST, but the tract-specific HMOA successfully identified these disparities (Mirchandani et al. 2021).

21.2.1.6 Limitations of Advanced Tractography—Or When to Use Classical DTI

Advanced tractography techniques have emerged as powerful tools for unravelling the intricacies of white matter fibre pathways within the brain, especially in fibre-dense areas. However, with this promise come limitations that necessitate careful consideration. The computational demands of some advanced tractography models can hinder their widespread use, particularly in clinical settings where quick results are essential. Additionally, these models depend heavily on high-quality diffusion MRI data; in situations where the data quality is subpar, the fidelity of the reconstruction can be disproportionately impacted, given the more exacting requirements of these techniques. Validating the results of advanced tractography proves challenging due to the complexity and user dependence of the algorithms involved (Schilling et al. 2019; Yendiki et al. 2022). This challenge raises questions about the clinical interpretability of advanced methods and has potentially hindered their uptake in the clinical world.

In contrast, DTI, though less intricate, remains widely used in the field, even in research studies (Rheault et al. 2025). The standardisation and widespread availability of DTI make it a pragmatic choice in multicenter studies or clinical trials. Additionally, DTI algorithms have been refined to align with the white matter anatomy as revealed by postmortem dissections, thereby increasing the likelihood of accurately representing the white matter's central structures. Ultimately, the choice between advanced tractography and classical DTI hinges upon the unique demands of each study or clinical application, ranging from the necessity of rapid results and technical feasibility to the importance of standardised protocols and interpretability in varied clinical or research contexts (Aydogan et al. 2025).

21.2.2 Acquiring Clinically Suitable DWI Data

To effectively utilise HARDI methods like spherical deconvolution, we will outline the essential components for conducting a successful DWI study using a clinical scanner. While additional recommendations could be made for more comprehensive research studies where time is less constrained, we focus on identifying the fundamental elements that can deliver high-quality data within the tight time frames typical of a busy clinical setting. Consequently, the acquisition protocol we propose is designed to be feasible on most modern clinical scanners within a time frame of 5–7 min.

1. Use a *b*-value of 1500 s/mm^2 or higher to allow the chosen diffusion model to reliably separate fibres that cross the same voxel at angles as small as 40°. Although higher *b*-values would be more desirable (e.g., 2000 s/mm^2), they may

be challenging for clinical systems, so we recommend the more moderate choice of 1500 s/mm^2 as most modern clinical scanners can acquire images at such *b*-values without sacrificing data quality.

2. Acquire a minimum of 60 diffusion gradient directions. This covers the minimum requirement for models like spherical deconvolution and allows some leeway for cases where some volumes (directions) are lost due to motion or other artefacts.
3. Incorporate multiple b0 images into the sequence, with about three captured at the beginning and then one following every 10–15 volumes. This approach is beneficial during pre-processing and model fitting for two key reasons: first, acquiring several b0 images at the start guarantees that at least one or two are of high quality, even if the participant moves a lot during the initial 20 s of the acquisition. Second, interspersing b0 images throughout the sequence provides reference images at various points in the series, enhancing the robustness of motion correction models and simplifying quality control.
4. Pair the main acquisition with additional b0 images acquired with reverse-polarity phase encoding to be used for distortion correction (see pre-processing section). These images should have all imaging parameters exactly matched to the main acquisition except for the phase-encoding direction. In theory, only one such image is required, but in practice, it is highly recommended to acquire at least three to ensure data viability in case of severe patient motion.
5. Use isotropic voxels of a minimum size of 2 × 2 × 2 mm^3, which is the most common resolution used in research studies until today.
6. Do not allow the scanner to automatically upsample the data, i.e., ensure that the voxel size is not changed by the scanner after the signal is collected, as is commonly done for some clinical structural sequences (see pre-processing section).
7. If the study involves patients who are unable to keep still for more than a few minutes at a time and good data cannot be acquired using the full-length MRI sequence, one could consider splitting the sequence into two sets of 30 different directions. This should be done as a last resort and only if no viable data can be acquired in one go, as it complicates data pre-processing. This will not be necessary on most modern clinical systems, which will have imaging acceleration features like multiband (or simultaneous multislice) that reduce the total acquisition time to approximately 4 min. If sequence splitting is done during data pre-processing, these exams can be combined later on to obtain the full set of 60 directions, for example, as done in Forkel et al. (2014). In this case, it is crucial to acquire a separate set of reverse phase-encoded b0 images for each of the two halves of the acquisition because a large head rotation during the short break between the two sets can result in different anatomical distortions (see pre-processing section on susceptibility distortions).

Other recommendations can be made for more advanced acquisitions that would typically result in longer exams but could be useful in a research context. For example, acquiring the diffusion signal using multiple *b*-values instead of only one allows for more advanced and complex modelling of the signal and of tissue microstructure (these types of acquisitions are known as multi-shell diffusion studies). Pushing the *b*-value even higher, beyond 3000 s/mm^2 can be quite useful for the reconstruction of very small fibre bundles that cross larger ones at small angles. This, of course, would benefit from a research 3T scanner, which would also help in acquiring higher resolution images without losing too much SNR. To avoid pulsatile effects caused by cardiac events, cardiac-gated data can be acquired, which is to say that the MRI sequence can be time-locked to the cardiac cycle to ensure that images are acquired only during R-R intervals. These types of modifications would usually elongate the total exam time and depend on the scanner's capabilities, so they would not apply to most hospital-based clinical studies, but they are worth considering for patient cohorts acquired at research centres.

21.2.3 Data Quality and Pre-processing

Multiple elements influence the quality of MRI data, encompassing characteristics of the patient (such as movement during the scan, variations in heart rate and brain pulsations, or breathing patterns), aspects of the MRI sequence itself (like eddy current distortions, thermal noise, and Gibbs ringing), and interactions between the patient's anatomy and the scanner's environment (e.g., susceptibility distortions). This section will discuss some of the primary artefacts typically addressed in DWI studies.

21.2.3.1 Head Motion Artefacts

Motion artefacts are a common issue in all imaging studies, not only in DWI, but they have a particularly significant impact in DWI. A standard DWI session lasts several minutes, during which it is crucial for the patient to remain as motionless as possible to reduce motion artefacts. A typical 2D EPI DWI sequence on a contemporary 1.5T or 3T scanner can complete a full brain scan (for one diffusion gradient direction) in 3–8 seconds, thanks to various scanning enhancements like imaging acceleration. This process of acquiring a single volume is repeated with varying diffusion weighting gradient directions until the entire set of images (e.g., 60 DWI volumes plus six b0 volumes) is captured. During this period, head motion can occur at any time and can compromise data quality within an individual volume (intra-volume motion) as well as affect the consistency of head position throughout the scan (inter-volume motion). See Fig. 21.6 for examples of these artefacts.

The fundamental step involves addressing inter-volume motion, which is when the brain's position changes within the field of view from one volume to another. This type of correction is relatively straightforward and is achieved through rigid body alignment, ensuring that each voxel in the image matrix consistently aligns with the same anatomical location across the entire series of images. However, it is important to note that each volume acquisition takes several seconds, and any movement during this period can lead to two types of artefacts. The first is signal dropout,

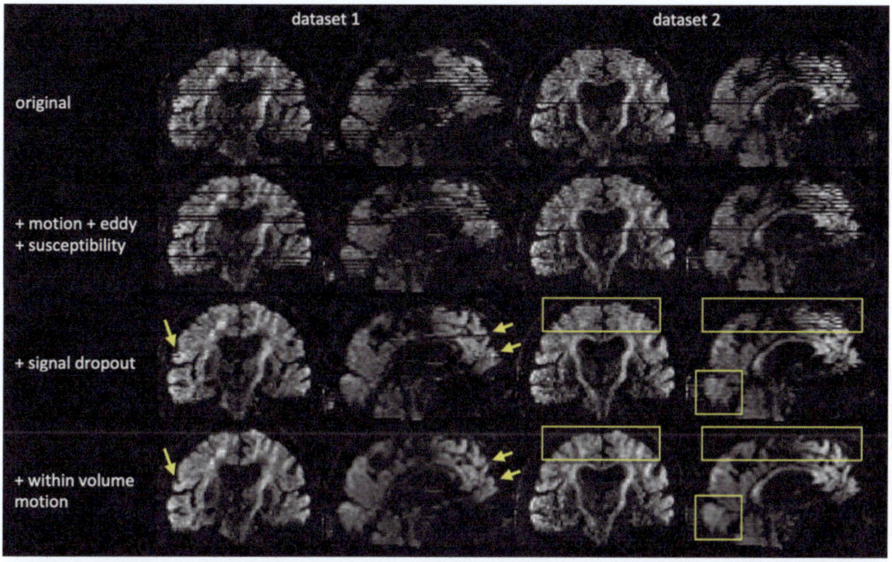

Fig. 21.6 Motion correction of diffusion images. In addition to the framewise displacement caused by subject motion, other motion artefacts are observed in DWI data, including signal dropout and within-volume misalignment. Here, two example datasets exhibit these artefacts. The top row represents the original data with all artefacts included; the second row represents data that have been corrected for framewise motion, eddy currents, and susceptibility distortions; the third row represents additional corrections for signal dropout; the last row represents a full correction that also accounts for within-volume slice misalignment. (Image adapted from Andersson et al. (2017) with permission (CC-BY))

which is caused by head motion that takes place while an imaging slice is being acquired, leading to a mismatch between the imaging parameters and the slice's new position. This results in the appearance of dark bands (slices) in the 3D image that can significantly disrupt model fitting if not properly corrected or if this happens frequently. The second artefact, resulting from intra-volume motion, appears as a 'zig-zag' pattern in the 3D volume when viewed perpendicular to the slice axis (such as a sagittal view of an axial acquisition). This occurs when the patient moves partway through acquiring a volume and remains in that new position, causing a misalignment in all slices imaged after that movement.

Some pre-processing tools take a model-based approach to correct for within-volume motion that results in signal dropout by exploiting the redundancy of the diffusion information in the data to regenerate the lost signal (Andersson et al. 2016). Others discard the slices with lost signals during the diffusion modelling step (Chang et al. 2005). Whichever approach is taken, accounting for signal dropout during pre-processing or modelling is a basic requirement, as it can have detrimental consequences during model fitting. Recent efforts have also been made to correct intra-volume motion with excellent results, albeit at a large computational cost (Andersson et al. 2017).

21.2.3.2 Eddy Current Distortions

DWI pulse sequences involve the rapid introduction and removal of magnetic diffusion weighting gradients, a process that induces weak secondary magnetic fields due to interactions with the scanner's hardware and that interferes with the proton spins. This results in imaging artefacts known as eddy current distortions (Fig. 21.7), which can compress, stretch, skew, or shift the imaged slice (Tournier et al. 2011). These distortions vary from one volume to the next as a function of the changing

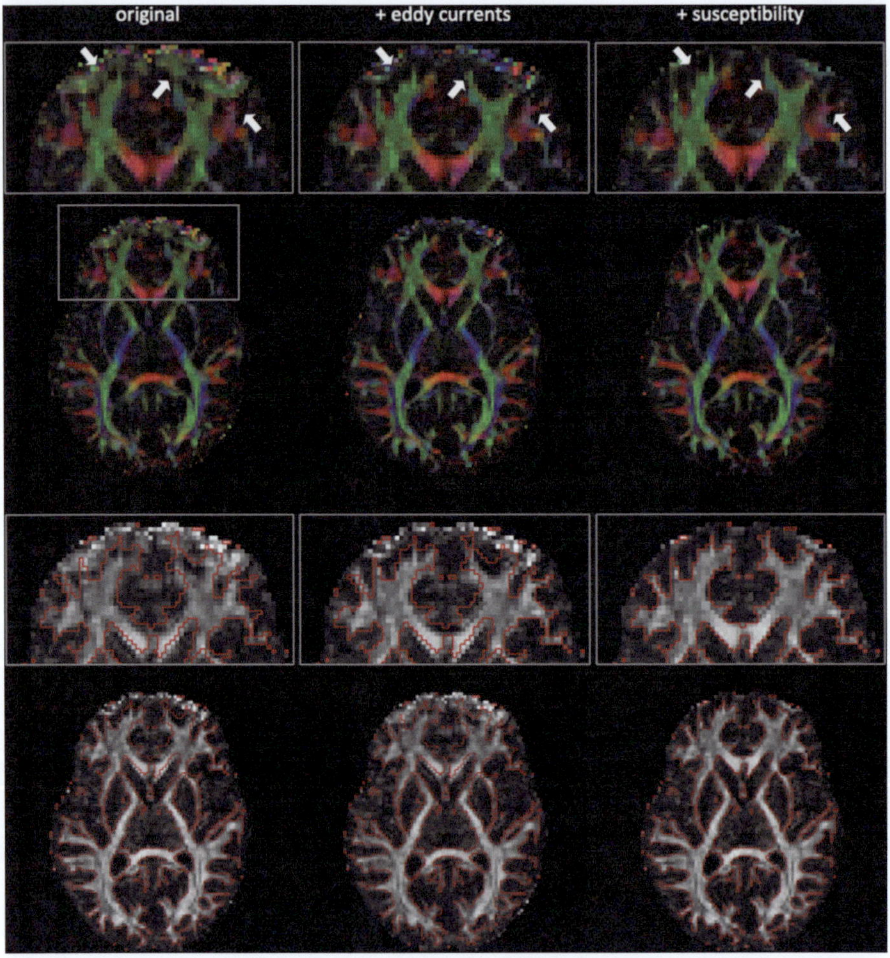

Fig. 21.7 The effects of eddy current and susceptibility distortions in DWI illustrated with greyscale and colour FA maps. The original data (left) suffers from both types of distortions, leading to a gross mismatch between the image and the subject's real anatomy and leading to false representations of local directional and microstructural information. Correcting for eddy current distortions (middle) improves the anatomical accuracy of the data and the robustness of the diffusion-derived metrics, but areas with high susceptibility distortions remain misaligned with the real anatomy (red contours). Only by including all corrections (right) does one obtain a dataset with high anatomical fidelity

diffusion gradients and cause an anatomical mismatch between volumes, above and beyond the one caused by motion. Therefore, they must be properly dealt with during data pre-processing. Importantly, eddy currents are only present in the diffusion-weighted images but not in the b0 images that are acquired without any diffusion-weighting gradients. So, a commonly used approach to correcting these distortions is registering the diffusion-weighted volumes to the b0 image, which acts as an undistorted reference.

21.2.3.3 Susceptibility Distortions

Different tissue types in and around the brain, such as grey and white matter, CSF, or air in the frontal sinuses, have different magnetic susceptibility properties. These tissue types are magnetised to different extents by the main magnetic field of the scanner, causing localised inhomogeneities in the effect of that magnetic field. This leads to localised and pronounced anatomical distortions that are common to EPI data, as in fMRI and DWI. The distortions cause a mismatch between the observed anatomy of the brain as imaged with EPI versus other MRI sequences that are robust to such effects, such as most structural T1-weighted or T2-weighted acquisitions. The most striking distortions occur in the orbitofrontal region due to the air-filled frontal, sphenoid, and ethmoid sinuses and in the inferior temporal region due to the air-filled mastoid cells of the mastoid process. That being said, susceptibility distortions usually affect the entire brain.

Susceptibility distortions are restricted to the phase-encoding axis of the EPI sequence, which is the axis along which each slice of the image is read. We can exploit this feature and acquire a few additional b0 images (i.e., without diffusion weighting) with the reverse phase encoding polarity, which will result in images with the same distortion magnitude but with the opposite distortion direction. The b0 images from the first acquisition and the reverse-phase b0 images can then be used to estimate the amount of distortion affecting each voxel and to correct the DWI data accordingly (Andersson and Sotiropoulos 2016; Andersson et al. 2003). As susceptibility distortions depend on the interaction of the patient's anatomy with the scanner's main magnetic field, it follows that the distortions will change when the patient substantially rotates their head. It is therefore crucial to ensure that minimal movement takes place during the short time window separating the main diffusion acquisition from that of the reverse b0 images. More recent methods have been proposed to correct for these types of distortions by combining information from the b0 images and a T1w image, doing away with the need for the reverse phase data (Schilling et al. 2020). See an example of these distortions in Fig. 21.7.

It is clear that this kind of correction is essential for DWI data, particularly when it is integrated with structural images for neurosurgical planning purposes. Regrettably, in many surgical departments, such corrections are often not implemented, primarily due to the absence of support for these corrective measures in the standard image processing software used in hospitals.

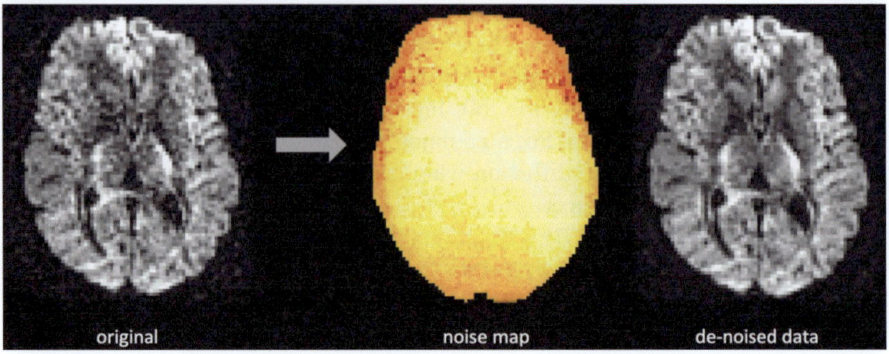

Fig. 21.8 Denoising diffusion data. The acquired signal can be noisy, and several approaches have been proposed to compensate for this problem during image pre-processing. These estimate the noise distribution from the data and yield a cleaner output that improves modelling

21.2.3.4 Thermal Noise

Noise is an inherent challenge in all signal measuring systems, ranging from cameras to MRI scanners. Thermal noise, for instance, is the signal detected when an MRI scanner operates without scanning any object—instead of capturing a completely black image with zeros in every voxel, the resulting image will display a somewhat random array of non-zero values. This occurs because various scanner components inadvertently add some signal to the image, even in the absence of a person in the scanner. DWI is particularly susceptible to this problem due to its dependence on signal attenuation for acquiring information. The issue becomes more pronounced with the use of higher *b*-values, a characteristic aspect of HARDI data. Separating noise from the actual signal in the acquired image makes image de-noising a complex task. However, several contemporary methods have been developed with impressive results, significantly enhancing both the overall image quality and the effectiveness of tractography (Fadnavis et al. 2020; Veraart et al. 2016a) (Fig. 21.8).

21.2.3.5 Gibbs Ringing Artefacts

A common image artefact observed in DWI at the interface between two regions with very different signal intensities is Gibbs ringing (Perrone et al. 2015; Veraart et al. 2016b). This artefact causes incorrect signal values around these border regions, e.g., it can appear as ‘ripples’ of rapidly fluctuating signal intensity, where the signal should otherwise be stable or smoothly varying.

21.2.3.6 A Robust Pre-processing Pipeline for Neurosurgery

Incorporating adjustments for all the previously mentioned artefacts is a crucial component of contemporary DWI pre-processing workflows. However, while some of these corrections are deemed essential, others are considered beneficial but not mandatory.

At a minimum, adjustments for inter-volume motion and eddy current distortions are imperative—data that has not undergone these corrections is generally deemed unsuitable for any form of analysis. In the case of intra-volume motion, if the employed tools do not incorporate signal dropout regeneration, this aspect needs to be addressed during model fitting. Similarly, for slice-wise motion correction, if it is impossible to correct it, then any volumes exhibiting obvious intra-volume motion artefacts that the pre-processing software cannot manage should be excluded.

Other steps can be added to the pipeline as needed depending on data quality. For instance, if the acquired images have a high SNR and a relatively small amount of Gibbs ringing, one could choose to omit these steps if faced with limited time or computational resources. However, it is recommended that a full correction be applied for all the possible sources of artefacts mentioned earlier (Ades-Aron et al. 2018).

Importantly, the pre-processing pipeline must follow a specific order. For instance, if denoising were to be performed, it must be the first step in the pipeline applied directly to the raw images. Otherwise, other steps can alter the noise information contained in the images, especially if these steps include data interpolation (this is the case for motion and geometric corrections). Therefore, a full pre-processing pipeline would follow this scheme, in order: (1) denoising; (2) Gibbs ringing correction; (3) motion and eddy current correction; (4) susceptibility correction. Depending on the tools used, step 4 is sometimes incorporated into step 3 to avoid multiple interpolations of the images. However, this is not always possible (see the section on susceptibility correction).

In essence, from foundational corrections to desired improvements and the incorporation of advanced refinements, the journey of acquiring and processing advanced tractography data involves a strategic orchestration of steps, each contributing to the fidelity and depth of insights derived from the data.

21.3 Clinical Integration of Advanced Tractography

As of the date of this chapter (January 2024), the most prominent clinically certified or approved systems for white matter tractography mainly revolve around DTI. In clinical practice, there are some 'push button' solutions that only address a subset of these correction steps or offer advanced modelling tools, such as those offered by the Medtronic StealthStation™ Surgical Navigation System or Imeka's Advanced Neuro Diagnostic Imaging. However, the more advanced tools for data pre-processing and modelling are, unfortunately, still confined to research use.

Tractography finds its primary applications in neurosurgery and epileptology, playing a pivotal role across three crucial stages: preoperative, intraoperative, and postoperative (Vergani et al. 2025; Sarubbo et al. 2025).

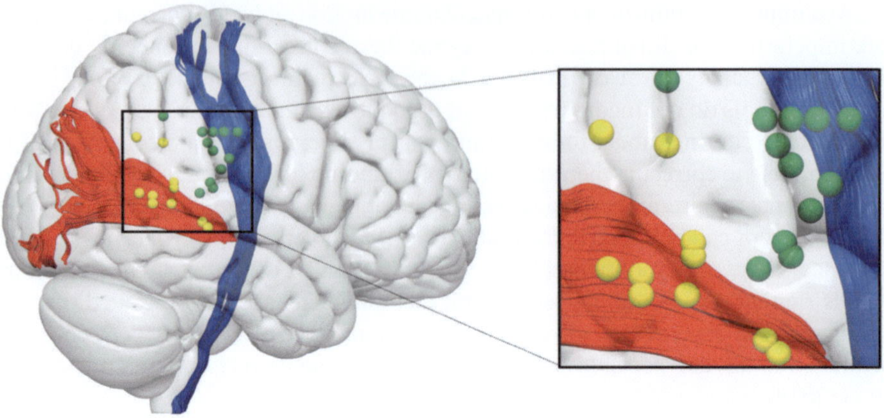

Fig. 21.9 Tractography and surgical mapping. Illustration depicting independent reconstruction tractography and mapping points, i.e., nMTS was not used as a seed (Chap. 10), showing positive subcortical responses associated with the optic radiations and corticospinal tract at the group level. The single-subject representation of the corticospinal tract is represented in blue, while the optic radiations are depicted in red. Yellow spheres denote positive subcortical responses for optic radiations induced by high-frequency bipolar stimulation, and green spheres signify positive subcortical responses for the corticospinal tract elicited through high-frequency monopolar stimulation at the group level. (Data based on Rajashekar et al. (2022))

21.3.1 Neurosurgical Preoperative Visualisation

Several instances within the clinical workflow benefit from tractography results in neurosurgery. Preoperatively, the 3D visual depiction of a network in relation to a lesion is a crucial tool for strategic surgical planning that aims to preserve functions postoperatively as well as possible. Such a priori 3D images have been shown to aid in planning entry zones to the tumour and, importantly, safe resection margins, especially when used by less experienced clinicians (Ille et al. 2021a). Advanced tractography is hence a great candidate for detailed preoperative anatomical guidance (Fig. 21.9). Nonetheless, high expertise in anatomy and tractography methods is needed to produce such tractograms. Moreover, the steps delineated earlier in this chapter are often deemed too time-consuming for the average clinical workflow, so the more time-efficient 'black box' solutions offered by built-in clinical software are often preferred. Future investigations will show if a more balanced consensus can be found.

In the preoperative phase, tractography's utility lies in meticulously assessing brain lesions. This assessment is then intricately compared with an in-depth evaluation of the patient's cognitive status, benchmarked against established norms. This comparative analysis is vital for the validation of tractography outcomes. A significant indicator of tractography's precision is when the imaged infiltration of white matter tracts corresponds with cognitive impairments, as predicted by the tracts' known functions (Forkel et al. 2022). Such congruence lends substantial credence to tractography's findings. These correlation studies

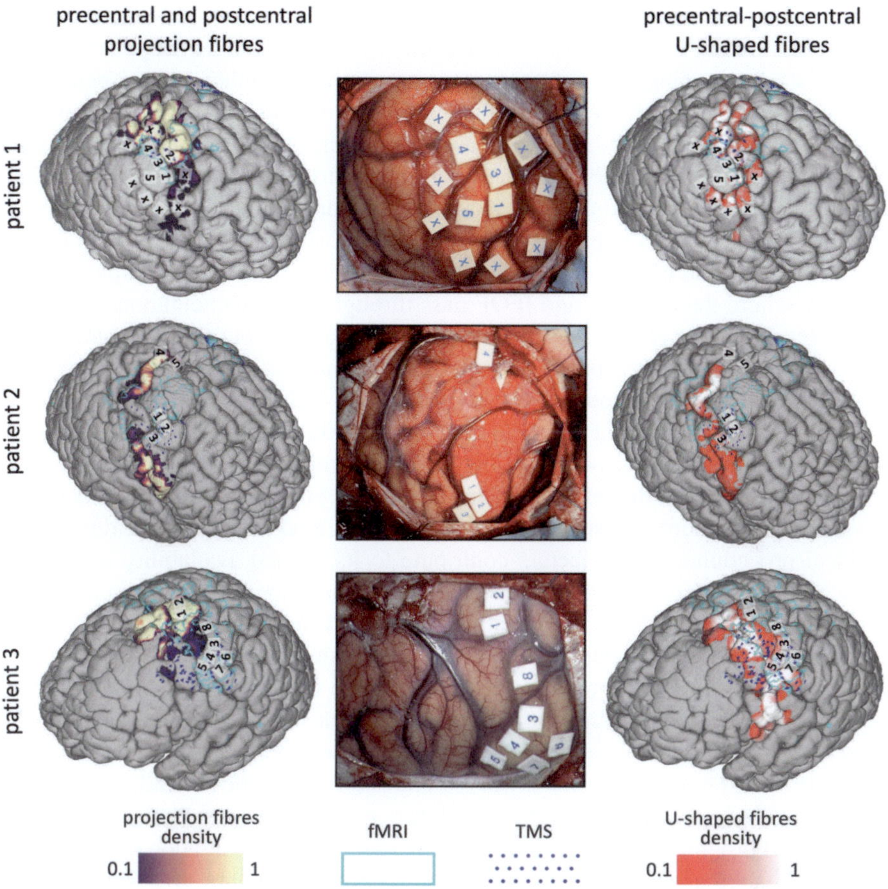

Fig. 21.10 Clinical examples of the integration of the structural and functional maps with intraoperative neuromonitoring data. The figure shows the relationship between functional nTMS and fMRI maps and tractography cortical projections in three tumour resection patients. Left and right panels display somatosensory projection and U-fibre terminations, respectively, with intraoperative cortical stimulation sites shown in the middle panel. (Figure taken from Beyh et al. (2024))

have been conducted with DTI-based tractography, highlighting its utility in clinical neuroimaging (Figs. 21.9 and 21.10).

21.3.2 Intraoperative Fibre Identification

Intraoperative direct electrical stimulation (DES) is the most direct way to interrogate the brain. However, caution has to be taken once the subcortical white matter is exposed: as already evident in preoperative tractograms, most areas are densely packed with multiple fibre bundles of interest. This makes it impossible to confirm that the exposed white matter during surgery presents solely and precisely the tract of interest. In the case of desired language preservation, for example, awake surgery cases often harbour lesions within or near the densely packed language network. It becomes a challenging task to disentangle tracts from each other without being able to verify them based on their respective cortical endpoints. This becomes particularly taxing for areas with tracts that have aligning trajectories, such as the arcuate fasciculus (AF) and the SLF along their dorsal trajectory, or the inferior longitudinal fasciculus (ILF) and inferior fronto-occipital fasciculus (IFOF) in the ventral stream (Weiller et al. 2021; Middlebrooks et al. 2017). An even greater risk of false identification lies in areas of tract crossroads, such as the temporo-parieto-occipital (TPO) junction where the AF, ILF, and IFOF pass through the same neighbourhood. Not only does this make identification and theoretical claims about the single tracts and their function ambitious, but it also renders this area a difficult and potentially unsafe entry zone for tumour removal (Tuncer et al. 2021).

Only a limited number of hospitals can conduct intraoperative tractography analysis, necessitating an intraoperative MRI (ioMRI) setup. This sophisticated arrangement permits the scanning of patients during surgery. However, implementing such a system is costly, demands specialised expertise, and tends to extend the duration of surgical procedures. These factors contribute to its limited availability and use in clinical settings. Even though there is a possibility for immediate intraoperative tractography through ioMRI, preoperatively prepared network analyses are often favoured and can be uploaded to the intraoperative neuronavigation system. However, it is relevant to note that preoperative segmentation of tracts is performed on data acquired in a supine position with an intact skull. That setup can be dramatically different from the surgical one where patients are placed in any position necessary to facilitate access to the tumour and where bone and tissue removal lead to tissue displacement (or brain shift). Once the craniotomy is completed and the resection commences and progresses, recurrent, reliable, and precise co-registration of imaging to the ever-changing neuroanatomy would be required. Swelling after opening and brain shift commonly cause anatomical distortions (Gerard et al. 2017), hence calling for continuously applied correction throughout the various stages of the operation. This can be done through software-based distortion corrections and elastic fusion techniques using new intraoperative images (Zhang et al. 2022; Ille et al. 2021b). Time restrictions often preclude the creation of a newly segmented tractogram based on the novel ioMR images, also due to the often lower resolution of ioMRI (Nimsky 2011; Maesawa et al. 2010). Hence, aligning the new ioMRI with the prepared tractogram and anatomy is recommended and shows good concordance with intraoperative testing of motor areas and function (Zhang et al. 2022).

21.3.3 Functional White Matter Mapping During Surgery

For motor functions, the stimulation of white matter tracts can be done with the patient asleep. For instance, motor areas and the CST can be localised by stimulating the brain and measuring corresponding MEPs on the patient's extremities (Bello et al. 2008; Zhu et al. 2012). For more complex functions, such as attention, facial recognition, and language, an awake and cooperative patient is required (De Witt Hamer et al. 2012; Ruis 2018). Under local sedation for the invasive procedure on the scalp and skull, the patient is woken up to full consciousness and asked to perform specific behavioural and cognitive tasks (Chap. 4). Stimulation of (sub-)cortical areas and/or white matter tracts using DES then disrupts local action potentials in those areas (Berger and Ojemann 1992; Duffau 2015; Vanderweyen et al. 2020). If a stimulated area or tract contributes to performing the assessed task, the patient's performance will be affected. The preoperatively segmented tractogram's validation is, then, based on the assumed function of the exposed white matter tract and its ability to disrupt the respective function when stimulated (e.g., Fig. 21.10). For example, after preoperative tracking of the IFOF with advanced tractography and successful registration of the 3D reconstruction to the patient's intraoperative neuroanatomy, it can be located during the resection of affected white matter. If stimulation of the IFOF disrupts the function being tested, then the IFOF's involvement in that function (whether directly or indirectly) is confirmed (Duffau et al. 2005). Studies report a high concordance between tractography and DES, where 82–97% of DES sites correspond to the expected locations derived from DTI-based preoperative tractography (Bello et al. 2007, 2008; Berman et al. 2007; Leclercq et al. 2010). Future investigations employing advanced tractography methods coupled with DES stimulation may improve this concordance by resolving some of the remaining discrepancies.

21.3.4 Postoperative Fibre Identification

In the postoperative stage, tractography is invaluable for analysing neuroanatomical alterations, such as changes in the lesion cavity, by contrasting these with preoperative imaging and tractography findings. This comparison is essential to discern the relationship between anatomical shifts and variations in patient behaviour. Behavioural evaluations are typically undertaken in the short term, within the initial three months following surgery, and over the longer term, at follow-up assessments at six, nine, and 12 months. These post-surgical evaluations, crucial for gauging the success and impact of neurosurgical procedures, have primarily been conducted utilising DTI-based tractography, emphasising its importance in ongoing patient monitoring and understanding surgical outcomes over time (Carrabba et al. 2022; Voets et al. 2021). As we highlighted earlier, though, such longitudinal monitoring would highly benefit from more robust and advanced methods that reveal information about each tract of interest based on fibre-specific metrics.

21.3.5 Integration of Tractography and Navigated TMS (nTMS)

Tractography results constantly face the challenge of balance between the over- and under-representation of fibres. Coupled with a functional mapping tool, subsets of tracts can be selected. This subset could be defined by functional involvement as delineated through fMRI activation (for a recent review, see Pasquini et al. 2023). A second option is stimulation mapping through DES or its preoperative counterpart, navigated transcranial magnetic stimulation (nTMS). nTMS pulses can be applied to the skull and reach the cortex (depth estimate: 20–30 mm) to interrogate an area's functionality (Chap. 10). However, while nTMS is the closest match to DES and is non-invasive in nature, it does not reach the depth of white matter tracts (the same applies to fMRI mapping). Nonetheless, cortical functional areas identified with nTMS can be used as regions of interest for tractography (see Chap. 10; Negwer et al. 2017; Sollmann et al. 2016). In line with high-resolution tractography results, nTMS-based cortical areas can be uploaded to intraoperative neuronavigation and co-registered to structural anatomical images of a patient to aid the resection workflow (Sollmann et al. 2018). Alternatively, as is shown in Fig. 21.10, the tractography and functional mapping can be done independently to reconstruct the full structural and functional network and then co-visualised to identify the relevant overlap. This approach results in a complete network map for the structural and functional networks and is therefore preferable over the partial reconstruction of functional seed-based tractography. As a future goal, projects may combine nTMS maps of cortical functionality with high-resolution advanced tractography to highlight tracts and identify the full structural and functional network implicated by the presence and/or resection of a brain tumour.

21.4 Concluding Remarks

This chapter has emphasised the transformative role of DWI and tractography in neurosurgery, offering unique insights into the brain's white matter tracts. While current clinical practice predominantly utilises DTI for its simplicity, there is significant potential for more advanced tractography methods, such as those based on the HARDI family of methods, particularly spherical deconvolution, to be integrated into standard clinical protocols.

The importance of tractography in preoperative, intraoperative, and postoperative stages highlights its impact on enhancing patient outcomes. Nevertheless, challenges such as the necessity for specialised sequences, preprocessing pipelines, and short analysis time remain barriers to its widespread adoption.

The future of tractography in clinical practice is promising, particularly with the potential integration with functional mapping tools like nTMS. This could lead to more precise and personalised neurosurgical interventions. As we continue to advance our understanding and technology, tractography stands on the brink of significantly revolutionising neurosurgical practice and improving long-term behavioural-cognitive outcomes. This chapter underscores the importance of these neuroimaging techniques' ongoing development and clinical integration.

Acknowledgements We thank the members of the CNSlab for discussions and feedback.

Funding This project received funding from the Dutch Research Council (NWO) Aspasia grant (SJF, 'Human individuality: phenotypes, cognition, and brain disorders') and the Donders Mohrmann Fellowship No. 2401515 (SJF, NEUROVARIABILITY).

References

Ades-Aron B, Veraart J, Kochunov P, McGuire S, Sherman P, Kellner E, Novikov DS, Fieremans E. Evaluation of the accuracy and precision of the diffusion parameter EStImation with Gibbs and NoisE removal pipeline. Neuroimage. 2018;183:532–43. https://doi.org/10.1016/J.NEUROIMAGE.2018.07.066.

Andersson JLR, Sotiropoulos SN. An integrated approach to correction for off-resonance effects and subject movement in diffusion MR imaging. Neuroimage. 2016;125:1063–78. https://doi.org/10.1016/j.neuroimage.2015.10.019.

Andersson JLR, Skare S, Ashburner J. How to correct susceptibility distortions in spin-echo echo-planar images: application to diffusion tensor imaging. Neuroimage. 2003;20(2):870–88. https://doi.org/10.1016/S1053-8119(03)00336-7.

Andersson JLR, Graham MS, Zsoldos E, Sotiropoulos SN. Incorporating outlier detection and replacement into a non-parametric framework for movement and distortion correction of diffusion MR images. Neuroimage. 2016;141:556–72. https://doi.org/10.1016/j.neuroimage.2016.06.058.

Andersson JLR, Graham MS, Drobnjak I, Zhang H, Filippini N, Bastiani M. Towards a comprehensive framework for movement and distortion correction of diffusion MR images: within volume movement. Neuroimage. 2017;152(March):450–66. https://doi.org/10.1016/j.neuroimage.2017.02.085.

Aydogan DB, Leemans A, Dubois J et al. DTI-based fiber tractography: why not?. Brain Struct Funct. 2025;230:84. https://doi.org/10.1007/s00429-025-02960-2.

Basser PJ, Mattiello J, LeBihan D. MR diffusion tensor spectroscopy and imaging. Biophysical Journal. 1994;66:259–67. https://doi.org/10.1016/S0006-3495(94)80775-1.

Behrens TEJ, Berg HJ, Jbabdi S, Rushworth MFS, Woolrich MW. Probabilistic diffusion tractography with multiple fibre orientations: what can we gain? Neuroimage. 2007;34(1):144–55. https://doi.org/10.1016/j.neuroimage.2006.09.018.

Bello L, Gallucci M, Fava M, Carrabba G, Giussani C, Acerbi F, Baratta P, et al. Intraoperative subcortical languagetract mapping guides surgical removalof gliomas involving speech areas. Neurosurgery. 2007;60(1):67–82.

Bello L, Gambini A, Castellano A, et al. Motor and language DTI fiber tracking combined with intraoperative subcortical mapping for surgical removal of gliomas. Neuroimage. 2008;39(1):369–82.

Berger MS, Ojemann GA. Intraoperative brain mapping techniques in neuro-oncology. Stereotact Funct Neurosurg. 1992;58(1–4):153–61.

Berman JI, Berger MS, Chung S, Nagarajan SS, Henry RG. Accuracy of diffusion tensor magnetic resonance imaging tractography assessed using intraoperative subcortical stimulation mapping and magnetic source imaging. J Neurosurg. 2007;107(3):488–94.

Beyh A, Howells H, Giampiccolo D, Cancemi D, Requejo FDS, Citro S, Keeble H, Lavrador JP, Bhangoo R, Ashkan K, Dell'Acqua F, Catani M, Vergani F. Connectivity defines the distinctive anatomy and function of the hand-knob area. Brain Commun. 2024;6(5):fcae261. https://doi.org/10.1093/braincomms/fcae261.

Carrabba G, Fiore G, Di Cristofori A, Bana C, Borellini L, Zarino B, Conte G, Triulzi F, Rocca A, Giussani C, Caroli M. Diffusion tensor imaging, intra-operative neurophysiological monitoring and small craniotomy: results in a consecutive series of 103 gliomas. Front Oncol. 2022;12:897147.

Chang LC, Jones DK, Pierpaoli C. RESTORE: robust estimation of tensors by outlier rejection. Magn Reson Med. 2005;53(5):1088–95. https://doi.org/10.1002/mrm.20426.

De Witt Hamer PC, Robles SG, Zwinderman AH, Duffau H, Berger MS. Impact of intraoperative stimulation brain mapping on glioma surgery outcome: a meta-analysis. J Clin Oncol. 2012;30(20):2559–65.

Dell'Acqua F, Catani M. Structural human brain networks: hot topics in diffusion tractography. Curr Opin Neurol. 2012;25(4):375–83. https://doi.org/10.1097/WCO.0b013e328355d544.

Dell'Acqua F, Tournier J-D. Modelling white matter with spherical deconvolution: how and why? NMR Biomed. 2019;32(4):e3945. https://doi.org/10.1002/nbm.3945.

Dell'Acqua F, Scifo P, Rizzo G, Catani M, Simmons A, Scotti G, Fazio F. A modified damped Richardson-Lucy algorithm to reduce isotropic background effects in spherical deconvolution. Neuroimage. 2010;49(2):1446–58. https://doi.org/10.1016/j.neuroimage.2009.09.033.

Dell'Acqua F, Simmons A, Williams SCR, Catani M. Can spherical deconvolution provide more information than fiber orientations? Hindrance modulated orientational anisotropy, a true-tract specific index to characterize white matter diffusion. Hum Brain Mapp. 2013;34(10):2464–83. https://doi.org/10.1002/hbm.22080.

Descoteaux M, Deriche R, Knösche TR, Anwander A. Deterministic and probabilistic tractography based on complex fibre orientation distributions. IEEE Trans Med Imaging. 2009;28(2):269–86. https://doi.org/10.1109/TMI.2008.2004424.

Duffau H. Stimulation mapping of white matter tracts to study brain functional connectivity. Nat Rev Neurol. 2015;11(5):255–65.

Duffau H, Gatignol P, Mandonnet E, Peruzzi P, Tzourio-Mazoyer N, Capelle L. New insights into the anatomo-functional connectivity of the semantic system: a study using cortico-subcortical electrostimulations. Brain. 2005;128(4):797–810.

Fadnavis S, Batson J, Garyfallidis E. Patch2Self: denoising diffusion MRI with self-supervised learning. In: Advances in neural information processing systems, 2020-Decem (NeurIPS); 2020.

Forkel S, Catani M. Structural neuroimaging. In: de Groot AMB, Hagoort P, editors. Research methods in psycholinguistics and the neurobiology of language: a practical guide. Wiley; 2017. p. 371. https://www.wiley.com/en-us/Research+Methods+in+Psycholinguistics+and+the+Neurobiology+of+Language%3A+A+Practical+Guide-p-9781119109846.

Forkel SJ, Thiebaut De Schotten M, Dell'Acqua F, Kalra L, Murphy DGM, Williams SCR, Catani M. Anatomical predictors of aphasia recovery: a tractography study of bilateral perisylvian language networks. Brain. 2014;137(7):2027–39. https://doi.org/10.1093/brain/awu113.

Forkel SJ, Labache L, Nachev P, Thiebaut de Schotten M, Hesling I. Stroke disconnectome decodes reading networks. Brain Struct Funct. 2022;227:2897–908. https://doi.org/10.1007/S00429-022-02575-X/FIGURES/3.

Forkel SJ, Bortolami C, Dulyan L, Barrett RLC, Beyh A. Dissecting white matter pathways: aneuroanatomical approach. In Dell'Acqua F, Descoteaux M, Leemans A (Eds.), Handbook of Diffusion MR Tractography. Academic Press; 2025. pp. 397–421. https://doi.org/10.1016/B978-0-12-818894-1.00012-4.

Gerard IJ, Kersten-Oertel M, Petrecca K, Sirhan D, Hall JA, Collins DL. Brain shift in neuronavigation of brain tumors: a review. Med Image Anal. 2017;35:403–20.

Ille S, Schwendner M, Zhang W, Schroeder A, Meyer B, Krieg SM. Tractography for subcortical resection of gliomas is highly accurate for motor and language function: ioMRI-based elastic fusion disproves the severity of brain shift. Cancers (Basel). 2021a;13(8):1787.

Ille S, Ohlerth AK, Colle D, Colle H, Dragoy O, Goodden J, Robe P, Rofes A, Mandonnet E, Robert E, Satoer D. Augmented reality for the virtual dissection of white matter pathways. Acta Neurochir. 2021b;163:895–903.

Jones DK, editor. Diffusion MRI: theory, methods, and applications. Oxford Academic; 2010. https://doi.org/10.1093/MED/9780195369779.001.0001.

Le Bihan D. Apparent diffusion coefficient and beyond : what diffusion MR imaging can tell us about tissue structure. Radiology. 2013;268(2):318–22. https://doi.org/10.1148/radiol.13130420.

Le Bihan D, Johansen-Berg H. Diffusion MRI at 25: exploring brain tissue structure and function. Neuroimage. 2012;61(2):324–41. https://doi.org/10.1016/J.NEUROIMAGE.2011.11.006.

Leclercq D, Duffau H, Delmaire C, Capelle L, Gatignol P, Ducros M, Chiras J, Lehéricy S. Comparison of diffusion tensor imaging tractography of language tracts and intraoperative subcortical stimulations. J Neurosurg. 2010;112(3):503–11.

Maesawa S, Fujii M, Nakahara N, Watanabe T, Wakabayashi T, Yoshida J. Intraoperative tractography and motor evoked potential (MEP) monitoring in surgery for gliomas around the corticospinal tract. World Neurosurg. 2010;74(1):153–61.

Middlebrooks EH, Yagmurlu K, Szaflarski JP, Rahman M, Bozkurt B. A contemporary framework of language processing in the human brain in the context of preoperative and intraoperative language mapping. Neuroradiology. 2017;59(1):69–87. https://doi.org/10.1007/s00234-016-1772-0.

Mirchandani AS, Beyh A, Lavrador JP, Howells H, Dell'Acqua F, Vergani F. Altered corticospinal microstructure and motor cortex excitability in gliomas: an advanced tractography and transcranial magnetic stimulation study. J Neurosurg. 2021;134(5):1368–76. https://doi.org/10.3171/2020.2.JNS192994.

Negwer C, Ille S, Hauck T, Sollmann N, Maurer S, Kirschke JS, Ringel F, Meyer B, Krieg SM. Visualization of subcortical language pathways by diffusion tensor imaging fiber tracking based on rTMS language mapping. Brain Imaging Behav. 2017;11:899–914.

Nimsky C. Intraoperative acquisition of fMRI and DTI. Neurosurg Clin N Am. 2011;22(2):269–77, ix.

Pasquini L, Peck KK, Jenabi M, Holodny A. Functional MRI in neuro-oncology: state of the art and future directions. Radiology. 2023;308(3):e222028.

Perrone D, Aelterman J, Pižurica A, Jeurissen B, Philips W, Leemans A. The effect of Gibbs ringing artifacts on measures derived from diffusion MRI. NeuroImage. 2015;120:441–55. https://doi.org/10.1016/J.NEUROIMAGE.2015.06.068.

Rajashekar D, Lavrador JP, Ghimire P, Keeble H, Harris L, Pereira N, Patel S, Beyh A, Gullan R, Ashkan K, Bhangoo R, Vergani F. Simultaneous motor and visual intraoperative neuromonitoring in asleep parietal lobe surgery: dual strip technique. J Pers Med. 2022;12(9):1478. https://doi.org/10.3390/JPM12091478.

Rheault F, Mayberg H, Thiebaut de Schotten M, Roebroeck A, Forkel SJ. The scientific value of tractography: accuracy vs Usefulness. BSAF. 2025;29:230(4):59. https://doi.org/10.1007/s00429-025-02921-9.

Ruis C. Monitoring cognition during awake brain surgery in adults: a systematic review. J Clin Exp Neuropsychol. 2018;40(10):1081–104.

Schilling KG, Nath V, Hansen C, Parvathaneni P, Blaber J, Gao Y, Neher P, Aydogan DB, Shi Y, Ocampo-Pineda M, Schiavi S, Daducci A, Girard G, Barakovic M, Rafael-Patino J, Romascano D, Rensonnet G, Pizzolato M, Bates A, et al. Limits to anatomical accuracy of diffusion tractography using modern approaches. Neuroimage. 2019;185:1–11. https://doi.org/10.1016/J.NEUROIMAGE.2018.10.029.

Schilling KG, Blaber J, Hansen C, Cai L, Rogers B, Anderson AW, Smith S, Kanakaraj P, Rex T, Resnick SM, Shafer AT, Cutting LE, Woodward N, Zald D, Landman BA. Distortion correction of diffusion weighted MRI without reverse phase-encoding scans or field-maps. PLoS One. 2020;15(7):e0236418. https://doi.org/10.1371/JOURNAL.PONE.0236418.

Sollmann N, Negwer C, Ille S, Maurer S, Hauck T, Kirschke JS, Ringel F, Meyer B, Krieg SM. Feasibility of nTMS-based DTI fiber tracking of language pathways in neurosurgical patients using a fractional anisotropy threshold. J Neurosci Methods. 2016;267:45–54.

Sollmann N, Kelm A, Ille S, Schröder A, Zimmer C, Ringel F, Meyer B, Krieg SM. Setup presentation and clinical outcome analysis of treating highly language-eloquent gliomas via preoperative navigated transcranial magnetic stimulation and tractography. Neurosurg Focus. 2018;44(6):E2.

Stejskal EO, Tanner JE. Spin diffusion measurements: spin echoes in the presence of a time-dependent field gradient. J Chem Phys. 1965;42(1):288–92. https://doi.org/10.1063/1.1695690.

Thiebaut de Schotten M, Dell'Acqua F, Forkel SJ, Simmons A, Vergani F, Murphy DGM, Catani M. A lateralized brain network for visuospatial attention. Nat Neurosci. 2011;14(10):1245–6. https://doi.org/10.1038/nn.2905.

Tournier JD, Mori S, Leemans A. Diffusion tensor imaging and beyond. Magn Reson Med. 2011;65(6):1532–56. https://doi.org/10.1002/mrm.22924.

Tournier JD, Calamante F, Connelly A. Robust determination of the fibre orientation distribution in diffusion MRI: Non-negativity constrained super-resolved spherical deconvolution. NeuroImage. 2007;35(4):1459–72. https://doi.org/10.1016/j.neuroimage.2007.02.016.

Tuch DS. Q-ball imaging. Magn Reson Med. 2004;52(6):1358–72. https://doi.org/10.1002/MRM.20279.
Tuncer MS, Salvati LF, Grittner U, Hardt J, Schilling R, Bährend I, Silva LL, Fekonja LS, Faust K, Vajkoczy P, Rosenstock T, Picht T. Towards a tractography-based risk stratification model for language area associated gliomas. Neuroimage Clin. 2021;29:102541. https://doi.org/10.1016/J.NICL.2020.102541.
Vanderweyen DC, Theaud G, Sidhu J, Rheault F, Sarubbo S, Descoteaux M, Fortin D. The role of diffusion tractography in refining glial tumor resection. Brain Struct Funct. 2020;225:1413–36.
Veraart J, Fieremans E, Novikov DS. Diffusion MRI noise mapping using random matrix theory. Magn Reson Med. 2016a;76(5):1582–93. https://doi.org/10.1002/mrm.26059.
Veraart J, Fieremans E, Jelescu IO, Knoll F, Novikov DS. Gibbs ringing in diffusion MRI. Magn Reson Med. 2016b;76(1):301–14. https://doi.org/10.1002/MRM.25866.
Vergani F, Sarubbo S, Yang JYM. DEBATE: is tractography accurate enough to be used to inform surgery? Brain Struct Funct. 2025;230:81. https://doi.org/10.1007/s00429-025-02948-y.
Sarubbo S, Vergani F, Yang JYM. Tractography in brain tumor surgery: current clinical impact and future challenges. Brain Struct Funct. 2025;230:93. https://doi.org/10.1007/s00429-025-02956-y.
Voets NL, Pretorius P, Birch MD, Apostolopoulos V, Stacey R, Plaha P. Diffusion tractography for awake craniotomy: accuracy and factors affecting specificity. J Neuro-Oncol. 2021;153(3):547–57.
Wedeen VJ, Hagmann P, Tseng WYI, Reese TG, Weisskoff RM. Mapping complex tissue architecture with diffusion spectrum magnetic resonance imaging. Magn Reson Med. 2005;54(6):1377–86. https://doi.org/10.1002/MRM.20642.
Weiller C, Reisert M, Peto I, Hennig J, Makris N, Petrides M, Rijntjes M, Egger K. The ventral pathway of the human brain: a continuous association tract system. Neuroimage. 2021;234:117977.
Yendiki A, Aggarwal M, Axer M, Howard AFD, van Walsum AM v C, Haber SN. Post mortem mapping of connectional anatomy for the validation of diffusion MRI. Neuroimage. 2022;256:119146. https://doi.org/10.1016/J.NEUROIMAGE.2022.119146.
Zhang W, Ille S, Schwendner M, Wiestler B, Meyer B, Krieg SM. Tracking motor and language eloquent white matter pathways with intraoperative fiber tracking versus preoperative tractography adjusted by intraoperative MRI–based elastic fusion. J Neurosurg. 2022;137(4):1114–23.
Zhu FP, Wu JS, Song YY, Yao CJ, Zhuang DX, Xu G, Tang WJ, Qin ZY, Mao Y, Zhou LF. Clinical application of motor pathway mapping using diffusion tensor imaging tractography and intraoperative direct subcortical stimulation in cerebral glioma surgery: a prospective cohort study. Neurosurgery. 2012;71(6):1170–84.

Online Tutorials

Lecture on methods to study white matter (https://youtu.be/oX2Pgti8olQ).
Lecture on tractography methods (https://youtu.be/DpCb7uVivAk).
Tutorial for the arcuate fasciculus (https://youtu.be/ypUSTD-5p1Q).
Tutorial for the SLF dissections (https://youtu.be/CojNe42RUjE).
Tutorial on tractography data visualisation (https://youtu.be/L4KPpm2e4ck).

Software and Data Used in This Chapter

http://trackvis.org/
http://www.bcblab.com/BCB/Opendata.html
https://fsl.fmrib.ox.ac.uk/fsl/fslwiki

https://www.exploredti.com/
https://www.humanconnectome.org/
https://www.mathworks.com/products/matlab.html
https://www.mr-startrack.com/
https://www.nitrc.org/projects/surfice/
https://www.stephanieforkel.com/opendata

Analysis and Creation of Structural Connectomes in Brain Tumor Patients: Methodological Challenges and Potential Solutions

22

Haosu Zhang and Lucius S. Fekonja

22.1 Introduction

22.1.1 Definition and Relevance of Connectome in Neuroscience and Neurosurgery

The term "connectome" refers to the comprehensive map of the structural and functional connections of the nervous system. A connectome is a comprehensive description of the brain's structural and functional connectivity (Sporns et al. 2005; Sporns 2011; Abd Hamid et al. 2011). Over the past decade, the concept of the connectome has become increasingly central in neuroscience, as it provides insights into the brain's complex organizational structures and helps link structural features with functional outcomes. For neurosurgeons, understanding the connectome is of utmost importance, as surgical interventions often require precise targeting to preserve essential structural and functional networks while addressing pathological regions. Knowledge of the connectome aids in the identification of critical pathways, minimizing postoperative deficits, and maximizing surgical outcomes. Thus, connectomes are particularly relevant to tumor neurosurgery because brain tumors can disrupt the brain's network structure and function. By mapping the connectome of a brain tumor patient, neurosurgeons may better understand how the tumor is

H. Zhang
Department of Neurosurgery, Heidelberg University Hospital, Heidelberg, Germany
e-mail: Haosu.Zhang@med.uni-heidelberg.de

L. S. Fekonja (✉)
Department of Neurosurgery, Charité—University Hospital, Berlin, Germany

Cluster of Excellence: "Matters of Activity. Image Space Material", Humboldt University, Berlin, Germany
e-mail: lucius.fekonja@charite.de

S. M. Krieg, T. Picht (eds.), *Navigated Transcranial Magnetic Stimulation in Neurosurgery*, https://doi.org/10.1007/978-3-031-97155-6_22

affecting the patient's brain and possibly apply this knowledge to their treatment strategy (Andres et al. 2011; Tuncer et al. 2022; Fekonja et al. 2022).

Creating a connectome requires a synthesis of neuroimaging and analysis methods. Diffusion-weighted magnetic resonance imaging (dMRI) serves as the basic modality for obtaining structural connectivity by modeling the movements of water molecules approximating axonal pathways (Hagmann et al. 2010; Jones 2003; Jeurissen et al. 2019). On the other hand, functional magnetic resonance imaging (fMRI) is employed to delineate functional connectivity by measuring blood oxygen level-dependent signals, highlighting brain regions with synchronized neural activity (Andres et al. 2011; Smith et al. 2013a).

In the context of brain tumor patients, several tools and software packages that were developed essentially for healthy data can also be used. These toolboxes enable the structural connectome construction on the basis of dMRI data to construct fiber orientation distributions (FODs) and enable tractography, allowing detailed visualization and analysis of white matter (WM) tracts or whole brain tractograms, even in the presence of pathological distortions due to tumors (Fekonja et al. 2022, 2020; Reisch et al. 2022; Zhang et al. 2022a).

In addition to the structural connectome derived from dMRI, functional connectomics derived from fMRI enable the identification of connected regions based on spatial signal fluctuations and facilitate the mapping of functional connectivity between these areas (Van Essen et al. 2012). These functional networks represent patterns of brain activity based on blood oxygenation levels between regions.

22.1.2 Overview of Connectome Creation and Analysis in Brain Tumor Patients

The creation of a connectome relies on the fusion of advanced imaging and analysis methods; cf. Figure 22.1. dMRI is commonly used to detect structural connectivity. dMRI non-invasively allows measuring the rate of tissue water diffusion (Tournier et al. 2011). On the other hand, rsfMRI is used to describe functional connectivity by measuring signals that depend on blood oxygenation and reveal brain regions with synchronized neuronal activity (van de Ven et al. 2004).

Currently, there exist software tools specifically developed for the processing and analysis of structural MRI data (Henschel et al. 2020; Fischl 2012). They facilitate the reconstruction of cortical surfaces, subcortical segmentation, and morphometric analysis. In brain tumor patients, where anatomical distortions are common, the robustness of these tools becomes particularly crucial for obtaining accurate representations of cortical structures (Fekonja et al. 2022).

For functional data derived from fMRI scans, fMRIprep stands out as a preprocessing pipeline designed to ensure data quality and consistency (Esteban et al. 2019). It standardizes the processing of fMRI data by integrating best practices, minimizing user input, and maximizing reproducibility, which is essential for ensuring the reliability of functional connectivity analyses, especially when evaluating the influence of tumors on neural networks.

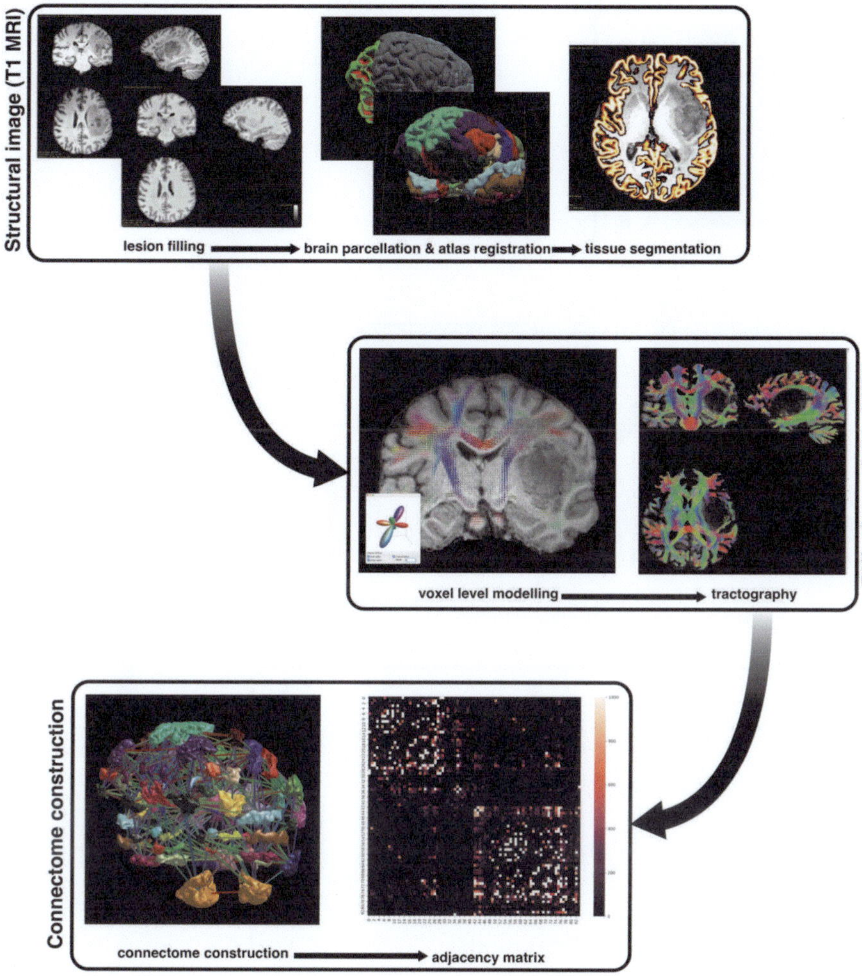

Fig. 22.1 Connectome construction workflow. The connectome construction includes various neuroimaging methods. T1 MRI data is used for brain tissue segmentation and parcellation. Diffusion-weighted magnetic resonance imaging (dMRI) data is used for tractography based on the source data, which contains the voxel-level modeling of the fiber populations. Furthermore, the tractography algorithm is guided by tissue segmentation to apply biologically realistic priors to the streamlines generated. Finally, the connectome matrix is constructed based on a whole-brain streamline file and a node parcellation image, representing the structural connections within the brain

Together, these methods and tools form a cohesive framework that allows us to construct and analyze the connectomes of brain tumor patients.

By examining altered connectivity patterns, researchers can identify networks that may be compromised by tumor growth and possibly related to functional deficits (Tuncer et al. 2022; Fekonja et al. 2022); cf. Figure 22.2. Additionally,

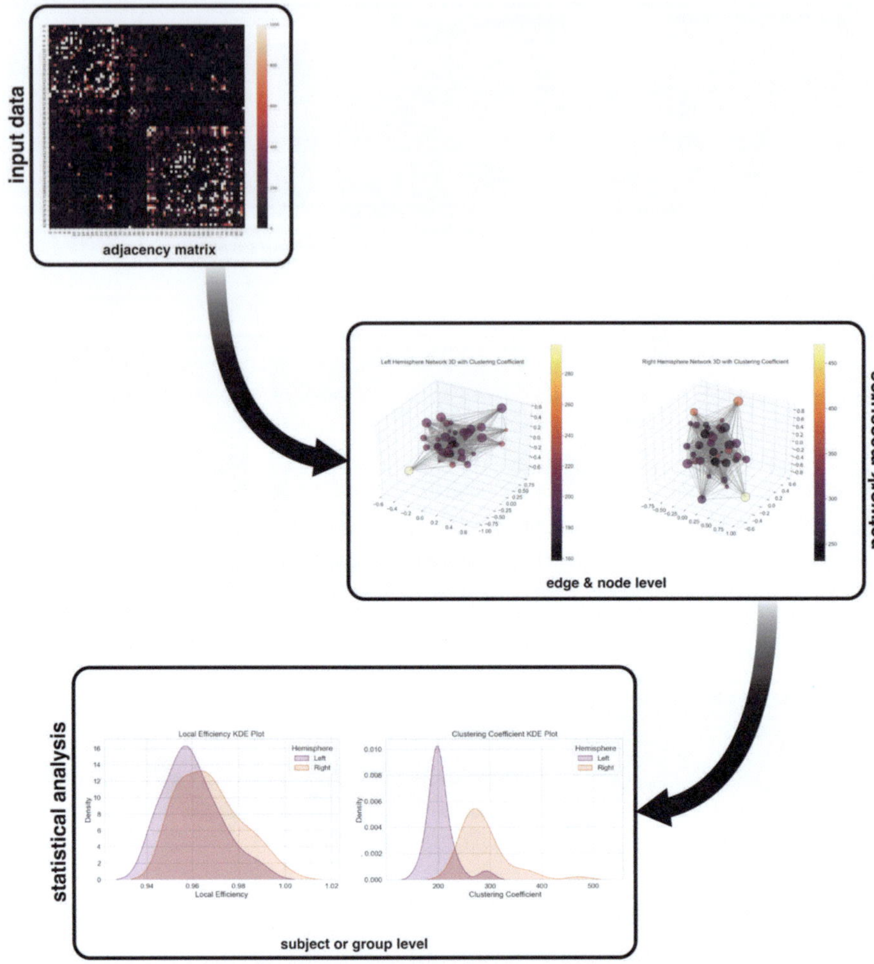

Fig. 22.2 Connectome analysis. The adjacency matrix, representing the connectome, can, e.g., be imported into a Python environment for in-depth analysis. Using graph theory allows us to assess the impact of the neural network and how the network adapts to tumor growth. A variety of network-specific analysis methods offer further insights into the unique patients' connectomes, aiding in the understanding of tumor-induced alterations in neural connectivity

connectome-based neuro-function localization aids surgeons in planning tumor resections (Henderson et al. 2020). The combination of preoperative mapping of eloquent regions by nTMS and their connections allows for preserving vital functions and can minimize postsurgical complications (Abd Hamid et al. 2011; Zhang et al. 2022a,b; Wang et al. 2021; Tuncer et al. 2021; Engelhardt et al. 2023).

Connectomes can be subjected to various analytical approaches to uncover meaningful insights to interpret the organization of cerebral function (Barabasi et al. 2023; Bassett and Sporns 2017). Graph theory, a prominent method in connectome analysis, quantifies the brain's network properties by the connectome's nodes and edges. This facilitates the characterization of network topology. Connectome alterations due to brain tumors can be identified by comparing healthy controls or tracking changes over time (Fekonja et al. 2022; Wei et al. 2023). These parameters from graph theory are now widely accepted and applied (Lawrence et al. 2014), for example, to study surgical processes in the network by assessing the robustness and efficiency of spatial and temporal signal transmission (Lawrence et al. 2014) or to model the patient's brain before and after surgery (Aerts et al. 2020, 2018).

Despite its potential, connectomics in relation to brain tumor patients is still partially impacted by different factors, such as variability in pathological diagnoses (Nenonen et al. 2015), patient heterogeneity (Nenonen et al. 2015; Veenith et al. 2013), and dynamic functional reorganization (Pallast et al. 2020). Complex analysis techniques, including multi-modal acquisitions and machine learning methods, pose the potential to possibly overcome these limitations, enhancing the accuracy of connectome-based assessments (Abd Hamid et al. 2011; Shams et al. 2022).

22.1.3 Challenges and Advancements in Methodology for Connectome Creation

The creation of a connectome in the presence of a brain tumor introduces specific challenges, largely because tumors distort and modify the inherent anatomical structure and imaging properties of the patient's brain:

1. Tumor-related difficulties in brain parcellation: The process of brain parcellation involves segmenting different brain regions, especially the gray matter (GM) and subcortical gray matter (SGM), based on structural MRI scans like a T1 image or, in certain cases, on T2-weighted ontrast. However, tumors not only cause anatomical shifts due to their mass effect but also might lead to edema, further disturbing the individual anatomy. Additionally, tumors can drastically alter the MRI signal intensities of involved voxels, hindering the segmentation algorithms' ability to distinguish between different tissue types. Consequently, the parcellation of GM and SGM can be compromised, leading to inaccurate or incomplete connectome representations (Jeurissen et al. 2019).
2. Overcoming parcellation challenges with lesion filling: Given the challenges introduced by tumors, lesion-filling methods have been developed to replace the tumor's anatomical space with healthy anatomy for better parcellation results (Radwan et al. 2021). Introducing artificial healthy anatomy provides a more uniform tissue representation, thereby facilitating tools like FreeSurfer in achieving error-free parcellations (Fischl 2012). Although this introduces non-native anatomy, the aim is to ensure that the connectome can be constructed without significant errors. Streamlines from the tractogram can then be accurately assigned to the appropriate brain parcels or nodes, preserving the integrity of the connectome.

3. Simple atlas registration as an alternative: A more rudimentary approach to overcome the tumor-induced parcellation challenge is to register an existing atlas directly to the patient's MRI. While this method is simpler and bypasses the direct challenges posed by tumors, it lacks the granularity and refinement offered, e.g., by tools like FreeSurfer (Fischl 2012). FreeSurfer constructs vertices, capturing the individual gyrification patterns of the cerebral cortex, and reconstructs a highly detailed model of the patient's brain (Fischl 2012). Thus, it often achieves a more accurate depiction of the cortical and subcortical structures. In contrast, direct atlas registration, being a more generalized approach, might not encapsulate the patient-specific nuances, thus potentially compromising the resolution and precision of the resultant connectome.
4. In functional connectomics, the challenge of inferring meaningful connectivity from resting-state fMRI data becomes particularly pronounced when applied to tumor-related analysis. While providing valuable insights, fMRI networks inherently exhibit susceptibility to generating spurious correlations because of the interfered global supply of blood and oxygen by tumors (Guyon et al. 2021; Waqar et al. 2021). The tumor-related alterations prompt the exploration of new techniques, such as dynamic functional connectivity analysis (Fu et al. 2019). This approach employs moving frames and steps to identify consistent network performance while excluding unstable alterations induced by the tumor microenvironment.
5. Advancing from DTI to CSD: One of the most notable advancements in the methodology for connectome creation has been the transition from Diffusion Tensor Imaging (DTI) to Constrained Spherical Deconvolution (CSD). While DTI has historically been the mainstay for analyzing diffusion MRI data, it models each voxel as containing a single fiber orientation, which can be oversimplified, as up to 90% of WM voxels include multiple fiber populations (Farquharson et al. 2013). CSD, on the other hand, can resolve multiple fiber orientations within a single voxel. This ability to identify complex intra-voxel structures translates into a more accurate representation of neural pathways.

The presence of tumors can distort the local fiber architecture, making it crucial to employ methodologies that can accurately resolve multiple fiber populations in proximity to or infiltrated by the tumor or edema. In such cases, DTI's inherent limitation in resolving crossing fibers may lead to oversimplified or even erroneous representations of tracts, potentially misguiding clinical decisions or therapeutic interventions. Conversely, CSD's ability to delineate multiple fiber orientations within each voxel offers a more precise depiction of neural pathways, even in the challenging milieu introduced by tumorous growths (Fekonja et al. 2022, 2020; Farquharson et al. 2013; Shams et al. 2023).

Tractography based on DTI often yields information that is consistently unreliable and can be misleading in a clinical context. By contrast, higher-order tractography models, utilizing identical diffusion-weighted datasets, evidently showcase fiber pathways with greater accuracy, offering enhanced estimations of safety margins that could be advantageous in neurosurgery (Farquharson et al. 2013).

22.2 Impact of MRI Sequences on Connectome Creation

22.2.1 Imbalance Between Clinically Available Data Quality and the Requirements of Advanced Methodology

There exists a noticeable disparity between the quality of dMRI data routinely available in clinical settings and the data quality prerequisites of state-of-the-art connectome construction methodologies.

Many clinical environments primarily utilize single-shell dMRI sequences due to their time efficiency and simplicity. However, the advanced algorithms and tools designed for rigorous connectome analysis often evolve with the technology and optimize their state-of-the-art models, usually adapted to high data quality, enabling the extraction of superior data.

This imbalance presents a challenge. The utilization of single-shell data with advanced methodologies can lead to suboptimal or even inaccurate connectome representations. On the other hand, acquiring multi-shell data in clinical scenarios might be constrained by factors such as longer scanning times, patient comfort, and available MRI hardware capabilities. Bridging this gap requires a concerted effort from both the clinical and research communities to prioritize the acquisition of high-quality dMRI data that aligns with the demands of sophisticated connectome analysis tools.

Furthermore, there's often a preference for MRI scans with larger voxel sizes in clinical settings. This means that each voxel represents a bigger volume of brain tissue, thereby encompassing more anatomical details within a single voxel. Such a choice often stems from economic and infrastructural constraints. In contrast, experimental datasets used in basic neuroscience research typically employ smaller voxel sizes, capturing more granular and detailed information about the underlying anatomy. This higher resolution is advantageous for advanced methodologies, offering a more intricate representation of neural structures and pathways. However, the disparity between the commonly larger voxel sizes in clinical settings and the high-resolution demands of cutting-edge acquisitions introduces challenges in achieving the most accurate and comprehensive neural mapping.

22.2.2 Overview of Different MRI Sequences: Single-Shell Versus Multi-shell

MRI sequences play a fundamental role in determining the quality and specificity of the data collected, which subsequently affects the integrity of the connectome generated. Regarding dMRI, two primary sequences are predominantly used: single-shell and multi-shell.

Single-shell dMRI sequences capture diffusion data at a single *b*-value (e.g., $b = 1000$ s/mm^2), which is a measure of the diffusion-sensitizing gradient strength. This type of sequence is simpler and quicker to acquire, making it frequently used in clinical settings. However, its simplicity can be a limitation when trying to unravel the fiber configurations and crossings present in the human brain.

Contrarily, multi-shell dMRI sequences gather data at multiple *b*-values, providing a richer depiction of the diffusion profiles across various tissue environments. The advantage of multi-shell sequences lies in their ability to more accurately discern complex fiber architectures and resolve tissue-dependent signals, which is vital for a comprehensive and accurate connectome representation. Such sequences are particularly useful in investigating microstructural changes and heterogeneity in neural tissues, which may be prevalent in brain tumor patients (Tournier et al. 2004, 2007; Jeurissen et al. 2014). However, it's important to note that multi-shell sequences have some trade-offs. Acquiring data at multiple diffusion weightings increases scan time (Jha et al. 2022), which can be a limitation, especially in clinical settings. Additionally, data analysis from multi-shell sequences can be more complex and require specialized software and expertise.

22.2.3 The Role of Angular Resolution

Angular resolution in dMRI refers to the number of gradient directions sampled per *b*-value during data acquisition. A higher angular resolution, for instance, acquiring data in 45 or more directions per *b*-value, offers a more comprehensive sampling of the diffusion profile within each voxel. This richness of information is crucial when delineating neural pathways, especially in areas where fibers cross, converge, or diverge. In the context of brain tumor patients, achieving a high angular resolution becomes even more pivotal. Due to their disruptive nature, tumors can further impair the structural clarity of already complex neural pathways. Therefore, employing dMRI sequences with high angular resolution is essential to ensure that the complexities and nuances of the surrounding neural architecture are accurately captured. In essence, a high angular resolution compensates for the structural ambiguities introduced by the tumor and aids in creating a detailed and reliable connectome representation, ensuring that tumor-related distortions do not overshadow the underlying neural connectivity patterns (Jeurissen et al. 2019; Hutter et al. 2018).

22.3 Challenges in Tissue Segmentation Due to Anatomical Shifts

22.3.1 Introduction to Tissue Segmentation in the Context of Connectome Creation

Tissue segmentation, especially during connectome creation, refers to the process of demarcating different anatomical structures in the brain based on MRI scans, primarily the T1-weighted image. A pivotal approach used in connectome construction is the anatomically constrained tractography (ACT) framework (Smith et al. 2012).

ACT relies on a segmented structural anatomy image, derived from the T1 scan, to guide and constrain the tractography process (Smith et al. 2012). The main goal is to ensure the biological plausibility of the streamlines generated, thereby rejecting those that don't adhere to known anatomical constraints. Subsequently, the SIFT2 (spherical-deconvolution informed filtering of tractograms 2) framework utilizes this anatomically informed tractography to further refine and filter the streamlines, ensuring a robust and accurate connectome representation (Smith et al., 2013b, 2015).

22.3.2 Examination of Problems Induced by Tumor-Related Changes in Intensities

Brain tumors introduce significant disruptions to the native architecture and imaging properties of the brain. One of the pronounced challenges arises in the tissue segmentation phase, particularly when generating the necessary structural anatomy image from the T1 scan. Tumorous regions can alter the inherent MRI signal intensities, leading to ambiguities in tissue classification. A common issue, in this case, is the misattribution of WM to GM, and vice versa. Such misclassifications, if not rectified, can compromise the efficacy of the ACT framework, leading to erroneous streamline generation and, consequently, an inaccurate representation of neural connectivity.

22.3.3 Exploration of Solutions to Overcome Segmentation Issues

To surmount the challenges posed by tumor-induced intensity alterations, several strategies have been devised. One prevalent approach involves the use of a lesion-filled T1 image; cf. above and Fig. 22.1, facilitating more accurate tissue segmentation. Alternatively, another strategy involves directly editing the tissue-segmentation image to incorporate the lesion mask. By doing so, priors are not set in the lesioned area, providing a more neutral landscape for the ACT framework to operate on. Both these methods aim to preserve the anatomical fidelity required by the ACT and, subsequently, the SIFT2 framework, ensuring that tumor-induced anatomical distortions do not interfere with the final connectome. Moreover, current methodological advances allow us to turn clinical, heterogeneous brain scans into high-resolution T1 images, which further aids connectome construction with clinical data.

22.4 Problems and Potential Solutions in Atlas Registration

22.4.1 Brief Overview of Atlas Registration and Its Importance in Connectome Creation

Atlas registration refers to the process of aligning individual brain images to a common reference or atlas, also known as image registration (Toga and Thompson 2001; Fu et al. 2020), which is a critical step in creating structural and functional connectomes (Tustison et al. 2021; Klein and Tourville 2012; Rolls et al. 2020). It is essential for comparing and combining data across different subjects, localizing brain regions, and facilitating the identification of the inter-subject common brain activity.

The importance of atlas registration in connectome creation cannot be overstated, especially when the study involves datasets from different subjects or hybrid analysis with multi-modal images. Aligning brain imaging data to a common template facilitates bringing the unclear variation under control, which offers a more comprehensive understanding of brain structures by capturing both structural and functional aspects of neural networks. However, brain tumors pose a challenge for accurate segmentation and registration (cf. above). Therefore, these approaches can introduce varying degrees of data distortion, and the lesion-filled images must be inspected carefully before subsequent use.

22.4.2 Limitations of Current Methods in Tumor Cases

While tools like FreeSurfer have been groundbreaking in neuroimaging, offering robust cortical reconstruction and parcellation from structural MRI data, they are not without limitations, especially in pathologies that result in pronounced anatomical deviations (Fischl 2012). When confronted with anatomical shifts, such as those stemming from large tumors or edema, default pipelines can falter, making them susceptible to inaccuracies when anatomical norms are disrupted. Specifically, when tumors are centrally located or intensively shift anatomy, they can alter the spatial relationships between major subcortical (GM, WM) brain structures, leading to parcellation or registration errors. This misalignment can, in turn, result in the erroneous assignment of cortical labels or, in some cases, the failure to assign any label at all (Fekonja et al. 2021).

One potential solution to mitigate such challenges is the incorporation of manual interventions or adjustments post-registration or post-parcellation. By providing human oversight and refinement in problematic regions, we can ensure a more accurate alignment. Additionally, integrating adaptive algorithms that are sensitive to atypical anatomies or employing complementary tools that offer alternative registration strategies, such as ANTs (Tustison et al. 2021) or in combination with lesion-filling methods, can further improve the fidelity of atlas-to-subject registration or parcellation in the presence of tumors or other significant anatomical anomalies (cf. Figure 22.1).

22.5 Structural Connectomes and Their Relationship with nTMS

The study of structural connectomes in relation to navigated transcranial magnetic stimulation (nTMS) in patients with brain tumors can reveal valuable insights into the complex relationships between tumor presence, connectivity alterations, and their clinical implications. The structural connectome in relation to nTMS adds to our understanding of the effects of brain tumors on neural networks and function. Here we outline the potential that may arise from such analyses:

1. Tumor-related disruptions in WM tracts: nTMS-related structural connectome analysis can highlight tumor-affected edges, nodes, networks and sub-networks in relation to effectively stimulated cortical areas.
2. Functional connectivity alterations: Tumor-related structural changes can influence functional connectivity networks. Investigating the relationship between nTMS-based structural connectomes and fMRI data may reveal how structural alterations impact functional connectivity patterns. This can help to understand the relationship between structural and functional connectomes in the brain and their alterations under the impacts of cerebral tumors.
3. Precise localization of functional areas: nTMS can efficiently localize functional cortical areas in relation to motor, language, or sensory functions (Krieg et al. 2016). It has proven comparable with the invasive gold standard, direct cortical stimulation. However, in relation to language function, methodological difficulties prevail (Bahrend et al. 2020). Nevertheless, combining this functional information with the structural or functional connectome data can elucidate how tumors impact the network, similar to the combination of nTMS and tractography (Reisch et al. 2022). This information can serve in neurosurgical planning and selecting appropriate neurosurgical strategies to potentially minimize postoperative deficits (Ille et al. 2021).
4. Predictive indicators: By applying, e.g., graph theory, one may identify specific connectivity patterns associated with surgical outcomes (Zhang et al. 2022b; Ille et al. 2022; Hallett et al. 2017). Thus, for example, the robustness and efficiency of nTMS-related networks may help to assess the functional outcome after surgery (Zhang et al. 2022b; Ille et al. 2022). Furthermore, it may possibly help to investigate the brain's compensation capacity to overcome tumor induced disturbances and to identify effective or ineffective compensatory mechanisms or reorganization (Engelhardt et al. 2023).

22.6 Conclusion and Outlook

The analysis and creation of structural connectomes in brain tumor patients, as explored in this chapter, present a unique pathway to understanding the complex interplay between brain tumors and neural network organization. This approach, while promising, confronts a variety of methodological challenges. These range from the anatomical distortions and disruptions introduced by the tumors to the advanced methods required for accurate connectome construction.

Despite these hurdles, connectome analysis in the context of brain tumor patients is vibrant and evolving. It is a domain marked by both technical innovation and its application to deepen our understanding of the brain in pathological states. As methodologies continue to advance and mature, they promise to unveil new perspectives and insights into the cerebral network under threat of tumor growth. nTMS emerges as a complementary tool in this endeavor, offering a unique window into additional functional aspects of connectomes. The future of this field holds the potential to uncover solutions and understandings previously unattainable, offering new perspectives in the complex realm of studying and treating brain tumor patients.

References

Abd Hamid AI, Yusoff AN, Mukari SZ, Mohamad M. Brain activation during addition and subtraction tasks in-noise and in-quiet. Malays J Med Sci. 2011;18(2):3–15.

Aerts H, Schirner M, Jeurissen B, Van Roost D, Achten E, Ritter P, et al. Modeling brain dynamics in brain tumor patients using the virtual brain. eNeuro. 2018;5(3):ENEURO.0083-18.2018.

Aerts H, Schirner M, Dhollander T, Jeurissen B, Achten E, Van Roost D, et al. Modeling brain dynamics after tumor resection using the virtual brain. Neuroimage. 2020;213:116738.

Andres M, Pelgrims B, Michaux N, Olivier E, Pesenti M. Role of distinct parietal areas in arithmetic: an fMRI-guided TMS study. Neuroimage. 2011;54(4):3048–56.

Bahrend I, Muench MR, Schneider H, Moshourab R, Dreyer FR, Vajkoczy P, et al. Incidence and linguistic quality of speech errors: a comparison of preoperative transcranial magnetic stimulation and intraoperative direct cortex stimulation. J Neurosurg. 2020;134(5):1409–18.

Barabasi DL, Bianconi G, Bullmore E, Burgess M, Chung S, Eliassi-Rad T, et al. Neuroscience needs network science. J Neurosci. 2023;43(34):5989–95.

Bassett DS, Sporns O. Network neuroscience. Nat Neurosci. 2017;20(3):353–64.

Engelhardt M, Grittner U, Krieg S, Picht T. Preconditioning of the motor network with repetitive navigated transcranial magnetic stimulation (rnTMS) to improve oncological and functional outcome in brain tumor surgery: a study protocol for a randomized, sham-controlled, triple-blind clinical trial. Trials. 2023;24(1):638.

Esteban O, Markiewicz CJ, Blair RW, Moodie CA, Isik AI, Erramuzpe A, et al. fMRIPrep: a robust preprocessing pipeline for functional MRI. Nat Methods. 2019;16(1):111–6.

Farquharson S, Tournier JD, Calamante F, Fabinyi G, Schneider-Kolsky M, Jackson GD, et al. White matter fiber tractography: why we need to move beyond DTI. J Neurosurg. 2013;118(6):1367–77.

Fekonja LS, Wang Z, Aydogan DB, Roine T, Engelhardt M, Dreyer FR, et al. Detecting corticospinal tract impairment in tumor patients with fiber density and tensor-based metrics. Front Oncol. 2020;10:622358.

Fekonja LS, Wang Z, Cacciola A, Roine T, Aydogan DB, Mewes D, et al. Network analysis shows decreased ipsilesional structural connectivity in glioma patients. medRxiv. 2021:2021.06.22.21259319.

Fekonja LS, Wang Z, Cacciola A, Roine T, Aydogan DB, Mewes D, et al. Network analysis shows decreased ipsilesional structural connectivity in glioma patients. Commun Biol. 2022;5(1):258.

Fischl B. FreeSurfer. Neuroimage. 2012;62(2):774–81.

Fu Z, Du Y, Calhoun VD. The dynamic functional network connectivity analysis framework. Engineering (Beijing). 2019;5(2):190–3.

Fu Y, Lei Y, Wang T, Curran WJ, Liu T, Yang X. Deep learning in medical image registration: a review. Phys Med Biol. 2020;65(20):20TR01.

Guyon J, Chapouly C, Andrique L, Bikfalvi A, Daubon T. The normal and brain tumor vasculature: morphological and functional characteristics and therapeutic targeting. Front Physiol. 2021;12:622615.

Hagmann P, Sporns O, Madan N, Cammoun L, Pienaar R, Wedeen VJ, et al. White matter maturation reshapes structural connectivity in the late developing human brain. Proc Natl Acad Sci U S A. 2010;107(44):19067–72.

Hallett M, Di Iorio R, Rossini PM, Park JE, Chen R, Celnik P, et al. Contribution of transcranial magnetic stimulation to assessment of brain connectivity and networks. Clin Neurophysiol. 2017;128(11):2125–39.

Henderson F, Abdullah KG, Verma R, Brem S. Tractography and the connectome in neurosurgical treatment of gliomas: the premise, the progress, and the potential. Neurosurg Focus. 2020;48(2):E6.

Henschel L, Conjeti S, Estrada S, Diers K, Fischl B, Reuter M. FastSurfer - a fast and accurate deep learning based neuroimaging pipeline. Neuroimage. 2020;219:117012.

Hutter J, Tournier JD, Price AN, Cordero-Grande L, Hughes EJ, Malik S, et al. Time-efficient and flexible design of optimized multishell HARDI diffusion. Magn Reson Med. 2018;79(3):1276–92.

Ille S, Kelm A, Schroeder A, Albers LE, Negwer C, Butenschoen VM, et al. Navigated repetitive transcranial magnetic stimulation improves the outcome of postsurgical paresis in glioma patients - a randomized, double-blinded trial. Brain Stimul. 2021;14(4):780–7.

Ille S, Zhang H, Sogerer L, Schwendner M, Schoder A, Meyer B, et al. Preoperative function-specific connectome analysis predicts surgery-related aphasia after glioma resection. Hum Brain Mapp. 2022;43(18):5408–20.

Jeurissen B, Tournier JD, Dhollander T, Connelly A, Sijbers J. Multi-tissue constrained spherical deconvolution for improved analysis of multi-shell diffusion MRI data. Neuroimage. 2014;103:411–26.

Jeurissen B, Descoteaux M, Mori S, Leemans A. Diffusion MRI fiber tractography of the brain. NMR Biomed. 2019;32(4):e3785.

Jha RR, Jaswal G, Bhavsar A, Nigam A. Single-shell to multi-shell dMRI transformation using spatial and volumetric multilevel hierarchical reconstruction framework. Magn Reson Imaging. 2022;87:133–56.

Jones DK. Determining and visualizing uncertainty in estimates of fiber orientation from diffusion tensor MRI. Magn Reson Med. 2003;49(1):7–12.

Klein A, Tourville J. 101 labeled brain images and a consistent human cortical labeling protocol. Front Neurosci. 2012;6:171.

Krieg SM, Sollmann N, Tanigawa N, Foerschler A, Meyer B, Ringel F. Cortical distribution of speech and language errors investigated by visual object naming and navigated transcranial magnetic stimulation. Brain Struct Funct. 2016;221(4):2259–86.

Lawrence AJ, Chung AW, Morris RG, Markus HS, Barrick TR. Structural network efficiency is associated with cognitive impairment in small-vessel disease. Neurology. 2014;83(4):304–11.

Nenonen M, Hakulinen U, Brander A, Ohman J, Dastidar P, Luoto TM. Possible confounding factors on cerebral diffusion tensor imaging measurements. Acta Radiol Open. 2015;4(2):2047981614546795.

Pallast N, Wieters F, Nill M, Fink GR, Aswendt M. Graph theoretical quantification of white matter reorganization after cortical stroke in mice. Neuroimage. 2020;217:116873.

Radwan AM, Emsell L, Blommaert J, Zhylka A, Kovacs S, Theys T, et al. Virtual brain grafting: enabling whole brain parcellation in the presence of large lesions. Neuroimage. 2021;229:117731.

Reisch K, Bottcher F, Tuncer MS, Schneider H, Vajkoczy P, Picht T, et al. Tractography-based navigated TMS language mapping protocol. Front Oncol. 2022;12:1008442.

Rolls ET, Huang CC, Lin CP, Feng J, Joliot M. Automated anatomical labelling atlas 3. Neuroimage. 2020;206:116189.

Shams B, Wang Z, Roine T, Aydogan DB, Vajkoczy P, Lippert C, et al. Machine learning-based prediction of motor status in glioma patients using diffusion MRI metrics along the corticospinal tract. Brain Commun. 2022;4(3):fcac141.

Shams B, Reisch K, Vajkoczy P, Lippert C, Picht T, Fekonja LS. Improved prediction of glioma-related aphasia by diffusion MRI metrics, machine learning, and automated fiber bundle segmentation. Hum Brain Mapp. 2023;44(12):4480–97.

Smith RE, Tournier JD, Calamante F, Connelly A. Anatomically-constrained tractography: improved diffusion MRI streamlines tractography through effective use of anatomical information. Neuroimage. 2012;62(3):1924–38.

Smith SM, Vidaurre D, Beckmann CF, Glasser MF, Jenkinson M, Miller KL, et al. Functional connectomics from resting-state fMRI. Trends Cogn Sci. 2013a;17(12):666–82.

Smith RE, Tournier JD, Calamante F, Connelly A. SIFT: spherical-deconvolution informed filtering of tractograms. Neuroimage. 2013b;67:298–312.

Smith RE, Tournier JD, Calamante F, Connelly A. SIFT2: enabling dense quantitative assessment of brain white matter connectivity using streamlines tractography. Neuroimage. 2015;119:338–51.

Sporns O. The human connectome: a complex network. Ann N Y Acad Sci. 2011;1224:109–25.

Sporns O, Tononi G, Kotter R. The human connectome: a structural description of the human brain. PLoS Comput Biol. 2005;1(4):e42.

Toga AW, Thompson PM. The role of image registration in brain mapping. Image Vis Comput. 2001;19(1–2):3–24.

Tournier JD, Calamante F, Gadian DG, Connelly A. Direct estimation of the fiber orientation density function from diffusion-weighted MRI data using spherical deconvolution. Neuroimage. 2004;23(3):1176–85.

Tournier JD, Calamante F, Connelly A. Robust determination of the fibre orientation distribution in diffusion MRI: non-negativity constrained super-resolved spherical deconvolution. Neuroimage. 2007;35(4):1459–72.

Tournier JD, Mori S, Leemans A. Diffusion tensor imaging and beyond. Magn Reson Med. 2011;65(6):1532–56.

Tuncer MS, Salvati LF, Grittner U, Hardt J, Schilling R, Bahrend I, et al. Towards a tractography-based risk stratification model for language area associated gliomas. Neuroimage Clin. 2021;29:102541.

Tuncer MS, Fekonja LS, Ott S, Pfnur A, Karbe AG, Engelhardt M, et al. Role of interhemispheric connectivity in recovery from postoperative supplementary motor area syndrome in glioma patients. J Neurosurg. 2022;139(2):324–33.

Tustison NJ, Cook PA, Holbrook AJ, Johnson HJ, Muschelli J, Devenyi GA, et al. The ANTsX ecosystem for quantitative biological and medical imaging. Sci Rep. 2021;11(1):9068.

van de Ven VG, Formisano E, Prvulovic D, Roeder CH, Linden DE. Functional connectivity as revealed by spatial independent component analysis of fMRI measurements during rest. Hum Brain Mapp. 2004;22(3):165–78.

Van Essen DC, Ugurbil K, Auerbach E, Barch D, Behrens TE, Bucholz R, et al. The Human Connectome Project: a data acquisition perspective. Neuroimage. 2012;62(4):2222–31.

Veenith TV, Carter E, Grossac J, Newcombe VF, Outtrim JG, Lupson V, et al. Inter subject variability and reproducibility of diffusion tensor imaging within and between different imaging sessions. PLoS One. 2013;8(6):e65941.

Wang Z, Dreyer F, Pulvermuller F, Ntemou E, Vajkoczy P, Fekonja LS, et al. Support vector machine based aphasia classification of transcranial magnetic stimulation language mapping in brain tumor patients. Neuroimage Clin. 2021;29:102536.

Waqar M, Lewis D, Agushi E, Gittins M, Jackson A, Coope D. Cerebral and tumoral blood flow in adult gliomas: a systematic review of results from magnetic resonance imaging. Br J Radiol. 2021;94(1125):20201450.

Wei Y, Li C, Cui Z, Mayrand RC, Zou J, Wong A, et al. Structural connectome quantifies tumour invasion and predicts survival in glioblastoma patients. Brain. 2023;146(4):1714–27.

Zhang W, Ille S, Schwendner M, Wiestler B, Meyer B, Krieg SM. Tracking motor and language eloquent white matter pathways with intraoperative fiber tracking versus preoperative tractography adjusted by intraoperative MRI-based elastic fusion. J Neurosurg. 2022a;137(4):1114–23.

Zhang H, Ille S, Sogerer L, Schwendner M, Schroder A, Meyer B, et al. Elucidating the structural-functional connectome of language in glioma-induced aphasia using nTMS and DTI. Hum Brain Mapp. 2022b;43(6):1836–49.

Brain Plasticity and Neurosurgery 23

Petro Julkunen and Jari Karhu

23.1 Brain Plasticity—Unmasking Existing Connections and/or Establishing New Ones

Neural plasticity is considered an intrinsic property of the human nervous system and does not necessarily represent behavioral gain. Network plasticity is the mechanism for development and learning, as well as a cause for maladaptive reorganization, such as epileptic phenomena in conjunction with brain tumors.

The human central nervous system is capable of short- and long-term change and adaptation throughout life (Kaas 1997; Pascual-Leone et al. 2005). Unmasking of existing connections, shifting synaptic weighting, and even sprouting of new dendritic connections and formation of new synapses are possible (Kaas 1997). These modifications can be driven by afferent input, which is often inseparable from efferent demands and the functional significance of the tasks. Despite the largely uncertain exact molecular and biophysical determinants, enough repetitions of a given task or stimulus in the human neuronal system are likely to give rise to long-standing modification in participating networks, leading to, for example, movement learning. Plastic changes seem to underlie the acquisition of new skills, the adaptation to new contexts, and the recovery of function after injury. The other critical issue regarding motor action is the question of what is represented in the motor cortex output maps: muscles, postures, or movements.

P. Julkunen (✉)
Department of Clinical Neurophysiology, Kuopio University Hospital, Kuopio, Finland

Department of Technical Physics, University of Eastern Finland, Kuopio, Finland
e-mail: petro.julkunen@uef.fi

J. Karhu
Satucon Inc, Kuopio, Finland
e-mail: jari.karhu@satucon.fi

S. M. Krieg, T. Picht (eds.), *Navigated Transcranial Magnetic Stimulation in Neurosurgery*, https://doi.org/10.1007/978-3-031-97155-6_23

Brain plasticity in adults can be observed, e.g., via fast induction in stroke (Rossini et al. 2003). Indications of brain plasticity in slow-growing lesions such as low-grade glioma (LGG) provide theoretical support for enabling surgery in areas essential to language or motor function that might otherwise be considered inoperable (Desmurget et al. 2007; Duffau 2005). Grey matter plasticity is accompanied by white matter plasticity of subcortical pathways affecting reorganization (Szalisznyo et al. 2013). Hence, major tumor resections without induction of functional loss in networks with preserved connectivity and good prognosis after stroke lesions with preserved motor tract functional connectivity demonstrate the different modes of plasticity, while it may be that the plasticity of the white matter is more limited than that of the grey matter (Di Pino et al. 2014; Ius et al. 2011). Consequently, brain plasticity may cause even large tumors to cause no neurological symptoms. Comparison of recovery between slow-growing lesions and acute injuries has suggested different reorganization patterns (Desmurget et al. 2007; Keidel et al. 2010). For recovery, a concept of "minimal common brain" has been introduced, which suggests that there exists a set of mechanisms or networks that are necessary for basic cognitive functions, so minimalistic that it is not sufficient for complex functions (Ius et al. 2011).

There have been several attempts to develop a taxonomy for adaptive/maladaptive plasticity in compromised brains. According to Cargnelutti et al. (2020), they rely on the duration of the altered brain state leading to (i) *regional plasticity*, which is relocation or substitution of a brain area, for example, in early blind patients or after early hemispherectomy; (ii) *functional plasticity*, which refers to moving function from a lesioned area to undamaged brain, i.e., from M1 to PMA; and (iii) *structural plasticity*, which refers to altered shape or number of neural connections for example in the hippocampus of London taxi drivers. For recovery and rehabilitation, plasticity often refers to training-induced neural changes: (i) molecular and microstructural changes, such as dendritic sprouting and molecular diffusion; (ii) *systems-level plasticity*, referring to reorganization within the intrinsic network, for example, in motor networks; and (iii) *between-regions plasticity*, e.g., alterations in effective connectivity participating in a given function or action (Cargnelutti et al. 2020).

Commonly, brain plasticity is considered to cover adaptive changes in the neural networks, including cellular, synaptic, and pathway changes, which are exhibited as functional reorganization (Smits et al. 2015) within anatomical constraints (Bulubas et al. 2018). From a neurosurgical context, the plasticity processes have a role in the prevention of and recovery from neurological and cognitive deficits by adjusting brain functions (Cargnelutti et al. 2020). In this chapter, we extend the definition to include those changes that have the appearance of plasticity but are potentially caused by mechanical effects.

23.1.1 Single-Cell Level Plasticity (Intrinsic Excitability)

At the single-cell level, synaptic plasticity refers to changes in the connections between neurons, whereas non-synaptic plasticity refers to changes in their intrinsic excitability. In general, the connections between network components are prone to synaptic plasticity, while component functions themselves (i.e., intrinsic excitability) of the neurons are prone to intrinsic plasticity.

Intrinsic excitability is the net sum of excitatory-inhibitory single-cell reactions either to synaptic input (Koch 1998) or exposure to external whole-cell stimulation, such as an electric field induced by TMS (Muller-Dahlhaus and Vlachos 2013). It may be attributed mostly to the balance and distribution of fast- and slow-adapting ion channels leading to adaptive changes in membrane excitability and conductance. When a neuron is stimulated by an external electric field, the geometry of the dendrites and axons in the stimulating field also has a profound effect on the overall excitability of a single neuron. Indeed, the same principle can be expanded to glial cells and all cells in the brain with sufficient length of neuronal projections in relation to field strength (Ruohonen and Karhu 2010).

23.1.2 Synaptic Plasticity

The connections between neural network components with anatomical proximity and/or connections to, e.g., injured or lesioned cortex are prone to synaptic plasticity, which is required for learning (and memory). This is the prerequisite for neuronal adaptation to injury or lesion and subsequent restoration of functions. For example, Koch coined and elucidated a terminology for "*synaptic strength*" (Koch 1998). The coupling strength of two neurons is described in terms of n = the number of presynaptic transmitter release sites, p = the probability of transmitter release, and q = some measure of postsynaptic response, such as current, voltage, or conductance change. Taken together, these measures give tools for obtaining the time-dependent response $R = npq$ for "*quantal*" handling of the synaptic efficacy in neural networks and a simplified method for the characterization of plastic network changes.

23.1.3 Hebbian Plasticity

Hebb described plasticity by the example of two adjacent neurons that could take part in firing each other with the efficiency of the firing cells increased as a consequence of some growth process or metabolic change (Hebb 1949). The original formulation is nowadays often described as "*what fires together, wires together.*" Nevertheless, Hebb's principle fits nicely together with the quantal description of synaptic plasticity. Moreover, it fulfills the basic empirical requirements for LTP (long-term potentiation), which is the best-known and most-studied neuronal learning—and adaptive—mechanism in the mammalian brain.

A healthy human brain is known to display adaptive plasticity. Learning new skills results in the adaptation of the neural networks involved in the developed or trained function (Adkins et al. 2006; Muellbacher et al. 2001; Pascual-Leone et al. 1995). This type of Hebbian adaptation has been observed in different types of groups, like musicians and athletes (Elbert et al. 1995; Pearce et al. 2000; Rosenkranz et al. 2007; Tyc et al. 2005; Vaalto et al. 2013). Musicians are a good model of use-dependent adaptation neuroplasticity, with, e.g., adaptive changes in Broca's area (Abdul-Kareem et al. 2011; Sluming et al. 2002) and the primary motor area (Bangert and Schlaug 2006; Vaalto et al. 2013). These types of adaptive changes in the brain may continue and become active when required, e.g., to enhance or restore brain functions. To understand the analogy behind adaptive neuroplasticity, neural network models may be used (Ng et al. 2024).

23.1.4 Modulation of a Neural Network

The plastic effects of lesions and surgery in neural networks, operating as part of the human connectome, involve the optimization of brain functions. Neural networks are organized to produce outputs based on inputs, with the "*hidden layer*" modulating responses (Fig. 23.1). A suitable neural function is achieved through a network with minimal size, optimized energy efficiency, and low error rates (Laughlin and Sejnowski 2003).

For instance, in the sensory-motor system, external stimuli activate the hidden layer, influencing neural impulses and generating outputs, such as motor actions. Depending on the input, the hidden layer of the network modulates the neural impulse to produce an output, and hence the output depends on the inputs of the neural network and the modulatory and controlling effects of the *hidden layer* (Fig. 23.1a). Adaptation occurs if the network fails to produce desired outputs from the inputs it receives. Thus, adaptation is required if the inputs or the hidden layer are impaired, and consequently, the output does not produce a suitable response (Fig. 23.1b). Hebbian theory (Hebb 1949) describes synaptic plasticity, allowing the formation of new connections to compensate for impairments. The biasing of the different components affecting the output could be adjusted at the general level, for instance, via modulation of excitation level, e.g., by modulation of firing rates or number of presynaptic connections. In addition, new connections to other existing neural networks may be formed. The extent and location of the impairments determine the type and extent of adaptation required. Hebbian plasticity is required for forming new connections, while intrinsic plasticity is required for regulation of the synaptic plasticity to which the component functions (i.e., intrinsic excitability) of the neurons are prone to.

The neuronal networks that govern the actual neural functions could have several layers (i.e., several hidden network levels), continuing from the outputs of the simple network presented in Fig. 23.1. As such, impaired parts of the neural networks likely have effects that cascade into multiple outputs of several layers of the neural functions. On the other hand, the compensating network components will be in

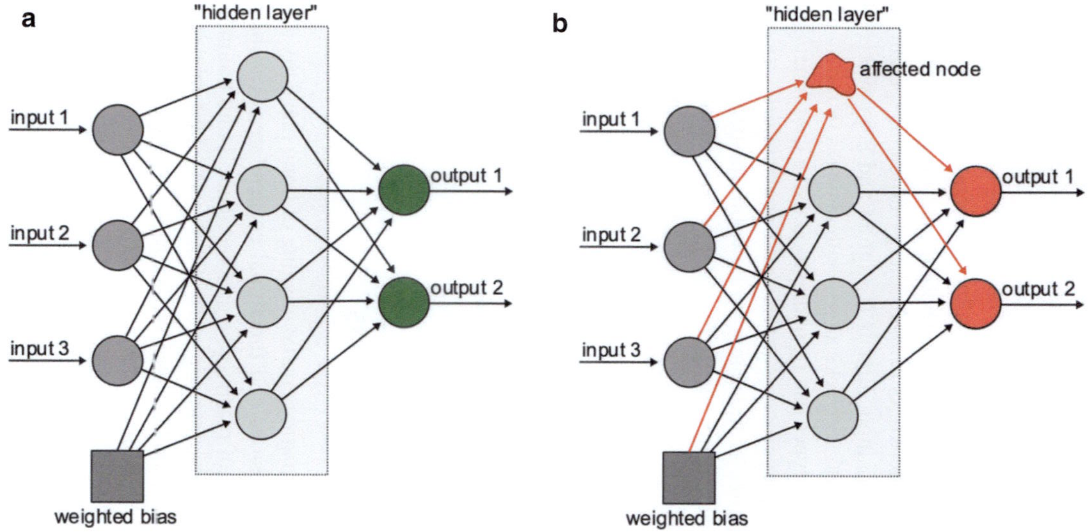

Fig. 23.1 Neural network example. Simplified neural network example of a normally functioning (**a**) and lesioned network (**a**) and resulting modulation of network function adapting to impairment in parts of the network. (**a**) The neural network represents a functional network of neurons in the brain, which has an input and an output, and the inputs are modulated by a "hidden layer" of excitatory and inhibitory neurons to produce a certain output (green circles). (**b**) If a lesion (e.g., tumor or stroke) impairs an input or part of the network (red node in the "hidden layer"), the output is affected and requires adaptation from the network to compensate for the impairment. A weighted bias from outside the network may control for the weight of the effect of each component in the network, for instance, via modulation of the general level of excitation or parallel/connected networks

higher numbers, and hence, a lower level of adaptation from the individual components may be required than in a small network. The plasticity required for a brain function recovery after a focal lesion or injury therefore involves not only the areas in the vicinity of the lesion but also requires the reorganization of all brain networks (Guggisberg et al. 2008; Szalisznyo et al. 2013). To understand the analogy of several layers and connections, neurons are suggested to be able to receive and deliver signals via thousands of synapses, thereby extensively processing inputs to implement all information operations in the nervous system (Laughlin and Sejnowski 2003). Consequently, resectable areas of the brain should be considered as components within a neural network, meaning that after their removal, the neural network should reorganize to eventually preserve behavioral function (Ius et al. 2011).

In an ideal case, neuronal networks provide energy-efficient, spatially compact, and accurate processing of the input signals (Laughlin and Sejnowski 2003). The continuous adaptation of neural networks aligns with changes in inputs and objectives. Differences in recovery types can be understood using cost functions, revealing distinctions between slow-growing lesions and acute injuries (Keidel et al. 2010). The intrinsic properties of the components within the neural networks may also be affected by maladaptive plasticity. In epilepsy, the components within the network activate synchronously with adjusted firing rates to cause changes in overall network function and excitability.

Understanding brain areas near resections as components in a neural network or as part of the human connectome aids in comprehending the required reorganization for the preservation of functions. Protecting connectivity minimizes the extent of required network reorganization. In practice, sufficiently intact subcortical white matter tracts are needed for clinically meaningful sparing and recovery of large-scale neural network functions.

23.2 Imaging Plasticity with nTMS

Multiple modalities of neuroimaging enable imaging of brain plasticity effects, and the interaction between the lesions and functional cortical areas can be revealed. Depending on the source type of plasticity, e.g., experience, brain trauma, continuous use, training, etc., the proper imaging method may differ in detecting anatomic changes, functional changes, and connectivity changes (Cargnelutti et al. 2020). Most often, the relative localization of the functionally crucial cortical sites is done presurgically to set surgical constraints and to aid planning of the procedure. NTMS is utilized in this task (Sollmann et al. 2021). The suitable responses are typically motor responses recorded from muscles using EMG or interruption responses in the speech recorded using real-time video recording. While it is likely possible to identify plasticity effects in the language-related brain areas, the main focus has been on the motor areas with muscle responses, as quantification of the induced responses is convenient via evoked responses like MEP or CSP (Foltys et al. 2003; Forster et al. 2012; Jussen et al. 2016; Mäkelä et al. 2013; Pascual-Leone et al. 1994; Pitkänen et al. 2015; Sollmann et al. 2021; Säisänen et al. 2010; Vaalto et al. 2013).

A cortical map can be constructed for stimulus locations accompanied by response size (Forster et al. 2012; Julkunen 2014; Kallioniemi and Julkunen 2016; Pitkänen et al. 2015; Sollmann et al. 2021) (Fig. 23.2). The produced cortical map is a momentary representation of the cortical functional organization. Therefore, the plasticity-induced effect before or after the mapping cannot be quantified without separate mapping data. For neurosurgery, the most important application of nTMS is to produce momentary cortical maps representative of certain neural functions. These cortical maps are alternatives to cortical maps produced by other methods such as fMRI, PET, single-photon emission computed tomography (SPECT), EEG, or DES. These methods may complement and contradict each other, and they could be further amended by the inclusion of connectivity information (Cargnelutti et al. 2020). As neural plasticity arises in several ways, it appears in different cortical maps in different ways. The accuracy of these methods is limited due to local neurovascular and metabolic coupling, physical properties of the tissue, and the

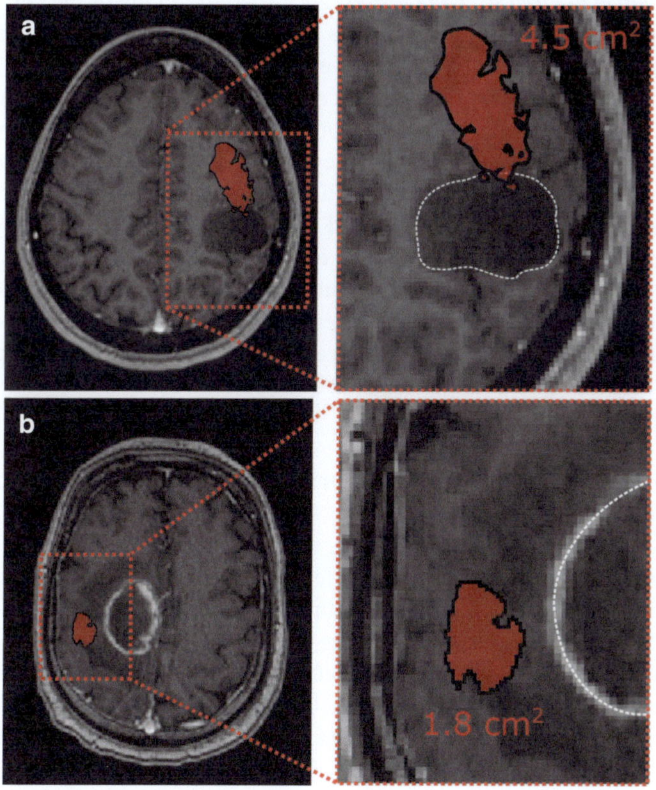

Fig. 23.2 Cortical map for hand muscles. Two examples (**a**, **b**) of an outlined cortical map of hand muscle function on the primary motor cortex overlaid on an axial MRI slice. The functional area is represented as red, and the affected anatomical structure is outlined with a white dashed line in the close-up on the right. Outlining and area estimation of the cortical map were performed using spline interpolation of MEP amplitudes (Julkunen 2014)

distinction of essential areas from modulatory areas, i.e. areas that need to be preserved and areas that can be resected without permanent harm, cannot be made with confidence (Ius et al. 2011).

23.2.1 NTMS Cortical Maps for Detecting and Accounting for Plasticity

NTMS is used to construct cortical maps as a clinical procedure, e.g., preceding surgery or radiotherapy (Conti et al. 2013; Kato et al. 2014; Lefaucheur and Picht 2016; Picht et al. 2009; Sollmann et al. 2021; Säisänen et al. 2010). To observe plasticity effects in the cortical maps, parameters measured from the maps are used for quantitative evaluation of, e.g., cortical hotspot location, center of gravity (CoG), map area, MEP volume, number of responses, rMT, and MEP amplitude. The hotspot location represents the location of the largest response within the cortical map (Reijonen et al. 2020a). *CoG* represents a spatial average of the cortical map of a function (Borghetti et al. 2008; Byrnes et al. 1998; Classen et al. 1998; Freund et al. 2011; Julkunen 2014; Kallioniemi et al. 2016; Wassermann et al. 1992) and can be used to detect relocation (Byrnes et al. 1998; Siebner and Rothwell 2003). *Map area* can be estimated based on response size distribution to compute streamline edges for the cortical map extent to evaluate the size of the functional cortical area (Julkunen 2014; Jussen et al. 2016; Pitkänen et al. 2015; Sollmann et al. 2021). The cortical map has also been evaluated using an *MEP volume* map by summing up all responses (Hetu et al. 2011; Kesar et al. 2012) or by counting the number of induced responses/active sites on a stimulus grid (Foltys et al. 2003; Gagne et al. 2011; Malcolm et al. 2006; Pascual-Leone et al. 1995). To study excitability changes, simple measures of response threshold or response amplitude can be conducted (Julkunen et al. 2016a; Pascual-Leone et al. 1995).

23.2.2 Physical Changes in the Cortex Affecting Brain Mapping with nTMS

From a physical perspective, plasticity effects can be expected to be visually apparent during presurgical mapping of functional cortical areas, as plasticity preceding the cortical mapping procedure may have reorganized the network by altering (1) the location of functional motor areas (*relocation*), (2) the extent of the functional motor areas (*resizing*), or (3) the excitability of the functional motor areas (*excitability*). Surgical operations may also either directly or indirectly provoke plasticity in similar ways.

The known and important physical factors that affect nTMS mapping include (1) TMS coil placement (position, rotation, and tilt), (2) the distance between the TMS coil and the targeted cortical surface, (3) the induced electric field direction with respect to the cortical neuronal organization, (4) the neuronal organization, (5) the strength of the stimulus, (6) stimulus characteristics, and (7) stimulus spacing (Danner et al. 2012; Julkunen et al. 2012; Kallioniemi et al. 2015; Reijonen et al. 2020b; Ruohonen and Karhu 2010; Schmidt et al. 2015; Sollmann et al. 2021). These factors provide the physical, methodological, and technical constraints for the accuracy of nTMS mapping to detect plasticity-induced changes in cortical representation. However, these factors do not account for the neural plasticity causing reorganization of the cortical functions. Instead, macroscopic lesions close to the stimulated area, such as tumors, cause physical effects that may exhibit changes in cortical map location, size, and excitability.

For instance, a tumor located in the vicinity of the primary motor cortex could, as a result of expansion, cause dislocation of the cortical structure, resulting in the appearance of relocation plasticity. On the other hand, an extracortically located tumor dislocating the cortex could increase the distance of the stimulated cortex to the coil, thus requiring greater stimulation power to achieve sufficient excitation in the cortex (Fig. 23.3c). This could cause the impression of reduced excitability and/or a wider area of cortical excitation in the immediately adjacent tissue. Instead, a subcortical tumor compressing the cortex from beneath could "push" the cortical surface from inside the sulcus toward the stimulation coil, hence reducing the required stimulation intensity to achieve cortical excitation and response (Fig. 23.3b, d). This could cause the impression of increased excitability. Compression and

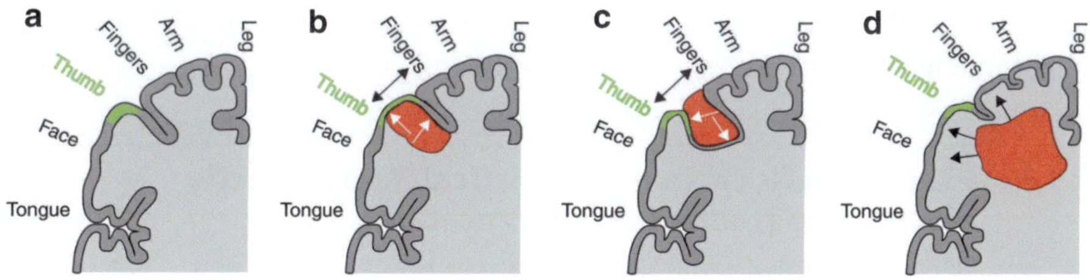

Fig. 23.3 Effect of lesion growth within the primary motor cortex. Simplified schematic images representing the effect of lesion growth within the primary motor cortex. In the schematic images, the coronal view is used for 2D visualization of the common homunculus for simplicity reasons. (**a**) Normal, intact brain in adult human subjects. The functional area of the thumb is highlighted in green. (**b**) A subcortical growth affecting cortical tissue geometry and causing mechanical dislocation of the muscle representation area by compression of the cortex from beneath. (**c**) An extracortical growth affecting tissue geometry and causing mechanical dislocation of the muscle representation by compression of the cortex from the outside. (**d**) A large subcortical growth causing subcortical tissue dislocation and resulting compression of the cortical structure. Vectors in the images indicate the direction of compression. Lesions are represented as red areas

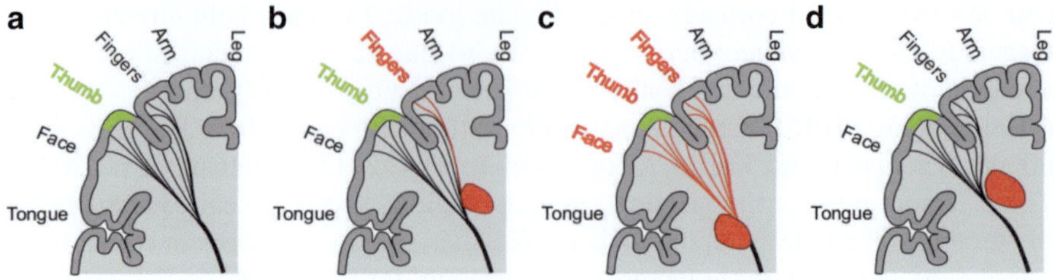

Fig. 23.4 Effect of lesion growth within the axonal motor pathway. Simplified schematic images representing the effect of lesion growth within the axonal motor pathway. In the schematic images, the coronal view is used for 2D visualization of the common homunculus for simplicity reasons. (**a**) Normal, intact brain in adult human subjects. The functional area of the thumb is highlighted in green with connected descending axonal pathways as black lines. (**b**) A subcortical lesion affecting parts of the descending motor pathway and partly impairing connectivity and motor function. Red lines indicate affected axonal pathways. (**c**) A subcortical lesion affecting a large portion of the descending motor pathway and impairing connectivity and motor function. (**d**) A subcortical lesion affecting parts of the descending motor pathway by compression and dislocating the motor pathway with potential effects on motor function and connectivity. Lesions are represented as red areas

stretching of the cortical tissue may also appear as changed cortical excitability, as the neuronal organization is altered, and consequently, the excitable volume of neurons is changed. This means that upon stimulus targeting, the spatial density of potentially activated neurons is lower.

Compression and stretching could further give the impression of a resized functional area. Changes in the curvature of the cortex may also affect the apparent excitability and hence affect the required stimulation power. Dislocation of a functional area could therefore be accompanied by changes in excitability. Similar types of changes may occur in the axonal pathways (Fig. 23.4).

23.3 Plasticity Effects in nTMS Cortical Maps Directly Relevant to Neurosurgical Procedures

23.3.1 Plasticity Preceding Surgery

To account for any plasticity-induced changes in normal brain function and anatomic brain areas, plasticity preceding surgery should be mapped. Here we consider the sources of plasticity to be: (1) *lesion-induced*, (2) *use-dependent*, or (3) *maladaptive plasticity*. The *lesion-induced* plasticity may be caused, e.g., by stroke or tumor, while *use-dependent* plasticity may be, e.g., due to muscle disuse, amputation, or training (Elbert and Rockstroh 2004). *Maladaptive plasticity* may be, e.g., due to focal cortical dysplasia (FCD) causing epilepsy or adaptation to neural network changes causing, e.g., pain or tinnitus (Langguth et al. 2005). The separation of these source types is not strict, and they may overlap. For instance, FCD may induce changes in lesion-induced and maladaptive types. FCD has been demonstrated to cause major reorganization of the motor function (Narayana et al. 2015)

(Fig. 23.5b). In addition, large lesions or traumas could have radical effects on the reorganization of cortical functions. Radical cortical reorganization has been demonstrated after partial hemispherectomy to treat refractory seizure disorders (Narayana et al. 2015) (Fig. 23.5c). The appearance of plasticity in this way is likely affected both by the dysfunctional hemisphere and the partial hemispherectomy.

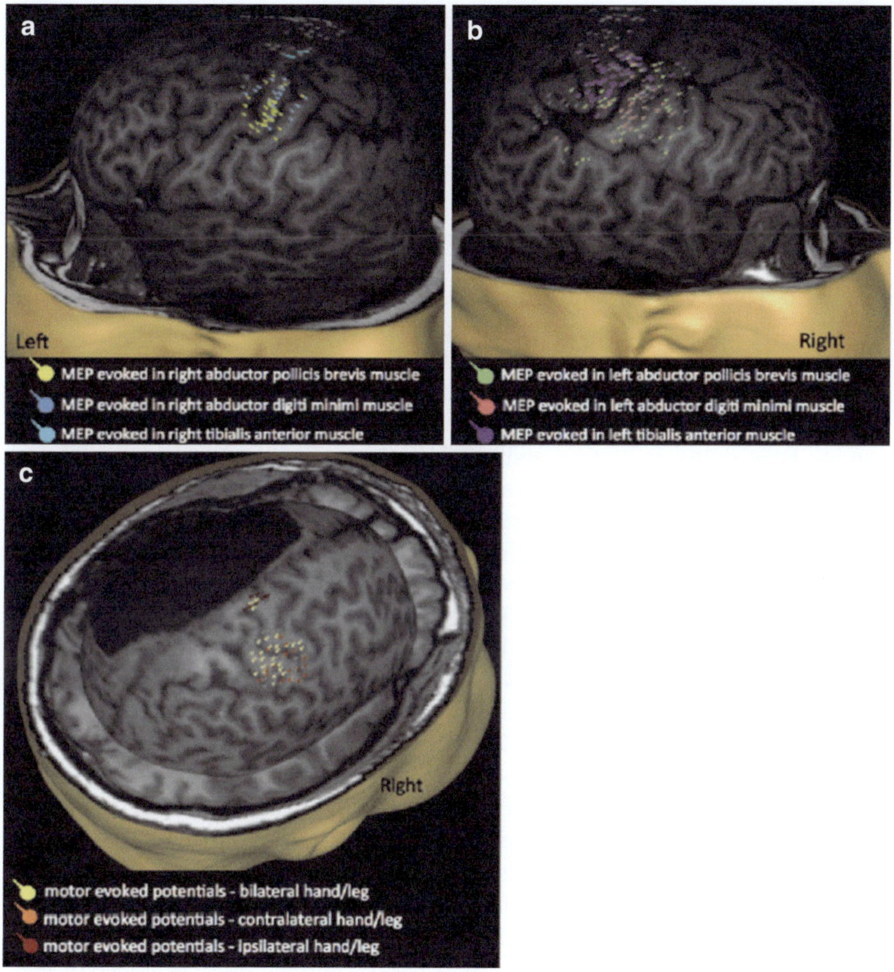

Fig. 23.5 Functional reorganization of the motor cortex. (**a**) nTMS motor mapping demonstrating the effect of cortical dysplasia on cortical functional reorganization in a 13-year-old girl. Normal organization of the left motor cortex with normal cortical localization of the right hand and leg. (**b**) Polymicrogyria and the vertical cleft extending from the posterior aspect of the right Sylvian fissure. The location and extent of the left-hand muscle representation in the right hemisphere is aberrant and localized over the area of polymicrogyria with the displaced location of the primary leg motor cortex. (**c**) nTMS motor mapping demonstrating cortical reorganization in a 16-year-old female patient who had suffered left hemisphere trauma at 32 weeks' gestation. Bilateral limb representation is noted in the right primary hand and leg motor area. (Modified from Narayana et al. (2015) with permission)

Brain plasticity in the context of neurosurgery does not need to be adaptive plasticity, but instead, the appearance of plasticity may be caused simply by the mechanical pressure of a lesion, which may cause changes in function and altered appearance in cortical maps (Conway et al. 2017). Once the source of mechanical load (e.g., tumor) is removed, normal functions may be regained, and little or no neural adaptation is required. Therefore, unlike use-dependent and maladaptive plasticity, lesion-induced plasticity is not necessarily associated with adaptation. Use-dependent plasticity manifests due to changes in the activation of the cortex and peripheral connections. It is easily demonstrated via immobilization of restrictions of movements or related muscle disuse, which may reduce the size of the functional motor area in a cortical map (Elbert and Rockstroh 2004; Liepert et al. 1995; Raffin and Siebner 2019), while training of skills may expand the functional motor area (Elbert et al. 1995; Elbert and Rockstroh 2004; Pascual-Leone et al. 1995; Raffin and Siebner 2019; Vaalto et al. 2013).

Maladaptive plasticity, like in FCD, is a harmful adaptation, which may cause epilepsy or pain by disturbing normal neural network function. The summation of multiple appearances of plasticity may complicate the identification of different sources of plasticity based purely on the cortical map. Hence, structural MRI may help by imaging the axonal pathways using tractography (please see the chapter on "Advanced Tractography"). Previously, lesion-induced impairment of normal function has been shown in cortical and subcortical structures and pathways (Papagno et al. 2011). The effects of lesion-induced plasticity are likely both relocating and resizing of the functional areas. A subcortical efferent lesion may cease a descending motor tract from functioning at different locations of the tract, whereas altered sensory pathways may change functional activation patterns feeding into motor functions and therefore induce plastic effects.

Previously recorded plasticity effects due to lesions in the brain are numerous, the most fundamental of which are stroke and tumors. Gliomas have been shown to cause relocation of the functional motor areas, as they tend to shift motor areas in their close vicinity (Conway et al. 2017; Takahashi et al. 2012). Similar observations have been made in language-related areas as a potential hemispheric shift (Duffau 2020; Krieg et al. 2013; Rösler et al. 2014). In addition, SMA appears to have a major role in motor cortex plasticity in high-grade glioma patients (Majos et al. 2015). Cortical maps of low-grade glioma patients have revealed various patterns of reorganization with brain functions remaining within the tumor, reorganizing around the tumor, spreading in the ipsilateral hemisphere, or even moving to the contralateral hemisphere (Desmurget et al. 2007). Further, the appearance of plasticity, as evidenced by previous electrophysiological studies, provides an understanding of the role of plasticity in low-grade gliomas (Ng and Duffau 2023; Sarubbo et al. 2020).

In epilepsy, the epileptogenic zone can often be detected with MRI as reorganized structures. The use of TMS cortical mapping has demonstrated the representational adaptations of the motor cortex in epilepsy when epileptogenic focus involves a motor area (Labyt et al. 2007). The adaptations include changes in excitability and apparent representation resizing, potentially due to modified inhibition and representation shift. FCDs, a common cause of intractable epilepsy, are known to reorganize the local network (Otsubo et al. 2005; Sisodiya et al. 2009). Intracranial arteriovenous malformations (AVMs) are also known to induce plasticity, the effects of which can be observed using nTMS (Kato et al. 2014). Previously, emphasis on right-sided language lateralization in AVM patients has been reported (Lehericy et al. 2002; Pouratian and Bookheimer 2010; Vikingstad et al. 2000). Cerebral palsy has been demonstrated in examples to relocate motor functions to enable ipsilateral activation of the primary motor tract with TMS (Pihko et al. 2014).

In stroke, the timing of creating the cortical map is crucial, as vast time-dependent changes tend to occur both in the acute and subacute phases, while milder changes may still occur during the chronic phase (Julkunen et al. 2016a, b; Mäkelä et al. 2015). Stroke-induced plastic changes may reveal extensive plasticity effects (Fig. 23.6). Unlike stroke, where plastic effects are rehabilitative and potentially recovering toward normal function, tumors, and lesions tend to have a progression that steers brain function away from normal function, and adaptive and non-adaptive processes may induce plasticity to occur in various ways, e.g., as relocated functional area, extended functional area, and altered excitability (Conway et al. 2017; Krieg et al. 2013). With nTMS, relocation of motor areas may be disguised also as altered excitability in the situation when a tumor of subcortical origin extends or "pushes" the cortex toward the skull, thus reducing coil-to-cortex distance, giving the appearance of a lower excitability threshold and diminished functional map due to suboptimal stimulus strength.

For neurosurgical applications of cortical mapping and to understand/account for plasticity effects, it may be of interest to determine whether the observed plastic changes prior to presurgical mapping are expected to continue after surgery, therefore potentially having an effect on long-term brain function.

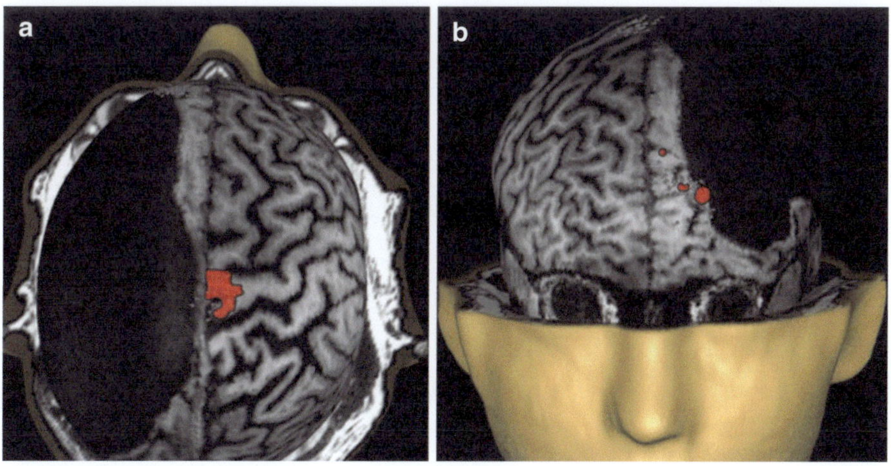

Fig. 23.6 Functional reorganization after stroke. (**a**) Left foot and (**b**) right foot muscle representation revealed by nTMS mapping in a 19-year-old male epilepsy patient with right-sided hemiparesis and an extensive perinatal vascular infarction in the left middle cerebral artery territory. Sites eliciting MEPs are indicated by red color. (Image data courtesy of Professor Jyrki Mäkelä et al. (2015))

23.3.2 Plasticity Following Surgery

Normalization of the plastic effects preceding surgery may occur after surgery. However, relocation after surgery may predominantly be observed as a shift toward the resection cavity, as has been reported in the case of gliomas (Conway et al. 2017). Potentially, this shift or lesion-induced relocation prior to surgery may not have induced adaptive changes, and removal of the source of mechanical tissue compression allows for a quick recovery. The vascularization of the cortex close to the resection cavity also plays a critical role. A report on extra-intracranial bypass surgery in occlusive cerebrovascular disease suggests a reversibly impaired cortical motor function in the ischemic brain, with cerebral revascularization leading to improved motor output observed as increased cortical motor excitability and resized motor representation (Jussen et al. 2016).

The reversible effects of the plasticity preceding surgery may occur as the original inductor is removed. In case of lesion-induced plasticity preceding surgery, the lesion removal may, in addition to the aforementioned relocation, allow for retaining neural network connections, enabling adaptation to normal network function. This is expected after tumor resection, as normalized excitability and functional recruitment, and most of all normalized brain function is expected. For instance, in the case of retained muscle function, use-dependent plasticity may cause recovery of the motor representation to be observed in the cortical maps (Fig. 23.7). Similar effects can be observed with language function. Obviously, the mechanical effect of the resection cavity needs to be accounted for (Conway et al. 2017).

As maladaptive plasticity may be caused, e.g., by changes in the input to the neural network, the brain may try to compensate for the lower-level input by

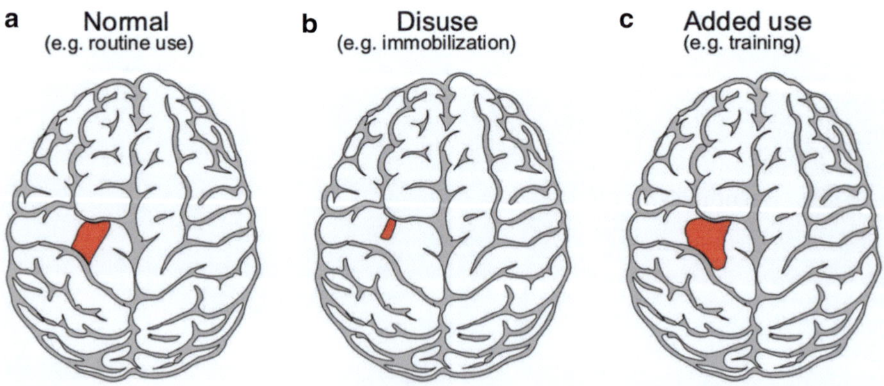

Fig. 23.7 Use-dependent plasticity. Schematic example of potential effects of use-dependent plasticity on the functional hand motor area of the primary motor cortex. (**a**) Normal representation area (as red). (**b**) Effects of long-term disuse of the muscle result in a reduction of the size of the functional motor area (Elbert and Rockstroh 2004; Liepert et al. 1995). (**c**) Added use, e.g., via training, may expand the functional representation area (Elbert et al. 1995; Elbert and Rockstroh 2004; Pascual-Leone et al. 1995)

increasing the excitability of the remaining neural network, which could result in false or suboptimal outputs. This type of maladaptive plasticity could be promoted by surgical procedures, i.e., both the resection itself as well as the vascular changes.

23.4 Final Discussions on Plasticity Effects in Cortical Maps

Even though the effects of different types of plasticity may be observed as a summation in the cortical map, it may be impossible to identify the different sources of plasticity due to various chemical and mechanical factors causing the observed plasticity effects, with the contribution of the preferred functional output. While some of the plasticity effects may be relevant to identifying the cortical map construction for neurosurgical applications, some effects of plasticity are irrelevant for the timescale used for the currently most important application of nTMS in neurosurgery, the preoperative mapping. Short-term plasticity is usually controlled and considered negligible for the mapping procedure, as the time between mapping and surgery/radiotherapy is commonly short, as it should be, especially in cases where plastic changes are expected to be rapid. Theoretically, an aggressive tumor growth affecting either the cortex or the subcortical tracts could cause some of the plasticity effects to occur between the mapping and surgery, and therefore the validity of the mapping could be compromised.

As the indications for neurosurgical operations vary and functional reorganization might show potentially rapid (transient) effects (Duffau 2006), localizing the brain functions and connected tracts needs to be performed individually. Even though there are functional limitations in each imaging technique, combinations of different methods will lead to the best results in terms of surgical indications and optimal EOR (Ius et al. 2011). The combination of DTI with nTMS enables the assessment of cortical function and connected white matter tracts (Conti et al. 2014; Frey et al. 2012; Kamada et al. 2007; Negwer et al. 2016; Sollmann et al. 2021).

Brain plasticity is often considered to be the normal ongoing state of the central nervous system throughout life (Pascual-Leone et al. 2005). However, the state and demand of plasticity are modulated heavily by lesions, trauma, or surgical interventions affecting the neural networks of the brain. The emphasis on the continuous ongoing state of plasticity is crucial with slow-growing tumors with low expansion rates necessitating continuous functional reorganization and implementation of compensatory networks (Desmurget et al. 2007; Ius et al. 2011; Ng and Duffau 2023). In addition, areas outside the damaged area may take over the impaired functions while facilitating recovery (Duffau 2006). The dynamics of the reorganization of brain networks occurring through adaptation in everyday life or after a lesion demonstrate the versatile redundancies that exist in the brain available for functional substitution (Bavelier and Neville 2002; Duffau et al. 2000; Ius et al. 2011; Rossini et al. 2003; Schieber and Hibbard 1993). For optimizing individual surgical outcomes, it is important to understand the potential sources and appearances of brain plasticity.

In conclusion, the relative localization of the functionally crucial cortical sites and the integrity of cortical and subcortical tracts is done presurgically to set surgical constraints and to aid in planning the procedure; nTMS is well suited for this task (Sollmann et al. 2021). The stratification of patients with the best neural reserve for protective and corrective neural plasticity is still very much under investigation. These attempts are inseparable from the attempts to utilize neuromodulatory techniques in optimizing the functional outcome of surgery. In the future, these techniques may significantly aid the oldest and most beneficial method to boost postoperative recovery and neural well-being: continuous (self)rehabilitation.

References

Abdul-Kareem IA, Stancak A, Parkes LM, Sluming V. Increased gray matter volume of left pars opercularis in male orchestral musicians correlate positively with years of musical performance. J Magn Reson Imaging. 2011;33:24–32.

Adkins DL, Boychuk J, Remple MS, Kleim JA. Motor training induces experience-specific patterns of plasticity across motor cortex and spinal cord. J Appl Physiol. 2006;1985(101):1776–82.

Bangert M, Schlaug G. Specialization of the specialized in features of external human brain morphology. Eur J Neurosci. 2006;24:1832–4.

Bavelier D, Neville HJ. Cross-modal plasticity: where and how? Nat Rev Neurosci. 2002;3:443–52.

Borghetti D, Sartucci F, Petacchi E, Guzzetta A, Piras MF, Murri L, Cioni G. Transcranial magnetic stimulation mapping: a model based on spline interpolation. Brain Res Bull. 2008;77:143–8.

Bulubas L, Sollmann N, Tanigawa N, Zimmer C, Meyer B, Krieg SM. Reorganization of motor representations in patients with brain lesions: a navigated transcranial magnetic stimulation study. Brain Topogr. 2018;31:288–99.

Byrnes ML, Thickbroom GW, Wilson SA, Sacco P, Shipman JM, Stell R, Mastaglia FL. The corticomotor representation of upper limb muscles in writer's cramp and changes following botulinum toxin injection. Brain. 1998;121(Pt 5):977–88.

Cargnelutti E, Ius T, Skrap M, Tomasino B. What do we know about pre- and postoperative plasticity in patients with glioma? A review of neuroimaging and intraoperative mapping studies. Neuroimage Clin. 2020;28:102435.

Classen J, Knorr U, Werhahn KJ, Schlaug G, Kunesch E, Cohen LG, Seitz RJ, Benecke R. Multimodal output mapping of human central motor representation on different spatial scales. J Physiol. 1998;512(Pt 1):163–79.

Conti A, Pontoriero A, Ricciardi GK, Granata F, Vinci S, Angileri FF, Pergolizzi S, Alafaci C, Rizzo V, Quartarone A, Germano A, Foroni RI, De Renzis C, Tomasello F. Integration of functional neuroimaging in CyberKnife radiosurgery: feasibility and dosimetric results. Neurosurg Focus. 2013;34:E5.

Conti A, Raffa G, Granata F, Rizzo V, Germano A, Tomasello F. Navigated transcranial magnetic stimulation for "somatotopic" tractography of the corticospinal tract. Neurosurgery. 2014;10(Suppl 4):542–54; discussion 554.

Conway N, Wildschuetz N, Moser T, Bulubas L, Sollmann N, Tanigawa N, Meyer B, Krieg SM. Cortical plasticity of motor-eloquent areas measured by navigated transcranial magnetic stimulation in patients with glioma. J Neurosurg. 2017;127:981–91.

Danner N, Könönen M, Säisänen L, Laitinen R, Mervaala E, Julkunen P. Effect of individual anatomy on resting motor threshold-computed electric field as a measure of cortical excitability. J Neurosci Methods. 2012;203:298–304.

Desmurget M, Bonnetblanc F, Duffau H. Contrasting acute and slow-growing lesions: a new door to brain plasticity. Brain. 2007;130:898–914.

Di Pino G, Pellegrino G, Assenza G, Capone F, Ferreri F, Formica D, Ranieri F, Tombini M, Ziemann U, Rothwell JC, Di Lazzaro V. Modulation of brain plasticity in stroke: a novel model for neurorehabilitation. Nat Rev Neurol. 2014;10:597–608.
Duffau H. Lessons from brain mapping in surgery for low-grade glioma: insights into associations between tumour and brain plasticity. Lancet Neurol. 2005;4:476–86.
Duffau H. Brain plasticity: from pathophysiological mechanisms to therapeutic applications. J Clin Neurosci. 2006;13:885–97.
Duffau H. Functional mapping before and after low-grade glioma surgery: a new way to decipher various spatiotemporal patterns of individual neuroplastic potential in brain tumor patients. Cancers (Basel). 2020;12:2611.
Duffau H, Sichez JP, Lehericy S. Intraoperative unmasking of brain redundant motor sites during resection of a precentral angioma: evidence using direct cortical stimulation. Ann Neurol. 2000;47:132–5.
Elbert T, Rockstroh B. Reorganization of human cerebral cortex: the range of changes following use and injury. Neuroscientist. 2004;10:129–41.
Elbert T, Pantev C, Wienbruch C, Rockstroh B, Taub E. Increased cortical representation of the fingers of the left hand in string players. Science. 1995;270:305–7.
Foltys H, Krings T, Meister IG, Sparing R, Boroojerdi B, Thron A, Topper R. Motor representation in patients rapidly recovering after stroke: a functional magnetic resonance imaging and transcranial magnetic stimulation study. Clin Neurophysiol. 2003;114:2404–15.
Forster MT, Senft C, Hattingen E, Lorei M, Seifert V, Szelenyi A. Motor cortex evaluation by nTMS after surgery of central region tumors: a feasibility study. Acta Neurochir (Wien). 2012;154:1351–9.
Freund P, Rothwell J, Craggs M, Thompson AJ, Bestmann S. Corticomotor representation to a human forearm muscle changes following cervical spinal cord injury. Eur J Neurosci. 2011;34:1839–46.
Frey D, Strack V, Wiener E, Jussen D, Vajkoczy P, Picht T. A new approach for corticospinal tract reconstruction based on navigated transcranial stimulation and standardized fractional anisotropy values. Neuroimage. 2012;62:1600–9.
Gagne M, Hetu S, Reilly KT, Mercier C. The map is not the territory: motor system reorganization in upper limb amputees. Hum Brain Mapp. 2011;32:509–19.
Guggisberg AG, Honma SM, Findlay AM, Dalal SS, Kirsch HE, Berger MS, Nagarajan SS. Mapping functional connectivity in patients with brain lesions. Ann Neurol. 2008;63:193–203.
Hebb DO. Organization of behavior. New York: Wiley; 1949.
Hetu S, Gagne M, Reilly KT, Mercier C. Short-term reliability of transcranial magnetic stimulation motor maps in upper limb amputees. J Clin Neurosci. 2011;18:728–30.
Ius T, Angelini E, Thiebaut de Schotten M, Mandonnet E, Duffau H. Evidence for potentials and limitations of brain plasticity using an atlas of functional resectability of WHO grade II gliomas: towards a "minimal common brain". Neuroimage. 2011;56:992–1000.
Julkunen P. Methods for estimating cortical motor representation size and location in navigated transcranial magnetic stimulation. J Neurosci Methods. 2014;232:125–33.
Julkunen P, Säisänen L, Danner N, Awiszus F, Könönen M. Within-subject effect of coil-to-cortex distance on cortical electric field threshold and motor evoked potentials in transcranial magnetic stimulation. J Neurosci Methods. 2012;206:158–64.
Julkunen P, Könönen M, Määttä S, Tarkka IM, Hiekkala SH, Säisänen L, Vanninen R, Karhu J, Jakala P. Longitudinal study on modulated corticospinal excitability throughout recovery in supratentorial stroke. Neurosci Lett. 2016a;617:88–93.
Julkunen P, Määttä S, Säisänen L, Kallioniemi E, Könönen M, Jäkälä P, Vanninen R, Vaalto S. Functional and structural cortical characteristics after restricted focal motor cortical infarction evaluated at chronic stage – indications from a preliminary study. Clin Neurophysiol. 2016b;127:2775–84.

Jussen D, Zdunczyk A, Schmidt S, Rosler J, Buchert R, Julkunen P, Karhu J, Brandt S, Picht T, Vajkoczy P. Motor plasticity after extra-intracranial bypass surgery in occlusive cerebrovascular disease. Neurology. 2016;87:27–35.

Kaas JH. Functional plasticity in adult cortex. Orlando: Academic Press; 1997.

Kallioniemi E, Julkunen P. Alternative stimulation intensities for mapping cortical motor area with navigated TMS. Brain Topogr. 2016;29:395–404.

Kallioniemi E, Könönen M, Säisänen L, Gröhn H, Julkunen P. Functional neuronal anisotropy assessed with neuronavigated transcranial magnetic stimulation. J Neurosci Methods. 2015;256:82–90.

Kallioniemi E, Pitkänen M, Könönen M, Vanninen R, Julkunen P. Localization of cortical primary motor area of the hand using navigated transcranial magnetic stimulation, BOLD and arterial spin labeling fMRI. J Neurosci Methods. 2016;273:138–48.

Kamada K, Todo T, Masutani Y, Aoki S, Ino K, Morita A, Saito N. Visualization of the frontotemporal language fibers by tractography combined with functional magnetic resonance imaging and magnetoencephalography. J Neurosurg. 2007;106:90–8.

Kato N, Schilt S, Schneider H, Frey D, Kufeld M, Vajkoczy P, Picht T. Functional brain mapping of patients with arteriovenous malformations using navigated transcranial magnetic stimulation: first experience in ten patients. Acta Neurochir (Wien). 2014;156:885–95.

Keidel JL, Welbourne SR, Lambon Ralph MA. Solving the paradox of the equipotential and modular brain: a neurocomputational model of stroke vs. slow-growing glioma. Neuropsychologia. 2010;48:1716–24.

Kesar TM, Sawaki L, Burdette JH, Cabrera MN, Kolaski K, Smith BP, O'Shea TM, Koman LA, Wittenberg GF. Motor cortical functional geometry in cerebral palsy and its relationship to disability. Clin Neurophysiol. 2012;123:1383–90.

Koch C. Biophysics of computation: information processing in single neurons. New York: Oxford University Press; 1998.

Krieg SM, Sollmann N, Hauck T, Ille S, Foerschler A, Meyer B, Ringel F. Functional language shift to the right hemisphere in patients with language-eloquent brain tumors. PLoS One. 2013;8:e75403.

Labyt E, Houdayer E, Cassim F, Bourriez JL, Derambure P, Devanne H. Motor representation areas in epileptic patients with focal motor seizures: a TMS study. Epilepsy Res. 2007;75:197–205.

Langguth B, Eichhammer P, Zowe M, Kleinjung T, Jacob P, Binder H, Sand P, Hajak G. Altered motor cortex excitability in tinnitus patients: a hint at crossmodal plasticity. Neurosci Lett. 2005;380:326–9.

Laughlin SB, Sejnowski TJ. Communication in neuronal networks. Science. 2003;301:1870–4.

Lefaucheur JP, Picht T. The value of preoperative functional cortical mapping using navigated TMS. Neurophysiol Clin. 2016;46:125–33.

Lehericy S, Biondi A, Sourour N, Vlaicu M, du Montcel ST, Cohen L, Vivas E, Capelle L, Faillot T, Casasco A, Le Bihan D, Marsault C. Arteriovenous brain malformations: is functional MR imaging reliable for studying language reorganization in patients? Initial observations. Radiology. 2002;223:672–82.

Liepert J, Tegenthoff M, Malin JP. Changes of cortical motor area size during immobilization. Electroencephalogr Clin Neurophysiol. 1995;97:382–6.

Majos A, Bryszewski B, Kosla KN, Pfaifer L, Jaskolski D, Stefanczyk L. Process of the functional reorganization of the cortical centers for movement in GBM patients: fMRI study. Clin Neuroradiol. 2015;27:71–9.

Mäkelä JP, Vitikainen AM, Lioumis P, Paetau R, Ahtola E, Kuusela L, Valanne L, Blomstedt G, Gaily E. Functional plasticity of the motor cortical structures demonstrated by navigated TMS in two patients with epilepsy. Brain Stimul. 2013;6:286–91.

Mäkelä JP, Lioumis P, Laaksonen K, Forss N, Tatlisumak T, Kaste M, Mustanoja S. Cortical excitability measured with nTMS and MEG during stroke recovery. Neural Plast. 2015;2015:309546.

Malcolm MP, Triggs WJ, Light KE, Shechtman O, Khandekar G, Gonzalez Rothi LJ. Reliability of motor cortex transcranial magnetic stimulation in four muscle representations. Clin Neurophysiol. 2006;117:1037–46.

Muellbacher W, Ziemann U, Boroojerdi B, Cohen L, Hallett M. Role of the human motor cortex in rapid motor learning. Exp Brain Res. 2001;136:431–8.
Muller-Dahlhaus F, Vlachos A. Unraveling the cellular and molecular mechanisms of repetitive magnetic stimulation. Front Mol Neurosci. 2013;6:50.
Narayana S, Papanicolaou AC, McGregor A, Boop FA, Wheless JW. Clinical applications of transcranial magnetic stimulation in pediatric neurology. J Child Neurol. 2015;30:1111–24.
Negwer C, Sollmann N, Ille S, Hauck T, Maurer S, Kirschke JS, Ringel F, Meyer B, Krieg SM. Language pathway tracking: comparing nTMS-based DTI fiber tracking with a cubic ROIs-based protocol. J Neurosurg. 2016:1–9.
Ng S, Duffau H. Brain plasticity profiling as a key support to therapeutic decision-making in low-grade glioma oncological strategies. Cancers (Basel). 2023;15:3698.
Ng S, Duffau H, Herbet G. Perspectives in human brain plasticity sparked by glioma invasion: from intraoperative (re)mappings to neural reconfigurations. Neural Regen Res. 2024;19:947–8.
Otsubo H, Iida K, Oishi M, Okuda C, Ochi A, Pang E, Weiss SK, Rutka JT, Chuang SH, Snead OC 3rd. Neurophysiologic findings of neuronal migration disorders: intrinsic epileptogenicity of focal cortical dysplasia on electroencephalography, electrocorticography, and magnetoencephalography. J Child Neurol. 2005;20:357–63.
Papagno C, Gallucci M, Casarotti A, Castellano A, Falini A, Fava E, Giussani C, Carrabba G, Bello L, Caramazza A. Connectivity constraints on cortical reorganization of neural circuits involved in object naming. Neuroimage. 2011;55:1306–13.
Pascual-Leone A, Grafman J, Hallett M. Modulation of cortical motor output maps during development of implicit and explicit knowledge. Science. 1994;263:1287–9.
Pascual-Leone A, Nguyet D, Cohen LG, Brasil-Neto JP, Cammarota A, Hallett M. Modulation of muscle responses evoked by transcranial magnetic stimulation during the acquisition of new fine motor skills. J Neurophysiol. 1995;74:1037–45.
Pascual-Leone A, Amedi A, Fregni F, Merabet LB. The plastic human brain cortex. Annu Rev Neurosci. 2005;28:377–401.
Pearce AJ, Thickbroom GW, Byrnes ML, Mastaglia FL. Functional reorganisation of the corticomotor projection to the hand in skilled racquet players. Exp Brain Res. 2000;130:238–43.
Picht T, Mularski S, Kuehn B, Vajkoczy P, Kombos T, Suess O. Navigated transcranial magnetic stimulation for preoperative functional diagnostics in brain tumor surgery. Neurosurgery. 2009;65:93–8; discussion 98–99.
Pihko E, Nevalainen P, Vaalto S, Laaksonen K, Mäenpää H, Valanne L, Lauronen L. Reactivity of sensorimotor oscillations is altered in children with hemiplegic cerebral palsy: a magnetoencephalographic study. Hum Brain Mapp. 2014;35:4105–17.
Pitkänen M, Kallioniemi E, Julkunen P. Extent and location of the excitatory and inhibitory cortical hand representation maps: a navigated transcranial magnetic stimulation study. Brain Topogr. 2015;28:657–65.
Pouratian N, Bookheimer SY. The reliability of neuroanatomy as a predictor of eloquence: a review. Neurosurg Focus. 2010;28:E3.
Raffin E, Siebner HR. Use-dependent plasticity in human primary motor hand area: synergistic interplay between training and immobilization. Cereb Cortex. 2019;29:356–71.
Reijonen J, Pitkänen M, Kallioniemi E, Mohammadi A, Ilmoniemi RJ, Julkunen P. Spatial extent of cortical motor hotspot in navigated transcranial magnetic stimulation. J Neurosci Methods. 2020a;346:108893.
Reijonen J, Säisänen L, Könönen M, Mohammadi A, Julkunen P. The effect of coil placement and orientation on the assessment of focal excitability in motor mapping with navigated transcranial magnetic stimulation. J Neurosci Methods. 2020b;331:108521.
Rosenkranz K, Williamon A, Rothwell JC. Motorcortical excitability and synaptic plasticity is enhanced in professional musicians. J Neurosci. 2007;27:5200–6.
Rösler J, Niraula B, Strack V, Zdunczyk A, Schilt S, Savolainen P, Lioumis P, Mäkelä J, Vajkoczy P, Frey D, Picht T. Language mapping in healthy volunteers and brain tumor patients with a novel navigated TMS system: evidence of tumor-induced plasticity. Clin Neurophysiol. 2014;125:526–36.

Rossini PM, Calautti C, Pauri F, Baron JC. Post-stroke plastic reorganisation in the adult brain. Lancet Neurol. 2003;2:493–502.
Ruohonen J, Karhu J. Navigated transcranial magnetic stimulation. Neurophysiol Clin. 2010;40:7–17.
Säisänen L, Könönen M, Julkunen P, Määttä S, Vanninen R, Immonen A, Jutila L, Kälviäinen R, Jääskeläinen JE, Mervaala E. Non-invasive preoperative localization of primary motor cortex in epilepsy surgery by navigated transcranial magnetic stimulation. Epilepsy Res. 2010;92:134–44.
Sarubbo S, Tate M, De Benedictis A, Merler S, Moritz-Gasser S, Herbet G, Duffau H. Mapping critical cortical hubs and white matter pathways by direct electrical stimulation: an original functional atlas of the human brain. Neuroimage. 2020;205:116237.
Schieber MH, Hibbard LS. How somatotopic is the motor cortex hand area? Science. 1993;261:489–92.
Schmidt S, Bathe-Peters R, Fleischmann R, Ronnefarth M, Scholz M, Brandt SA. Nonphysiological factors in navigated TMS studies; confounding covariates and valid intracortical estimates. Hum Brain Mapp. 2015;36:40–9.
Siebner HR, Rothwell J. Transcranial magnetic stimulation: new insights into representational cortical plasticity. Exp Brain Res. 2003;148:1–16.
Sisodiya SM, Fauser S, Cross JH, Thom M. Focal cortical dysplasia type II: biological features and clinical perspectives. Lancet Neurol. 2009;8:830–43.
Sluming V, Barrick T, Howard M, Cezayirli E, Mayes A, Roberts N. Voxel-based morphometry reveals increased gray matter density in Broca's area in male symphony orchestra musicians. Neuroimage. 2002;17:1613–22.
Smits A, Zetterling M, Lundin M, Melin B, Fahlstrom M, Grabowska A, Larsson EM, Berntsson SG. Neurological impairment linked with cortico-subcortical infiltration of diffuse low-grade gliomas at initial diagnosis supports early brain plasticity. Front Neurol. 2015;6:137.
Sollmann N, Krieg SM, Säisänen L, Julkunen P. Mapping of motor function with neuronavigated transcranial magnetic stimulation: a review on clinical application in brain tumors and methods for ensuring feasible accuracy. Brain Sci. 2021;11:897.
Szalisznyo K, Silverstein DN, Duffau H, Smits A. Pathological neural attractor dynamics in slowly growing gliomas supports an optimal time frame for white matter plasticity. PLoS One. 2013;8:e69798.
Takahashi S, Jussen D, Vajkoczy P, Picht T. Plastic relocation of motor cortex in a patient with LGG (low grade glioma) confirmed by NBS (navigated brain stimulation). Acta Neurochir (Wien). 2012;154:2003–8; discussion 2008.
Tyc F, Boyadjian A, Devanne H. Motor cortex plasticity induced by extensive training revealed by transcranial magnetic stimulation in human. Eur J Neurosci. 2005;21:259–66.
Vaalto S, Julkunen P, Säisänen L, Könönen M, Määttä S, Karhu J. Long-term plasticity may be manifested as reduction or expansion of cortical representations of actively used muscles in motor skill specialists. Neuroreport. 2013;24:596–600.
Vikingstad EM, Cao Y, Thomas AJ, Johnson AF, Malik GM, Welch KM. Language hemispheric dominance in patients with congenital lesions of eloquent brain. Neurosurgery. 2000;47:562–70.
Wassermann EM, McShane LM, Hallett M, Cohen LG. Noninvasive mapping of muscle representations in human motor cortex. Electroencephalogr Clin Neurophysiol. 1992;85:1–8.

Outlook on the Potential of nTMS in Neurosurgery

24

Clara Starkweather, Mitchel Berger, and Phiroz E. Tarapore

The purpose of this chapter is to anticipate and motivate the next generation of nTMS applications over the next several years. Recent studies have highlighted the potential for nTMS to develop in two major areas that will benefit neurosurgical patients. First, on the diagnostic frontier, nTMS may be used to map functional organization with respect to motor and language function, as well as other brain functions such as visuospatial and affective processing. Second, on the therapeutic frontier, nTMS may be used both to rehabilitate patients with surgery-related deficits and to optimize patients preoperatively. The treatment protocols for postoperative acquired deficits will likely be similar to those applied post-stroke, which have now accrued substantial evidence. The induction of functional reorganization, so-called prehabilitation, could potentially make highly eloquent tumors safer for resection. Before this application is ready for clinical use, we must answer several proximal research questions.

C. Starkweather (✉)
Department of Neurological Surgery, San Francisco, CA, USA
e-mail: Clara.Starkweather@ucsf.edu

M. Berger
UCSF Brain Tumor Center, San Francisco, CA, USA
e-mail: Mitchel.Berger@ucsf.edu

P. E. Tarapore
Department of Neurological Surgery, University of California at San Francisco, San Francisco, CA, USA

Veterans Affairs Medical Center, Department of Veterans Affairs, San Francisco, CA, USA
e-mail: Phiroz.Tarapore@ucsf.edu

S. M. Krieg, T. Picht (eds.), *Navigated Transcranial Magnetic Stimulation in Neurosurgery*, https://doi.org/10.1007/978-3-031-97155-6_24

24.1 Diagnostic Use for Detecting Functional Reorganization

24.1.1 Current Knowledge of Lesion-Induced Functional Reorganization

Intraparenchymal brain tumors promote functional reorganization, which means that critical cortical functions may be relocated from one brain region to another (Duffau 2006; Ius et al. 2011; Krieg et al. 2013a; Kawashima et al. 2013; Southwell et al. 2016). Particularly in the case of slow-growing tumors, cortical function does not reside in the typical, anatomically defined brain region. These patients' brains are different from those of healthy people because the tumor itself induces cortical reorganization such that eloquent functions may rely on other brain areas. There is evidence of this reorganization from redo surgical cases, which revealed differences in brain function between the first and second surgeries. For example, Duffau and colleagues reported that they were not initially able to resect gliomas completely during the first surgery because intraoperative mapping by direct electrical stimulation (DES) showed eloquent motor or language function within the region of the tumor (Robles et al. 2008). They followed these patients postoperatively, waiting for functional reorganization to take place prior to re-operation. On repeat surgery, they were able to resect the tumor residual because DES showed no further function in the tumor area. Similarly, our group also observed functional reorganization when examining 18 patients who underwent repeated awake surgery with intraoperative DES mapping. We found a loss of function in previously motor- or language-positive brain areas in six out of 18 patients and a gain of function in another patient who showed new essential motor and language function in previously negative areas. These findings provide direct evidence that eloquent brain functions can be supplanted by other brain areas in patients with brain tumors.

While this capability is profoundly exciting, it is important to distinguish between cortical and subcortical functional reorganization. The potential for functional reorganization is not equal in all brain regions. Injury to the cortex can potentially recover, while subcortical white matter lesions are often irreversible (Herbet et al. 2016). The impairment of subcortical tracts usually results in severe functional impairment without significant capacity for functional recovery (Galluzzi et al. 2008). This finding applies to glioma patients as well as stroke patients. Subcortical tracts are far less likely than cortical areas to show functional reorganization if infiltrated by a tumor (Herbet et al. 2016). In contrast, cortical regions display substantial functional reorganization. Stroke patients, for instance, have the potential to change the organization of their motor cortex within 6 months after the ischemic event, leading to improved motor function (Freundlieb et al. 2015). Glioma patients can preserve motor abilities despite tumor infiltration within their M1, as described above (Robles et al. 2008; Duffau 2005). Primary unimodal cortex, such as the precentral gyrus, is less capable of reorganization than higher-order cortical regions such as the left angular gyrus or dorsal superior temporal gyrus. One reason for this difference may be that cortices with a distinct function, such as the precentral gyrus or Wernicke's area, have a critical role within their functional network; thus, the capacity for functional reorganization is limited, and the absence of functional reorganization produces severe deficits. In contrast, other brain regions with less distinct functional determination participate in wider cortical networks with more redundancy. Some areas represent an intermediate level of functional integration, such as supplemental areas responsible for the fine-tuning and coordination of functional processing. Two such regions are the SMA of the SFG (motor function) or the left-sided MFG and IFG (language function). Because language is organized as a complex and adaptive network including multiple subcortical pathways connecting the same cortical areas, language function is more likely to exhibit functional reorganization than motor function (Ius et al. 2011; Herbet et al. 2016; Duffau 2005, 2014; De Witt Hamer et al. 2013).

There are various theories on the mechanisms of cortical neuroplasticity. The two most accepted theories are addressed here. In the first theory, adjacent neurons in the corresponding cortical layer are recruited, as has been shown for traumatic brain and spinal cord injury patients (Harris et al. 2013; Di Rienzo et al. 2014). Such recruitment can also occur via existing contralateral connections (Levy et al. 2008). The second theory involves the disinhibition of inhibitory interneurons (Duffau 2014). Functional synapses are eliminated during development because they represent redundant connections ("synaptic pruning") (Schuldiner and Yaron 2015). In the setting of injury, during which some of these primary connections are disrupted, previously inhibited redundant circuits are recruited in order to improve impaired function. Thus, injury or functional impairment can promote the reactivation of these latent compensatory tracts (Gaucher et al. 2013; Duffau 2001). While recruitment takes considerable time to develop, the unmasking of latent redundant circuits can occur immediately. In glioma patients, there is evidence for both mechanisms. Short-term functional reorganization has been shown in the precentral gyrus intraoperatively within the ipsilateral cortex, while recruitment of the contralateral SMA was shown to develop during long-term follow-up and corresponds with improved functional recovery after SMA resection (Duffau 2001; Krainik et al. 2004). Similar ipsi- and contralateral functional reorganization has also been reported for language function (Krieg et al. 2013a, 2014; Sarubbo et al. 2012). Despite the relative paucity of data on functional reorganization, some factors affecting functional reorganization have been identified. These factors include age, growth kinetics of the lesion, tumor location, affected function (network vs. unimodal), and gender (Galluzzi et al. 2008; Keidel et al. 2010; Charras et al. 2015; Kuo et al. 2006). Although data showing the high potential of functional reorganization for neurosurgical patients is growing, functional reorganization cannot be relied upon to restore function after surgery—preservation of function is far preferable to restoration of function, and the best outcomes require safe surgery that includes intraoperative DES mapping when appropriate.

Following functional reorganization with nTMS opens a new horizon in neurosurgical oncology, since a tumor in a presumed eloquent location, be it in the motor cortex, language cortex, or subcortical structures, may still be considered safe for surgical removal (Kawashima et al. 2013; Takahashi et al. 2012; Krieg et al. 2013b). The preoperative localization of the functional cortex may make more patients eligible for resection. While this localization would benefit relatively few patients with metastases, in which surgery is usually feasible irrespective of location, patients suffering from high-grade gliomas and even more so from low-grade gliomas could benefit significantly, as the preoperative nTMS map demonstrates the exact relationship between tumor and eloquent cortex (Picht et al. 2012). More importantly, in cases of a highly eloquent tumor location, longitudinal follow-up with annual nTMS mappings could identify at some future time point when the risk of surgical resection becomes acceptable.

24.1.2 Detecting Functional Reorganization by nTMS Motor Mapping

Within the last few years, reports have increasingly demonstrated that nTMS can identify functional reorganization and contribute significantly to the management of neuro-oncology patients. In one case report, a patient suffering from an LGG within the hand knob of the precentral gyrus was initially deemed to have a nonresectable lesion. Preoperative nTMS mapping revealed motor areas to be outside the anatomically presumed motor eloquent cortex (Takahashi et al. 2012). This case was later supported by data from two other cases that showed comparable observations in two epilepsy patients (Mäkelä et al. 2013). These initial case reports are supported by larger investigations into the utility of nTMS in detecting relocation of motor function. In one study, the 3D data of nTMS motor maps of 100 brain tumor patients were fused (Bulubas et al. 2016) (Fig. 24.1a). The authors found an extensive distribution of primary motor function with short-latency motor evoked potentials (MEPs) far beyond the precentral gyrus, but also different distributions of motor

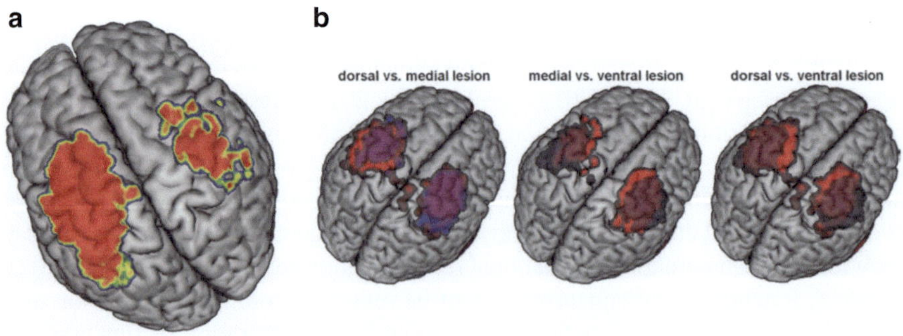

Fig. 24.1 From Bulubas et al. (**a**) Extensive motor areas as identified via nTMS. This heat map is the result of the normalization and fusion of the motor mapping data of 100 brain tumor patients. The motor area, in which MEPs were elicited, goes far anterior of the Rolandic region. This effect is even more prominent in the (dominant) left hemisphere. (**b**) Motor areas of tumors are located inside the precentral gyrus. Patients with a tumor inside the precentral gyrus were divided into three groups depending on the tumor location (dorsal, medial, and ventral parts of the precentral gyrus). For each group, the motor mapping data were normalized and fused. Even between these groups, a difference in the pattern of motor function location can be observed

function depending on the location of the parenchymal brain tumor, even within the precentral gyrus (Fig. 24.1b).

Another recent study investigated the frequency and spatial pattern of functional reorganization in 22 glioma patients (Conway et al. 2017). After spatial normalization to account for brain shift, the authors reported an average shift of the center of gravity (CoG) of M1 of 9.7 ± 1.5 mm (mean ± SEM) on the anteroposterior axis, a value outside the margin of error of the nTMS system and thus suggestive of significant change (Ruohonen and Karhu 2010). These findings are supported by another study on healthy volunteers analyzing the optimal parameters for characterizing motor cortex plasticity with nTMS. This study found that resting motor threshold (rMT), CoG, and mean MEPs are reliable parameters showing high test-retest reliability for monitoring functional reorganization during longer follow-ups (Kraus and Gharabaghi 2016). Thus, the currently available data shows that (1) cortical functional reorganization occurs in LGG but also in HGG patients and (2) nTMS is able to detect such functional reorganization noninvasively.

24.1.3 Detecting Functional Reorganization by nrTMS Language Mapping

The right hemisphere has been shown to participate in language function not only in healthy participants (Vigneau et al. 2011; Schuhmann et al. 2012; Devlin and Watkins 2007; Brennan and Pylkkänen 2012) but also in patients after left hemispheric stroke (Baumgaertner et al. 2013; Baum et al. 2012) or brain tumor (Briganti et al. 2012; Perrone-Bertolotti et al. 2012; Bonelli et al. 2012; Wang et al. 2013; Thiel et al. 2005, 2006). These reports used a variety of methods, including neuropsychological assessment, non-navigated TMS, and fMRI. The fMRI technique in particular can be confounded by intracerebral tumors and ischemic lesions (Giussani et al. 2010; Hou et al. 2006; Ille et al. 2015). Published nTMS-based studies of language laterality have demonstrated that right-handed patients harboring left-sided perisylvian brain tumors can experience a shift of language function to the right hemisphere, in contradistinction to healthy subjects (Krieg et al. 2013a). These results were later confirmed by a second nTMS research group (Rösler et al. 2014). These nrTMS studies corroborated studies using other modalities, as well as previous intraoperative data (Duffau 2006; Robles et al. 2008). We have also seen that functional reorganization not only causes a shift of language function to the contralateral hemisphere but also to other parts of the ipsilateral hemisphere (Kawashima et al. 2013; Krieg et al. 2014). Another study of 80 patients undergoing resection of left-sided perisylvian brain tumors showed that patients with more right-sided language have a lower risk of aphasia 5 days after tumor resection (Ille et al. 2016). Finally, nrTMS language mapping allows for the longitudinal reexamination of patients with known tumors in high-risk regions. By following patients at regular intervals, nrTMS-based language mapping alerts the clinician when a previously unresectable tumor has become resectable due to the relocation of critical language sites (Krieg et al. 2014) (Fig. 24.2).

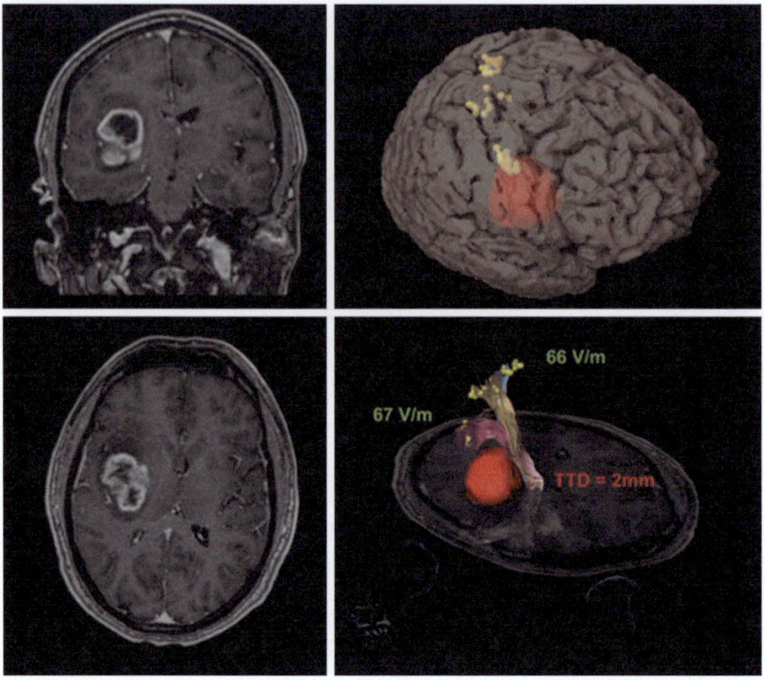

Fig. 24.2 From Rosenstock et al. nTMS motor mapping was performed to map the patient's cortical motor representation. In particular, seeds from nTMS motor positive sites (yellow) were used for tractography in order to map the corticospinal tract relative to the tumor (red)

Similar to risk stratification for motor eloquent tumors, recent studies have also highlighted the yield of nTMS in risk stratification for the resection of language eloquent tumors. Specialized fiber tracking protocols have been developed for mapping the arcuate fasciculus (Zoli et al. 2021) in tumor-infiltrated brains. Furthermore, the integrity of the inferior frontal-occipital fasciculus and superior longitudinal fasciculus within tumor-infiltrated tissue has been shown to be predictive of postoperative language deficits (Ius et al. 2021), demonstrating the utility of detailed preoperative language tract mapping. Similar to work in motor-eloquent tumors in which the corticospinal tract was mapped using seeds localized using nTMS (Rosenstock et al. 2021), a recent study created preoperative language connectomes for each patient by combining diffusion tractography with nTMS (Ille et al. 2022). Anatomic areas were identified as language-positive regions if patients' naming during nTMS stimulation was degraded compared to sessions without stimulation. The authors found that a machine learning model trained on pre-operative connectomes predicted the presence of post-operative aphasia, suggesting that some features of these connectomes may make individuals' language networks more or less robust to surgical resection. While this work introduces the potential for nTMS in preoperative language mapping, it raises several questions for further investigation: the identity of the abstract predictive features detected by the machine learning algorithm, whether these highly trained features would hold in a multicenter study, and whether individualized connectomics could be used to guide the surgical resection itself.

24.1.4 Mapping of Other Neurocognitive and Affective Functions

Performing nTMS maps of other neurocognitive functions, such as arithmetic processing or visuospatial attention, can help to avoid common postoperative neuropsychological deficits. For example, one study describes a protocol in which patients with right parietal lobe tumors underwent mapping of the visuospatial cortico-subcortical network by using an nTMS-implemented version of the Hooper Visual Organization Test to identify critical anatomic points to avoid during surgical resection (Raffa et al. 2021). Such tasks can (and should) be tailored to the individual patient, based on the skillset that each patient wishes to preserve. It still remains to be seen how useful the nTMS maps of these higher-level neurocognitive functions will be in the clinical management of brain tumor patients. Evaluating and optimizing these studies will be a major issue in the development of nTMS.

TMS has demonstrated efficacy in the treatment of psychiatric illnesses such as major depressive disorder (Perera et al. 2016) and obsessive-compulsive disorder (Carmi et al. 2018). The efficacy of TMS in psychiatric disorders may be leveraged to map appropriate affective targets and identify appropriate candidates for psychiatric neurosurgical procedures. As an example, consider that stimulation of the subgenual cingulate and nearby white matter pathways results in sustained remission in four out of six patients with treatment-resistant depression (Mayberg et al. 2005). This pioneering work showcases one challenge in the widespread implementation of deep brain stimulation for depression, which is that these interventions historically work in only a subset of patients (Dougherty et al. 2015; Holtzheimer et al. 2017; Bergfeld et al. 2016). As such, it is difficult to justify an invasive procedure to implant a chronic indwelling device if only a fraction of patients derive benefit. One could, however, envision a treatment pipeline in which responders are identified preoperatively by stimulating the subgenual cingulate (or other region of interest) with noninvasive nTMS and evaluating their response to treatment in the region of interest. Such studies and identification of partial responders may soon be feasible given coil designs that are capable of reaching deeper targets (Smith et al. 2021; Wu et al. 2018). In this way, a non-invasive screening process using nTMS could be implemented to map efficacy on potential affective targets and identify candidates most likely to benefit from psychiatric neurosurgical procedures.

24.2 Therapeutic Use: Inducing Functional Reorganization with nRTMS

24.2.1 General Considerations

Besides the therapeutic applications for chronic pain, tinnitus, and depression, nrTMS has been found to enhance recovery of aphasia and paresis after stroke (Takeuchi et al. 2009; Takeuchi and Izumi 2012; Abo et al. 2014; Kim et al. 2006), which is perhaps the most studied, comparable application to its use in the recovery of post-surgical deficits. One mechanism for this effect is thought to be the induction of functional reorganization (Abo et al. 2014; Kim et al. 2006; Du et al. 2016). This theory is supported by various laboratory investigations showing that rTMS can induce plasticity on the synaptic and cellular levels (Lenz et al. 2016; Korchounov and Ziemann 2011). Preliminary studies on the use of nTMS have yielded promising results. For example, Abo et al. (2014) conducted a randomized multicenter study that demonstrated significantly improved upper limb mobility following a 15-day rehabilitative period including repetitive TMS plus occupational therapy versus movement therapy alone. However, future studies would benefit from a more rigorous trial design incorporating convincing sham or control conditions, which has proven difficult given the salient auditory feedback and sensations associated with TMS delivery. New device designs incorporating quieter coils (Zeng et al. 2022) will be useful in designing trials that may provide higher-quality evidence in favor of TMS in stroke recovery, as well as coil designs that incorporate sham trials (Memarian Sorkhabi and Denison 2022).

24.2.2 Treatment of Surgery-Related Deficits

Preliminary studies of nrTMS-based treatment after stroke have demonstrated an impressive improvement in motor deficits (Takeuchi et al. 2009; Takeuchi and Izumi 2012; Abo et al. 2014; Kim et al. 2006) (also see Chap. 19). Due to the fact that the majority of postoperative deficits also result from ischemic lesions, it is reasonable to investigate whether nrTMS can induce reorganization and facilitate recovery in neurosurgical patients with postoperative deficits (Krieg et al. 2012, 2013c; Gempt et al. 2013; Obermueller et al. 2014). In an unpublished feasibility study, brain tumor patients with postoperative impairment of motor or language function tolerated 7 days of nrTMS therapy, even if started on the first postoperative day.

Larger clinical trials on the postoperative use of nrTMS treatment for surgery-related deficits are ongoing and represent a promising application of this technology. One recent randomized double-blinded trial by Ille et al. (2021) asked whether patients suffering from acute upper limb paresis following surgery would benefit from nrTMS. This study demonstrated that nTMS over the course of 7 days following surgery significantly improved outcomes measured by the Fugl-Meyer Assessment at 3-month follow-up. Importantly, patients selected for this study had a new subcortical infarct within the corticospinal tract on their postoperative MRI scan. Although subcortical lesions are typically thought to invoke less plasticity and window for recuperation compared to cortical lesions, this study shows that even postoperative subcortical insults have their potential for recovery augmented by nrTMS.

24.2.3 Prehabilitation: Preoperative Induction of Functional Reorganization

Rather than waiting for a tumor to induce functional reorganization, prehabilitation uses nrTMS to try and move functionally eloquent brain regions away from a planned surgical site (Fig. 24.3). To date, there are only three published case reports on this technique; nevertheless, these reports provide proof of concept and valuable data for the noninvasive induction of functional reorganization by nrTMS. In all three cases, using a protocol of 7–12 sessions of nrTMS therapy for 10–20 min each day, the investigators were able to modify the organization of language networks (Barwood et al. 2011; Andoh and Martinot 2008; Barcia et al. 2012). Barcia et al. reported on a glioma patient who underwent incomplete resection due to intraoperatively identified eloquent brain tissue within the tumor. The patient underwent intralesional treatment with nrTMS. After each session, the patient experienced worsened language function, but this effect lessened over the course of 12 sessions. This change was interpreted as a functional reorganization of language function away from the previously identified brain area (Barcia et al. 2012). Two other groups obtained comparable results (Barwood et al. 2011; Andoh and Martinot 2008). Of note, nrTMS is used in each of these three reports and is the only reported noninvasive method for inducing functional reorganization. Importantly, an ongoing sham-controlled, triple-blinded clinical trial will test whether motor pre-habilitation can

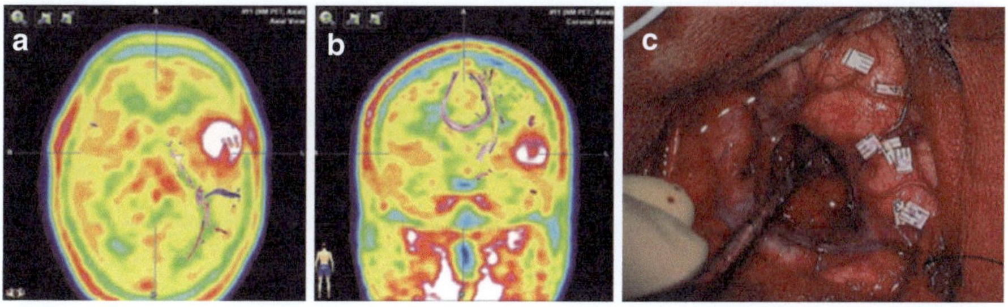

Fig. 24.3 Candidate for preoperative prehabilitation by nrTMS treatment. These axial (**a**) and coronal (**b**) slides show the fusion of a preoperative PET scan and preoperative nTMS/nrTMS motor and language mapping in a bilingual patient suffering from an anaplastic astrocytoma WHO III in the triangular part of the left IFG. The nrTMS mapping shows a language-involved cortex right inside the upper part of the tumor for both languages, which was later confirmed during awake surgery by DES mapping (**c**). Prehabilitation could be applied to move language function outside the tumor. Language mapping of the language-negative cortex could monitor the success of prehabilitation and therefore confirm the time point of optimal second resective surgery going for GTR

successfully induce functional reorganization of the motor network (Engelhardt et al. 2023).

Even using invasive techniques, there is only one reported series of prehabilitation. Barcia and colleagues induced functional reorganization in five out of five patients by applying high-frequency stimulation via implanted subdural grid electrodes for 27–37 days (Rivera-Rivera et al. 2017). Strikingly, the authors showed that language function shifted to the contralateral hemisphere (as assessed with fMRI) after stimulation. Nevertheless, the reported rate of severe complications was 60% (two patients with brain infection, and one patient with severe intracranial hemorrhage). This rate is unacceptably high for the technique to be applied broadly. The nrTMS treatment, which is non-invasive and carries minimal risk, may have the same capability but without ethical concerns.

Interestingly, despite substantial interest in exploring whether nrTMS can be used to induce plasticity and shifts in key brain functions in preoperative settings, there is a dearth of recent literature on this topic, perhaps indicating technical challenges in realizing this strategy. One recent approach that may be helpful toward optimizing pre-operative stimulation protocols is TMS-EEG. This strategy allows researchers to titrate stimulation intensity in a closed-loop fashion by using EEG to monitor changes in brain electrical activity during TMS (Cruciani et al. 2023). Despite technical challenges and ongoing research in implementing preoperative nrTMS, prehabilitation is a fascinating topic for further study that could revolutionize our understanding of resectable neurosurgical lesions.

24.3 Conclusion

With each passing year and each additional study, the application of nTMS in neurosurgery becomes increasingly exciting. The future may allow us to map neurological functions other than motor and language; it may also enable us to follow functional reorganization and use this information to optimize the timing of an intervention in neuro-oncological pathology. Moreover, the therapeutic use of nrTMS for prehabilitation and treatment of postoperative deficits may be a transformative new technique for clinical neurosurgery.

References

Abo M, Kakuda W, Momosaki R, Harashima H, Kojima M, Watanabe S, et al. Randomized, multicenter, comparative study of NEURO versus CIMT in poststroke patients with upper limb hemiparesis: the NEURO-VERIFY study. Int J Stroke. 2014;9(5):607–12.

Andoh J, Martinot JL. Interhemispheric compensation: a hypothesis of TMS-induced effects on language-related areas. Eur Psychiatry. 2008;23(4):281–8.

Barcia JA, Sanz A, González-Hidalgo M, De Las Heras C, Alonso-Lera P, Díaz P, et al. rTMS stimulation to induce plastic changes at the language motor area in a patient with a left recidivant brain tumor affecting Broca's area. Neurocase. 2012;18(2):132–8.

Barwood CHS, Murdoch BE, Whelan BM, Lloyd D, Riek S, O'Sullivan J, et al. The effects of low frequency Repetitive Transcranial Magnetic Stimulation (rTMS) and sham condition rTMS on behavioural language in chronic non-fluent aphasia: short term outcomes. NeuroRehabilitation. 2011;28(2):113–28.

Baum SH, Martin RC, Hamilton AC, Beauchamp MS. Multisensory speech perception without the left superior temporal sulcus. NeuroImage. 2012;62(3):1825–32.

Baumgaertner A, Hartwigsen G, Roman Siebner H. Right-hemispheric processing of non-linguistic word features: implications for mapping language recovery after stroke. Hum Brain Mapp. 2013;34(6):1293–305.

Bergfeld IO, Mantione M, Hoogendoorn MLC, Ruhé HG, Notten P, van Laarhoven J, et al. Deep brain stimulation of the ventral anterior limb of the internal capsule for treatment-resistant depression: a randomized clinical trial. JAMA Psychiatry. 2016;73(5):456.

Bonelli SB, Thompson PJ, Yogarajah M, Vollmar C, Powell RHW, Symms MR, et al. Imaging language networks before and after anterior temporal lobe resection: results of a longitudinal fMRI study: *Imaging Language Networks in TLE*. Epilepsia. 2012;53(4):639–50.

Brennan J, Pylkkänen L. The time-course and spatial distribution of brain activity associated with sentence processing. NeuroImage. 2012;60(2):1139–48.

Briganti C, Sestieri C, Mattei PA, Esposito R, Galzio RJ, Tartaro A, et al. Reorganization of functional connectivity of the language network in patients with brain gliomas. Am J Neuroradiol. 2012;33(10):1983–90.

Bulubas L, Sabih J, Wohlschlaeger A, Sollmann N, Hauck T, Ille S, et al. Motor areas of the frontal cortex in patients with motor eloquent brain lesions. J Neurosurg. 2016;125(6):1431–42.

Carmi L, Alyagon U, Barnea-Ygael N, Zohar J, Dar R, Zangen A. Clinical and electrophysiological outcomes of deep TMS over the medial prefrontal and anterior cingulate cortices in OCD patients. Brain Stimul. 2018;11(1):158–65.

Charras P, Herbet G, Deverdun J, De Champfleur NM, Duffau H, Bartolomeo P, et al. Functional reorganization of the attentional networks in low-grade glioma patients: a longitudinal study. Cortex. 2015;63:27–41.

Conway N, Wildschuetz N, Moser T, Bulubas L, Sollmann N, Tanigawa N, et al. Cortical plasticity of motor-eloquent areas measured by navigated transcranial magnetic stimulation in patients with glioma. J Neurosurg. 2017;127(5):981–91.

Cruciani A, Mancuso M, Sveva V, Maccarrone D, Todisco A, Motolese F, et al. Using TMS-EEG to assess the effects of neuromodulation techniques: a narrative review. Front Hum Neurosci. 2023;17:1247104.

De Witt Hamer PC, Hendriks EJ, Mandonnet E, Barkhof F, Zwinderman AH, Duffau H. Resection probability maps for quality assessment of glioma surgery without brain location bias. Elder JB, editor. PLoS One. 2013;8(9):e73353.

Devlin JT, Watkins KE. Stimulating language: insights from TMS. Brain. 2007;130(3):610–22.

Di Rienzo F, Guillot A, Mateo S, Daligault S, Delpuech C, Rode G, et al. Neuroplasticity of prehensile neural networks after quadriplegia. Neuroscience. 2014;274:82–92.

Dougherty DD, Rezai AR, Carpenter LL, Howland RH, Bhati MT, O'Reardon JP, et al. A randomized sham-controlled trial of deep brain stimulation of the ventral capsule/ventral striatum for chronic treatment-resistant depression. Biol Psychiatry. 2015;78(4):240–8.

Du J, Tian L, Liu W, Hu J, Xu G, Ma M, et al. Effects of repetitive transcranial magnetic stimulation on motor recovery and motor cortex excitability in patients with stroke: a randomized controlled trial. Eur J Neurol. 2016;23(11):1666–72.

Duffau H. Acute functional reorganisation of the human motor cortex during resection of central lesions: a study using intraoperative brain mapping. J Neurol Neurosurg Psychiatry. 2001;70(4):506–13.

Duffau H. Lessons from brain mapping in surgery for low-grade glioma: insights into associations between tumour and brain plasticity. Lancet Neurol. 2005;4(8):476–86.

Duffau H. New concepts in surgery of WHO grade II gliomas: functional brain mapping, connectionism and plasticity – a review. J Neuro-Oncol. 2006;79(1):77–115.

Duffau H. The huge plastic potential of adult brain and the role of connectomics: new insights provided by serial mappings in glioma surgery. Cortex. 2014;58:325–37.

Engelhardt M, Grittner U, Krieg S, Picht T. Preconditioning of the motor network with repetitive navigated transcranial magnetic stimulation (rnTMS) to improve oncological and functional outcome in brain tumor surgery: a study protocol for a randomized, sham-controlled, triple-blind clinical trial. Trials. 2023;24(1):638.

Freundlieb N, Philipp S, Drabik A, Gerloff C, Forkert ND, Hummel FC. Ipsilesional motor area size correlates with functional recovery after stroke: a 6-month follow-up longitudinal TMS motor mapping study. Restor Neurol Neurosci. 2015;33(2):221–31.

Galluzzi S, Lanni C, Pantoni L, Filippi M, Frisoni GB. White matter lesions in the elderly: pathophysiological hypothesis on the effect on brain plasticity and reserve. J Neurol Sci. 2008;273(1–2):3–9.

Gaucher Q, Huetz C, Gourevitch B, Edeline JM. Cortical inhibition reduces information redundancy at presentation of communication sounds in the primary auditory cortex. J Neurosci. 2013;33(26):10713–28.

Gempt J, Krieg SM, Hüttinger S, Buchmann N, Ryang YM, Shiban E, et al. Postoperative ischemic changes after glioma resection identified by diffusion-weighted magnetic resonance imaging and their association with intraoperative motor evoked potentials: clinical article. J Neurosurg. 2013;119(4):829–36.

Giussani C, Roux FE, Ojemann J, Sganzerla EP, Pirillo D, Papagno C. Is preoperative functional magnetic resonance imaging reliable for language areas mapping in brain tumor surgery? Review of language functional magnetic resonance imaging and direct cortical stimulation correlation studies. Neurosurgery. 2010;66(1):113–20.

Harris NG, Nogueira MSM, Verley DR, Sutton RL. Chondroitinase enhances cortical map plasticity and increases functionally active sprouting axons after brain injury. J Neurotrauma. 2013;30(14):1257–69.

Herbet G, Maheu M, Costi E, Lafargue G, Duffau H. Mapping neuroplastic potential in brain-damaged patients. Brain. 2016;139(3):829–44.

Holtzheimer PE, Husain MM, Lisanby SH, Taylor SF, Whitworth LA, McClintock S, et al. Subcallosal cingulate deep brain stimulation for treatment-resistant depression: a multisite, randomised, sham-controlled trial. Lancet Psychiatry. 2017;4(11):839–49.

Hou BL, Bradbury M, Peck KK, Petrovich NM, Gutin PH, Holodny AI. Effect of brain tumor neovasculature defined by rCBV on BOLD fMRI activation volume in the primary motor cortex. NeuroImage. 2006;32(2):489–97.

Ille S, Sollmann N, Hauck T, Maurer S, Tanigawa N, Obermueller T, et al. Impairment of preoperative language mapping by lesion location: a functional magnetic resonance imaging, navigated transcranial magnetic stimulation, and direct cortical stimulation study. J Neurosurg. 2015;123(2):314–24.

Ille S, Kulchytska N, Sollmann N, Wittig R, Beurskens E, Butenschoen VM, et al. Hemispheric language dominance measured by repetitive navigated transcranial magnetic stimulation and postoperative course of language function in brain tumor patients. Neuropsychologia. 2016;91:50–60.

Ille S, Kelm A, Schroeder A, Albers LE, Negwer C, Butenschoen VM, et al. Navigated repetitive transcranial magnetic stimulation improves the outcome of postsurgical paresis in glioma patients – a randomized, double-blinded trial. Brain Stimul. 2021;14(4):780–7.

Ille S, Zhang H, Sogerer L, Schwendner M, Schöder A, Meyer B, et al. Preoperative function-specific connectome analysis predicts surgery-related aphasia after glioma resection. Hum Brain Mapp. 2022;43(18):5408–20.

Ius T, Angelini E, Thiebaut De Schotten M, Mandonnet E, Duffau H. Evidence for potentials and limitations of brain plasticity using an atlas of functional resectability of WHO grade II gliomas: towards a "minimal common brain". NeuroImage. 2011;56(3):992–1000.

Ius T, Somma T, Baiano C, Guarracino I, Pauletto G, Nilo A, et al. Risk assessment by pre-surgical tractography in left hemisphere low-grade gliomas. Front Neurol. 2021;12:648432.

Kawashima A, Krieg SM, Faust K, Schneider H, Vajkoczy P, Picht T. Plastic reshaping of cortical language areas evaluated by navigated transcranial magnetic stimulation in a surgical case of glioblastoma multiforme. Clin Neurol Neurosurg. 2013;115(10):2226–9.

Keidel JL, Welbourne SR, Lambon Ralph MA. Solving the paradox of the equipotential and modular brain: a neurocomputational model of stroke vs. slow-growing glioma. Neuropsychologia. 2010;48(6):1716–24.

Kim YH, You SH, Ko MH, Park JW, Lee KH, Jang SH, et al. Repetitive transcranial magnetic stimulation–induced corticomotor excitability and associated motor skill acquisition in chronic stroke. Stroke. 2006;37(6):1471–6.

Korchounov A, Ziemann U. Neuromodulatory neurotransmitters influence LTP-like plasticity in human cortex: a pharmaco-TMS study. Neuropsychopharmacology. 2011;36(9):1894–902.

Krainik A, Duffau H, Capelle L, Cornu P, Boch AL, Mangin JF, et al. Role of the healthy hemisphere in recovery after resection of the supplementary motor area. Neurology. 2004;62(8):1323–32.

Kraus D, Gharabaghi A. Neuromuscular plasticity: disentangling stable and variable motor maps in the human sensorimotor cortex. Neural Plast. 2016;2016:1–13.

Krieg SM, Shiban E, Droese D, Gempt J, Buchmann N, Pape H, et al. Predictive value and safety of intraoperative neurophysiological monitoring with motor evoked potentials in glioma surgery. Neurosurgery. 2012;70(5):1060–71.

Krieg SM, Sollmann N, Hauck T, Ille S, Foerschler A, Meyer B, et al. Functional language shift to the right hemisphere in patients with language-eloquent brain tumors. Kassubek J, editor. PLoS One. 2013a;8(9):e75403.

Krieg SM, Schnurbus L, Shiban E, Droese D, Obermueller T, Buchmann N, et al. Surgery of highly eloquent gliomas primarily assessed as non-resectable: risks and benefits in a cohort study. BMC Cancer. 2013b;13(1):51.

Krieg SM, Schäffner M, Shiban E, Droese D, Obermüller T, Gempt J, et al. Reliability of intraoperative neurophysiological monitoring using motor evoked potentials during resection of metastases in motor-eloquent brain regions: clinical article. J Neurosurg. 2013c;118(6):1269–78.

Krieg SM, Sollmann N, Hauck T, Ille S, Meyer B, Ringel F. Repeated mapping of cortical language sites by preoperative navigated transcranial magnetic stimulation compared to repeated intraoperative DCS mapping in awake craniotomy. BMC Neurosci. 2014;15(1):20.

Kuo MF, Paulus W, Nitsche MA. Sex differences in cortical neuroplasticity in humans. Neuroreport. 2006;17:1703–7.

Lenz M, Galanis C, Müller-Dahlhaus F, Opitz A, Wierenga CJ, Szabó G, et al. Repetitive magnetic stimulation induces plasticity of inhibitory synapses. Nat Commun. 2016;7(1):10020.

Levy R, Ruland S, Weinand M, Lowry D, Dafer R, Bakay R. Cortical stimulation for the rehabilitation of patients with hemiparetic stroke: a multicenter feasibility study of safety and efficacy. J Neurosurg. 2008;108(4):707–14.

Mäkelä JP, Vitikainen AM, Lioumis P, Paetau R, Ahtola E, Kuusela L, et al. Functional plasticity of the motor cortical structures demonstrated by navigated TMS in two patients with epilepsy. Brain Stimul. 2013;6(3):286–91.

Mayberg HS, Lozano AM, Voon V, McNeely HE, Seminowicz D, Hamani C, et al. Deep brain stimulation for treatment-resistant depression. Neuron. 2005;45(5):651–60.

Memarian Sorkhabi M, Denison T. A neurostimulator system for real, sham, and multi-target transcranial magnetic stimulation. J Neural Eng. 2022;19(2):026035.

Obermueller T, Schaeffner M, Gerhardt J, Meyer B, Ringel F, Krieg SM. Risks of postoperative paresis in motor eloquently and non-eloquently located brain metastases. BMC Cancer. 2014;14(1):21.

Perera T, George MS, Grammer G, Janicak PG, Pascual-Leone A, Wirecki TS. The Clinical TMS Society consensus review and treatment recommendations for TMS therapy for major depressive disorder. Brain Stimul. 2016;9(3):336–46.

Perrone-Bertolotti M, Zoubrinetzky R, Yvert G, Le Bas JF, Baciu M. Functional MRI and neuropsychological evidence for language plasticity before and after surgery in one patient with left temporal lobe epilepsy. Epilepsy Behav. 2012;23(1):81–6.

Picht T, Schulz J, Hanna M, Schmidt S, Suess O, Vajkoczy P. Assessment of the influence of navigated transcranial magnetic stimulation on surgical planning for tumors in or near the motor cortex. Neurosurgery. 2012;70(5):1248–57.

Raffa G, Quattropani MC, Marzano G, Curcio A, Rizzo V, Sebestyén G, et al. Mapping and preserving the visuospatial network by repetitive nTMS and DTI tractography in patients with right parietal lobe tumors. Front Oncol. 2021;11:677172.

Rivera-Rivera PA, Rios-Lago M, Sanchez-Casarrubios S, Salazar O, Yus M, González-Hidalgo M, et al. Cortical plasticity catalyzed by prehabilitation enables extensive resection of brain tumors in eloquent areas. J Neurosurg. 2017;126(4):1323–33.

Robles SG, Gatignol P, Lehéricy S, Duffau H. Long-term brain plasticity allowing a multistage surgical approach to World Health Organization Grade II gliomas in eloquent areas: report of 2 cases. J Neurosurg. 2008;109(4):615–24.

Rosenstock T, Tuncer MS, Münch MR, Vajkoczy P, Picht T, Faust K. Preoperative nTMS and intraoperative neurophysiology – a comparative analysis in patients with motor-eloquent glioma. Front Oncol. 2021;11:676626.

Rösler J, Niraula B, Strack V, Zdunczyk A, Schilt S, Savolainen P, et al. Language mapping in healthy volunteers and brain tumor patients with a novel navigated TMS system: evidence of tumor-induced plasticity. Clin Neurophysiol. 2014;125(3):526–36.

Ruohonen J, Karhu J. Navigated transcranial magnetic stimulation. Neurophysiol Clin Neurophysiol. 2010;40(1):7–17.

Sarubbo S, Latini F, Sette E, Milani P, Granieri E, Fainardi E, et al. Is the resection of gliomas in Wernicke's area reliable?: Wernicke's area resection. Acta Neurochir (Wien). 2012;154(9):1653–62.

Schuhmann T, Schiller NO, Goebel R, Sack AT. Speaking of which: dissecting the neurocognitive network of language production in picture naming. Cereb Cortex. 2012;22(3):701–9.

Schuldiner O, Yaron A. Mechanisms of developmental neurite pruning. Cell Mol Life Sci. 2015;72(1):101–19.

Smith MC, Li A, Sievenpiper DF. A multifunction dense array system with reconfigurable depth of penetration. IEEE J Electromagn RF Microw Med Biol. 2021;5(1):35–45.

Southwell DG, Hervey-Jumper SL, Perry DW, Berger MS. Intraoperative mapping during repeat awake craniotomy reveals the functional plasticity of adult cortex. J Neurosurg. 2016;124(5):1460–9.

Takahashi S, Jussen D, Vajkoczy P, Picht T. Plastic relocation of motor cortex in a patient with LGG (low grade glioma) confirmed by NBS (navigated brain stimulation). Acta Neurochir (Wien). 2012;154(11):2003–8.

Takeuchi N, Izumi SI. Noninvasive brain stimulation for motor recovery after stroke: mechanisms and future views. Stroke Res Treat. 2012;2012:1–10.

Takeuchi N, Tada T, Toshima M, Matsuo Y, Ikoma K. Repetitive transcranial magnetic stimulation over bilateral hemispheres enhances motor function and training effect of paretic hand in patients after stroke. J Rehabil Med. 2009;41(13):1049–54.

Thiel A, Habedank B, Winhuisen L, Herholz K, Kessler J, Haupt WF, et al. Essential language function of the right hemisphere in brain tumor patients. Ann Neurol. 2005;57(1):128–31.

Thiel A, Habedank B, Herholz K, Kessler J, Winhuisen L, Haupt WF, et al. From the left to the right: how the brain compensates progressive loss of language function. Brain Lang. 2006;98(1):57–65.

Vigneau M, Beaucousin V, Hervé PY, Jobard G, Petit L, Crivello F, et al. What is right-hemisphere contribution to phonological, lexico-semantic, and sentence processing? NeuroImage. 2011;54(1):577–93.

Wang L, Chen D, Yang X, Olson JJ, Gopinath K, Fan T, et al. Group independent component analysis and functional MRI examination of changes in language areas associated with brain tumors at different locations. Zhan W, editor. PLoS One. 2013;8(3):e59657.

Wu YX, Yu HY, Liu ZW. Numerical investigation of the magnetic and electric field distributions produced by biconical transcranial magnetic stimulation coil for optimal design. IEEE Trans Magn. 2018;54(11):1–5.

Zeng Z, Koponen LM, Hamdan R, Li Z, Goetz SM, Peterchev AV. Modular multilevel TMS device with wide output range and ultrabrief pulse capability for sound reduction. J Neural Eng. 2022;19(2):026008.

Zoli M, Talozzi L, Martinoni M, Manners DN, Badaloni F, Testa C, et al. From neurosurgical planning to histopathological brain tumor characterization: potentialities of arcuate fasciculus along-tract diffusion tensor imaging tractography measures. Front Neurol. 2021;12:633209.

MIX
Papier aus verantwortungsvollen Quellen
Paper from responsible sources
FSC® C105338

If you have any concerns about our products,
you can contact us on
ProductSafety@springernature.com

In case Publisher is established outside the EU,
the EU authorized representative is:
Springer Nature Customer Service Center GmbH
Europaplatz 3, 69115 Heidelberg, Germany

Printed by Libri Plureos GmbH
in Hamburg, Germany